Introductory Algebra:

An Applied Approach

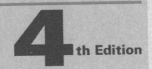

4th Edition

Introductory Algebra
An Applied Approach

Richard N. Aufmann
Palomar College

Vernon C. Barker
Palomar College

HOUGHTON MIFFLIN COMPANY Boston   Toronto
Dallas   Geneva, Illinois   Palo Alto
Princeton, New Jersey

Cover design: Len Massiglia, LMA Communications

Cover image: Herman Cherry, REPEAT 1989.
       Oil on linen, 33″ × 22″
       Courtesy Luise Ross Gallery, NY
       Photo credit: Regina Cherry

Senior Sponsoring Editor: *Maureen O'Connor*
Senior Associate Editor: *Robert Hupp*
Project Editor: *Maria A. Morelli*
Editorial Assistant: *So-Chung Shinn*
Production/Design Coordinator: *Carol Merrigan*
Senior Manufacturing Coordinator: *Priscilla Bailey*

Printed in the U.S.A.

ISBN:
Text: 0-395-70831-1
Instructor's Annotated Edition: 0-395-71218-1

123456789-B-98 97 96 95 94

Content and Format © 1995 HMCo.

Contents

Preface *xiii*

To the Student *xxi*

Index of Applications *xxiii*

1 Real Numbers 1

Section 1.1 Introduction to Integers 3
    Objective A   *To use inequality symbols with integers* 3
    Objective B   *To use opposites and absolute value* 4

Section 1.2 Addition and Subtraction of Integers 7
    Objective A   *To add integers* 7
    Objective B   *To subtract integers* 8
    Objective C   *To solve application problems* 10

Section 1.3 Multiplication and Division of Integers 15
    Objective A   *To multiply integers* 15
    Objective B   *To divide integers* 16
    Objective C   *To solve application problems* 18

Section 1.4 Operations with Rational Numbers 21
    Objective A   *To write a rational number in simplest form and as a decimal* 21
    Objective B   *To convert between percents, fractions, and decimals* 23
    Objective C   *To add or subtract rational numbers* 24
    Objective D   *To multiply or divide rational numbers* 26
    Objective E   *To solve application problems* 28

Section 1.5 Exponents and the Order of Operations Agreement 35
    Objective A   *To evaluate exponential expressions* 35
    Objective B   *To use the Order of Operations Agreement to simplify expressions* 36

    Project in Mathematics: Using the $\boxed{+/-}$ Key on a Scientific Calculator 41
    Chapter Summary 41   Chapter Review 43   Chapter Test 45

2 Variable Expressions 47

Section 2.1 Evaluating Variable Expressions 49
    Objective A   *To evaluate a variable expression* 49

Section 2.2 Simplifying Variable Expressions 53
    Objective A   *To simplify a variable expression using the Properties of Addition* 53
    Objective B   *To simplify a variable expression using the Properties of Multiplication* 55
    Objective C   *To simplify a variable expression using the Distributive Property* 56
    Objective D   *To simplify general variable expressions* 58

Section 2.3   Translating Verbal Expressions into Variable Expressions   **63**

Objective A   To translate a verbal expression into a variable expression, given the variable   63

Objective B   To translate a verbal expression into a variable expression and then simplify   64

Objective C   To translate application problems   66

Project in Mathematics: Prime and Composite Numbers 71   Chapter Summary 72
Chapter Review 73   Chapter Test 75   Cumulative Review 77

3   Solving Equations   79

Section 3.1   Introduction to Equations   **81**

Objective A   To determine whether a given number is a solution of an equation in one variable   81

Objective B   To solve an equation of the form $x + a = b$ 82

Objective C   To solve an equation of the form $ax = b$ 83

Objective D   To solve application problems   85

Section 3.2   General Equations—Part I   **93**

Objective A   To solve an equation of the form $ax + b = c$ 93

Objective B   To solve application problems   94

Section 3.3   General Equations—Part II   **99**

Objective A   To solve an equation of the form $ax + b = cx + d$ 99

Objective B   To solve an equation containing parentheses   100

Objective C   To solve application problems   102

Section 3.4   Translating Sentences into Equations   **107**

Objective A   To translate a number problem into an equation and solve   107

Objective B   To solve application problems   110

Project in Mathematics: Nielsen Ratings 115   Chapter Summary 116
Chapter Review 117   Chapter Test 119   Cumulative Review 121

4   Solving Equations: Applications   123

Section 4.1   Markup and Discount   **125**

Objective A   To solve markup problems   125

Objective B   To solve discount problems   126

Section 4.2   Investment Problems   **129**

Objective A   To solve investment problems   129

Section 4.3   Mixture Problems   **133**

Objective A   To solve value mixture problems   133

Objective B   To solve percent mixture problems   135

Section 4.4   Uniform Motion Problems   **141**

Objective A   To solve uniform motion problems   141

Section 4.5   Geometry Problems   **145**
   Objective A   *To solve perimeter problems*   145
   Objective B   *To solve problems involving the angles of a triangle*   146

   Project in Mathematics: Finding Counter Examples to Conjectures   149
   Chapter Summary 150   Chapter Review 151   Chapter Test 153
   Cumulative Review 155

5   Polynomials   157

Section 5.1   Addition and Subtraction of Polynomials   **159**
   Objective A   *To add polynomials*   159
   Objective B   *To subtract polynomials*   160

Section 5.2   Multiplication of Monomials   **163**
   Objective A   *To multiply monomials*   163
   Objective B   *To simplify powers of monomials*   164

Section 5.3   Multiplication of Polynomials   **167**
   Objective A   *To multiply a polynomial by a monomial*   167
   Objective B   *To multiply two polynomials*   167
   Objective C   *To multiply two binomials*   168
   Objective D   *To multiply binomials that have special products*   169
   Objective E   *To solve application problems*   170

Section 5.4   Integer Exponents and Scientific Notation   **175**
   Objective A   *To divide monomials*   175
   Objective B   *To write a number in scientific notation*   180

Section 5.5   Division of Polynomials   **185**
   Objective A   *To divide a polynomial by a monomial*   185
   Objective B   *To divide polynomials*   185

   Project in Mathematics: Intensity of Illumination 189   Chapter Summary 190
   Chapter Review 191   Chapter Test 193   Cumulative Review 195

6   Factoring   197

Section 6.1   Common Factors   **199**
   Objective A   *To factor a monomial from a polynomial*   199
   Objective B   *To factor by grouping*   201

Section 6.2   Factoring Polynomials of the Form $x^2 + bx + c$ **205**
   Objective A   *To factor a trinomial of the form $x^2 + bx + c$*   205
   Objective B   *To factor completely*   207

Section 6.3   Factoring Polynomials of the Form $ax^2 + bx + c$ **213**
   Objective A   *To factor a trinomial of the form $ax^2 + bx + c$ by using
      trial factors*   213
   Objective B   *To factor a trinomial of the form $ax^2 + bx + c$ by grouping*   215

Section 6.4   **Special Factoring**   **221**

Objective A   *To factor the difference of two squares and perfect-square trinomials*   221

Objective B   *To factor completely*   223

Section 6.5   **Solving Equations**   **229**

Objective A   *To solve equations by factoring*   229

Objective B   *To solve application problems*   231

Project in Mathematics: Evaluating Polynomials Using a Calculator 237
Chapter Summary 238   Chapter Review 239   Chapter Test 241
Cumulative Review 243

7   Rational Expressions   **245**

Section 7.1   **Multiplication and Division of Rational Expressions**   **247**

Objective A   *To simplify a rational expression*   247

Objective B   *To multiply rational expressions*   248

Objective C   *To divide rational expressions*   250

Section 7.2   **Expressing Fractions in Terms of the Least Common Multiple (LCM)**   **255**

Objective A   *To find the least common multiple (LCM) of two or more polynomials*   255

Objective B   *To express two fractions in terms of the LCM of their denominators*   256

Section 7.3   **Addition and Subtraction of Rational Expressions**   **259**

Objective A   *To add or subtract rational expressions with like denominators*   259

Objective B   *To add or subtract rational expressions with unlike denominators*   260

Section 7.4   **Complex Fractions**   **267**

Objective A   *To simplify a complex fraction*   267

Section 7.5   **Solving Equations Containing Fractions**   **271**

Objective A   *To solve an equation containing fractions*   271

Section 7.6   **Ratio and Proportion**   **275**

Objective A   *To solve a proportion*   275

Objective B   *To solve application problems*   276

Section 7.7   **Literal Equations**   **279**

Objective A   *To solve a literal equation for one of the variables*   279

Section 7.8   **Application Problems**   **283**

Objective A   *To solve work problems*   283

Objective B   *To solve uniform motion problems*   285

Project in Mathematics: Continued Fractions 291   Chapter Summary 292
Chapter Review 293   Chapter Test 295   Cumulative Review 297

8   Linear Equations in Two Variables   **299**

Section 8.1   **The Rectangular Coordinate System**   **301**

Objective A   *To graph points in a rectangular coordinate system*   301

Objective B   To determine ordered-pair solutions of an equation in
two variables   303
Objective C   To determine whether a set of ordered pairs is a function   305
Objective D   To evaluate a function written in functional notation   308

Section 8.2   Linear Equations in Two Variables   **313**
Objective A   To graph an equation of the form $y = mx + b$ 313
Objective B   To graph an equation of the form $Ax + By = C$ 315
Objective C   To solve application problems   318

Section 8.3   Intercepts and Slopes of Straight Lines   **323**
Objective A   To find the x- and y-intercepts of a straight line   323
Objective B   To find the slope of a straight line   324
Objective C   To graph a line using the slope and the y-intercept   327

Section 8.4   Equations of Straight Lines   **333**
Objective A   To find the equation of a line given a point and the slope   333
Objective B   To find the equation of a line given two points   334
Objective C   To solve application problems   336

Project in Mathematics: Graphing Linear Equations with a Graphing Utility 341
Chapter Summary 342   Chapter Review 343   Chapter Test 345
Cumulative Review 347

9   Systems of Linear Equations   349

Section 9.1   Solving Systems of Linear Equations by Graphing   **351**
Objective A   To solve a system of linear equations by graphing   351

Section 9.2   Solving Systems of Linear Equations by the Substitution Method   **359**
Objective A   To solve a system of linear equations by the
substitution method   359

Section 9.3   Solving Systems of Equations by the Addition Method   **365**
Objective A   To solve a system of linear equations by the addition method   365

Section 9.4   Application Problems in Two Variables   **371**
Objective A   To solve rate-of-wind or -current problems   371
Objective B   To solve application problems using two variables   372

Project in Mathematics: Calculator Solution of a System of Linear Equations 377
Chapter Summary 378   Chapter Review 379   Chapter Test 381
Cumulative Review 383

10   Inequalities   385

Section 10.1   Sets   **387**
Objective A   To write a set using the roster method   387
Objective B   To write a set using set-builder notation   388
Objective C   To graph an inequality on the number line   389

Section 10.2   The Addition and Multiplication Properties of Inequalities   **393**
Objective A   To solve an inequality using the Addition Property of inequalities   393

Objective B   *To solve an inequality using the Multiplication Property of Inequalities* 394

Objective C   *To solve application problems* 396

Section 10.3   General Inequalities   **401**

Objective A   *To solve general inequalities* 401

Objective B   *To solve application problems* 402

Section 10.4   Graphing Linear Inequalities   **405**

Objective A   *To graph an inequality in two variables* 405

Project in Mathematics: Measurements as Approximations 409
Chapter Summary 410   Chapter Review 411   Chapter Test 413
Cumulative Review 415

11 Radical Expressions   417

Section 11.1   Introduction to Radical Expressions   **419**

Objective A   *To simplify numerical radical expressions* 419

Objective B   *To simplify variable radical expressions* 421

Section 11.2   Addition and Subtraction of Radical Expressions   **425**

Objective A   *To add and subtract radical expressions* 425

Section 11.3   Multiplication and Division of Radical Expressions   **429**

Objective A   *To multiply radical expressions* 429

Objective B   *To divide radical expressions* 430

Section 11.4   Solving Equations Containing Radical Expressions   **435**

Objective A   *To solve an equation containing a radical expression* 435

Objective B   *To solve application problems* 437

Project in Mathematics: Distance to the Horizon 441   Chapter Summary 442
Chapter Review 443   Chapter Test 445   Cumulative Review 447

12 Quadratic Equations   449

Section 12.1   Solving Quadratic Equations by Factoring or by Taking Square Roots   **451**

Objective A   *To solve a quadratic equation by factoring* 451

Objective B   *To solve a quadratic equation by taking square roots* 453

Section 12.2   Solving Quadratic Equations by Completing the Square   **457**

Objective A   *To solve a quadratic equation by completing the square* 457

Section 12.3   Solving Quadratic Equations by Using the Quadratic Formula   **463**

Objective A   *To solve a quadratic equation by using the quadratic formula* 463

Section 12.4   Graphing Quadratic Equations in Two Variables   **467**

Objective A   *To graph a quadratic equation of the form $y = ax^2 + bx + c$* 467

Section 12.5   Application Problems   **471**

Objective A   *To solve application problems* **471**

Projects in Mathematics: Geometric Construction of Completing the Square 475
Graphing a Quadratic Function by Using a Graphing Calculator 476
Chapter Summary 476   Chapter Review 477   Chapter Test 479
Cumulative Review 481   Final Exam 483

Content and Format © 1995 HMCo.

Solutions to "You Try It" A1

Answers to Odd-Numbered Exercises A39

Appendix: Guidelines for Using Graphing Calculators A13
          Table of Square Roots A83
          Table of Symbols A84
          Table of Measurement Abbreviations A84

Glossary G1

Index I1

Preface

. . . .

The fourth edition of *Introductory Algebra: An Applied Approach* provides mathematically sound and comprehensive coverage of the topics considered essential in an introductory algebra course. The text has been designed not only to meet the needs of the traditional college student but also to serve the needs of returning students who need to strengthen their math proficiency.

In this new edition of *Introductory Algebra: An Applied Approach*, careful attention has been given to implementing the standards suggested by NCTM. Each chapter begins with a mathematical vignette in which there may be a historical note, an application, or an interesting issue related to mathematics. At the end of each section are "Applying the Concepts" exercises that include writing, synthesis, critical thinking, and challenge problems. The chapter ends with "Project in Mathematics," which can be used for cooperative learning activities.

INSTRUCTIONAL FEATURES

Interactive Approach

Introductory Algebra: An Applied Approach uses an interactive style that provides students with an opportunity to try a skill as it is presented. Each section is divided into objectives, and every objective contains one or more sets of matched-pair examples. The first example in each set is worked out; the second example, called "You Try It," is for the student to work. By solving this problem, students practice concepts as they are presented in the text. There are *complete* worked-out solutions to these examples at the end of the book. By comparing their solution to the solution in the appendix, students are able to obtain immediate feedback on and reinforcement of the concept.

Emphasis on Problem-Solving Strategies

Introductory Algebra: An Applied Approach features a carefully developed approach to problem solving that emphasizes developing strategies to solve problems. Students are encouraged to develop their own strategies, to draw diagrams, and to write their strategies as part of their solution to a problem. In each case, model strategies are presented as guides for students to follow as they attempt the You Try It problem. Having students provide strategies is a natural way to incorporate writing into the math curriculum.

Emphasis on Applications

The traditional approach to teaching algebra covers only the straightforward manipulation of numbers and variables and thereby fails to teach students the practical value of algebra. By contrast, *Introductory Algebra: An Applied Approach* contains an extensive collection of contemporary application problems. Wherever appropriate, the last objective of a section presents applications that require the student to use the skills covered in that section to solve practical problems. This carefully integrated applied approach generates students' awareness of the value of algebra as a real-life tool.

Completely Integrated Learning System
Organized by Objectives

Each chapter begins with a list of the learning objectives included within that chapter. Each of the objectives is then restated in the chapter to remind the student of the current topic of discussion. The same objectives that organize the text are used as the structure for exercises, testing programs, and the Computer Tutor. For each objective in the text there is a corresponding computer tutorial and a corresponding set of test questions.

AN INTERACTIVE APPROACH

Instructors have long realized the need for a text that requires students to use a skill as it is being taught. *Introductory Algebra: An Applied Approach* uses an interactive technique that meets this need. Every objective, including the one shown below, contains at least one pair of examples. One of the examples is worked. The second example in the pair (called "You Try It" in this edition) is not worked so that students may interact with the text by solving it. To provide immediate feedback, a complete solution to this example is provided in the Answer Section. The benefit of this interactive style is that students can check that a new skill has been learned before attempting a homework assignment.

An explanatory passage begins each skill objective.

Paired examples follow the explanatory passage.

The interactive key is the You Try It. It has not been worked so that the student may practice the skill, referring to the worked example at the left if necessary.

Reference to the Answer Section allows the student to check solutions immediately.

3.2 General Equations—Part I

> **Objective A** **To solve an equation of the form $ax + b = c$** ....

In solving an equation of the form $ax + b = c$, the goal is to rewrite the equation in the form *variable = constant*. This requires the application of both the Addition and the Multiplication Properties of Equations.

➡ Solve: $\frac{3}{4}x - 2 = -11$

The goal is to write the equation in the form *variable = constant*.

$$\frac{3}{4}x - 2 = -11$$

$$\frac{3}{4}x - 2 + 2 = -11 + 2 \qquad \bullet \text{ Add 2 to each side of the equation.}$$

$$\frac{3}{4}x = -9 \qquad \bullet \text{ Simplify.}$$

$$\frac{4}{3} \cdot \frac{3}{4}x = \frac{4}{3}(-9) \qquad \bullet \text{ Multiply each side of the equation by } \frac{4}{3}.$$

$$x = -12 \qquad \bullet \text{ The equation is of the form } variable = constant.$$

The solution is -12.

LOOK CLOSELY

Check: $\dfrac{3}{4}x - 2 = -11$

$$\frac{3}{4}(-12) - 2 \,\Big|\, -11$$

$$-11 = -11$$
A true equation

Example 1 Solve: $3x - 7 = -5$	**You Try It 1** Solve: $5x + 7 = 10$
Solution $\begin{aligned} 3x - 7 &= -5 \\ 3x - 7 + 7 &= -5 + 7 \\ 3x &= 2 \\ \frac{3x}{3} &= \frac{2}{3} \\ x &= \frac{2}{3} \end{aligned}$ The solution is $\frac{2}{3}$.	**Your solution**
Example 2 Solve: $5 = 9 - 2x$	**You Try It 2** Solve: $2 = 11 + 3x$
Solution $\begin{aligned} 5 &= 9 - 2x \\ 5 - 9 &= 9 - 9 - 2x \\ -4 &= -2x \\ \frac{-4}{-2} &= \frac{-2x}{-2} \\ 2 &= x \end{aligned}$ The solution is 2.	**Your solution**

Solutions on p. A7

The traditional teaching approach neglects the difficulties that students have in making the transition from arithmetic to algebra, specifically from concrete arithmetic to symbolic algebra. This text recognizes the formidable task the student faces by introducing variables in a very natural way—through applications of mathematics. A benefit of this approach is that the student becomes aware of the value of algebra as a real-life tool.

The solution of an application problem in this text is always accompanied by two parts: **Strategy** and **Solution.** The strategy is a written description of the steps necessary to solve the problem; the solution is the implementation of the strategy. This format provides students with a structure for problem solving. It also encourages students to write strategies for solving problems, which fosters organizing problem-solving strategies in a logical way.

110   Chapter 3 / Solving Equations

Objective B   To solve application problems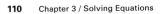

Vast number of application problems.

Example 4
A Fahrenheit temperature of 68° is 32° more than $\frac{9}{5}$ the Celsius temperature. Find the Celsius temperature.

Strategy
Write and solve an equation.

A strategy which the student may use in solving an application problem is stated.

Solution

$$\boxed{68} \text{ is } \boxed{\text{32 degrees more than } \frac{9}{5} \text{ the Celsius temperature}}$$

$$68 = \frac{9}{5}C + 32$$

$$68 - 32 = \frac{9}{5}C + 32 - 32$$

$$36 = \frac{9}{5}C$$

$$\frac{5}{9} \cdot 36 = \frac{5}{9} \cdot \frac{9}{5}C$$

$$20 = C \quad \text{The Celsius temperature is 20°.}$$

This strategy is used in the solution of the worked example.

Example 5
A board 20 ft long is cut into two pieces. Five times the length of the smaller piece is 2 ft more than twice the length of the longer piece. Find the length of each piece.

Strategy
Make a drawing.

Write and solve an equation.

Students are encouraged to write a strategy for the application problem they try to solve in the You Try It.

Solution

$$\boxed{\text{Five times the smaller piece}} \text{ is } \boxed{\text{two feet more than twice the longer}}$$

$$5x = 2(20 - x) + 2$$
$$5x = 40 - 2x + 2$$
$$5x = 42 - 2x$$
$$5x + 2x = 42 - 2x + 2x$$
$$7x = 42$$
$$\frac{7x}{7} = \frac{42}{7}$$
$$x = 6 \quad \text{The shorter piece is 6 ft.}$$
$$20 - x = 14 \quad \text{The longer piece is 14 ft.}$$

You Try It 4
A Celsius temperature of 20° is $\frac{5}{9}$ of the difference between the Fahrenheit temperature and 32. Find the Fahrenheit temperature.

Your strategy

Your solution

You Try It 5
A company makes 140 televisions per day. Three times the number of black and white TV's made equals 20 less than the number of color TV's made. Find the number of color TV's made each day.

Your strategy

Your solution

Solutions on pp. A9–A10

OBJECTIVE-SPECIFIC APPROACH

Many texts in mathematics are not organized in a manner that facilitates management of learning. Typically, students are left to wander through a maze of apparently unrelated lessons, exercise sets, and tests. *Introductory Algebra: An Applied Approach* solves this problem by organizing all lessons, exercise sets, computer tutorials and tests around a carefully constructed hierarchy of objectives. The advantage of this objective-by-objective organization is that it enables the student who is uncertain at any step in the learning process to refer easily to the original presentation and review that material.

The objective-specific approach also gives the instructor greater control over the management of student progress. The Computerized Test Generator and the printed Test Bank are organized by the same objectives as the text. These items are accompanied by answers to the test items, which allows the instructor to quickly determine those objectives for which a student may need additional instruction.

The Computer Tutor is also organized around the objectives of the text. As a result, supplemental instruction is available for any objectives that are troublesome for a student.

A numbered objective statement names the topic of each lesson.

2.2 Simplifying Variable Expressions

Objective A   *To simplify a variable expression using the Properties of Addition* ....

The exercise sets correspond to the objectives in the text.

2.2 Exercises

Objective A

Simplify.

1. $6x + 8x$ **2.** $12x + 13x$ **3.** $9a - 4a$ **4.** $12a - 3a$

The answers to the Chapter Test show the objective to study if the student incorrectly answers an item.

CHAPTER TEST *pages 75–76*

1. 22 [2.1A]   **2.** 3 [2.1A]   **3.** $5x$ [2.2A]   **4.** y^2 [2.2A]   **5.** $-9x - 7y$ [2.2A]
6. $-2x - 5y$ [2.2C]   **7.** $2x$ [2.2B]   **8.** $3x$ [2.2B]   **9.** $36y$ [2.2B]   **10.** $-10a$ [2.2B]
11. $15 - 35b$ [2.2C]   **12.** $-4x + 8$ [2.2C]   **13.** $-6x^2 + 21y^2$ [2.2C]

The answers to the Cumulative Review exercises also show the objective that relates to the exercise.

CUMULATIVE REVIEW *pages 77–78*

1. -7 [1.2A]   **2.** 5 [1.2B]   **3.** 24 [1.3A]   **4.** -5 [1.3B]   **5.** 1.25 [1.4A]
6. $\frac{11}{48}$ [1.4C]   **7.** $\frac{1}{6}$ [1.4D]   **8.** $\frac{1}{4}$ [1.4D]   **9.** 75% [1.4B]   **10.** -5 [1.5B]
11. $\frac{53}{48}$ [1.5B]   **12.** 16 [2.1A]   **13.** $5x^2$ [2.2A]   **14.** $-7a - 10b$ [2.2A]   **15.** $6a$ [2.2B]

ADDITIONAL LEARNING AIDS

Project in Mathematics

The Project in Mathematics feature occurs at the end of each chapter. These projects can be used as extra credit or cooperative learning activities. The projects cover various aspects of mathematics including the use of calculators, extended applications, additional problem solving strategies, and other topics related to mathematics.

Chapter Summary

The Chapter Summary includes Key Words and Essential Rules that were covered in the chapter. The chapter summary is a self-contained reference guide for the student preparing for a test.

Study Skills

To The Student, provides suggestions for using this text and approaches to creating good study habits.

Computer Tutor

The Computer Tutor is a networkable, interactive, algorithmically driven software package. This powerful ancillary features full-color graphics, extensive hints, animated solution steps, a glossary, and a comprehensive class management system.

Glossary

New to this edition, a Glossary at the back of the book includes definitions of terms used in the text.

EXERCISES

End-of-Section Exercises

Introductory Algebra: An Applied Approach contains more than **6000** exercises. At the end of each section there are exercise sets that are keyed to the corresponding learning objective. The exercises are carefully developed to ensure that students can apply the concepts in the section to a variety of problem situations.

Applying the Concepts Exercises

The End-of-Section Exercises are followed by Applying the Concepts Exercises. These sections contain a variety of exercise types which include:

- problems that offer students more challenges
- problems that require students to determine if a statement is always true, sometimes true, or never true
- problems that ask students to determine incorrect procedures
- problems that require a calculator to determine a solution

Writing Exercises

Within the "Applying the Concepts" exercises, there are writing exercises denoted by **[W]**. These exercises ask students to write about a topic in the section or to research and report on a related topic.

Chapter Review Exercises

Review exercises are found at the end of each chapter. These exercises are selected to help students integrate all of the topics presented in the chapter. The answers to all review exercises are given in the answer section.

Cumulative Review Exercises

Cumulative Review Exercises, which appear at the end of each chapter beginning with Chapter 2, help students maintain skills learned in previous chapters. The answers to all Cumulative Review exercises are given in the answer section. Along with the answer, there is a reference to the objective that pertains to each exercise.

NEW TO THIS EDITION

Topical Coverage

Introductory Algebra: An Applied Approach retains its strong committment to applications of mathematics. However, at the request of users of earlier editions, we have moved some of the applications from Chapter 4, *Applications of First-Degree Equations*, to appropriate places in other chapters. This provides for a more balanced approach to application problems throughout the text.

We have added many new application problems to the text. Some of these problems incorporate short writing exercises. These exercises ask students to explain, in the context of the application, the significance of their answer.

Since the first edition of this text, we have emphasized to students that they should write the strategy they will use to solve a word problem. In our classes, we have found that students will frequently not include a diagram to aid them in their attempt to solve a problem. With this edition, we have included the "Draw a diagram" instruction in those applications where it is appropriate.

In this edition of *Introductory Algebra: An Applied Approach*, we have separated division of monomials and division of polynomials in Chapter 5 into two sections. This provides more emphasis on the operations of negative exponents and polynomial division.

In Chapter 8, *Linear Equations in Two Variables*, we have added an introduction to functions and to function notation.

Margin Notes

Margin notes are interspersed throughout the text. These notes are called *Look Closely* and *Point of Interest*. The *Look Closely* note warns students that a procedure may be particularly involved or reminds students that there are certain checks of their work that should be performed. The *Point of Interest* note is an interesting sidelight of the topic being discussed. In addition, there are *Instructor Notes*, which are printed only in the Instructor's Annotated Edition.

Applying the Concepts

The End-of-Section exercises are followed by Applying the Concept Exercises. These exercises include challenge problems and applications of the concepts presented in the section in a way that requires a student to combine several skills to solve a problem.

Computer Tutor

The Computer Tutor has been completely revised. It is now an algorithmically based tutor that includes color and animation. The algorithmic feature essentially provides an infinite number of practice problems for students to attempt. The algorithms have been carefully crafted to present a variety of problem types from easy to difficult. A complete solution to each problem is available.

There is an interactive feature of the Computer Tutor that requires students to respond to questions about the topic in the current lesson. In this way, students can assess their understanding of concepts as they are presented. There is a Glossary which can be accessed at any time so that students can look up words whose definitions they may have forgotten.

When the student completes a lesson, a printed report is available. This optional report gives the student's name, the objective studied, the number of problems attempted, the number of problems correct, and the percent correct.

Projects in Mathematics

Through the Projects in Mathematics feature, some of the standards suggested by the NCTM can be implemented. These Projects offer an opportunity for students to explore several topics relating to calculators, applications of math, logic, problem solving, and related subjects.

Graphing Calculators

Graphing calculators are incorporated as an optional feature at appropriate places throughout the text. Some of the Projects in Mathematics discuss the use of these calculators.

Appendix A is a special section with guidelines for graphing on various models of graphing calculators.

Glossary

The glossary at the back of the book includes definitions to key terms referenced to the section in which they were first defined.

SUPPLEMENTS FOR THE INSTRUCTOR

Instructor's Annotated Edition

The Instructor's Annotated Edition is an exact replica of the student text except that answers to all exercises are given in the text. Also, there are Instructor Notes in margin that offer suggestions for presenting the material in that objective.

Instructor's Resource Manual with Chapter Tests

The Instructor's Resource Manual contains the printed testing program, which is the first of three sources of testing material. Four printed tests, two free response and two multiple choice, are provided for each chapter. In addition, there are cumulative tests after Chapters 4, 8, and 11, and a final exam. The Instructor's Resource manual also includes suggestions for course sequencing and outlines for the answers to the writing exercises.

Computerized Test Generator

The Computerized Test Generator is the second source of testing material. The data base contains more than 3000 test items. The Test Generator is designed to provide an unlimited number of tests for each chapter, cumulative chapter tests, and a final exam. It is available for the IBM PC and compatible computers and the Macintosh.

Printed Test Bank

The Printed Test Bank, the third component of the testing material, is a printout of all items in the Computerized Test Generator. Instructors who do not have access to a computer can use the test bank to select items to include on a test being prepared by hand.

Solutions Manual

The Solutions Manual contains worked-out solutions for all end-of-section exercises, critical thinking exercises, chapter review exercises, and cumulative review exercises.

Videotapes

The videotape series contains lessons to accompany *Introductory Algebra: An Applied Approach*. These lessons are closely tied to specific sections of the text. Each tape begins with footage of a real-life application, which is solved during the lesson.

SUPPLEMENTS FOR THE STUDENT

Student Solutions Manual

The Student Solutions Manual contains the complete solutions to all odd-numbered exercises in the text.

Computer Tutor

The Computer Tutor is an interactive instructional computer program for student use. Each objective of the text is supported by a tutorial in the Computer Tutor. These tutorials contain an interactive lesson which covers the material in the objective. Following the lesson are randomly generated exercises for the student to attempt. A record of the student's progress is available.

The Computer Tutor can be used in several ways: (1) to cover material the student missed because of an absence; (2) to reinforce instruction on a concept that the student has not yet mastered; (3) to review material in preparation for exams. This tutorial is available for the IBM PC and compatible computers running Windows and the Macintosh.

ACKNOWLEDGMENTS

The authors would like to thank the people who have reviewed this manuscript and provided many valuable suggestions:

Nancy E. Carpenter, *Johnson County Community College, KS;* **Ellen Casey,** *Massachusetts Bay Community College, MA;* **Barbara Cribbs,** *Stark Technical College, OH;* **Therese Cummings,** *William Rainey Harper College, IL;* **Richard A. Davis,** *Community College of Allegheny County, PA;* **Irene Doo,** *Austin Community College, TX;* **Robert Eicken,** *Illinois Central College, IL;* **Cheryl Hobneck,** *Illinois Valley Community College, IL;* **Laura L. Hoye,** *Trident Technical College, SC;* **Paul Hrabovsky,** *Indiana University of Pennsylvania, PA;* **Alice Kristufek,** *Slippery Rock University, PA;* **Harvey A. LeBoff,** *Middlesex Community College, MA;* **Randy Leifson,** *Pierce College, WA;* **Doris Lewis,** *University of Toledo, OH;* **Sarah Jawe Martin,** *Virginia Western Community College, VA;* **Hector Mendez,** *University of Georgia, GA;* **Linda Mudge,** *Illinois Valley Community College, IL;* **Doris Nice,** *University of Wisconsin/Parkside, WI;* **Ghee Lip Ong,** *Delta-Ouachita Vocational Technical Institute, LA;* **William Peters,** *San Diego Mesa College, CA;* **Edith M. Ruben,** *New River Community College, VA;* **Louis Shanin,** *Victor Valley College, CA;* **Judy Staver,** *Florida Community College at Jacksonville, FL;* **Loretta Sullivan,** *University of Detroit Mercy, MI;* **Gary Sundberg,** *Little Hoop Community College, ND;* **Sharon Testone,** *Onondaga Community College, NY;* **Jennie Thompson,** *Leeward Community College, HI;* **Raymond Weaver,** *Community College of Allegheny County, PA*

To the Student

Many students feel that they will never understand math while others appear to do very well with little effort. Often times what makes the difference is that successful students take an active role in the learning process.

Learning mathematics requires your *active* participation. Although doing homework is one way you can actively participate, it is not the only way. First, you must attend class regularly and become an active participant. Secondly, you must become actively involved with the textbook.

Introductory Algebra: An Applied Approach was written and designed with you in mind as a participant. Here are some suggestions on how to use the features of this textbook.

There are 12 chapters in this text. Each chapter is divided into sections and each section is subdivided into learning objectives. Each learning objective is labeled with a letter from A–E.

First, read each objective statement carefully so you will understand the learning goal that is being presented. Next, read the objective material carefully, being sure to note each bold word. These words indicate important concepts that you should familiarize yourself with. Study each in-text example carefully, noting the techniques and strategies used to solve the example.

You will then come to the key learning feature of this text, the *boxed examples*. These examples have been designed to aid you in a very specific way. Notice that in each example box, the example on the left is completely worked out and the "You Try It" example on the right is not. *You* are expected to work the right-hand example (in the space provided) in order to immediately test your understanding of the material you have just studied.

You should study the worked-out example carefully by working through each step presented. This allows you to focus on each step and reinforces the technique for solving that type of problem. You can then use the worked-out example as a model for solving similar problems.

Solve the "You Try It" example using the problem-solving techniques that you have just studied. When you have completed your solution, check your work by turning to the page in the appendix where the complete solution can be found. The page number on which the solution appears is printed at the bottom of the example box in the right-hand corner. By checking your solution, you will know immediately whether or not you fully understand the skill just studied.

When you have completed studying an objective, do all of the exercises in the exercise set that correspond with that objective. The exercises will be labeled with the same letter as the objective. Algebra is a subject that needs to be learned in small sections and practiced continually in order to be mastered. Doing all of the exercises in each exercise set will help you master the problem solving techniques necessary for success.

Once you have completed the exercises to an objective, you should check your answers to the odd-numbered exercises with those found in the back of the book.

After completing a chapter, read the Chapter Summary. This summary highlights the important topics covered in the chapter. Following the Chapter Summary are Chapter Review Exercises, a Chapter Test, and a Cumulative

Review (beginning with Chapter 2). Doing the review exercises is an important way of testing your understanding of the chapter. The answer to each review exercise is in an appendix at the back of the book. Each answer is followed by a reference that tells which objective that exercise was taken from. For example, (4.2B) means Section 4.2, Objective B. After checking your answers, restudy any objective that you missed. It may be very helpful to retry some of the exercises for that objective to reinforce your problem-solving techniques.

The Chapter Test should be used to prepare for an exam. We suggest that you try the Chapter Test a few days before your actual exam. Take the test in a quiet place and try to complete the test in the same amount of time you will be allowed for your exam. When taking the Chapter Test, practice the strategies of successful test takers: 1) scan the entire test to get a feel for the questions; 2) read the directions carefully; 3) work the problems that are easiest for you first; and perhaps most importantly, 4) try to stay calm.

When you have completed the Chapter Test, check your answers. If you missed a question, review the material in that objective and rework some of the exercises from that objective. This will strengthen your ability to perform the skills in that objective.

The Cumulative Review allows you to refresh the skills you have learned in previous chapters. This is very important in mathematics. By consistently reviewing previous materials, you will retain the previous skills as you build new ones.

Remember, to be successful, attend class regularly; read the textbook carefully; actively participate in class; work with your textbook using the "You Try It" examples for immediate feedback and reinforcement of each skill; do all the homework assignments; review constantly; and work carefully.

Index of Applications

Agriculture
Maximum area of irrigation, 440
Mixture in a silo, 380

Architecture and Fine Arts
Eye-pleasing proportions, 113
Graph of architect fee and square feet, 322
Graph of hit records and number of concerts, 311
Height of the Eiffel Tower, 118
Tickets to community theater, 382

Astronomy
Mass of earth, 183
Parsec, 183
Sulfur in the atmosphere, 184
Temperature on the Moon, 10
Temperature in stratosphere, 10
Weight and distance in space, 444

Business
Airline overbooking, 92
Break-even point, 105, 106, 282
Cost, 125, 151, 153
Cost of commercial, 91
Cost of mailing, 380
Cost of software programs, 375
Cost of wheat and rye flour, 375
Depreciation of office machine, 485
Discount, 117
Discount price, 128
Discount rate, 126, 127, 128, 151, 153
Graph of depreciation, 322
Graph of the cost of a transatlantic phone call, 331
Graph of the revenue of Texas Instruments, 340
Markup, 128
Markup rate, 125, 127, 153, 156, 198, 485
Profit-Loss chart, 10, 14, 28, 33
Purchasing fluorescent and incandescent bulbs, 374
Purchasing orange and grapefruit trees, 374
Regular price, 126, 128, 151
Sale price, 127, 128, 151, 153, 156, 348
Selling price, 127, 396, 448, 482
Simple interest, 117
Straight line depreciation, 97
Value mixture, 133, 134, 137, 138, 140, 152, 153, 154, 156, 373, 485

Chemistry
Amount of oxygen, 91
Avogadro's number, 183
Percent mixture, 135, 136, 139, 140, 152, 153, 154, 156, 198, 296, 298, 384, 448, 485
Temperature of mixture, 120, 122

Computers
Access time, 120
Fee for computer bulletin board, 404
Graph of the age of computer and down time, 344
Licensing fee for computer program, 404
Number of pixils, 113
Speed of a computer, 113
Time to develop software, 400

Construction
Adding dye to base paint, 486
Area of a fountain, 440
Area of a garage, 122
Area of ceramic tile, 276
Cutting a board, 110, 120, 122, 246
Estimating amount of paint needed, 278
Graph of area of counter top and cost, 318
Graph showing size and cost of a house, 336
Height of a building, 438
Length of guy wire, 438
Lighting for billboards, 278

Consumer Economics
Charges of a computer bulletin board, 344
Comparison cost for transportation, 404
Cost of a book, 117
Cost of electricity, 113
Cost of life insurance, 298
Early reservation discount, 91
Graph of rental car fee and miles driven, 322
Graph of the depreciation of a used Macintosh, 326
Graph of the number of ATMs, 339
Income tax, 399
Making syrup for desserts, 277
Median price of a home, 86
Monthly payment on a car, 276

Most economical rental car, 402, 404, 416
Radius of a pizza, 474
Real estate taxes, 348
Total gasoline tax, 34

Consumption of Resources
Graph of number of miles driven and tread depth, 339

Economics
Congressional budget, 86
Cost for prime and non-prime time, 376
Cumulative return for S&P, 358
Currency exchange, 33, 46
Input-output analysis, 350
Per share earnings, 44
National debt, 183

Education
Graph of hours studying and grade, 305
Graph of midterm and final exam, 307
Graph of reading scores and history grade, 336
Graph of the increase in cost of tuition, 346
Lowest score to attain points, 411, 482
Multiple choice exam, 20
Number of research assistants, 113
Percent grade, 85
Percent of student in fine arts, 91
Range of test scores for an A grade, 396
Requirement for a B grade, 92, 400

Environment and Biological Sciences
Estimating number of fish, 278, 416
Estimating number of elk, 278
Relating height and length of the humerus, 97
Requirement for grade A hamburger, 404
Requirement to be labeled orange juice, 404

Geography
Area of the earth, 184
Circumference of the earth, 246

Daily high temperature, 18, 20
Daily low temperature, 18, 20, 45
Difference in elevation, 13
Difference in temperature, 14
Tsunami wave, 444
Wind-chill temperature, 13, 14

Geometry
Angles of a triangle, 146, 147, 148, 152, 154, 348, 416, 485
Area of a softball diamond, 174
Area of a circle, 170, 174, 194, 238
Area of a rectangle, 170, 173, 174, 192
Area of a square, 173, 192, 198
Area of a triangle, 440
Base of a triangle, 246, 298, 473
Complementary angles, 106
Distance to horizon, 441
Distance to memorial, 466
Height and width of TV screen, 440
Height of a triangle, 473
Legs of a right triangle, 474
Length and width of a rectangle, 198, 234, 237, 238, 242, 244, 474, 480, 486
Maximum length of a rectangle, 402
Measure of angles, 98, 113, 114
Perimeter of a rectangle, 145, 147, 148, 152, 154, 156
Perimeter of a triangle, 145, 147, 148, 152, 154, 156, 440
Perimeter of a square, 113
Pythagorean Theorem, 440
Radius of a circle, 238
Side of a square, 238, 242, 384
Supplementary angles, 106, 376
Width of a border, 238
Width of a frame, 242
Width of a rectangle, 414, 472

Health and Physical Fitness
Amount of medication, 276
Amount spent on drugs, 28
Daily allowance of vitamin C, 400
Graph of heart rate and medication, 307
Graph of the height and weight of students, 343
Minimum height to ride roller coaster, 413
Recommended cholesterol levels, 400
Recommended number of air vents, 277

History
Problem from Chinese manuscript, 376

Home Maintenance
Area of lawn sprinkler, 296
Length and width of patio, 113
Minimum length of a garden, 412
Repair bill for washing machine, 155
Speed of lazer printer, 278

Investments
Amount invested, 130, 131, 132, 151, 153, 156, 246, 376, 384, 485
Annual percentage rate, 456
Annual simple interest, 129
Dividend per share from two stocks, 376
Graph of the value of an investment and time, 318
Graph of the decline in price of a stock, 332
Graph of the increase in number of mutual funds, 346
Income from an investment, 276
Number of shares of stock, 380, 482
Moving average price of stock, 20

Manufacturing and Industry
Allowable circumference for ball bearing, 413
Cost per unit, 119, 122
Decrease in production, 86
Manufacture of TVs, 110
Number of pickups, 92
Quality control, 277, 278
Units of production, 94

Mathematics
Calculations of Pi, 386
Coin problem, 376
Consecutive integers, 109, 112, 120, 198, 233, 482, 234, 235, 246, 446, 448, 473, 474, 486
Continued fractions, 291
Contradiction, 114
Distance of a screen, 242
Dividing profits from car wash, 278
Early Egyptian fractions, 2
Egyptian arithmetic operations, 158
Geometric construction of completing the square, 475
Geometric proofs of algebraic identities, 200
Magic squares, 300
Mersenne primes, 80
Perfect numbers, 206

Physics
Ages of gold and silver coins, 376
Boiling and freezing points, 43

Charge on an electron, 183
Conversion, Centigrade-Fahrenheit, 110
Distance between two planes, 466
Falling object, 97, 166, 234, 235, 346, 448
Golden ratio, 434
Graph of distance and time, 318
Graph of the pressure on a diver, 332
Height of periscope, 438, 439
Intensity of illumination, 189
Kinetic energy, 456
Length of pendulum, 438, 440, 446
Length of light wave, 184
Lever system, 102, 105, 118
Pressure in the ocean, 94
Radius of sharpest turn, 444
Rate of current problem, 372, 375, 380, 384, 474, 480, 486
Rate of wind problem, 370, 372, 375, 380, 382, 384, 474, 478, 482, 486
Stopping distance, 456
Time to fly a certain distance, 120
Total resistance, 280
Uniform motion, 141, 142, 143, 144, 152, 154, 156, 198, 246, 285, 286, 289, 290, 294, 296, 298, 384, 471, 474, 482, 485
Work problems, 284, 287, 288, 290, 294, 296, 298, 348, 448, 472, 474, 486

Social Science
Characteristics of children, 33
Exit poll, 277
Graph of the decline in price of a home, 332
Graph of distance from fire station and damage, 345
Graph of the number of employed and retired, 340
Lottery tickets, 278
Nielsen ratings, 115
Number to override a veto, 91
Number of phone calls, 98
Survey of consumers, 34

Sports
Circumference of official baseball, 399
Distance of pitcher's mound, 439
Earned run average, 294
Graph of distance ran and time, 325
Graph of number of aces and games won, 311
Graph of runs scored and won or lost, 311

Content and Format © 1995 HMCo.

Increasing playoff games, 91
Length and width of batter's box, 473
Length and width of a pool, 473
Length and width of a tennis court, 473
Number of field goals, 113
Number of foul shots made, 278
Number of hits to increase earnings, 403

Number of three point baskets, 113
Qualifying for a tournament, 399
Schedules for sports teams, 237
Scoring of a basketball team, 376
Scoring of a football team, 376
Seating of a football stadium, 91
Time baseball will stay in air, 462
Time basketball is in the air, 462
Wins-losses for the NBA, 33

Wages and Earned Income
Alternate methods of payment, 403
Consulting fee, 118
Earning a commission, 400
Monthly salary, 400, 414
Percent increase in hourly wage, 86

Introductory Algebra
An Applied Approach

Real Numbers

Objectives

Section 1.1
To use inequality symbols with integers
To use opposites and absolute value

Section 1.2
To add integers
To subtract integers
To solve application problems

Section 1.3
To multiply integers
To divide integers
To solve application problems

Section 1.4
To write a rational number in simplest form
  and as a decimal
To convert between percents, fractions, and
  decimals
To add or subtract rational numbers
To multiply or divide rational numbers
To solve application problems

Section 1.5
To evaluate exponential expressions
To use the Order of Operations Agreement to
  simplify expressions

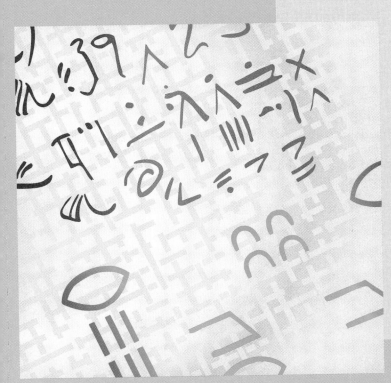

Early Egyptian Fractions

One of the earliest written documents of mathematics is the Rhind Papyrus. This tablet was found in Egypt in 1858, but it is estimated that the writings date back to 1650 B.C.

The Rhind Papyrus contains over 80 problems. A study of these problems has enabled mathematicians and scientists to understand some of the methods by which the early Egyptians used mathematics.

Evidence gained from the Papyrus shows that the Egyptian method of calculating with fractions was much different from the methods used today. All fractions were represented in terms of what are called "unit fractions." A unit fraction is a fraction in which the numerator is one. This fraction was symbolized (using modern numbers) with a bar over the number.

For example, $\overline{3} = \dfrac{1}{3}; \qquad \overline{15} = \dfrac{1}{15}$

The early Egyptians also tended to deal with powers of two $(2, 4, 8, 16, \ldots)$. As a result, representing fractions with 2 in the numerator in terms of unit fractions was an important matter. The Rhind Papyrus has a table giving the equivalent unit fractions for all fractions with odd denominators from 5 to 101 and 2 as the numerator. Some of these are listed below.

$$\frac{2}{5} = \overline{3}\ \overline{15} \qquad \left(\frac{2}{5} = \frac{1}{3} + \frac{1}{15}\right)$$

$$\frac{2}{7} = \overline{4}\ \overline{28}$$

$$\frac{2}{11} = \overline{6}\ \overline{66}$$

$$\frac{2}{19} = \overline{12}\ \overline{76}\ \overline{114}$$

Content and Format © 1995 HMCo.

1.1 Introduction to Integers

Objective A   ***To use inequality symbols with integers*** ....

It seems to be a human characteristic to put similar items in the same place. For instance, a biologist places similar animals in groups called *phyla,* and a chemist places atoms with similar characteristics in a *periodic table.*

Mathematicians likewise place objects with similar properties in *sets* and use braces to surround the objects in the set. The numbers that we use to count objects, such as the number of people at a baseball game or the number of horses on a ranch, have similar characteristics. These numbers are the *natural numbers.*

$$\textbf{Natural numbers} = \{1, 2, 3, 4, 5, 6, 7, 8, 9, 10, 11, \ldots\}$$

The natural numbers alone do not provide all the numbers that are useful in applications. For instance, a meteorologist needs numbers below zero and above zero.

$$\textbf{Integers} = \{\ldots, -5, -4, -3, -2, -1, 0, 1, 2, 3, 4, 5, \ldots\}$$

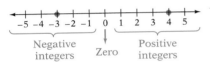

Fig. 1

Each integer can be shown on a **number line**. The **graph** of an integer is shown by placing a heavy dot on the number line directly above the number. The graph of -3 and that of 4 are shown on the number line in Fig. 1.

The integers to the left of zero are **negative integers**. The integers to the right of zero are **positive integers**. Zero is neither a positive nor a negative integer.

Consider the sentences below.

> The quarterback threw the football and the receiver caught *it.*
> An accountant purchased a calculator and placed *it* in a briefcase.

In the first sentence, *it* means football; in the second sentence, *it* means calculator. In language, the word *it* can stand for many different objects. Similarly, in mathematics, a letter of the alphabet can be used to stand for a number. Such a letter is called a **variable**. Variables are used in the next definition.

> **Definition of Inequality Symbols**
>
> If *a* and *b* are two numbers and *a* is to the left of *b* on the number line, then *a* is **less than** *b*. This is written $a < b$.
> If *a* and *b* are two numbers and *a* is to the right of *b* on the number line, then *a* is **greater than** *b*. This is written $a > b$.

There are also inequality symbols for **less than or equal to** ($\leq$) and **greater than or equal to** ($\geq$). For instance,

$$6 \leq 6 \text{ because } 6 = 6. \qquad 7 \leq 15 \text{ because } 7 < 15.$$

It is convenient to use a variable to represent, or stand for, any one of the elements of a set. For instance, the statement "*x* is an *element of* the set $\{0, 2, 4, 6\}$" means that *x* can be replaced by 0, 2, 4, or 6. The symbol for "is an element of" is $\in$; the symbol for "is not an element of" is $\notin$. For example,

$$2 \in \{0, 2, 4, 6\} \qquad 6 \in \{0, 2, 4, 6\} \qquad 7 \notin \{0, 2, 4, 6\}$$

Example 1
Let $x \in \{-6, -2, 0\}$. For which values of x is the inequality $x \le -2$ a true statement?

Solution
Replace x by each element of the set and determine whether the inequality is true.

$$x \le -2$$
$$-6 \le -2 \quad \text{True.} \quad -6 < -2$$
$$-2 \le -2 \quad \text{True.} \quad -2 = -2$$
$$0 \le -2 \quad \text{False}$$

The inequality is true for -6 and -2.

You Try It 1
Let $y \in \{-5, -1, 5\}$. For which values of y is the inequality $y > -1$ a true statement?

Your solution

$x = -5$

$-5 > -1$ no

$+7 > -1$ ye

Solution on p. A1

Objective B   To use opposites and absolute value

Fig. 2

LOOK CLOSELY
From Fig. 2, the distance from 0 to 5 is 5; $|5| = 5$. The distance from 0 to -5 is 5; $|-5| = 5$.

POINT OF INTEREST
The definition of absolute value that we have given in the box is written in what is called "rhetorical style." That is, it is written without the use of variables. This is how *all* mathematics was written prior to the Renaissance. During that period from the 14th to the 16th century, the idea of expressing a variable symbolically was developed.

On the number line, the numbers 5 and -5 are the same distance from zero but on opposite sides of zero. The numbers 5 and -5 are called **opposites** or **additive inverses** of each other. (See Fig. 2.)

The opposite (or additive inverse) of 5 is -5. The opposite of -5 is 5. The symbol for opposite is $-$.

$-(5)$ means the opposite of *positive* 5. $-(5) = -5$
$-(-5)$ means the opposite of *negative* 5. $-(-5) = 5$

The **absolute value** of a number is its distance from zero on the number line. The symbol for absolute value is two vertical bars, $|\ |$.

> **Absolute Value**
> The absolute value of a positive number is the number itself. The absolute value of zero is zero. The absolute value of a negative number is the opposite of the negative number.

➡ Evaluate: $-|-12|$

$$-|-12| = -12$$

• The absolute value sign does not affect the negative sign in front of the absolute value sign.

Example 2
Let $a \in \{-12, 0, 4\}$. Find $-a$, the opposite of a, for each element of the set.

Solution
Replace a by each element of the set.

$$-a$$
$$-(-12) = 12$$
$$-(0) = 0 \qquad \text{• Zero is neither positive}$$
$$-(4) = -4 \qquad \text{nor negative.}$$

You Try It 2
Let $z \in \{-11, 0, 8\}$. Evaluate $|z|$ for each element of the set.

Your solution

$z =$

$|-1| = 11$

$|0| = 0$

$|8| = 0$

Solution on p. A1

1.1 Exercises

Objective A

Place the correct symbol, $<$ or $>$, between the two numbers.

1. 8 -6 **2.** -14 16   **3.** -12 1   **4.** 35   28   **5.** 42   19

6. -42 27   **7.** 0 -31 **8.** -17 0   **9.** 53 -46 **10.** -27 -39

Answer True or False.

11. $-13 > 0$ **12.** $-20 > 3$ **13.** $12 > -31$ **14.** $9 > 7$ **15.** $-5 > -2$

16. $-44 > -21$ **17.** $-4 > -120$ **18.** $0 > -8$ **19.** $-1 > 0$ **20.** $-10 > -88$

21. Let $a \in \{-7, 0, 2, 5\}$. For which values of a is the inequality $a > 2$ a true statement?

22. Let $b \in \{-8, 0, 7, 15\}$. For which values of b is the inequality $b > 7$ a true statement?

23. Let $x \in \{-23, -18, -8, 0\}$. For which values of x is the inequality $x < -8$ a true statement?

24. Let $w \in \{-33, -24, -10, 0\}$. For which values of w is the inequality $w < -10$ a true statement?

25. Let $a \in \{-33, -15, 21, 37\}$. For which values of a is the inequality $a > -10$ a true statement?

26. Let $v \in \{-27, -14, 14, 27\}$. For which values of v is the inequality $v > -15$ a true statement?

27. Let $b \in \{-52, -46, 0, 46, 52\}$. For which values of b is the inequality $0 < b$ a true statement?

28. Let $y \in \{-12, -9, 0, 12, 34\}$. For which values of y is the inequality $0 < y$ a true statement?

29. Let $n \in \{-23, -1, 0, 4, 29\}$. For which values of n is the inequality $-6 > n$ a true statement?

30. Let $m \in \{-33, -11, 0, 12, 45\}$. For which values of m is the inequality $-15 > m$ a true statement?

Objective B

Find the additive inverse.

31. 4   **32.** 8   **33.** −9   **34.** −12   **35.** −28   **36.** −36

Evaluate.

37. −(−14)   **38.** −(−40)   **39.** −(77)   **40.** −(39)   **41.** −(0)   **42.** −(−13)

43. |−74| **44.** |−96| **45.** −|−82| **46.** −|−53| **47.** −|81| **48.** −|38|

Place the correct symbol, < or >, between the values of the two numbers.

49. |−83| |58| **50.** |22| |−19| **51.** |43| |−52| **52.** |−71| |−92|

53. |−68| |−42| **54.** |12| |−31| **55.** |−45| |−61| **56.** |−28| |43|

57. Let $p \in \{-19, 0, 28\}$. Evaluate $-p$ for each element of the set.

58. Let $q \in \{-34, 0, 31\}$. Evaluate $-q$ for each element of the set.

59. Let $x \in \{-45, 0, 17\}$. Evaluate $-|x|$ for each element of the set.

60. Let $y \in \{-91, 0, 48\}$. Evaluate $-|y|$ for each element of the set.

APPLYING THE CONCEPTS

61. Let $z \in \{-15, -10, 0, 5, 10\}$. For which values of z is the inequality $-z > -|z|$ a true statement?

62. Let $y \in \{-42, -31, 0, 17, 48\}$. For which values of y is the inequality $-y < -|y|$ a true statement?

63. If x represents a negative integer, then $-x$ represents a ___+___ integer.

64. If x represents a positive integer, then $-x$ represents a ___−___ integer.

65. If a and b are integers and $a < b$, is the inequality $|a| < |b|$ always true, sometimes true, or never true?

66. If x is an integer, is the inequality $|x| < -3$ always true, sometimes true, or never true?

67. Give examples of some games that use negative integers in the scoring.

68. [W] In your own words, explain the meaning of the absolute value of a number and the additive inverse of a number.

69. [W] Give some examples of English words that are used as variables.

1.2 Addition and Subtraction of Integers

Objective A *To add integers*

A number can be represented anywhere along the number line by an arrow. A positive number is represented by an arrow pointing to the right, and a negative number is represented by an arrow pointing to the left. The size of the number is represented by the length of the arrow.

Addition of integers can be shown on the number line. To add integers, start at zero and draw an arrow representing the first number. At the tip of the first arrow, draw a second arrow representing the second number. The sum is below the tip of second arrow.

$4 + 2 = 6$

$-4 + (-2) = -6$

$-4 + 2 = -2$

$4 + (-2) = 2$

The pattern for the addition of integers shown on the number line can be summarized in the following rule.

> **Addition of Integers**
>
> - *Numbers have the same sign*
>   To add two numbers with the same sign, add the absolute values of the numbers. Then attach the sign of the addends.
>
> - *Numbers have different signs*
>   To add two numbers with different signs, find the absolute value of each number. Then subtract the smaller of these numbers from the larger one. Attach the sign of the number with the larger absolute value.

⟹ Add: $(-9) + 8$

$|-9| = 9$ $|8| = 8$ • The signs are different. Find the absolute value of each number.

$9 - 8 = 1$ • Subtract the smaller number from the larger.

$(-9) + 8 = -1$ • Attach the sign of the number with larger absolute value. Because $|-9| > |8|$, use the sign of -9.

⟹ Add: $(-23) + 47 + (-18) + (-5)$

To add more than two numbers, add the first two numbers. Then add the sum to the third number. Continue until all the numbers are added.

$$(-23) + 47 + (-18) + 5 = 24 + (-18) + 5$$
$$= 6 + 5$$
$$= 11$$

Example 1   Add: $(-52) + (-39)$

Solution   The signs are the same. Add the absolute values of each number: $52 + 39 = 91$

Attach the sign of the addends: $(-52) + (-39) = -91$

Example 2   Add: $37 + (-52) + (-21) + (-7)$

Solution $37 + (-52) + (-21) + (-7)$
$= -15 + (-21) + (-7)$
$= -36 + (-7)$
$= -43$

You Try It 1   Add: $100 + (-43)$

Your solution

You Try It 2   Add: $(-51) + 42 + 17 + (-102)$

Your solution

Solutions on p. A1

Objective B   To subtract integers

Look at the two expressions below and note that each expression equals the same number.

$8 - 3 = 5$ 8 minus 3 is 5

$8 + (-3) = 5$ 8 plus the opposite of 3 is 5

This example suggests that to subtract two numbers, we add the opposite of the second number to the first number.

first number	−	second number	=	first number	+	the opposite of the second number	
40	−	60	=	40	+	(-60)	$= -20$
−40	−	60	=	−40	+	(-60)	$= -100$
−40	−	(-60)	=	−40	+	60	$= 20$
40	−	(-60)	=	40	+	60	$= 100$

➡ Subtract: $-21 - (-40)$

Change this sign to plus.

$-21 - (-40) = -21 + 40 = 19$

Change -40 to the opposite of -40.

+40
-21
19

• Rewrite each subtraction as addition of the opposite. Then add.

➡ Subtract: $15 - 51$

Change this sign to plus.

$15 - 51 = 15 + (-51) = -36$

Change 51 to the opposite of 51.

• Rewrite each subtraction as addition of the opposite. Then add.

➡ Subtract: $-12 - (-21) - 15$

$$-12 - (-21) - 15 = -12 + 21 + (-15)$$
$$= 9 + (-15)$$
$$= -6$$

-12 + 21

• Rewrite each subtraction as addition of the opposite. Then add.

Example 3   Subtract: $-11 - 15$

Solution $-11 - 15 = -11 + (-15)$
$\qquad\qquad\quad = -26$

Example 4   Subtract:
$-14 - 18 - (-21) - 4$

Solution $-14 - 18 - (-21) - 4$
$\qquad = -14 + (-18) + 21 + (-4)$
$\qquad = -32 + 21 + (-4)$
$\qquad = -11 + (-4)$
$\qquad = -15$

You Try It 3   Subtract: $19 - (-32)$

Your solution

You Try It 4   Subtract:
$-9 - (-12) - 17 - 4$

Your solution

Solutions on p. A1

Objective C To solve application problems

Positive and negative numbers are used to express the profitability of a company. A profit is recorded as a positive number; a loss is recorded as a negative number.

➡ The bar chart at the right shows the annual profit and loss, to the nearest ten million dollars, for National Semiconductor for the years 1991, 1992, and 1993.

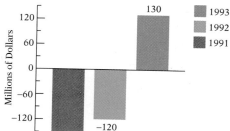

a. Determine the total profit for the three years.

b. Determine the difference between the profit in 1993 and the profit in 1992.

Strategy

a. To determine the total profit for the three years, add the profits and losses for the three years.

b. To determine the difference between the profit in 1993 and that in 1992, subtract the profits for the two years.

Solution

a. Total profit = $-150 + (-120) + 130 = -140$
The company had a loss of $140 million.

b. Difference in profit between 1993 and 1992 = $130 - (-120) = 250$
There was a $250 million difference in profit between 1992 and 1993.

Other applications of positive and negative numbers are shown in the examples below.

To solve an application problem, first read the problem carefully. The *Strategy* involves identifying the quantity to be found and planning the steps that are necessary to find the quantity. The *Solution* involves performing each operation stated in the Strategy and writing the answer.

Example 5
The average temperature on the sunlit side of the moon is approximately 215°F. On the dark side, it is approximately −250°F. Find the difference between these average temperatures.

Strategy
To find the difference, subtract the average temperature on the dark side of the moon (−250°) from the average temperature on the sunlit side (215°).

Solution
$215 - (-250)$
$215 + 250$
465
The difference is 465°F.

You Try It 5
The average temperature throughout the earth's stratosphere is −70°F. The average temperature on the earth's surface is 57°F. Find the difference between these average temperatures.

Your strategy

Your solution

Solution on p. A1

Content and Format © 1995 HMCo.

1.2 Exercises

Objective A

Add.

1. $-3 + (-8)$

2. $-6 + (-9)$

3. $-8 + 3$

4. $-9 + 2$

$$\begin{array}{r} -9 \\ + \;\; +2 \\ \hline -7 \end{array}$$

5. $-3 + (-80)$

6. $-12 + (-1)$

7. $-23 + (-23)$

8. $-12 + (-12)$

9. $16 + (-16)$

10. $-17 + 17$

11. $48 + (-53)$

12. $19 + (-41)$

13. $-17 + (-3) + 29$

14. $13 + 62 + (-38)$

15. $-3 + (-8) + 12$

16. $-27 + (-42) + (-18)$

17. $13 + (-22) + 4 + (-5)$

18. $-14 + (-3) + 7 + (-21)$

19. $-22 + 20 + 2 + (-18)$

20. $-6 + (-8) + 14 + (-4)$

21. $-16 + (-17) + (-18) + 10$

22. $-25 + (-31) + 24 + 19$

23. $26 + (-15) + (-11) + (-12)$

24. $-32 + 40 + (-8) + (-19)$

25. $-17 + (-18) + 45 + (-10)$

26. $23 + (-15) + 9 + (-15)$

27. $46 + (-17) + (-13) + (-50)$

28. $-37 + (-17) + (-12) + (-15)$

29. $-14 + (-15) + (-11) + 40$

30. $28 + (-19) + (-8) + (-1)$

31. $-23 + (-22) + (-21) + 5$

32. $-31 + 9 + (-16) + (-15)$

33. $72 + (-22) + (-14) + (-9)$

34. $-22 + (-17) + 58 + 29$

Objective B

Subtract.

35. $16 - 8$

36. $12 - 3$

37. $7 - 14$

38. $6 - 9$

39. $-7 - 2$

40. $-9 - 4$

41. $7 - (-2)$

42. $3 - (-4)$

43. $-6 - (-3)$

44. $-4 - (-2)$

45. $6 - (-12)$

46. $-12 - 16$

47. $-4 - 3 - 2$

48. $4 - 5 - 12$

49. $12 - (-7) - 8$

50. $-12 - (-3) - (-15)$

51. $-19 - (-19) - 18$

52. $-8 - (-8) - 14$

53. $-17 - (-8) - (-9)$

54. $7 - 8 - (-1)$

55. $-30 - (-65) - 29 - 4$

56. $42 - (-82) - 65 - 7$

57. $-16 - 47 - 63 - 12$

58. $42 - (-30) - 65 - (-11)$

59. $-47 - (-67) - 13 - 15$

60. $-18 - 49 - (-84) - 27$

61. $-19 - 17 - (-36) - 12$

62. $48 - 19 - 29 - 51$

63. $21 - (-14) - 43 - 12$

64. $17 - (-17) - 14 - 21$

Content and Format © 1995 HMCo.

Objective C   *Application Problems*

The elevation, or height of places on earth is measured in relation to sea level—that is, the average level of the ocean's surface. The following table shows height above sea level as a positive number and shows depth below sea level as a negative number.

Place	Elevation (in feet)
Mt. Everest	29,028
Mt. Aconcagua	23,035
Mt. McKinley	20,320
Mt. Kilimanjaro	19,340
Salinas Grandes	−131
Death Valley	−282
Qattara Depression	−436
Dead Sea	−1286

65. Use the table to find the difference in elevation between Mt. McKinley and Death Valley (the highest and lowest points in North America).

66. Use the table to find the difference in elevation between Mt. Kilimanjaro and the Qattara Depression (the highest and lowest points in Africa).

67. Use the table to find the difference in elevation between Mt. Everest and the Dead Sea (the highest and lowest points in Asia).

68. Use the table to find the difference in elevation between Mt. Aconcagua and Salinas Grandes (the highest and lowest points in South America).

A meteorologist may report a wind-chill temperature. This is the equivalent temperature, including the effects of wind and temperature, that a person would feel in calm air conditions. The table below gives the wind-chill temperature for various wind speeds and temperatures. For instance, when the temperature is 5°F and the wind in blowing at 15 mph, the wind-chill temperature is −25°F. Use this table for Exercises 69 to 72.

Wind Chill Factors

		15°F	10°F	5°F	0°F	−5°F	−10°F	−15°F
		Calm Air Temperature (Fahrenheit)						
Miles	5	12	7	0	−5	−10	−15	−21
per	10	−3	−9	−15	−22	−27	−34	−40
Hour	15	−11	−18	−25	−31	−38	−45	−51
	20	−17	−24	−31	−39	−46	−53	−60

69. What is the difference between a calm air temperature of −10°F and the wind-chill temperature when the wind is blowing 15 mph?

70. What is the difference between a calm air temperature of 10°F and the wind-chill temperature when the wind is blowing 10 mph?

71. When the wind speed increases from 5 mph to 10 mph, does the wind-chill temperature decrease by the same number of degrees for each of the temperatures shown?

72. For a wind speed of 15 mph, when the calm air temperature decreases by 5°F, does the wind-chill temperature decrease by the same number of degrees for each of the temperatures shown? If not, determine the largest decrease in temperature.

73. On January 22, 1943, the temperature at Spearfish, South Dakota, rose from −4°F to 45°F in two minutes. How many degrees did the temperature rise during those two minutes?

74. In a 24-hour period in January of 1916, the temperature in Browning, Montana, dropped from 44°F to −56°F. How many degrees did the temperature drop during that time?

The bar chart at the right shows the profit and loss, in millions of dollars, for General Motors for the years 1990 through 1993. Use this chart for Exercises 75–78.

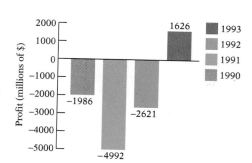

75. What was the total profit or loss for the four years shown on the graph?

76. What was the difference between the profit or loss in 1993 and that in 1992?

77. What was the difference between the profit or loss in 1990 and that in 1991?

78. What was the difference between the profit or loss in 1993 and that in 1991?

APPLYING THE CONCEPTS

79. If $x \in \{-3, 0, 4\}$, for which value of x is the expression $3 - x$ the greatest?

80. If $z \in \{-5, 0, 6\}$, for which value of z is the expression $4 - z$ the smallest?

81. If a and b are integers, is the expression $|a + b| = |a| + |b|$ always true, sometimes true, or never true?

82. If a and b are integers, is the expression $|a - b| = |a| - |b|$ always true, sometimes true, or never true?

83. Is the difference between two integers always smaller than either one of the numbers in the difference? If not, give an example for which the difference between two integers is greater than either integer.

84. If a is an integer, is $-a$ always a negative integer? If not, give an example wherein $-a$ is a positive integer.

1.3 Multiplication and Division of Integers

Objective A   To multiply integers

Multiplication is the repeated addition of the same number. The product 3×5 is shown on the number line below.

5 is added 3 times
$$3 \times 5 = 5 + 5 + 5 = 15$$

POINT OF INTEREST
The cross $\times$ was first used as a symbol for multiplication in 1631 in a book titled *The Key to Mathematics*. Also in that year, another book, *Practice of the Analytical Art*, advocated the use of a dot to indicate multiplication.

To indicate multiplication, several different symbols are used.

$$3 \times 5 = 15 \qquad 3 \cdot 5 = 15 \qquad (3)(5) = 15 \qquad 5(3) = 15 \qquad (3)5 = 15$$

Note that when parentheses are used and there is no arithmetic operation symbol, you are to assume that the operation is multiplication. Each number in a product is called a **factor**. For instance, 3 and 5 are factors of the product $3 \cdot 5 = 15$.

Now consider the product of a positive and a negative number.

-5 is added 3 times
$$3(-5) = (-5) + (-5) + (-5) = -15$$

• Multiplication is repeated addition.

This suggests that the product of a positive number and a negative number is negative. Here are a few more examples.

$$4 \times (-7) = -28 \qquad (-6)5 = -30 \qquad (-5) \cdot 7 = -35$$

To find the product of two negative numbers, look at the pattern at the right. As -5 multiplies a sequence of decreasing integers, the products increase by 5.

These numbers decrease by 1.

These numbers increase by 5

$$-5 \times 3 = -15$$
$$-5 \times 2 = -10$$
$$-5 \times 1 = -5$$
$$-5 \times 0 = 0$$
$$-5 \times (-1) = 5$$
$$-5 \times (-2) = 10$$
$$-5 \times (-3) = 15$$

The pattern can be continued by requiring that the product of two negative numbers be positive.

Multiplication of Integers

• *Integers with the same sign*
  To multiply two numbers with the same sign, multiply the absolute values of the numbers. The product is positive.
• *Integers with different signs*
  To multiply two numbers with different signs, multiply the absolute values of the numbers. The product is negative.

➡ Multiply: $-2(5)(-7)(-4)$

$$-2(5)(-7)(-4) = -10(-7)(-4)$$
$$= 70(-4) = -280$$

• To multiply more than two numbers, multiply the first two. Then multiply the product by the third number. Continue until all the numbers are multiplied.

Consider the products shown at the right. Note that when there is an even number of negative factors, the product is positive. When there is an odd number of negative factors, the product is negative.

$$(-3)(-5) = 15$$
$$(-2)(-5)(-6) = -60$$
$$(-4)(-3)(-5)(-7) = 420$$
$$(-3)(-3)(-5)(-4)(-5) = -900$$
$$(-6)(-3)(-4)(-2)(-10)(-5) = 7200$$

This idea can be summarized by the following useful rule: **The product of an even number of negative factors is positive; the product of an odd number of negative factors is negative.**

Example 1   Multiply: $(-3)4(-5)$

Solution $(-3)4(-5) = (-12)(-5) = 60$

Example 2   Multiply: $12(-4)(-3)(-5)$

Solution $12(-4)(-3)(-5) = (-48)(-3)(-5)$
$$= 144(-5) = -720$$

You Try It 1   Multiply: $8(-9)10$

Your solution

You Try It 2   Multiply: $(-2)3(-8)7$

Your solution

Solutions on p. A1

Objective B   To divide integers

For every division problem there is a related multiplication problem.

$$\frac{8}{2} = 4 \qquad \text{because} \qquad 2 \cdot 4 = 8$$

Division   Related multiplication

This fact and the rules for multiplying integers can be used to illustrate the rules for dividing integers.

Note in the following examples that the quotient of two numbers with the same sign is positive.

$$\frac{12}{3} = 4 \text{ because } 4 \cdot 3 = 12 \qquad \frac{-12}{-3} = 4 \text{ because } 4 \cdot (-3) = -12$$

The next two examples illustrate that the quotient of two numbers with different signs is negative.

$$\frac{12}{-3} = -4 \text{ because } (-4)(-3) = 12 \qquad \frac{-12}{3} = -4 \text{ because } (-4) \cdot 3 = -12$$

Division of Integers

- *Integers with the same sign*
  To divide two numbers with the same sign, divide the absolute values of the numbers. The quotient is positive.
- *Integers with different signs*
  To divide two numbers with different signs, divide the absolute values of the numbers. The quotient is negative.

➡ Simplify: $-\dfrac{-56}{7}$

$$-\dfrac{-56}{7} = -\left(\dfrac{-56}{7}\right) = -(-8) = 8$$

Note that $\dfrac{-12}{3} = -4$, $\dfrac{12}{-3} = -4$, and $-\dfrac{12}{3} = -4$. This suggests the following rule.

If a and b are integers, and $b \neq 0$, then $\dfrac{-a}{b} = \dfrac{a}{-b} = -\dfrac{a}{b}$.

Properties of Zero and One in Division

- Zero divided by any number other than zero is zero.

$\dfrac{0}{a} = 0$ because $0 \cdot a = 0$ For example, $\dfrac{0}{7} = 0$ because $0 \cdot 7 = 0$.

- Division by zero is not defined.

To understand that division by zero is not permitted, suppose that $\dfrac{4}{0}$ were equal to n, where n is some number. Because each division problem has a related multiplication problem, $\dfrac{4}{0} = n$ means $n \cdot 0 = 4$. But $n \cdot 0 = 4$ is impossible because any number times 0 is 0. Therefore, division by 0 is not defined.

- Any number other than zero divided by itself is 1.

$\dfrac{a}{a} = 1$, $a \neq 0$ For example, $\dfrac{-8}{-8} = 1$.

- Any number divided by one is the number.

$\dfrac{a}{1} = a$ For example, $\dfrac{9}{1} = 9$.

Example 3   Divide: $(-120) \div (-8)$

Solution $(-120) \div (-8) = 15$

You Try It 3   Divide: $(-135) \div (-9)$

Your solution

Example 4   Divide: $\dfrac{95}{-5}$

Solution $\dfrac{95}{-5} = -19$

You Try It 4   Divide: $\dfrac{-72}{4}$

Your solution

Example 5   Divide: $-\dfrac{-81}{3}$

Solution $-\dfrac{-81}{3} = -(-27) = 27$

You Try It 5   Divide: $-\dfrac{36}{-12}$

Your solution

Solutions on pp. A1–A2

Objective C   To solve application problems

In many courses, your course grade depends on the *average* of all your test scores. You compute the average by calculating the sum of all your test scores and then dividing that result by the number of tests. Statisticians call this average an **arithmetic mean.** Besides its application to finding the average of your test scores, the arithmetic mean is used in many other situations.

Stock market analysts calculate the **moving average** of a stock. This is the arithmetic mean of the changes in the value of a stock for a given number of days. To illustrate the procedure, we will calculate the five-day moving average of a stock. In actual practice, a stock market analyst may use 15 days, 30 days, or some other number.

The table below shows the amount of increase or decrease, in cents, from the closing price of a stock for a 10-day period.

Day 1	Day 2	Day 3	Day 4	Day 5	Day 6	Day 7	Day 8	Day 9	Day 10
+50	−175	+225	0	−275	−75	−50	+50	−475	−50

To calculate the five-day moving average of this stock, determine the average of the stock for days 1 through 5, days 2 through 6, days 3 through 7, and so on.

Days 1–5	*Days 2–6*	*Days 3–7*	*Days 4–8*	*Days 5–9*	*Days 6–10*
+50	−175	+225	0	−275	−75
−175	+225	0	−275	−75	−50
+225	0	−275	−75	−50	+50
0	−275	−75	−50	+50	−475
−275	−75	−50	+50	−475	−50
Sum = −175	Sum = −300	Sum = −175	Sum = −350	Sum = −825	Sum = −600

$$\text{Ave} = \frac{-175}{5} = -35 \quad \text{Ave} = \frac{-300}{5} = -60 \quad \text{Ave} = \frac{-175}{5} = -35 \quad \text{Ave} = \frac{-350}{5} = -70 \quad \text{Ave} = \frac{-825}{5} = -165 \quad \text{Ave} = \frac{-600}{5} = -120$$

The five-day moving average is the list of means: $-35, -60, -35, -70, -165,$ and -120. If the list tends to increase, the price of the stock is showing an upward trend; if it decreases, the price of the stock is showing a downward trend. These trends help an analyst recommend stocks. Other applications of averages are shown in the examples below.

Example 6
The daily high temperatures (in degrees Celsius) for six days in Anchorage, Alaska, were: $-14°, 3°, 0°, -8°, 2°, -1°$.

Find the average daily high temperature.

Strategy
To find the average daily high temperature:

• Add the six temperature readings.
• Divide the sum by 6.

Solution
$-14 + 3 + 0 + (-8) + 2 + (-1) = -18$
$-18 \div 6 = -3$ • The average daily high temperature was −3°C.

You Try It 6
The daily low temperatures (in degrees Celsius) during one week were recorded as follows: $-6°, -7°, 0°, -5°, -8°, -1°, -1°$.

Find the average daily low temperature.

Your strategy

Your solution

Solution on p. A2

1.3 Exercises

Objective A

Multiply.

1. $(14)3$ **2.** $(17)6$ **3.** $-7 \cdot 4$ **4.** $-8 \cdot 7$ **5.** $(-12)(-5)$ **6.** $(-13)(-9)$

7. $-11(23)$ **8.** $-8(21)$ **9.** $(-17)14$ **10.** $(-15)12$ **11.** $6(-19)$ **12.** $17(-13)$

13. $7(5)(-3)$ **14.** $(-3)(-2)8$ **15.** $9(-7)(-4)$ **16.** $(-2)(6)(-4)$

17. $16(-3)5$ **18.** $20(-4)3$ **19.** $-4(-3)8$ **20.** $-5(-9)6$

21. $-3(-8)(-9)$ **22.** $-7(-6)(-5)$ **23.** $(-9)7(5)$ **24.** $(-8)7(10)$

25. $7(-2)(5)(-6)$ **26.** $(-3)7(-2)8$ **27.** $-9(-4)(-8)(-10)$ **28.** $-11(-3)(-5)(-2)$

29. $7(9)(-11)4$ **30.** $-12(-4)7(-2)$ **31.** $(-14)9(-11)0$ **32.** $(-13)(15)(-19)0$

Objective B

Divide.

33. $12 \div (-6)$ **34.** $18 \div (-3)$ **35.** $(-72) \div (-9)$ **36.** $(-64) \div (-8)$ **37.** $-42 \div 6$

38. $(-56) \div 8$ **39.** $(-144) \div 12$ **40.** $(-93) \div (-3)$ **41.** $48 \div (-8)$ **42.** $57 \div (-3)$

43. $\dfrac{-49}{7}$ **44.** $\dfrac{-45}{5}$ **45.** $\dfrac{-44}{-4}$ **46.** $\dfrac{-36}{-9}$ **47.** $\dfrac{98}{-7}$

48. $\dfrac{85}{-5}$ **49.** $-\dfrac{-120}{8}$ **50.** $-\dfrac{-72}{4}$ **51.** $-\dfrac{-80}{-5}$ **52.** $-\dfrac{-114}{-6}$

53. $0 \div (-9)$ **54.** $0 \div (-14)$ **55.** $\dfrac{-261}{9}$ **56.** $\dfrac{-128}{4}$ **57.** $9 \div 0$

58. $(-21) \div 0$ **59.** $\dfrac{132}{-12}$ **60.** $\dfrac{250}{-25}$ **61.** $\dfrac{0}{0}$ **62.** $\dfrac{-58}{0}$

Objective C   *Application Problems*

Solve.

63. The high temperatures for a six-day period in Barrow, Alaska, were −23°F, −29°F, −21°F, −28°F, −28°F, and −27°F. Calculate the average daily high temperature.

64. The low temperatures for a ten-day period in a midwestern city were −4°F, −9°F, −5°F, −2°F, 4°F, −1°F, −1°F, −2°F, −2°F and 2°F. Calculate the average daily low temperature for this city.

65. The value of AT&T on September 28, 1993 was $59.50. The table below shows the amount of increase or decrease, to the nearest 25 cents, from the September 28 closing price of the stock for a 10-day period.

Day 1	Day 2	Day 3	Day 4	Day 5	Day 6	Day 7	Day 8	Day 9	Day 10
−25	−25	−25	−50	−50	0	−25	0	+25	−25

Calculate the five-day moving average for this stock.

66. The value of IBM on September 28, 1993 was $41.75. The table below shows the amount of increase or decrease, to the nearest 25 cents, from the September 28 closing price of the stock for a 10-day period.

Day 1	Day 2	Day 3	Day 4	Day 5	Day 6	Day 7	Day 8	Day 9	Day 10
+25	+50	+200	+50	−50	+25	0	+25	−25	−25

Calculate the five-day moving average for this stock.

67. To discourage random guessing on a multiple-choice exam, a professor assigns 5 points for a correct answer, −2 points for an incorrect answer, and 0 points for leaving the question blank. What is the score for a student who had 20 correct answers, had 13 incorrect answers, and left 7 questions blank?

68. To discourage random guessing on a multiple-choice exam, a professor assigns 7 points for a correct answer, −3 points for an incorrect answer, and −1 point for leaving the question blank. What is the score for a student who had 17 correct answers, had 8 incorrect answers, and left 2 questions blank?

APPLYING THE CONCEPTS

69. If $x \in \{-5, 0, 6\}$, for which value of x does the expression $2x$ have the smallest value?

70. If $x \in \{-6, -2, 7\}$, for which value of x does the expression $-3x$ have the greatest value?

71. Explain why $0 \div 0$ is not defined.
[W]

72. If $-4x$ equals a positive integer, is x a positive or a negative integer?
[W] Explain your answer.

73. If $5y$ equals a negative integer, is y a positive or a negative integer?
[W] Explain your answer.

Operations with Rational Numbers

Objective A   *To write a rational number in simplest form and as a decimal* ············

POINT OF INTEREST
As early as A.D. 630 the
Hindu mathematician
Brahmagupta wrote a
fraction as one number over
another spearted by a space.
The Arab mathematician
al Hassar (around A.D 1050)
was the first to show a
fraction with the horizontal
bar separating the numerator
and denominator.

A *rational number* is the quotient of two integers. A rational number written in this way is commonly called a fraction. Here are some examples of rational numbers.

$$\frac{3}{4}, \frac{-4}{9}, \frac{15}{-4}, \frac{8}{1}, -\frac{5}{6}$$

> **Rational Numbers**
>
> A **rational number** is a number that can be written in the form $\frac{a}{b}$, where a and b are integers and $b \neq 0$.

Because an integer can be written as the quotient of the integer and 1, every integer is a rational number. For instance,

$$\frac{6}{1} = 6 \qquad \frac{-8}{1} = -8$$

A fraction is in **simplest form** when there are no common factors in the numerator and the denominator. The fractions $\frac{4}{6}$ and $\frac{2}{3}$ are equivalent fractions because they represent the same part of a whole. However, the fraction $\frac{2}{3}$ is in simplest form because there are no common factors (other than 1) in the numerator and denominator.

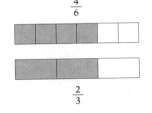

To write a fraction in simplest form, eliminate the common factors from the numerator and denominator by using the fact that $1 \cdot \frac{a}{b} = \frac{a}{b}$.

$$\frac{4}{6} = \frac{2 \cdot 2}{2 \cdot 3} = \frac{2}{2} \cdot \frac{2}{3} = 1 \cdot \frac{2}{3} = \frac{2}{3}$$

The process of eliminating common factors is usually written as shown at the right

$$\frac{4}{6} = \frac{\overset{1}{2} \cdot 2}{\underset{1}{2} \cdot 3} = \frac{2}{3}$$

If you have difficulty determining the common factors, write the numerator and denominator in terms of prime factors. (Recall that a prime number is a number divisible only by itself and 1. The first ten prime numbers are 2, 3, 5, 7, 11, 13, 17, 19, 23, and 29.)

➡ Write $\frac{18}{30}$ in simplest form.

$$\frac{18}{30} = \frac{\overset{1}{2} \cdot \overset{1}{3} \cdot 3}{\underset{1}{2} \cdot \underset{1}{3} \cdot 5} = \frac{3}{5}$$

A number written in **decimal notation** is also a rational number.

three-tenths $0.3 = \dfrac{3}{10}$ forty-three thousandths $0.043 = \dfrac{43}{1000}$

A rational number written as a fraction can be written in decimal notation by dividing the numerator of the fraction by the denominator. Think of the fraction bar as "divided by."

➡ Write $\dfrac{5}{8}$ as a decimal.

$$
\begin{array}{r}
0.625 \\
8\overline{)5.000} \\
-4\,8 \\
\hline
20 \\
-16 \\
\hline
40 \\
-40 \\
\hline
0
\end{array}
$$

• Divide the numerator, 5, by the denominator, 8.

• Dividing the numerator by the denominator resulted in a remainder of 0. The decimal 0.625 is called a **terminating decimal**.

$$\dfrac{5}{8} = 0.625$$

➡ Write $\dfrac{4}{11}$ in decimal notation.

$$
\begin{array}{r}
0.3636\,3 \\
11\overline{)4.0000} \\
-3\,3 \\
\hline
70 \\
-66 \\
\hline
40 \\
-33 \\
\hline
70 \\
-66 \\
\hline
4\,0
\end{array}
$$

• Divide the numerator, 4, by the denominator, 11.

• No matter how long we continue to divide, the remainder is never zero. The decimal $0.\overline{36}$ is a **repeating decimal**. The bar over the 36 indicates that these digits repeat.

$$\dfrac{4}{11} = 0.\overline{36}$$

Every rational number can be written as a terminating or a repeating decimal. Some numbers—for example, $\sqrt{7}$ and π—have decimal representations that never terminate or repeat. These numbers are called **irrational numbers.**

$$\sqrt{7} \approx 2.6457513\ldots \qquad \pi \approx 3.1415927\ldots$$

Real Numbers

The rational numbers and the irrational numbers taken together are called the **real numbers.**

Example 1   Write $\frac{20}{42}$ in simplest form.

Solution $\frac{20}{42} = \frac{\overset{1}{\cancel{2}} \cdot 2 \cdot 5}{\cancel{2} \cdot 3 \cdot 7} = \frac{10}{21}$

You Try It 1   Write $\frac{4}{9}$ as a decimal. Place a bar over the repeating digits of the decimal.

Your solution

Solution on p. A2

Objective B   ***To convert between percents, fractions, and decimals***

"A population growth rate of 3%," "a manufacturer's discount of 25%," and "an 8% increase in pay" are typical examples of the many ways in which percent is used in applied problems. **Percent** means "parts of 100." Thus 27% means 27 parts of 100.

In applied problems involving a percent, it may be necessary to rewrite a percent as a fraction or decimal, or to rewrite a fraction or decimal as a percent.

To write a percent as a fraction, remove the percent sign and multiply by $\frac{1}{100}$.

➡ Write 27% as a fraction.

$$27\% = 27\left(\frac{1}{100}\right) = \frac{27}{100}$$ • **Remove the percent sign and multiply by $\frac{1}{100}$.**

To write a percent as a decimal, remove the percent sign and multiply by 0.01.

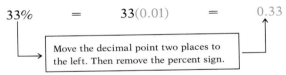

$$33\% \quad = \quad 33(0.01) \quad = \quad 0.33$$

Move the decimal point two places to the left. Then remove the percent sign.

A fraction or decimal can be written as a percent by multiplying by 100%. For example, $\frac{5}{8}$ is changed to a percent as follows.

To write a decimal as a percent, multiply by 100%.

$$\frac{5}{8} = \frac{5}{8}(100\%) = \frac{500}{8}\% = 62.5\%, \text{ or } 62\frac{1}{2}\%$$

$$0.82 \quad = \quad 0.82(100\%) \quad = \quad 82\%$$

Move the decimal point two places to the right. Then write the percent sign.

Example 2
Write 130% as a fraction and as a decimal.

Solution

$$130\% = 130\left(\frac{1}{100}\right) = \frac{130}{100} = 1\frac{3}{10}$$

$$130\% = 130(0.01) = 1.30$$

You Try It 2
Write 125% as a fraction and as a decimal.

Your solution

Solution on p. A2

Example 3

Write $33\frac{1}{3}\%$ as a fraction.

Solution

$$33\frac{1}{3}\% = 33\frac{1}{3}\left(\frac{1}{100}\right) = \frac{100}{3}\left(\frac{1}{100}\right) = \frac{1}{3}$$

You Try It 3

Write $16\frac{2}{3}\%$ as a fraction.

Your solution

Example 4

Write 0.25% as a decimal.

Solution

$$0.25\% = 0.25(0.01) = 0.0025$$

You Try It 4

Write 0.5% as a decimal.

Your solution

Example 5

Write $\frac{5}{6}$ as a percent.

Solution

$$\frac{5}{6} = \frac{5}{6}(100\%) = \frac{500}{6}\% = 83\frac{1}{3}\%.$$

You Try It 5

Write 0.043 as a percent.

Your solution

Solutions on p. A2

Objective C   To add or subtract rational numbers

Four of the 8 squares are shaded as purple. This is $\frac{4}{8}$ of the entire rectangle. Three of the 8 squares are shaded green. This is $\frac{3}{8}$ of the entire rectangle. So, 7 of the 8 squares, or $\frac{7}{8}$, of the entire rectangle is shaded.

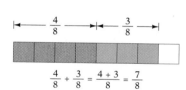

$$\frac{4}{8} + \frac{3}{8} = \frac{4+3}{8} = \frac{7}{8}$$

Addition of Fractions

To add two fractions with the same denominator, add the numerators and place the sum over the common denominator.

$$\frac{a}{c} + \frac{b}{c} = \frac{a+b}{c}$$

To add fractions with different denominators, first rewrite the fractions as an equivalent fraction with a common denominator. Then add the fractions.

The common denominator is the **least common multiple** (LCM) of the denominators. This is the smallest number that is a multiple of each of the denominators. It can be found by first writing each denominator as a product of prime factors. The LCM must contain the factors of each denominator.

$$6 = 2 \cdot 3$$
$$10 = 2 \cdot 5 \qquad \text{LCM} = 2 \cdot 3 \cdot 5 = 30$$

Factors of 10

Factors of 6

Content and Format © 1995 HMCo.

➡ Add: $-\dfrac{5}{6} + \dfrac{3}{10}$

The LCM of 6 and 10 is 30. This is frequently called the **least common denominator.** Rewrite the fractions as equivalent fractions with denominator 30. Then add the fractions.

$$-\frac{5}{6} + \frac{3}{10} = -\frac{25}{30} + \frac{9}{30} = \frac{-25 + 9}{30} = \frac{-16}{30} = -\frac{8}{15}$$

● $-\dfrac{5}{6} = -\dfrac{25}{30},\ \dfrac{3}{10} = \dfrac{9}{30}$

To subtract fractions, subtract the numerators and place the difference over the common denominator.

➡ Subtract: $-\dfrac{4}{9} - \left(-\dfrac{7}{12}\right)$

The LCM of 9 and 12 is 36. Rewrite the fractions as equivalent fractions with denominator 36. Then subtract the fractions.

$$-\frac{4}{9} - \left(-\frac{7}{12}\right) = -\frac{16}{36} - \left(-\frac{21}{36}\right) = \frac{-16 - (-21)}{36} = \frac{-16 + 21}{36} = \frac{5}{36}$$

To add or subtract decimals, write the numbers so that the decimal points are in a vertical line. Then proceed as in the addition or subtraction of integers. Write the decimal point in the answer directly below the decimal points in the problem.

➡ Add: $-114.039 + 84.76$

$|-114.039| = 114.039$

$|84.76| = 84.76$

$$\begin{array}{r} 114.039 \\ -\ 84.76 \\ \hline 29.279 \end{array}$$

$-114.039 + 84.76 = -29.279$

● The signs are different. Find the absolute value of each number.

● Subtract the smaller of these numbers from the larger.

● Attach the sign of the number with larger absolute value. Because $|-114.039| > |84.76|$, use the sign of -114.039.

Example 6 Simplify: $\dfrac{5}{16} - \dfrac{7}{40}$

Solution The LCM of 16 and 40 is 80.

$$\frac{5}{16} - \frac{7}{40} = \frac{25}{80} - \frac{14}{80} = \frac{25 - 14}{80} = \frac{11}{80}$$

You Try It 6 Simplify: $\dfrac{5}{9} - \dfrac{11}{12}$

Your solution

Example 7 Simplify: $-\dfrac{3}{4} + \dfrac{1}{6} - \dfrac{5}{8}$

Solution The LCM of 4, 6, and 8 is 24.

$$-\frac{3}{4} + \frac{1}{6} - \frac{5}{8} = -\frac{18}{24} + \frac{4}{24} - \frac{15}{24}$$
$$= \frac{-18 + 4 - 15}{24}$$
$$= \frac{-29}{24} = -\frac{29}{24}$$

You Try It 7 Simplify: $-\dfrac{7}{8} - \dfrac{5}{6} + \dfrac{3}{4}$

Your solution

Solutions on p. A2

Example 8 Simplify: 42.987 − 98.61

Solution

42.987 − 98.61 = 42.987 + (−98.61)

= −55.623

You Try It 8 Simplify: 16.127 − 67.91

Your solution

Solution on p. A2

Objective D **To multiply or divide rational numbers** ....

The product of two fractions is the product of the numerators divided by the product of the denominators.

➡ Simplify: $\frac{3}{8} \times \frac{12}{17}$

$$\frac{3}{8} \times \frac{12}{17} = \frac{3 \cdot 12}{8 \cdot 17}$$

• Multiply the numerators. Multiply the denominators.

$$= \frac{3 \cdot \cancel{2} \cdot \cancel{2} \cdot 3}{2 \cdot \cancel{2} \cdot \cancel{2} \cdot 17}$$

• Write the prime factorization of each factor. Divide by the common factors.

$$= \frac{9}{34}$$

• Multiply the factors in the numerator and denominator.

To divide fractions, invert the divisor. Then multiply the fractions.

➡ Simplify: $\frac{3}{10} \div \left(-\frac{18}{25}\right)$

The signs are different. The quotient is negative.

$$\frac{3}{10} \div \left(-\frac{18}{25}\right) = -\left(\frac{3}{10} \div \frac{18}{25}\right) = -\left(\frac{3}{10} \cdot \frac{25}{18}\right) = -\left(\frac{3 \cdot 25}{10 \cdot 18}\right)$$

$$= -\left(\frac{\cancel{3} \cdot \cancel{5} \cdot 5}{2 \cdot \cancel{5} \cdot 2 \cdot \cancel{3} \cdot 3}\right) = -\frac{5}{12}$$

To multiply decimals, multiply as with integers. Write the decimal point in the product so that the number of decimal places in the product equals the sum of the decimal places in the factors.

➡ Simplify: −6.89 × 0.00035

$$\begin{array}{r} 6.89 \\ \times\, 0.00035 \\ \hline 3445 \\ 2067 \\ \hline 0.0024115 \end{array}$$

2 decimal places
5 decimal places

7 decimal places

• Multiply the absolute values.

−6.89 × 0.00035 = −0.0024115

• The signs are different. The product is negative.

To divide decimals, move the decimal point in the divisor far enough to make the divisor a whole number. Move the decimal point in the dividend the same number of places to the right. Place the decimal point in the quotient directly over the decimal point in the dividend. Then divide as with whole numbers.

LOOK CLOSELY

The symbol ≈ is used to indicate that the quotient is an approximate value that has been rounded off.

➡ Simplify 1.32 ÷ 0.27. Round to the nearest tenth.

$$
\begin{array}{r}
4.88 \approx 4.9 \\
0.27\overline{)132.00} \\
-108 \\
\hline
240 \\
-216 \\
\hline
240 \\
-216 \\
\hline
24
\end{array}
$$

• Move the decimal point 2 places to the right in the divisor and then in the dividend. Place the decimal point in the quotient.

Example 9 Simplify: $\frac{2}{3} \times \left(-\frac{9}{10}\right)$

Solution The product is negative.

$$\frac{2}{3} \times \left(-\frac{9}{10}\right) = -\left(\frac{2}{3} \times \frac{9}{10}\right) = -\frac{2 \cdot 9}{3 \cdot 10}$$

$$= -\frac{\overset{1}{\cancel{2}} \cdot \overset{1}{\cancel{3}} \cdot 3}{\underset{1}{\cancel{3}} \cdot \underset{1}{\cancel{2}} \cdot 5} = -\frac{3}{5}$$

You Try It 9 Simplify: $-\frac{7}{12} \times \frac{9}{14}$

Your solution

Example 10 Simplify: $-\frac{5}{8} \div \left(-\frac{5}{40}\right)$

Solution The quotient is positive.

$$-\frac{5}{8} \div \left(-\frac{5}{40}\right) = \frac{5}{8} \div \frac{5}{40} = \frac{5}{8} \times \frac{40}{5} = \frac{5 \cdot 40}{8 \cdot 5}$$

$$= \frac{\overset{1}{\cancel{5}} \cdot \overset{1}{\cancel{2}} \cdot \overset{1}{\cancel{2}} \cdot \overset{1}{\cancel{2}} \cdot 5}{\underset{1}{\cancel{2}} \cdot \underset{1}{\cancel{2}} \cdot \underset{1}{\cancel{2}} \cdot \underset{1}{\cancel{5}}} = \frac{5}{1} = 5$$

You Try It 10 Simplify: $-\frac{3}{8} \div \left(-\frac{5}{12}\right)$

Your solution

Example 11 Simplify: -4.29×8.2

Solution The product is negative.

$$
\begin{array}{r}
4.29 \\
\times 8.2 \\
\hline
858 \\
3432 \\
\hline
35.178
\end{array}
$$

$-4.29 \times 8.2 = -35.178$

You Try It 11 Simplify: -5.44×3.8

Your solution

Example 12 Simplify $-0.0792 \div (-0.42)$. Round to the nearest hundredth.

Solution

$$
\begin{array}{r}
0.188 \approx 0.19 \\
0.42\overline{)0.07.920} \\
-42 \\
\hline
372 \\
-336 \\
\hline
360 \\
-336 \\
\hline
24
\end{array}
$$

You Try It 12 Simplify $-0.394 \div 1.7$. Round to the nearest hundredth.

Your solution

Solutions on p. A3

Objective E   *To solve application problems* ....

An article in *Business Week* magazine reported that the U.S. budget deficit for 1993 was $236.4 billion. The number 236.4 billion means

$$236.4 \underbrace{\times\ 1,000,000,000}_{\text{billion}} = 236,400,000,000$$

Numbers such as 236.4 billion are used in many instances because they are easy to read and offer an approximation to the actual number. Such numbers are used in Example 13 and You Try It 13.

One of the applications of percent is to express a portion of a total as a percent. For instance, a recent survey of 450 mall shoppers found that 270 preferred the mall closest to their home even though it did not have the same store variety as a mall farther from home. The percent of shoppers who preferred the mall closest to home can be found by converting a fraction to a percent.

$$\frac{\text{portion preferring close to home}}{\text{total number surveyed}} = \frac{270}{450}$$

$$= 0.60 = 60\%$$

Example 13
The graph below shows the profit and loss of Hartmarx for the years 1990 through 1994.

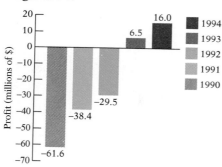

What was the difference between the profit in 1994 and the loss in 1990?

Strategy
To find the difference between the profit in 1994 and the loss in 1990, read the graph. Subtract the loss in 1990 (−$61.6 million) from the profit in 1994 ($16.0 million).

Solution
16.0 − (−61.6) = 77.6
The difference between the profit in 1994 and the loss in 1990 was $77.6 million.

You Try It 13
The circle graph below shows the amount spent in the United States for the most commonly prescribed drugs.

What percent of the total amount spent was spent for hypertension drugs? The hypertension drugs are Procardia, Cardizem, and Vasotec.

Your strategy

Your solution

Solution on p. A3

Content and Format © 1995 HMCo.

1.4 Exercises

Objective A

Write each fraction in simplest form.

1. $\dfrac{7}{21}$ **2.** $\dfrac{10}{15}$ **3.** $\dfrac{8}{22}$ **4.** $\dfrac{8}{60}$ **5.** $\dfrac{50}{75}$

6. $\dfrac{20}{44}$ **7.** $\dfrac{12}{8}$ **8.** $\dfrac{36}{9}$ **9.** $\dfrac{0}{36}$ **10.** $\dfrac{12}{18}$

11. $\dfrac{60}{100}$ **12.** $\dfrac{14}{45}$ **13.** $\dfrac{44}{60}$ **14.** $\dfrac{19}{51}$ **15.** $\dfrac{23}{46}$

16. $\dfrac{32}{48}$ **17.** $\dfrac{15}{80}$ **18.** $\dfrac{35}{135}$ **19.** $\dfrac{12}{20}$ **20.** $\dfrac{24}{100}$

Write as a decimal. Place a bar over the repeating digits of a repeating decimal.

21. $\dfrac{4}{5}$ **22.** $\dfrac{1}{6}$ **23.** $\dfrac{5}{6}$ **24.** $\dfrac{1}{8}$ **25.** $\dfrac{7}{8}$

26. $\dfrac{2}{9}$ **27.** $\dfrac{8}{9}$ **28.** $\dfrac{5}{11}$ **29.** $\dfrac{7}{12}$ **30.** $\dfrac{11}{12}$

31. $\dfrac{9}{16}$ **32.** $\dfrac{15}{16}$ **33.** $\dfrac{7}{18}$ **34.** $\dfrac{17}{18}$ **35.** $\dfrac{1}{20}$

36. $\dfrac{6}{25}$ **37.** $\dfrac{14}{25}$ **38.** $\dfrac{7}{30}$ **39.** $\dfrac{9}{40}$ **40.** $\dfrac{21}{40}$

Objective B

Write as a fraction and a decimal.

41. 75%   **42.** 40%   **43.** 64%   **44.** 88%   **45.** 125%

46. 160%   **47.** 19%   **48.** 87%   **49.** 5%   **50.** 450%

51. 380%   **52.** 8%   **53.** 2%   **54.** 105%   **55.** 82%

Write as a fraction.

56. $11\dfrac{1}{9}\%$ **57.** $4\dfrac{2}{7}\%$ **58.** $12\dfrac{1}{2}\%$ **59.** $37\dfrac{1}{2}\%$ **60.** $66\dfrac{2}{3}\%$

61. $\dfrac{1}{4}\%$ **62.** $\dfrac{1}{2}\%$ **63.** $6\dfrac{1}{4}\%$ **64.** $83\dfrac{1}{3}\%$ **65.** $5\dfrac{3}{4}\%$

Write as a decimal.

66. 7.3%   **67.** 9.1%   **68.** 15.8%   **69.** 16.7%   **70.** 0.3%

71. 0.9%   **72.** 9.9%   **73.** 9.15%   **74.** 121.2%   **75.** 18.23%

76. 62.14%   **77.** 0.15%   **78.** 0.27%   **79.** 0.02%   **80.** 0.08%

Write as a percent.

81. 0.15   **82.** 0.37   **83.** 0.05   **84.** 0.02   **85.** 0.175

86. 0.125   **87.** 1.15   **88.** 1.36   **89.** 0.008   **90.** 0.004

91. $\dfrac{27}{50}$ **92.** $\dfrac{83}{100}$ **93.** $\dfrac{1}{3}$ **94.** $\dfrac{3}{8}$ **95.** $\dfrac{5}{11}$

96. $\dfrac{4}{9}$ **97.** $\dfrac{7}{8}$ **98.** $\dfrac{9}{20}$ **99.** $1\dfrac{2}{3}$ **100.** $2\dfrac{1}{2}$

Objective C

Simplify.

101. $\dfrac{2}{3} + \dfrac{5}{12}$ **102.** $\dfrac{1}{2} + \dfrac{3}{8}$ **103.** $\dfrac{5}{8} - \dfrac{5}{6}$ **104.** $\dfrac{1}{9} - \dfrac{5}{27}$

105. $-\dfrac{5}{12} - \dfrac{3}{8}$ **106.** $-\dfrac{5}{6} - \dfrac{5}{9}$ **107.** $-\dfrac{6}{13} + \dfrac{17}{26}$ **108.** $-\dfrac{7}{12} + \dfrac{5}{8}$

109. $\dfrac{5}{8} - \left(-\dfrac{3}{4}\right)$ **110.** $\dfrac{3}{5} - \dfrac{11}{12}$ **111.** $\dfrac{11}{12} - \dfrac{5}{6}$ **112.** $-\dfrac{2}{3} - \left(-\dfrac{11}{18}\right)$

113. $-\dfrac{5}{8} - \left(-\dfrac{11}{12}\right)$ **114.** $\dfrac{1}{3} + \dfrac{5}{6} - \dfrac{2}{9}$ **115.** $\dfrac{1}{2} - \dfrac{2}{3} + \dfrac{1}{6}$ **116.** $-\dfrac{3}{8} - \dfrac{5}{12} - \dfrac{3}{16}$

117. $-\dfrac{5}{16} + \dfrac{3}{4} - \dfrac{7}{8}$ **118.** $\dfrac{1}{2} - \dfrac{3}{8} - \left(-\dfrac{1}{4}\right)$ **119.** $\dfrac{3}{4} - \left(-\dfrac{7}{12}\right) - \dfrac{7}{8}$ **120.** $\dfrac{1}{3} - \dfrac{1}{4} - \dfrac{1}{5}$

121. $\dfrac{2}{3} - \dfrac{1}{2} + \dfrac{5}{6}$ **122.** $\dfrac{5}{16} + \dfrac{1}{8} - \dfrac{1}{2}$ **123.** $\dfrac{5}{8} - \left(-\dfrac{5}{12}\right) + \dfrac{1}{3}$ **124.** $\dfrac{1}{8} - \dfrac{11}{12} + \dfrac{1}{2}$

125. $7.56 + 0.462$ **126.** $1.09 + 6.2$ **127.** $-32.1 - 6.7$ **128.** $5.13 - 8.179$

129. $-13.092 + 6.9$ **130.** $2.54 - 3.6$ **131.** $5.43 + 7.925$ **132.** $-16.92 - 6.925$

133. $-3.87 + 8.546$ **134.** $6.9027 - 17.692$ **135.** $2.09 - 6.72 - 5.4$

136. $-18.39 + 4.9 - 23.7$ **137.** $19 - (-3.72) - 82.75$ **138.** $-3.07 - (-2.97) - 17.4$

139. $16.4 - (-3.09) - 7.93$ **140.** $-3.09 - 4.6 - (-27.3)$ **141.** $2.66 - (-4.66) - 8.2$

142. $-0.34 - (-4.35) - 3.2$ **143.** $7.5 - 12.8 - (-0.57)$ **144.** $2.8 - (-3.44) + 2.3$

Objective D

Simplify.

145. $\dfrac{1}{2} \times \left(-\dfrac{3}{4}\right)$ **146.** $-\dfrac{2}{9} \times \left(-\dfrac{3}{14}\right)$ **147.** $\left(-\dfrac{3}{8}\right)\left(-\dfrac{4}{15}\right)$

148. $\left(-\dfrac{3}{4}\right)\left(-\dfrac{8}{27}\right)$ **149.** $-\dfrac{1}{2}\left(\dfrac{8}{9}\right)$ **150.** $\dfrac{5}{12}\left(-\dfrac{8}{15}\right)$

151. $\dfrac{5}{8}\left(-\dfrac{7}{12}\right)\dfrac{16}{25}$ **152.** $\left(\dfrac{5}{12}\right)\left(-\dfrac{8}{15}\right)\left(-\dfrac{1}{3}\right)$ **153.** $\dfrac{1}{2}\left(-\dfrac{3}{4}\right)\left(-\dfrac{5}{8}\right)$

154. $\dfrac{3}{8} \div \dfrac{1}{4}$ **155.** $\dfrac{5}{6} \div \left(-\dfrac{3}{4}\right)$ **156.** $-\dfrac{5}{12} \div \dfrac{15}{32}$

157. $-\dfrac{7}{8} \div \dfrac{4}{21}$ **158.** $\dfrac{7}{10} \div \dfrac{2}{5}$ **159.** $-\dfrac{15}{64} \div \left(-\dfrac{3}{40}\right)$

160. $\dfrac{1}{8} \div \left(-\dfrac{5}{12}\right)$ **161.** $-\dfrac{4}{9} \div \left(-\dfrac{2}{3}\right)$ **162.** $-\dfrac{6}{11} \div \dfrac{4}{9}$

163. $1.2(3.47)$ **164.** $(-0.8)6.2$ **165.** $(-1.89)(-2.3)$

166. $(6.9)(-4.2)$ **167.** $1.06(-3.8)$ **168.** $-2.7(-3.5)$

169. $1.2(-0.5)(3.7)$ **170.** $-2.4(6.1)(0.9)$ **171.** $2.3(-0.6)(0.8)$

172. $-1.2(-0.55)(1.9)$ **173.** $0.44(-2.3)(-0.5)$ **174.** $-3.4(-22.1)(-0.5)$

175. $1.8(0.33)(-0.4)$ **176.** $4.5(-0.22)(-0.8)$ **177.** $-24.7 \div 0.09$

Simplify. Round to the nearest hundredth.

178. $-1.27 \div (-1.7)$

179. $9.07 \div (-3.5)$

180. $0.0976 \div 0.042$

181. $-6.904 \div 1.35$

182. $-7.894 \div (-2.06)$

183. $-354.2086 \div 0.1719$

184. $78.564 \div (-2.337)$

Objective E   _Application Problems_

185. The table at the right shows the number of wins and losses for the Midwest League of Western Conference of the National Basketball Association for 1994.

Team	Won	Lost
Houston	58	24
San Antonio	55	27
Utah	53	29
Denver	42	40
Minnesota	20	62
Dallas	13	69

   a. Which team came closest to winning $\frac{2}{3}$ of its games?

   b. Which teams lost more than $\frac{3}{5}$ of their games?

   c. What percent of their games did Dallas lose?

186. The results of a survey conducted in part by Nintendo of 5000 children are shown in the circle graph at the right. In the survey, there children were asked which characteristic they would most like to have.

   a. What percent of the total number of students surveyed selected smart as the characteristic they would most like to have?

   b. What percent of the total number of students surveyed did not choose wealthy as the characteristic they would most like to have?

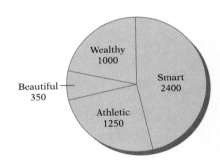

187. The net profits and losses for Rose Stores for the years 1990 through 1994 are shown in the graph at the right.

   a. What is the difference between the profit or loss in 1990 and that in 1992?

   b. What is the difference between the profit or loss in 1993 and that in 1991?

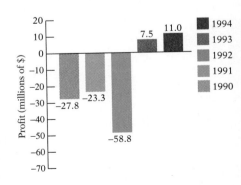

188. When a U.S. company does business with another country, it is necessary to convert U.S. currency into the currency of the other nation. These _exchange rates_ are determined by various factors. The exchange rates for some currencies are shown in the table at the right.

   a. If you sold goods worth $1.2 million U.S. dollars to Australia, how many Australian dollars would you receive?

   b. How many U.S. dollars, to the nearest thousandths, are equivalent to 1 Irish punt?

Foreign Currency per
U.S. Dollar

Australian dollar	1.5186
Belgian franc	34.80
Irish punt	0.6840

189. According the Federal Highway Administration, the average car is driven approximately 10,300 miles per year and uses approximately 495 gallons of gas. Assuming that the average cost of gasoline is $1.097 per gallon, which includes 44.3¢ for all taxes, and that there are 1.53 million cars on the road, determine how much total tax is paid for gasoline in one year. (You may not need all the data given in this problem.)

APPLYING THE CONCEPTS

190. Use a calculator to determine the decimal representation of $\frac{17}{99}$, $\frac{45}{99}$, and $\frac{73}{99}$. Make a conjecture as to the decimal representation of $\frac{83}{99}$. Does your conjecture work for $\frac{33}{99}$? What about $\frac{1}{99}$?

191. If the same positive integer is added to both the numerator and the denominator of $\frac{2}{5}$, is the new fraction less than, equal to, or greater than $\frac{2}{5}$?

192. A magic square is one in which the numbers in every row, column, and diagonal sum to the same number. Complete the magic square at the right.

$\frac{2}{3}$		
	$\frac{1}{6}$	$\frac{5}{6}$
		$-\frac{1}{3}$

193. If a and b are rational numbers and $a < b$, is it always possible to
[W] find a rational number c such that $a < c < b$? If not, explain why. If so, show how to find one.

194. For each part below, find a rational number, r, that satisfies the condition.
  a. $r^2 < r$,   **b.** $r^2 = r$, and **c.** $r^2 > r$

195. In a survey of consumers, approximately 43% said they would be willing to pay between $1000 and $2000 dollars more for a new car if the car had an EPA rating of 80 mpg. If your car now gets 28 mpg and you drive approximately 10,000 miles per year, in how many months would your savings on gasoline pay for the increased cost of such a car? Assume the average cost for gasoline is $1.06 per gallon.

196. Find three natural numbers a, b, and c such that $\frac{1}{a} + \frac{1}{b} + \frac{1}{c}$ is a natural number.

1.5 Exponents and the Order of Operations Agreement

Objective A   ***To evaluate exponential expressions*** ....

Repeated multiplication of the same factor can be written using an exponent.

$$2 \cdot 2 \cdot 2 \cdot 2 \cdot 2 = 2^5 \leftarrow \textbf{exponent}$$
$$\uparrow\!\!\text{---} \textbf{base}$$

$$a \cdot a \cdot a \cdot a = a^4 \leftarrow \textbf{exponent}$$
$$\uparrow\!\!\text{---} \textbf{base}$$

The **exponent** indicates how many times the factor, called the **base**, occurs in the multiplication. The multiplication $2 \cdot 2 \cdot 2 \cdot 2 \cdot 2$ is in **factored form.** The exponential expression 2^5 is in **exponential form.**

2^1 is read "the first power of 2" or just 2. Usually the exponent 1 is not written.

2^2 is read "the second power of 2" or "2 squared."

2^3 is read "the third power of 2" or "2 cubed."

2^4 is read "the fourth power of 2."

a^4 is read "the fourth power of *a*."

There is a geometric interpretation of the first three natural-number powers.

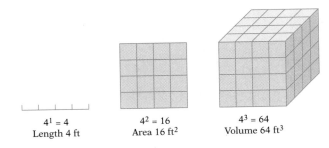

$4^1 = 4$
Length 4 ft

$4^2 = 16$
Area 16 ft^2

$4^3 = 64$
Volume 64 ft^3

To evaluate an exponential expression, write each factor as many times as indicated by the exponent. Then multiply.

➡ Evaluate $(-2)^4$.

$$(-2)^4 = (-2)(-2)(-2)(-2)$$
$$= 16$$

• Write (−2) as a factor 4 times.
• Multiply.

➡ Evaluate -2^4.

$$-2^4 = -(2 \cdot 2 \cdot 2 \cdot 2)$$
$$= -16$$

• Write 2 as a factor 4 times.
• Multiply.

From these last two examples, note the difference between $(-2)^4$ and -2^4.

$$(-2)^4 = (-2)(-2)(-2)(-2) = 16$$
$$-2^4 = -(2^4) = -(2 \cdot 2 \cdot 2 \cdot 2) = -16$$

Example 1   Evaluate -5^3.

Solution $-5^3 = -(5 \cdot 5 \cdot 5) = -125$

You Try It 1   Evaluate -6^3.

Your solution

Example 2   Evaluate $(-4)^4$.

Solution $(-4)^4 = (-4)(-4)(-4)(-4)$
$= 256$

You Try It 2   Evaluate $(-3)^4$.

Your solution

Example 3   Evaluate $(-3)^2 \cdot 2^3$

Solution $(-3)^2 \cdot 2^3 = (-3)(-3) \cdot (2)(2)(2)$
$= 9 \cdot 8 = 72$

You Try It 3   Evaluate $(3^3)(-2)^3$.

Your solution

Example 4   Evaluate $\left(-\dfrac{2}{3}\right)^3$.

Solution $\left(-\dfrac{2}{3}\right)^3 = \left(-\dfrac{2}{3}\right)\left(-\dfrac{2}{3}\right)\left(-\dfrac{2}{3}\right)$
$= -\dfrac{2 \cdot 2 \cdot 2}{3 \cdot 3 \cdot 3} = -\dfrac{8}{27}$

You Try It 4   Evaluate $\left(-\dfrac{2}{5}\right)^2$.

Your solution

Example 5   Evaluate $-4(0.7)^2$.

Solution $-4(0.7)^2 = -4(0.7)(0.7)$
$= -2.8(0.7) = -1.96$

You Try It 5   Evaluate $-3(0.3)^3$.

Your solution

Solutions on p. A4

Objective B   ***To use the Order of Operations Agreement to simplify expressions***

Let's evaluate $2 + 3 \cdot 5$.

There are two arithmetic operations, addition and multiplication, in this expression. The operations could be performed in different orders.

Multiply first. $2 + \underbrace{3 \cdot 5}$ Add first. $\underbrace{2 + 3} \cdot 5$

Then add. $\underbrace{2 + 15}$ Then multiply. $\underbrace{5 \cdot 5}$

17 25

In order to prevent there from being more than one answer for a numerical expression, an Order of Operations Agreement has been established.

The Order of Operations Agreement

Step 1   Perform operations inside grouping symbols. Grouping symbols include parentheses ( ), brackets [ ], braces { }, and the fraction bar.

Step 2   Simplify exponential expressions.

Step 3   Do multiplication and division as they occur from left to right.

Step 4   Do addition and subtraction as they occur from left to right.

➡ Evaluate $12 - 24(8 - 5) \div 2^2$.

$$12 - 24(8 - 5) \div 2^2 = 12 - 24(3) \div 2^2$$ • Perform operations inside grouping symbols.

$$= 12 - 24(3) \div 4$$ • Simplify exponential expressions.

$$= 12 - 72 \div 4$$ • Do multiplication and division as they occur from left to right.

$$= 12 - 18$$ • Do addition and subtraction as they occur from left to right.

$$= -6$$

One or more of the above steps may not be needed to evaluate an expression. In that case, proceed to the next step in the Order of Operations Agreement.

➡ Evaluate $\dfrac{4 + 8}{2 + 1} - (3 - 1) + 2$.

$$\dfrac{4 + 8}{2 + 1} - (3 - 1) + 2 = \dfrac{12}{3} - 2 + 2$$ • Perform operations inside grouping symbols.

$$= 4 - 2 + 2$$ • Do multiplication and division as they occur from left to right.

$$= 2 + 2$$ • Do addition and subtraction as they occur from left to right.

$$= 4$$

When an expression has grouping symbols inside grouping symbols, perform the operations inside the inner grouping symbols first.

➡ Evaluate $6 \div [4 - (6 - 8)] + 2^2$.

$$6 \div [4 - (6 - 8)] + 2^2 = 6 \div [4 - (-2)] + 2^2$$ • Perform operations inside grouping symbols.

$$= 6 \div 6 + 2^2$$ • Simplify exponential expressions.

$$= 6 \div 6 + 4$$ • Do multiplication and division as they occur from left to right.

$$= 1 + 4$$ • Do addition and subtraction as they occur from left to right.

$$= 5$$

Example 6
Evaluate $4 - 3[4 - 2(6 - 3)] \div 2$.

Solution
$$
\begin{aligned}
4 - 3[4 - 2(6 - 3)] \div 2 &= 4 - 3[4 - 2 \cdot 3] \div 2 \\
&= 4 - 3[4 - 6] \div 2 \\
&= 4 - 3[-2] \div 2 \\
&= 4 + 6 \div 2 \\
&= 4 + 3 \\
&= 7
\end{aligned}
$$

You Try It 6
Evaluate $18 - 5[8 - 2(2 - 5)] \div 10$.

Your solution

Example 7
Evaluate $27 \div (5 - 2)^2 + (-3)^2 \cdot 4$.

Solution
$$
\begin{aligned}
27 \div (5 - 2)^2 + (-3)^2 \cdot 4 &= 27 \div 3^2 + (-3)^2 \cdot 4 \\
&= 27 \div 9 + 9 \cdot 4 \\
&= 3 + 9 \cdot 4 \\
&= 3 + 36 \\
&= 39
\end{aligned}
$$

You Try It 7
Evaluate $36 \div (8 - 5)^2 - (-3)^2 \cdot 2$.

Your solution

Example 8
Evaluate $(1.75 - 1.3)^2 \div 0.025 + 6.1$.

Solution
$$
\begin{aligned}
(1.75 - 1.3)^2 &\div 0.025 + 6.1 \\
&= (0.45)^2 \div 0.025 + 6.1 \\
&= 0.2025 \div 0.025 + 6.1 \\
&= 8.1 + 6.1 \\
&= 14.2
\end{aligned}
$$

You Try It 8
Evaluate $(6.97 - 4.72)^2 \times 4.5 \div 0.05$.

Your solution

Example 9
Evaluate $\dfrac{5}{8} - \left(\dfrac{2}{5} - \dfrac{1}{2}\right) \div \left(\dfrac{2}{3}\right)^2$.

Solution
$$
\begin{aligned}
\frac{5}{8} - \left(\frac{2}{5} - \frac{1}{2}\right) \div \left(\frac{2}{3}\right)^2 &= \frac{5}{8} - \left(-\frac{1}{10}\right) \div \left(\frac{2}{3}\right)^2 \\
&= \frac{5}{8} + \frac{1}{10} \div \frac{4}{9} \\
&= \frac{5}{8} + \frac{1}{10} \cdot \frac{9}{4} \\
&= \frac{5}{8} + \frac{9}{40} \\
&= \frac{25}{40} + \frac{9}{40} \\
&= \frac{34}{40} = \frac{17}{20}
\end{aligned}
$$

You Try It 9
Evaluate $\dfrac{5}{8} \div \left(\dfrac{1}{3} - \dfrac{3}{4}\right) + \dfrac{7}{12}$.

Your solution

Solutions on p. A4

1.5 Exercises

- -

Objective A

Evaluate.

1. 6^2

2. 7^4

3. -7^2

4. -4^3

5. $(-3)^2$

6. $(-2)^3$

7. $(-3)^4$

8. $(-5)^3$

9. $\left(\dfrac{1}{2}\right)^2$

10. $\left(-\dfrac{3}{4}\right)^3$

11. $(0.3)^2$

12. $(1.5)^3$

13. $\left(\dfrac{2}{3}\right)^2 \cdot 3^3$

14. $\left(-\dfrac{1}{2}\right)^3 \cdot 8$

15. $(0.3)^3 \cdot 2^3$

16. $(0.5)^2 \cdot 3^3$

17. $(-3) \cdot 2^2$

18. $(-5) \cdot 3^4$

19. $(-2) \cdot (-2)^3$

20. $(-2) \cdot (-2)^2$

21. $2^3 \cdot 3^3 \cdot (-4)$

22. $(-3)^3 \cdot 5^2 \cdot 10$

23. $(-7) \cdot 4^2 \cdot 3^2$

24. $(-2) \cdot 2^3 \cdot (-3)^2$

25. $\left(\dfrac{2}{3}\right)^2 \cdot \dfrac{1}{4} \cdot 3^3$

26. $\left(\dfrac{3}{4}\right)^2 \cdot (-4) \cdot 2^3$

27. $8^2 \cdot (-3)^5 \cdot 5$

Objective B

Evaluate by using the Order of Operations Agreement.

28. $4 - 8 \div 2$

29. $2^2 \cdot 3 - 3$

30. $2(3 - 4) - (-3)^2$

31. $16 - 32 \div 2^3$

32. $24 - 18 \div 3 + 2$

33. $8 - (-3)^2 - (-2)$

34. $8 - 2(3)^2$

35. $16 - 16 \cdot 2 \div 4$

36. $12 + 16 \div 4 \cdot 2$

Evaluate by using the Order of Operations Agreement.

37. $16 - 2 \cdot 4^2$

38. $27 - 18 \div (-3^2)$

39. $4 + 12 \div 3 \cdot 2$

40. $16 + 15 \div (-5) - 2$

41. $14 - 2^2 - (4 - 7)$

42. $3 - 2[8 - (3 - 2)]$

43. $-2^2 + 4[16 \div (3 - 5)]$

44. $6 + \dfrac{16 - 4}{2^2 + 2} - 2$

45. $24 \div \dfrac{3^2}{8 - 5} - (-5)$

46. $96 \div 2[12 + (6 - 2)] - 3^2$

47. $4[16 - (7 - 1)] \div 10$

48. $18 \div 2 - 4^2 - (-3)^2$

49. $18 \div (9 - 2^3) + (-3)$

50. $16 - 3(8 - 3)^2 \div 5$

51. $4(-8) \div [2(7 - 3)^2]$

52. $\dfrac{(-19) + (-2)}{6^2 - 30} \div (2 - 4)$

53. $16 - 4 \cdot \dfrac{3^3 - 7}{2^3 + 2} - (-2)^2$

54. $(0.2)^2 \cdot (-0.5) + 1.72$

55. $0.3(1.7 - 4.8) + (1.2)^2$

56. $(1.8)^2 - 2.52 \div 1.8$

57. $(1.65 - 1.05)^2 \div 0.4 + 0.8$

58. $\dfrac{3}{8} \div \left(\dfrac{5}{6} + \dfrac{2}{3} \right)$

59. $\left(\dfrac{5}{12} - \dfrac{9}{16} \right) \dfrac{3}{7}$

60. $\left(\dfrac{3}{4} \right)^2 - \left(\dfrac{1}{2} \right)^3 \div \dfrac{3}{5}$

APPLYING THE CONCEPTS

61.   Find two fractions between $\left(\dfrac{2}{3} \right)$ and $\left(\dfrac{3}{4} \right)$. (There is more than one answer to this question.)

62.   The following was offered as the simplification of $6 + 2(4 - 9)$.
[W]

$$6 + 2(4 - 9) = 6 + 2(-5)$$
$$= 6(-5)$$
$$= -30$$

If this is a correct simplification, write yes for the answer. If it is incorrect, write no and explain the incorrect step.

63.   The following was offered as the simplification of $2 \cdot 3^3$.
[W]

$$2 \cdot 3^3 = 6^3 = 216$$

If this is a correct simplification, write yes for the answer. If it is incorrect, write no and explain the incorrect step.

Project in Mathematics

**Using the [+/−] Key
on a Scientific
Calculator**

Using your calculator to simplify numerical expressions sometimes requires use of the [+/−] key or, on some calculators, the negative key, which is frequently shown as [(−)]. These keys change the sign of the number currently in the display. To enter −4:

- For those calculators with [+/−], press 4 and then [+/−].
- For those calculators with [(−)], press [(−)] and then 4.

Here are the keystrokes for evaluating the expression $3(-4) - (-5)$.

Calculators with [+/−] key: 3 [×] 4 [+/−] [−] 5 [+/−] [=]

Calculators with [(−)] key: 3 [×] [(−)] 4 [−] [(−)] 5 [=]

This example illustrates that calculators make a distinction between negative and minus. To perform the operation $3 - (-3)$, you cannot enter 3 [−] [−] 3. This would result in 0, which is not the correct answer. You must enter

3 [−] 3 [+/−] [=]   or   3 [−] [(−)] 3 [=]

Chapter Summary

Key Words

The *natural numbers* are $1, 2, 3, 4, 5, 6, 7, \ldots$.

The *integers* are $\ldots, -4, -3, -2, -1, 0, 1, 2, 3, 4, \ldots$.

A number a is *less than* another number b, written $a < b$, if a is to the left of b on the number line.

A number a is *greater than* another number b, written $a > b$, if a is to the right of b on the number line.

The symbol $\leq$ means *is less than or equal to*.

The symbol $\geq$ means *is greater than or equal to*.

The *absolute value* of a number is its distance from zero on a number line.

A *rational number* is a number of the form $\frac{a}{b}$, where a and b are integers and b is not equal to zero. A rational number written in this form is commonly called a *fraction*.

Percent means parts of 100.

An *irrational number* is a number that has a decimal representation that never terminates or repeats.

An expression of the form a^n is in *exponential form*, where a is the base and n is the exponent.

Essential Rules   *Addition of Integers with the Same Sign*

To add two numbers with the same sign, add the absolute values of the numbers. Then attach the sign of the addends.

Addition of Integers with Different Signs

To add two numbers with different signs, find the absolute value of each number. Then subtract the smaller of these from the larger one. Attach the sign of the number with larger absolute value.

Subtraction of Integers

To subtract two integers, add the opposite of the second integer to the first integer.

Multiplication of Integers with the Same Sign

To multiply two numbers with the same sign, multiply the absolute values of the numbers. The product is positive.

Multiplication of Integers with Different Signs

To multiply two numbers with different signs, multiply the absolute values of the numbers. The product is negative.

Division of Integers with the Same Sign

The quotient of two numbers with the same sign is positive.

Division of Integers with Different Signs

The quotient of two numbers with different signs is negative.

Convert Percent to Decimal

To convert a percent to a decimal, remove the percent sign and multiply by 0.01.

Convert Percent to Fraction

To convert a percent to a fraction, remove the percent sign and multiply by $\frac{1}{100}$.

Convert Decimal to Percent

To convert a decimal to a percent, multiply by 100%.

Convert Fraction to Percent

To convert a fraction to a percent, multiply by 100%.

Order of Operations Agreement

Step 1   Perform operations inside grouping symbols. The grouping symbols are parentheses, brackets, braces, absolute value, and the fraction bar.

Step 2   Simplify exponential expressions.

Step 3   Do multiplication and division as they occur from left to right.

Step 4   Do addition and subtraction as they occur from left to right.

Chapter Review

SECTION 1.1

1. Place the correct symbol, $<$ or $>$, between the the two numbers. $-4 \quad 2$

2. Let $x \in \{-4, 0, 11\}$. For what values of x is the inequality $x < -1$ a true statement?

3. Find the additive inverse of -4.

4. Evaluate $-|-5|$.

5. Place the correct symbol, $<$ or $>$, between the two numbers. $-|6| \quad |-10|$

6. Let $y \in \{-7, 0, 9\}$. Evaluate $-|y|$ for each element of the set.

SECTION 1.2

7. Add: $-13 + 7$

8. Add: $-3 + (-12) + 6 + (-4)$

9. Subtract: $9 - 13$

10. Subtract: $16 - (-3) - 18$

11. To discourage random guessing on a multiple-choice exam, a professor assigns 6 points for a correct answer, -4 points for an incorrect answer, and -2 points for leaving a question blank. What is the score for a student who had 21 correct answers, had 5 incorrect answers, and left 4 questions blank?

SECTION 1.3

12. Multiply: $(-6)(7)$

13. Multiply: $-9(-9)$

14. Divide: $-32 \div (-4)$

15. Divide: $-100 \div 5$

16. The temperature at which mercury boils (that is, its boiling point) is 357°C. The temperature at which mercury freezes (its freezing point) is -39°C. Find the difference between the boiling point and the freezing point of mercury.

SECTION 1.4

17. Write $\dfrac{7}{25}$ as a decimal.

18. Write $\dfrac{2}{15}$ as a decimal. Place a bar over the repeating digits of the decimal.

19. Simplify: $\dfrac{1}{3} - \dfrac{1}{6} + \dfrac{5}{12}$

20. Simplify: $5.17 - 6.238$

21. Simplify: $-\dfrac{18}{35} \div \dfrac{27}{28}$

22. Simplify: $4.32 \cdot (-1.07)$

23. Write $79\dfrac{1}{2}\%$ as a fraction.

24. Write 6.2% as a decimal.

25. Write $\dfrac{5}{8}$ as a percent. Round to the nearest tenth of a percent.

26. Write $\dfrac{19}{35}$ as a percent. Write the remainder in fractional form.

27. The quarterly earnings per share of Wendy's stock are shown at the right.
   a. Find the percent increase in earnings per share for the third quarter 1993 over the third quarter 1992.

   b. Find the percent decrease in earnings from the fourth quarter 1992 to the first quarter 1993.

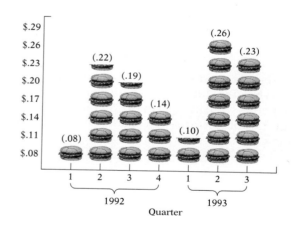

SECTION 1.5

28. Evaluate -5^2.

29. Evaluate $\left(-\dfrac{2}{3}\right)^4$.

30. Evaluate $5 - 2^2 + 9$.

31. Evaluate $15 \cdot (6 - 4)^2$.

32. Evaluate $\dfrac{5^2 + 11}{2^2 + 5} \div (2^3 - 2^2)$.

33. Evaluate $-3^2 + 4[18 + (12 - 20)]$.

Chapter Test

1. Place the correct symbol, $<$ or $>$, between the two numbers. $-2 \quad -40$

2. Let $y \in \{-5, -3, 0, 4\}$. For what values of y is the inequality $2 < y$ a true statement?

3. Find the opposite of -4.

4. Evaluate $-|-4|$.

5. Write 45% as a fraction and as a decimal.

6. Add: $-22 + 14 + (-8)$

7. Subtract: $16 - 30$

8. Subtract: $16 - (-30) - 42$

9. Multiply: -4×12

10. Multiply: $-5 \times (-6) \times 3$

11. Write $37\frac{1}{2}\%$ as a fraction.

12. Divide: $-561 \div (-33)$

13. Evaluate $16 \div 2[8 - 3(4 - 2)] + 1$.

14. The daily low temperature readings (in degrees Fahrenheit) for a three-day period were as follows: $-7°$, $9°$, $-8°$. Find the average low temperature for the three-day period.

15. Write 1.025 as a percent.

16. Write $\frac{7}{9}$ as a decimal. Place a bar over the repeating digits of the decimal.

17. Simplify: $-\frac{2}{5} + \frac{7}{15}$

18. Write $\frac{5}{6}$ as a percent. Write the remainder in fractional form.

19. Simplify: $\frac{5}{12} \div \left(-\frac{5}{6}\right)$

20. Simplify: $6.02 \times (-0.89)$

21. Evaluate $(-3^3) \cdot 2^2$.

22. Evaluate $\frac{3}{4} \cdot (4)^2$.

23. Evaluate $3^2 - 4 + 20 \div 5$.

24. Evaluate $\frac{-10 + 2}{2 + (-4)} \div 2 + 6$.

25. The exchange rates for some currencies are shown at the right. If you are in Sweden and purchase a car for 64,000 krona, what is the value of the purchase in U.S. dollars?

Foreign Currency per U.S. Dollar

Hong Kong dollar	7.7350
Swedish krona	8.0414
Japanese yen	103.85

2

Variable Expressions

Objectives

Section 2.1

To evaluate a variable expression

Section 2.2

To simplify a variable expression using the Properties of Addition

To simplify a variable expression using the Properties of Multiplication

To simplify a variable expression using the Distributive Property

To simplify general variable expressions

Section 2.3

To translate a verbal expression into a variable expression, given the variable

To translate a verbal expression into a variable expression and then simplify

To translate application problems

History of Variables

Prior to the 16th century, unknown quantities were represented by words. In Latin, the language in which most scholarly works were written, the word *res*, meaning "thing," was used. In Germany the word *zahl*, meaning "number," was used. In Italy the word *cosa*, also meaning "thing," was used.

Then in 1637, René Descartes, a French mathematician, began using the letters x, y, and z to represent variables. It is interesting to note, upon examining Descartes's work, that toward the end of the book the letters y and z were no longer used and x became the choice for a variable.

One explanation of why the letters y and z appeared less frequently has to do with the nature of printing presses during Descartes's time. A printer had a large tray that contained all the letters of the alphabet. There were many copies of each letter, especially those letters that are used frequently. For example, there were more e's than q's. Because the letters y and z do not occur frequently in French, a printer would have few of these letters on hand. Consequently, when Descartes started using these letters as variables, it quickly depleted the printer's supply and x's had to be used instead.

Today, x is used by most nations as the standard letter for a single unknown. In fact, x rays were so named because the scientists who discovered them did not know what they were and thus labeled them the "unknown rays" or x rays.

Evaluating Variable Expressions

Objective A   *To evaluate a variable expression* ....

Often we discuss a quantity without knowing its exact value—for example, the price of gold next month, the cost of a new automobile next year, or the tuition cost for next semester. Recall that a letter of the alphabet is used to stand for a quantity that is unknown or that can change, or *vary*. The letter is called a variable. An expression that contains one or more variables is called a **variable expression.**

A variable expression is shown at the right. The expression can be rewritten by writing subtraction as the addition of the opposite.

$$3x^2 - 5y + 2xy - x - 7$$

$$3x^2 + (-5y) + 2xy + (-x) + (-7)$$

Note that the expression has 5 addends. The **terms** of a variable expression are the addends of the expression. The expression has 5 terms.

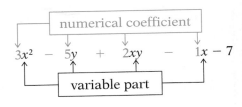

The terms $3x^2$, $-5y$, $2xy$, and $-x$ are **variable terms.**

The term -7 is a **constant term,** or simply a **constant.**

Each variable term is composed of a **numerical coefficient** and a **variable part** (the variable or variables and their exponents).

When the numerical coefficient is 1 or -1, the 1 is usually not written ($x = 1x$ and $-x = -1x$).

Variable expressions occur naturally in science. In a physics lab, a student may discover that a weight of 1 pound will stretch a spring $\frac{1}{2}$ inch. Two pounds will stretch the spring 1 inch. By experimenting, the student can discover that the distance the spring will stretch is found by multiplying the weight by $\frac{1}{2}$. By letting W represent the weight attached to the spring, the student can represent the distance the spring stretches by the variable expression $\frac{1}{2}W$.

With a weight of W pounds, the spring will stretch $\frac{1}{2} \cdot W = \frac{1}{2}W$ inches.

With a weight of 10 pounds, the spring will stretch $\frac{1}{2} \cdot 10 = 5$ inches. The number 10 is called the **value** of the variable W.

With a weight of 3 pounds, the spring will stretch $\frac{1}{2} \cdot 3 = 1\frac{1}{2}$ inches.

Replacing each variable by its value and then simplifying the resulting numerical expression is called **evaluating the variable expression.**

Evaluate $ab - b^2$ when $a = 2$ and $b = -3$.

Replace each variable in the expression by its value. Then use the Order of Operations Agreement to simplify the resulting numerical expression.

$ab - b^2$

$2(-3) - (-3)^2 = -6 - 9$

$$= -15$$

When $a = 2$ and $b = -3$, the value of $ab - b^2$ is -15.

Example 1   Name the variable terms of the expression $2a^2 - 5a + 7$.

Solution $2a^2$ and $5a$

You Try It 1   Name the constant term of the expression $6n^2 + 3n - 4$.

Your solution

Example 2   Evaluate $x^2 - 3xy$ when $x = 3$ and $y = -4$.

Solution $x^2 - 3xy$
$3^2 - 3(3)(-4) = 9 - 3(3)(-4)$
$= 9 - 9(-4)$
$= 9 - (-36)$
$= 45$

You Try It 2   Evaluate $2xy + y^2$ when $x = -4$ and $y = 2$.

Your solution $x = (-4)$ $y = (2)$

$2(-4)(2) + (2)^2$
$2(-4)(2) + 4 =$
$-8 - 16 + 4 =$
12

Example 3   Evaluate $\dfrac{a^2 - b^2}{a - b}$ when $a = 3$ and $b = -4$.

Solution $\dfrac{a^2 - b^2}{a - b}$

$\dfrac{3^2 - (-4)^2}{3 - (-4)} = \dfrac{9 - 16}{3 - (-4)}$

$\dfrac{-7}{7} = -1$

You Try It 3   Evaluate $\dfrac{a^2 + b^2}{a + b}$ when $a = 5$ and $b = -3$.

Your solution

Example 4   Evaluate $x^2 - 3(x - y) - z^2$ when $x = 2$, $y = -1$, and $z = 3$.

Solution $x^2 - 3(x - y) - z^2$
$= 2^2 - 3[2 - (-1)] - 3^2$
$= 2^2 - 3(3) - 3^2$
$= 4 - 3(3) - 9$
$= 4 - 9 - 9$
$= -5 - 9$
$= -14$

You Try It 4   Evaluate $x^3 - 2(x + y) + z^2$ when $x = 2$, $y = -4$, and $z = -3$.

Your solution

Solutions on p. A5

Content and Format © 1995 HMCo.

2.1 Exercises

Objective A

Name the terms of the variable expression. Then underline the constant term.

1. $2x^2 + 5x - 8$

2. $-3n^2 - 4n + 7$

3. $6 - a^4$

Name the variable terms of the expression. Then underline the variable part of each term.

4. $9b^2 - 4ab + a^2$

5. $7x^2y + 6xy^2 + 10$

6. $5 - 8n - 3n^2$

Name the coefficients of the variable terms.

7. $x^2 - 9x + 2$

8. $12a^2 - 8ab - b^2$

9. $n^3 - 4n^2 - n + 9$

Evaluate the variable expression when $a = 2$, $b = 3$, and $c = -4$.

10. $3a + 2b$

11. $a - 2c$

12. $-a^2$

13. $2c^2$

14. $-3a + 4b$

15. $3b - 3c$

16. $b^2 - 3$

17. $-3c + 4$

18. $16 \div (2c)$

19. $6b \div (-a)$

20. $bc \div (2a)$

21. $b^2 - 4ac$

22. $a^2 - b^2$

23. $b^2 - c^2$

24. $(a + b)^2$

25. $a^2 + b^2$

26. $2a - (c + a)^2$

27. $(b - a)^2 + 4c$

28. $b^2 - \dfrac{ac}{8}$

29. $\dfrac{5ab}{6} - 3cb$

30. $(b - 2a)^2 + bc$

Evaluate the variable expression when $a = -2$, $b = 4$, $c = -1$, and $d = 3$.

31. $\dfrac{b + c}{d}$

32. $\dfrac{d - b}{c}$

33. $\dfrac{2d + b}{-a}$

34. $\dfrac{b + 2d}{b}$

35. $\dfrac{b - d}{c - a}$

36. $\dfrac{2c - d}{-ad}$

37. $(b + d)^2 - 4a$

38. $(d - a)^2 - 3c$

39. $(d - a)^2 \div 5$

40. $3(b - a) - bc$

41. $\dfrac{b - 2a}{bc^2 - d}$

42. $\dfrac{b^2 - a}{ad + 3c}$

43. $\dfrac{1}{3}d^2 - \dfrac{3}{8}b^2$

44. $\dfrac{5}{8}a^4 - c^2$

45. $\dfrac{-4bc}{2a - b}$

46. $-\dfrac{3}{4}b + \dfrac{1}{2}(ac + bd)$

47. $-\dfrac{2}{3}d - \dfrac{1}{5}(bd - ac)$

48. $(b - a)^2 - (d - c)^2$

49. $(b + c)^2 + (a + d)^2$

50. $4ac + (2a)^2$

51. $3dc - (4c)^2$

APPLYING THE CONCEPTS

52. Choose any number. Evaluate the expressions $6a^2 + 2a - 10$
[W] and $2a(3a - 4) + 10(a - 1)$. Now choose a different number and evaluate the expressions again. Repeat this two more times with different numbers. What conclusions might you draw from your evaluations?

53. Explain in your own words the meaning of "evaluate an algebraic
[W] expression."

Evaluate the following expressions for $x = 2$, $y = 3$, and $z = -2$.

54. $3^x - x^3$

55. $2^y - y^2$

56. z^y

57. z^x

58. $x^x - y^y$

59. $y^{(x^2)}$

60. For each of the following, determine the first natural number x, greater than 1, for which the second expression is larger than the first.
    a. $x^3, 3^x$ **b.** $x^4, 4^x$ **c.** $x^5, 5^x$ **d.** $x^6, 6^x$

61. On the basis of your answer to Exercise 60, make a conjecture that appears to be true about the expressions x^n and n^x, where $n = 3, 4, 5, 6, 7, \ldots$ and x is a natural number greater than 1.

Simplifying Variable Expressions

Objective A   ***To simplify a variable expression using the Properties of Addition***

Like terms of a variable expression are terms with the same variable part. (Because $x^2 = x \cdot x$, x^2 and x are not like terms.)

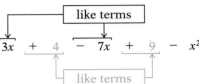

Constant terms are like terms. 4 and 9 are like terms.

To simplify a variable expression, use the Distributive Property to combine like terms by adding the numerical coefficients. The variable part remains unchanged.

Distributive Property

If a, b and c are real numbers, then $a(b + c) = ab + ac$.

The Distributive Property can also be written as $ba + ca = (b + c)a$. This form is used to simplify a variable expression.

➡ Simplify: $2x + 3x$
Use the Distributive Property to add the numerical coefficients of the like variable terms. This is called **combining like terms.**

$$2x + 3x = (2 + 3)x$$ ● Use the Distributive Property.
$$= 5x$$

➡ Simplify: $5y - 11y$
$$5y - 11y = (5 - 11)y$$ ● Use the Distributive Property. This step is usually done mentally.
$$= -6y$$

➡ Simplify: $5 + 7p$
The terms 5 and $7p$ are not like terms. The expression $5 + 7p$ is in simplest form.

LOOK CLOSELY

Simplifying an expression means to combine like terms. A constant term (5) and a variable term (7p) are not like terms and therefore cannot be combined.

The Associative Property of Addition

If a, b, and c are real numbers, then $(a + b) + c = a + (b + c)$.

When three or more terms are added, the terms can be grouped (with parentheses, for example) in any order. The sum is the same. For example,

$$(3x + 5x) + 9x = 3x + (5x + 9x)$$
$$8x + 9x = 3x + 14x$$
$$17x = 17x$$

> **The Commutative Property of Addition**
>
> If a and b are real numbers, then $a + b = b + a$.

When two like terms are added, the terms can be added in either order. The sum is the same. For example,

$$2x + (-4x) = -4x + 2x$$
$$-2x = -2x$$

> **The Addition Property of Zero**
>
> If a is a real number, then $a + 0 = 0 + a = a$.

The sum of a term and zero is the term. For example,

$$5x + 0 = 0 + 5x = 5x$$

> **The Inverse Property of Addition**
>
> If a is a real number, then $a + (-a) = (-a) + a = 0$.

The sum of a term and its opposite is zero. The opposite of a number is called its **additive inverse.**

$$7x + (-7x) = -7x + 7x = 0$$

➡ Simplify: $8x + 4y - 8x + y$

Use the Commutative and Associative Properties of Addition to rearrange and group like terms. Then combine like terms.

$$8x + 4y - 8x + y = \boxed{(8x - 8x) + (4y + y)} \quad \bullet \text{ This step is usually done mentally.}$$
$$= 0 + 5y$$
$$= 5y$$

➡ Simplify: $4x^2 + 5x - 6x^2 - 2x + 1$

Use the Commutative and Associative Properties of Addition to rearrange and group like terms. Then combine like terms.

$$4x^2 + 5x - 6x^2 - 2x + 1 = (4x^2 - 6x^2) + (5x - 2x) + 1$$
$$= -2x^2 + 3x + 1$$

Example 1 Simplify: $3x + 4y - 10x + 7y$

Solution

$3x + 4y - 10x + 7y = -7x + 11y$

You Try It 1 Simplify: $3a - 2b - 5a + 6b$

Your solution

Example 2 Simplify: $x^2 - 7 + 4x^2 - 16$

Solution

$x^2 - 7 + 4x^2 - 16 = 5x^2 - 23$

You Try It 2 Simplify: $-3y^2 + 7 + 8y^2 - 14$

Your solution

$5y^2 - 7 + 14$

Solutions on p. A5

Content and Format © 1995 HMCo.

Objective B   ***To simplify a variable expression using the Properties of Multiplication*** ....

In simplifying variable expressions, the following Properties of Multiplication are used.

> **The Associative Property of Multiplication**
>
> If *a, b* and *c* are real numbers, then $(a \cdot b) \cdot c = a \cdot (b \cdot c)$.

When three or more factors are multiplied, the factors can be grouped in any order. The product is the same. For example,

$$2(3x) = (2 \cdot 3)x = 6x$$

> **The Commutative Property of Mulitplication**
>
> If *a* and *b* are real numbers, then $a \cdot b = b \cdot a$.

Two factors can be multiplied in either order. The product is the same. For example,

$$(2x) \cdot 3 = 3 \cdot (2x) = 6x$$

> **The Multiplication Property of One**
>
> If *a* is a real number, then $a \cdot 1 = 1 \cdot a = a$.

The product of a term and one is the term. For example,

$$(8x)(1) = (1)(8x) = 8x$$

> **The Inverse Property of Multiplication**
>
> If *a* is a real number, and *a* is not equal to zero, then
>
> $a \cdot \dfrac{1}{a} = \dfrac{1}{a} \cdot a = 1.$

$\dfrac{1}{a}$ is called the **reciprocal** of *a*. $\dfrac{1}{a}$ is also called the **multiplicative inverse** of *a*.

The product of a number and its reciprocal is one. For example,

$$7 \cdot \frac{1}{7} = \frac{1}{7} \cdot 7 = 1$$

The multiplication properties just discussed are used to simplify variable expressions.

➡ Simplify: $2(-x)$
   Use the Associative Property of Multiplication to group factors.

$$2(-x) = 2(-1 \cdot x)$$
$$= [2(-1)]x$$
$$= -2x$$

➡ Simplify: $\frac{3}{2}\left(\frac{2x}{3}\right)$

Use the Associative Property of Multiplication to group factors.

$$\frac{3}{2}\left(\frac{2x}{3}\right) = \frac{3}{2}\left(\frac{2}{3}x\right)$$

• Note that $\frac{2x}{3} = \frac{2}{3}x$.

$$= \left(\frac{3}{2} \cdot \frac{2}{3}\right)x$$

• The steps in the dashed box are usually done mentally.

$$= 1 \cdot x$$
$$= x$$

➡ Simplify: $(16x)2$

Use the Commutative and Associative Properties of Multiplication to rearrange and group factors.

$$(16x)2 = 2(16x)$$
$$= (2 \cdot 16)x$$

• The steps in the dashed box are usually done mentally.

$$= 32x$$

Example 3   Simplify: $-2(3x^2)$

Solution $-2(3x^2) = -6x^2$

You Try It 3   Simplify: $-5(4y^2)$

Your solution $-5(4y^2)$
$-20y^2$

Example 4   Simplify: $-5(-10x)$

Solution $-5(-10x) = 50x$

You Try It 4   Simplify: $-7(-2a)$

Your solution $14a$

Example 5   Simplify: $(6x)(-4)$

Solution $(6x)(-4) = -24x$

You Try It 5   Simplify: $(-5x)(-2)$

Your solution $10x$

Solutions on p. A5

Objective C   **To simplify a variable expression using the Distributive Property**

Recall that the Distributive Property states that if a, b, and c are real numbers, then

$$a(b + c) = ab + ac$$

The Distributive Property is used to remove parentheses from a variable expression.

➡ Simplify: $3(2x + 7)$

$$3(2x + 7) = 3(2x) + 3(7)$$

• Use the Distributive Property. Do this step mentally.

$$= 6x + 21$$

Content and Format © 1995 HMCo.

➡ Simplify: $-5(4x + 6)$

$$-5(4x + 6) = \boxed{-5(4x) + (-5) \cdot 6}$$

$$= -20x - 30$$

• Use the Distributive Property.
  Do this step mentally.

➡ Simplify: $-(2x - 4)$

$$-(2x - 4) = \boxed{\begin{array}{c} -1(2x - 4) \\ -1(2x) - (-1)(4) \end{array}}$$

$$= -2x + 4$$

• Use the Distributive Property.
  Do these steps mentally.

Note: When a negative sign immediately precedes the parentheses, the sign of each term inside the parentheses is changed.

➡ Simplify: $-\frac{1}{2}(8x - 12y)$

$$-\frac{1}{2}(8x - 12y) = -\frac{1}{2}(8x) + \left(-\frac{1}{2}\right)(-12y)$$

$$= -4x + 6y$$

• Use the Distributive Property.
  Do this step mentally.

An extension of the Distributive Property is used when an expression contains more than two terms.

➡ Simplify: $3(4x - 2y - z)$

$$3(4x - 2y - z) = \boxed{3(4x) - 3(2y) - 3(z)}$$

$$= 12x - 6y - 3z$$

• Use the Distributive Property
  Do this step mentally.

Example 6
Simplify: $7(4 + 2x)$

Solution
$7(4 + 2x) = 28 + 14x$

You Try It 6
Simplify: $5(3 + 7b)$

$15 + 35b$

Your solution

Example 7
Simplify: $(2x - 6)2$

Solution
$(2x - 6)2 = 4x - 12$

You Try It 7
Simplify: $(3a - 1)5$

$15a - 5$

Your solution

Example 8
Simplify: $-3(-5a + 7b)$

Solution
$-3(-5a + 7b) = 15a - 21b$

You Try It 8
Simplify: $-8(-2a + 7b)$

$+16a - 56b$

Your solution

Solutions on p. A5

Example 9 Simplify: $3(x^2 - x - 5)$

Solution $3(x^2 - x - 5) = 3x^2 - 3x - 15$

You Try It 9 Simplify: $(12 - x + 8)$

Your solution $20 - x$

Example 10 Simplify: $-2(x^2 + 5x - 4)$

Solution
$-2(x^2 + 5x - 4) = -2x^2 - 10x + 8$

You Try It 10 Simplify: $3(-a^2 - 6a + 7)$

Your solution $3a^2 - 18a + 21$

Solutions on p. A5

Objective D   *To simplify general variable expressions*

When simplifying variable expressions, use the Distributive Property to remove parentheses and brackets used as grouping symbols.

➡ Simplify: $4(x - y) - 2(-3x + 6y)$

$4(x - y) - 2(-3x + 6y) = 4x - 4y + 6x - 12y$ • Use the Distributive Property.
$= 10x - 16y$ • Combine like terms.

Example 11 Simplify: $2x - 3(2x - 7y)$

Solution $2x - 3(2x - 7y) = 2x - 6x + 21y$
$= -4x + 21y$

You Try It 11 Simplify: $3y - 2(y - 7x)$

Your solution $3y - 2y + 14x$

y

Example 12 Simplify: $7(x - 2y) - (-x - 2y)$

Solution
$7(x - 2y) - (-x - 2y) = 7x - 14y + x + 2y$
$= 8x - 12y$

You Try It 12 Simplify:
$-2(x - 2y) - (-x + 3y)$

Your solution

Example 13 Simplify: $-2(-3x + 7y) - 14x$

Solution
$-2(-3x + 7y) - 14x = 6x - 14y - 14x$
$= -8x - 14y$

You Try It 13 Simplify:
$-5(-2y - 3x) + 4y$

Your solution

Example 14 Simplify:
$2x - 3[2x - 3(x + 7)]$

Solution
$2x - 3[2x - 3(x + 7)] = 2x - 3[2x - 3x - 21]$
$= 2x - 3[-x - 21]$
$= 2x + 3x + 63$
$= 5x + 63$

You Try It 14 Simplify:
$3y - 2[x - 4(2 - 3y)]$

Your solution

Solutions on p. A5

Content and Format © 1995 HMCo.

a − b − c

2.2 Exercises

Objective A

Simplify.

1. $6x + 8x$

2. $12x + 13x$

3. $9a - 4a$

4. $12a - 3a$

5. $4y + (-10y)$

6. $8y + (-6y)$

7. $-3b - 7$

8. $-12y - 3$

9. $-12a + 17a$

10. $-3a + 12a$

11. $5ab - 7ab$

12. $9ab - 3ab$

13. $-12xy + 17xy$

14. $-15xy + 3xy$

15. $-3ab + 3ab$

16. $-7ab + 7ab$

17. $-\dfrac{1}{2}x - \dfrac{1}{3}x$

18. $-\dfrac{2}{5}y + \dfrac{3}{10}y$

19. $\dfrac{3}{8}x^2 - \dfrac{5}{12}x^2$

20. $\dfrac{2}{3}y^2 - \dfrac{4}{9}y^2$

21. $3x + 5x + 3x$

22. $8x + 5x + 7x$

23. $5a - 3a + 5a$

24. $10a - 17a + 3a$

25. $-5x^2 - 12x^2 + 3x^2$

26. $-y^2 - 8y^2 + 7y^2$

27. $7x + (-8x) + 3y$

28. $8y + (-10x) + 8x$

29. $7x - 3y + 10x$

30. $8y + 8x - 8y$

31. $3a + (-7b) - 5a + b$

32. $-5b + 7a - 7b + 12a$

33. $3x + (-8y) - 10x + 4x$

34. $3y + (-12x) - 7y + 2y$

35. $x^2 - 7x + (-5x^2) + 5x$

36. $3x^2 + 5x - 10x^2 - 10x$

Objective B

Simplify.

37. $4(3x)$
38. $12(5x)$
39. $-3(7a)$
40. $-2(5a)$
41. $-2(-3y)$

42. $-5(-6y)$
43. $(4x)2$
44. $(6x)12$
45. $(3a)(-2)$
46. $(7a)(-4)$

47. $(-3b)(-4)$
48. $(-12b)(-9)$
49. $-5(3x^2)$
50. $-8(7x^2)$
51. $\frac{1}{3}(3x^2)$

52. $\frac{1}{6}(6x^2)$
53. $\frac{1}{5}(5a)$
54. $\frac{1}{8}(8x)$
55. $-\frac{1}{2}(-2x)$
56. $-\frac{1}{4}(-4a)$

57. $-\frac{1}{7}(-7n)$
58. $-\frac{1}{9}(-9b)$
59. $(3x)\left(\frac{1}{3}\right)$
60. $(12x)\left(\frac{1}{12}\right)$
61. $(-6y)\left(-\frac{1}{6}\right)$

62. $(-10n)\left(-\frac{1}{10}\right)$
63. $\frac{1}{3}(9x)$
64. $\frac{1}{7}(14x)$
65. $-\frac{1}{5}(10x)$
66. $-\frac{1}{8}(16x)$

67. $-\frac{2}{3}(12a^2)$
68. $-\frac{5}{8}(24a^2)$
69. $-\frac{1}{2}(-16y)$
70. $-\frac{3}{4}(-8y)$
71. $(16y)\left(\frac{1}{4}\right)$

72. $(33y)\left(\frac{1}{11}\right)$
73. $(-6x)\left(\frac{1}{3}\right)$
74. $(-10x)\left(\frac{1}{5}\right)$
75. $(-8a)\left(-\frac{3}{4}\right)$
76. $(21y)\left(-\frac{3}{7}\right)$

Objective C

Simplify.

77. $-(x+2)$
78. $-(x+7)$
79. $2(4x-3)$
80. $5(2x-7)$

81. $-2(a+7)$
82. $-5(a+16)$
83. $-3(2y-8)$
84. $-5(3y-7)$

85. $(5 - 3b)7$ **86.** $(10 - 7b)2$ **87.** $\frac{1}{3}(6 - 15y)$ **88.** $\frac{1}{2}(-8x + 4y)$

89. $3(5x^2 + 2x)$ **90.** $6(3x^2 + 2x)$ **91.** $-2(-y + 9)$ **92.** $-5(-2x + 7)$

93. $(-3x - 6)5$ **94.** $(-2x + 7)7$ **95.** $2(-3x^2 - 14)$ **96.** $5(-6x^2 - 3)$

97. $-3(2y^2 - 7)$ **98.** $-8(3y^2 - 12)$ **99.** $3(x^2 - y^2)$ **100.** $5(x^2 + y^2)$

101. $-\frac{2}{3}(6x - 18y)$ **102.** $-\frac{1}{2}(x - 4y)$ **103.** $-(6a^2 - 7b^2)$

104. $3(x^2 + 2x - 6)$ **105.** $4(x^2 - 3x + 5)$ **106.** $-2(y^2 - 2y + 4)$

107. $\frac{1}{2}(2x - 6y + 8)$ **108.** $-\frac{1}{3}(6x - 9y + 1)$ **109.** $4(-3a^2 - 5a + 7)$

110. $-5(-2x^2 - 3x + 7)$ **111.** $-3(-4x^2 + 3x - 4)$ **112.** $3(2x^2 + xy - 3y^2)$

113. $5(2x^2 - 4xy - y^2)$ **114.** $-(3a^2 + 5a - 4)$ **115.** $-(8b^2 - 6b + 9)$

Objective D

Simplify.

116. $4x - 2(3x + 8)$ **117.** $6a - (5a + 7)$ **118.** $9 - 3(4y + 6)$

119. $10 - (11x - 3)$ **120.** $5n - (7 - 2n)$ **121.** $8 - (12 + 4y)$

122. $3(x + 2) - 5(x - 7)$ **123.** $2(x - 4) - 4(x + 2)$ **124.** $12(y - 2) + 3(7 - 3y)$

125. $6(2y - 7) - (3 - 2y)$ **126.** $3(a - b) - (a + b)$ **127.** $2(a + 2b) - (a - 3b)$

128. $4[x - 2(x - 3)]$ **129.** $2[x + 2(x + 7)]$ **130.** $-2[3x + 2(4 - x)]$

131. $-5[2x + 3(5 - x)]$ **132.** $-3[2x - (x + 7)]$ **133.** $-2[3x - (5x - 2)]$

134. $2x - 3[x - (4 - x)]$ **135.** $-7x + 3[x - (3 - 2x)]$ **136.** $-5x - 2[2x - 4(x + 7)] - 6$

APPLYING THE CONCEPTS

137.   Determine whether the statement is true or false. If the statement is false, give an example that illustrates that it is false.
   a.   Division is a commutative operation.
   b.   Division is an associative operation
   c.   Subtraction is an associative operation.
   d.   Subtraction is a commutative operation.
   e.   Addition is a commutative operation.

138.   Is the statement "any number divided by itself is one" a true statement? If not, for what number or numbers is the statement not true?

139.   Does every number have an additive inverse? If not, which real numbers do not have an additive inverse?

140.   Does every number have a multiplicative inverse? If not, which real numbers do not have a multiplicative inverse?

141.   In your own words, explain the distributive property.
[W]

142.   Explain why division by zero is not allowed.
[W]

143.   Give examples of two operations that occur in everyday experience
[W]   that are not commutative (for example, putting on socks and then shoes).

144.   Find the additive inverse of $a - b$.

145.   Define an operation $\otimes$ as $a \otimes b = (a \cdot b) - (a + b)$.
   For example, $7 \otimes 5 = (7 \cdot 5) - (7 + 5) = 35 - 12 = 23$.
   a.   Is $\otimes$ a commutative operation? Support your answer.

   b.   Is $\otimes$ an associative operation? Support your answer.

Translating Verbal Expressions into Variable Expressions

Objective A   *To translate a verbal expression into a variable expression, given the variable*

One of the major skills required in applied mathematics is to translate a verbal expression into a variable expression. This requires recognizing the verbal phrases that translate into mathematical operations. A partial list of the verbal phrases used to indicate the different mathematical operations follows.

Addition	added to	6 added to y	$y + 6$
	more than	8 more than x	$x + 8$
	the sum of	the sum of x and z	$x + z$
	increased by	t increased by 9	$t + 9$
	the total of	the total of 5 and y	$5 + y$
Subtraction	minus	x minus 2	$x - 2$
	less than	7 less than t	$t - 7$
	decreased by	m decreased by 3	$m - 3$
	the difference between	the difference between y and 4	$y - 4$
Multiplication	times	10 times t	$10t$
	of	one half of x	$\frac{1}{2}x$
	the product of	the product of y and z	yz
	multiplied by	y multiplied by 11	$11y$
Division	divided by	x divided by 12	$\frac{x}{12}$
	the quotient of	the quotient of y and z	$\frac{y}{z}$
	the ratio of	the ratio of t to 9	$\frac{t}{9}$
Power	the square of	the square of x	x^2
	the cube of	the cube of a	a^3

➡ Translate "14 less than the cube of x" into a variable expression.

Identify the words that indicate the mathematical operations.

14 *less than* the *cube* of x

Use the identified operations to write the variable expression.

$x^3 - 14$

➡ Translate "the difference between the square of x and the sum of y and z" into a variable expression.

$$x^2 - (y + z)$$

Identify words that indicate the mathematical operations.	the difference between the square of x and the sum of y and z
Use the identified operations to write the variable expression.	$x^2 - (y + z)$

Example 1

$3n + n$

Translate "the total of 3 times n and n" into a variable expression.

Solution

the total of 3 times n and n
$3n + n$

You Try It 1

$2n - \frac{1}{3}n$ 2

Translate "the difference between twice n and one-third of n" into a variable expression.

Your solution

Example 2

Translate "m decreased by the sum of n and 12" into a variable expression.

Solution

m decreased by the sum of n and 12
$m - (n + 12)$

You Try It 2

$\dfrac{b-7}{15}$

Translate "the quotient of 7 less than b and 15" into a variable expression.

Your solution

Solutions on p. A6

Objective B   **To translate a verbal expression into a variable expression and then simplify** ....

In most applications that involve translating phrases into variable expressions, the variable to be used is not given. To translate these phrases, a variable must be assigned to an unknown quantity before the variable expression can be written.

➡ Translate "the sum of two consecutive integers" into a variable expression.

Assign a variable to one of the unknown quantities.	the first integer: n
Use the assigned variable to write an expression for any other unknown quantity.	the next consecutive integer: $n + 1$
Use the assigned variable to write the variable expression.	$n + (n + 1)$

(handwritten: $x \frac{2}{-12}$)

Translate "the quotient of twice a number and the difference between the number and twelve" into a variable expression.

Assign a variable to one of the unknown quantities.	the unknown number: n
Use the assigned variable to write an expression for any other unknown quantity.	twice a number: $2n$ the difference between the number and twelve: $n - 12$
Use the assigned variable to write the variable expression.	$\dfrac{2n}{n - 12}$

Example 3
Translate "a number added to the product of four and the square of the number" into a variable expression.

Solution
the unknown number: n
the square of the number: n^2
the product of four and the square of the number: $4n^2$
$n + 4n^2$

Example 4
Translate "the sum of the squares of two and a number" into a variable expression. Then simplify.

Solution
the unknown number: x
$2^2 + x^2$
$4 + x^2$

Example 5
Translate "four times the sum of half of a number and fourteen" into a variable expression. Then simplify.

Solution
the unknown number: n

half of the number: $\dfrac{1}{2}n$

the sum of half of the number and fourteen: $\dfrac{1}{2}n + 14$

$4\left(\dfrac{1}{2}n + 14\right)$

$2n + 56$

You Try It 3
Translate "negative 4 multiplied by the total of ten and the cube of a number" into a variable expression.

Your solution
(handwritten: $(-4 \cdot)(10 + x^3)$)
(handwritten: $-48 - 4x^3$)
(handwritten: $-4x^3 - 40$)

You Try It 4
Translate "the sum of three consecutive integers" into a variable expression. Then simplify.

Your solution
(handwritten: x, $x + 1$, $x + 2$)
(handwritten: $x + (x+1) + (x+2)$)
(handwritten: $3x + 3$)

You Try It 5
Translate "five times the difference between a number and sixty" into a variable expression. Then simplify.

Your solution
(handwritten: $(5)(x - 60) =$)
(handwritten: $5x - 300$)

Solutions on p. A6

Objective C **To translate application problems** ....

Many of the applications of mathematics require that you identify the unknown quantity, assign a variable to that quantity, and then attempt to express the other unknown quantities in terms of the variable.

➡ A deluxe car stereo costs $10 more than twice the cost of the standard car stereo. Express the cost of the deluxe car stereo <u>in terms of</u> the cost of the standard car stereo.

Assign a variable to the cost of the standard car stereo.	the cost of the standard car stereo: C
Express the cost of the deluxe car stereo in terms of C.	the cost of the deluxe car stereo is $10 more than twice that of the standard car stereo: $2C + 10$

Example 6

The length of a swimming pool is 4 feet less than two times the width. Express the length of the pool in terms of the width.

Solution

the width of the pool: w
the length is 4 feet less than two times the width: $2w - 4$

Example 7

A banker divided $5000 between two accounts, one paying 10% annual interest and the second paying 8% annual interest. Express the amount invested in the 10% account in terms of the amount invested in the 8% account.

Solution

the amount invested at 8%: x
the amount invested at 10%: $5000 - x$

You Try It 6

The speed of a new jet plane is twice the speed of an older model. Express the speed of the new model <u>in terms of the speed of</u> the old model.

Your solution

$n = 2.0$

new $J = 2J$
old $J = J$

You Try It 7

A guitar string 6 feet long was cut into two pieces. Express the length of the shorter piece in terms of the length of the longer piece.

$\frac{6}{2} =$

$6 - x$

x

$6 - x$

6

Your solution

Solutions on p. A6

total − part

Content and Format © 1995 HMCo.

2.3 Exercises

· ·

Objective A

Translate into a variable expression.

1. the sum of 8 and y

2. a less than 16

3. t increased by 10

4. p decreased by 7

5. z added to 14

6. q multiplied by 13

7. 20 less than the square of x

8. 6 times the difference between m and seven

9. the sum of three-fourths of n and 12

10. b decreased by the product of 2 and b

11. 8 increased by the quotient of n and 4

12. the product of -8 and y

13. the product of 3 and the total of y and 7

14. 8 divided by the difference between x and 6

15. the product of t and the sum of t and 16

16. the quotient of 6 less than n and twice n

17. 15 more than one half of the square of x

18. 19 less than the product of n and -2

19. the total of 5 times the cube of n and the square of n

20. the ratio of 9 more than m to m

$$\frac{9+m}{m}$$

21. r decreased by the quotient of r and 3

22. four-fifths of the sum of w and 10

23. the difference between the square of x and the total of x and 17

24. s increased by the quotient of 4 and s

$$s + \left(\frac{4}{s}\right)$$

25. the product of 9 and the total of z and 4

26. n increased by the difference between 10 times n and 9

Objective B

Translate into a variable expression. Then simplify.

27. twelve minus a number

28. a number divided by eighteen

29. two-thirds of a number

30. twenty more than a number

31. the quotient of twice a number and nine

32. ten times the difference between a number and fifty

33. eight less than the product of eleven and a number

34. the sum of five-eighths of a number and six

35. nine less than the total of a number and two

36. the difference between a number and three more than the number

37. the quotient of seven and the total of five and a number

38. four times the sum of a number and nineteen

39. five increased by one half of the sum of a number and three

40. the quotient of fifteen and the sum of a number and twelve

41. a number added to the difference between twice the number and four

42. the product of two-thirds and the sum of a number and seven

43. the product of five less than a number and seven

44. the difference between forty and the quotient of a number and twenty.

45. the quotient of five more than twice a number and the number

46. the sum of the square of a number and twice the number

47. a number decreased by the difference between three times the number and eight

48. the sum of eight more than a number and one-third of the number

Translate into a variable expression. Then simplify.

49. a number added to the product of three and the number

50. a number increased by the total of the number and nine

51. five more than the sum of a number and six

52. a number decreased by the difference between eight and the number

53. a number minus the sum of the number and ten

54. the difference between one-third of a number and five-eighths of the number

55. the sum of one-sixth of a number and four-ninths of the number

56. two more than the total of a number and five

57. the sum of a number divided by three and the number $\frac{n}{3} + n$

58. twice the sum of six times a number and seven

Objective C   Application Problems

59. A propeller-driven plane flies at a rate that is half that of a jet plane. Express the rate of the propeller plane in terms of the rate of the jet plane.

60. The length of a football field is 30 yards more than the width. Express the length of the field in terms of the width.

omit

61. The diameter of a basketball is approximately 4 times the diameter of a baseball. Express the diameter of the baseball in terms of the diameter of the basketball.   omit

62. An 18-carat gold chain has 6 grams less than twice the amount of gold in a 12-carat gold chain. Express the gold in the 18-carat gold chain in terms of the amount of gold in the 12-carat gold chain.

x = lowest bracket

63. The highest percent income tax bracket is 3 percent more than twice the lowest percent bracket. Express the highest income tax bracket in terms of the lowest bracket. $H = 3 + 2x$

64. A batter can swing a bat at a rate that is two-thirds the rate at which a pitcher can throw the ball. Express the rate of the ball in terms of the rate of the bat.

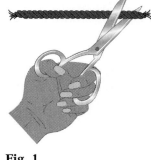

65. Twenty gallons of crude oil were poured into two containers of different size. Express the amount of crude oil poured into the smaller container in terms of the amount poured into the larger container.

66. A rope was cut into two pieces in such a way that the length of the longer piece was 2 feet less than 3 times the length of the shorter piece. Express the length of the longer piece in terms of the length of the shorter piece. (Fig. 1)

Fig. 1

67. A new design model of a laser printer can print pages at a rate that is seven pages more than one-half the speed of the older model. Express the speed of the newer model in terms of that of the older model.

68. Two cars are traveling in opposite directions and at different rates. Two hours later the cars are 200 miles apart. Express the distance traveled by the slower car in terms of the distance traveled by the faster car. (Fig. 2)

Fig. 2

APPLYING THE CONCEPTS

69. A wire whose length is given as x inches is bent into a square. Express the length of a side of the square in terms of x.

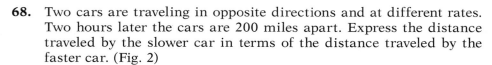

Fig. 3

70. The chemical formula for glucose (sugar) is $C_6H_{12}O_6$. This formula means that there are twelve hydrogen atoms for every six carbon atoms and six oxygen atoms in each molecule of glucose. If x represents the number of atoms of oxygen in a pound of sugar, express the number of hydrogen atoms in the pound of sugar. (Fig. 3)

71. A block-and-tackle system is designed so that pulling five feet on one end of a rope will move a weight on the other end a distance of three feet. If x represents the distance the rope is pulled, express the distance the weight moves in terms of x. (Fig. 4)

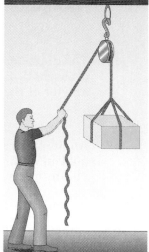

72. Translate the expressions $5x + 8$ and $5(x + 8)$ into phrases.
[W]

73. In your own words, explain how variables are used.
[W]

74. Explain the similarities and the differences between the expressions "the difference between x and 5" and "5 less than x."
[W]

Fig. 4

Project in Mathematics

Prime and Composite Numbers

A **prime number** is a natural number greater than 1 whose only natural-number factors are itself and 1. The number 11 is a prime number because the only natural-number factors of 11 are 11 and 1.

Eratosthenes, a Greek philosopher and astronomer who lived from 270 to 190 B.C., devised a method of identifying prime numbers. It is called the **Sieve of Eratosthenes.** The procedure is illustrated below.

~~1~~	(2)	(3)	~~4~~	(5)	~~6~~	(7)	~~8~~	~~9~~	~~10~~
(11)	~~12~~	(13)	~~14~~	~~15~~	~~16~~	(17)	~~18~~	(19)	~~20~~
~~21~~	~~22~~	(23)	~~24~~	~~25~~	~~26~~	~~27~~	~~28~~	(29)	~~30~~
(31)	~~32~~	~~33~~	~~34~~	~~35~~	~~36~~	(37)	~~38~~	~~39~~	~~40~~
(41)	~~42~~	(43)	~~44~~	~~45~~	~~46~~	(47)	~~48~~	~~49~~	~~50~~
~~51~~	~~52~~	(53)	~~54~~	~~55~~	~~56~~	~~57~~	~~58~~	(59)	~~60~~
(61)	~~62~~	~~63~~	~~64~~	~~65~~	~~66~~	(67)	~~68~~	~~69~~	~~70~~
(71)	~~72~~	(73)	~~74~~	~~75~~	~~76~~	~~77~~	~~78~~	(79)	~~80~~
~~81~~	~~82~~	(83)	~~84~~	~~85~~	~~86~~	~~87~~	~~88~~	(89)	~~90~~
~~91~~	~~92~~	~~93~~	~~94~~	~~95~~	~~96~~	(97)	~~98~~	~~99~~	~~100~~

List at the natural numbers from 1 to 100. Cross out the number 1, because it is not a prime number. The number 2 is prime; circle it. Cross out all the other multiples of 2 (4, 6, 8, . . .), because they are not prime. The number 3 is prime; circle it. Cross out all the other multiples of 3 (6, 9, 12, . . .) that are not already crossed out. The number 4, the next consecutive number in the list, has already been crossed out. The number 5 is prime; circle it. Cross out all the other multiples of 5 that are not already crossed out. Continue in this manner until all the prime numbers less than 100 are circled.

A **composite number** is a natural number greater than 1 that has a natural-number factor other than itself and 1. The number 21 is a composite number because it has factors of 3 and 7. All the numbers crossed out in the table above, except the number 1, are composite numbers.

Exercises:

Solve.

1. Use the Sieve of Eratosthenes to find the prime numbers between 100 and 200.
2. How many prime numbers are even numbers?
3. **a.** List two prime numbers that are consecutive natural numbers.
   b. Can there by any other pairs of prime numbers that are consecutive natural numbers?
4. Find the "twin primes" between 1 and 200. Twin primes are two prime numbers whose difference is 2. For instance, 3 and 5 are twin primes; 5 and 7 are also twin primes.

Chapter Summary

Key Words A *variable* is a letter that is used to stand for a quantity that is unknown.

A *variable expression* is an expression that contains one or more variables.

The *terms* of a variable expression are the addends of the expression.

A *variable term* is composed of a numerical coefficient and a variable part.

Like terms of a variable expression are terms that have the same variable part.

The *additive inverse* of a number is the opposite of the number.

The *multiplicative inverse* of a number is the reciprocal of the number.

Essential Rules *The Associative Property of Addition*
If a, b, and c are real numbers, then $(a + b) + c = a + (b + c)$.

The Commutative Property of Addition
If a and b are real numbers, then $a + b = b + a$.

The Addition Property of Zero
If a is a real number, then $a + 0 = 0 + a = a$.

The Additive Inverse Property
If a is a real number, than $a + (-a) = (-a) + a = 0$.

The Associative Property of Multiplication
If a, b, and c are real numbers, then $(ab)c = a(bc)$.

The Commutative Property of Multiplication
If a and b are real numbers, then $ab = ba$.

The Multiplication Property of One
If a is a real number, then $1 \cdot a = a \cdot 1 = a$.

The Inverse Property of Multiplication
If a is a non-zero real number, then $a\left(\dfrac{1}{a}\right) = \left(\dfrac{1}{a}\right)a = 1$.

The Distributive Property
If a, b, and c are real numbers, then $a(b + c) = ab + ac$.

Chapter Review

SECTION 2.1

1. Evaluate $a^2 - b^2$ when $a = 3$ and $b = 4$.

2. Evaluate $(b - a)^2 + c$ when $a = -2$, $b = 3$, and $c = 4$.

3. Evaluate $2bc \div (a + 7)$ when $a = 3$, $b = -5$, and $c = 4$.

4. Evaluate $(5c - 4a)^2 - b$ when $a = -1$, $b = 2$, and $c = 1$.

SECTION 2.2

5. Simplify: $7x + 4x$

6. Simplify: $12y - 17y$

7. Simplify: $6a - 4b + 2a$

8. Simplify: $4x - 3x^2 + 2x - x^2$

9. Simplify: $5c + (-2d) - 3d - (-4c)$

10. Simplify: $-9r + 2s - 6s + 12s$

11. Simplify: $5(4x)$

12. Simplify: $-3(-12y)$

13. Simplify: $\frac{1}{4}(-24a)$

14. Simplify: $-6(7x^2)$

15. Simplify: $(-50n)\left(\frac{1}{10}\right)$

16. Simplify: $5(2x - 7)$

17. Simplify: $2(6y^2 + 4y - 5)$

18. Simplify: $-9(7 + 4x)$

19. Simplify: $3(x^2 - 8x - 7)$

20. Simplify: $(7a^2 - 2a + 3)4$

21. Simplify: $18 - (4x - 2)$

22. Simplify: $-4(2x - 9) + 5(3x + 2)$

23. Simplify: $6(8y - 3) - 8(3y - 6)$

24. Simplify: $5[2 - 3(6x - 1)]$

$$5[2 - 18x + 3]$$
$$5[2 - 18x + 3]$$
$$5[5 - 18x] = 25 - 90x$$

SECTION 2.3

25. Translate "the product of 4 and x" into a variable expression.

26. Translate "6 less than x" into a variable expression.

27. Translate "two thirds of the total of x and 10" into a variable expression.

28. Translate "a number plus twice the number" into a variable expression. Then simplify.

29. Translate "three times a number plus the product of five and one less than the number" into a variable expression. Then simplify.

30. Translate "the difference between twice a number and one half of the number" into a variable expression. Then simplify.

31. A candy bar contains eight more calories than twice the number of calories in an apple. Express the number of calories in a candy bar in terms of the number of calories in an apple.

$x =$ apple

$8 + 2x =$ candy cal

32. A club treasurer has some five-dollar bills and some ten-dollar bills. The treasurer has a total of 35 bills. Express the number of five-dollar bills in terms of the number of ten-dollar bills.

Total = part

$x = \$10.$

$\$5 =$

$35 - x =$ five

33. The diameter of #8 copper wire is two mils less than 13 times the diameter of #30 copper wire. Express the diameter of #8 copper wire in terms of the diameter of #30 copper wire.

34. A baseball card collection contains five times as many National League players' cards as American League players' cards. Express the number of National League players' cards in terms of the number of American League players' cards.

Chapter Test

1. Evaluate $b^2 - 3ab$ when $a = 3$ and $b = -2$.

2. Evaluate $\dfrac{-2ab}{2b - a}$ when $a = -4$ and $b = 6$.

3. Simplify: $3x - 5x + 7x$

4. Simplify: $-7y^2 + 6y^2 - (-2y^2)$

5. Simplify: $3x - 7y - 12x$

6. Simplify: $3x + (-12y) - 5x - (-7y)$

7. Simplify: $\dfrac{1}{5}(10x)$

8. Simplify: $(12x)\left(\dfrac{1}{4}\right)$

9. Simplify: $(-3)(-12y)$

10. Simplify: $\dfrac{2}{3}(-15a)$

11. Simplify: $5(3 - 7b)$

12. Simplify: $-2(2x - 4)$

13. Simplify: $-3(2x^2 - 7y^2)$

14. Simplify: $-5(2x^2 - 3x + 6)$

15. Simplify: $2x - 3(x - 2)$

16. Simplify: $5(2x + 4) - 3(x - 6)$

17. Simplify: $2x + 3[4 - (3x - 7)]$

18. Simplify: $-2[x - 2(x - y)] + 5y$

19. Translate "the difference of the squares of a and b" into a variable expression.

20. Translate "the sum of a number and twice the square of the number" into a variable expression.

21. Translate "three less than the quotient of six and a number" into a variable expression.

22. Translate "b decreased by the product of b and 7" into a variable expression. Then simplify.

$7b - b$ $6 - 7b$

23. Translate "10 times the difference between x and 3" into a variable expression. Then simplify.

24. The speed of a pitcher's fastball is twice the speed of the catcher's return throw. Express the speed of the fastball in terms of the speed of the return throw.

25. A wire is cut into two lengths. The length of the longer piece is three inches less than four times the length of the shorter piece. Express the length of the longer piece in terms of the length of the shorter piece.

Cumulative Review

1. Add: $-4 + 7 + (-10)$

2. Subtract: $-16 \div (-25) - 4$

3. Multiply: $(-2)(3)(-4)$

4. Divide: $(-60) \div 12$

5. Write $1\frac{1}{4}$ as a decimal.

6. Simplify: $\frac{7}{12} - \frac{11}{16} - \left(-\frac{1}{3}\right)$

7. Simplify: $\frac{5}{12} \div \left(2\frac{1}{2}\right)$

8. Simplify: $\left(-\frac{9}{16}\right) \cdot \left(\frac{8}{27}\right) \cdot \left(-\frac{3}{2}\right)$

9. Write $\frac{3}{4}$ as a percent.

10. Simplify: $-2^5 \div (3 - 5)^2 - (-3)$

11. Simplify: $\left(-\frac{3}{4}\right)^2 - \left(\frac{3}{8} - \frac{11}{12}\right)$

12. Evaluate $a^2 - 3b$ when $a = 2$ and $b = -4$.

13. Simplify: $-2x^2 - (-3x^2) + 4x^2$

14. Simplify: $5a - 10b - 12a$

15. Simplify: $\frac{1}{2}(12a)$

16. Simplify: $\left(-\frac{5}{6}\right)(-36b)$

17. Simplify: $3(8 - 2x)$

18. Simplify: $-2(-3y + 9)$

19. Write $37\frac{1}{2}\%$ as a fraction.

20. Write 1.05% as a decimal.

21. Simplify: $-4(2x^2 - 3y^2)$

22. Simplify: $-3(3y^2 - 3y - 7)$

23. Simplify: $-3x - 2(2x - 7)$

24. Simplify: $4(3x - 2) - 7(x + 5)$

25. Simplify: $2x + 3[x - 2(4 - 2x)]$

26. Simplify: $3[2x - 3(x - 2y)] + 3y$

27. Translate "the sum of one-half of b and b" into a variable expression.

28. Translate "ten divided by the difference between y and two" into a variable expression.

29. Translate "the difference between eight and the quotient of a number and twelve" into a variable expression.

30. Translate "the sum of a number and two more than the number" into a variable expression. Then simplify.

31. Translate and simplify "twelve more than the product of three plus a number and five."

32. A "triple speed" CD-ROM drive spins three times faster than a normal CD-ROM drive. Express the speed of the "triple speed" CD-ROM drive in terms of that of the normal CD-ROM drive.

Chapter

3

Solving Equations

Objectives

Section 3.1

To determine whether a given number is a
  solution of an equation in one variable
To solve an equation of the form $x + a = b$
To solve an equation of the form $ax = b$
To solve application problems

Section 3.2

To solve an equation of the form $ax + b = c$
To solve application problems

Section 3.3

To solve an equation of the form
  $ax + b = cx + d$
To solve an equation containing parentheses
To solve application problems

Section 3.4

To translate a number problem into an
  equation and solve
To solve application problems

Mersenne Primes

A prime number that can be written in the form $2^n - 1$, where n is also prime, is called a Mersenne prime. The table below shows some Mersenne primes.

$$3 = 2^2 - 1$$
$$7 = 2^3 - 1$$
$$31 = 2^5 - 1$$
$$127 = 2^7 - 1$$

Not every prime number is a Mersenne prime. For example, 5 is a prime number but not a Mersenne prime. Also, not all numbers in the form $2^n - 1$, where n is prime, yield a prime number. For example, $2^{11} - 1 = 2047$ is not a prime number.

The search for Mersenne primes has been quite extensive, especially since the advent of the computer. One reason for the extensive research into large prime numbers (not only Mersenne primes) has to do with cryptology.

Cryptology is the study of making or breaking secret codes. One method for making a code that is difficult to break is called public key cryptology. For this method to work, it is necessary to use very large prime numbers. To keep anyone from breaking the code, each prime should have at least 200 digits.

Today the largest known Mersenne prime is $2^{216091} - 1$. This number has 65,050 digits in its representation.

Another Mersenne prime got special recognition in a postage-meter stamp! It is the number $2^{11213} - 1$. This number has 3276 digits in its representation.

Content and Format © 1995 HMCo.

 **Introduction to Equations**

Objective A   *To determine whether a given number is a solution of an equation in one variable*

POINT OF INTEREST
One of the most famous equations ever stated is $E = mc^2$. This equation, stated by Albert Einstein, shows that there is a relationship between mass m and energy E. As a side note, the chemical element einsteinium was named in honor of Einstein.

An **equation** expresses the equality of two mathematical expressions. The expressions can be either numerical or variable expressions.

$$\left. \begin{array}{l} 9 + 3 = 12 \\ 3x - 2 = 10 \\ y^2 + 4 = 2y - 1 \\ z = 2 \end{array} \right\} \text{Equations}$$

The equation at the right is true if the variable is replaced by 5.

$$x + 8 = 13$$
$$5 + 8 = 13 \qquad \text{A true equation}$$

The equation is false if the variable is replaced by 7.

$$7 + 8 = 13 \qquad \text{A false equation}$$

A **solution** of an equation is a number that, when substituted for the variable, results in a true equation. 5 is a solution of the equation $x + 8 = 13$. 7 is not a solution of the equation $x + 8 = 13$.

⇒ Is -2 a solution of $2x + 5 = x^2 - 3$?
To determine whether -2 is a solution of $2x + 5 = x^2 - 3$, replace x by -2. Then simplify and compare the results. If the results are equal, -2 is a solution of the equation. If the results are not equal, -2 is not a solution of the equation.

$$\begin{array}{c|c} \multicolumn{2}{c}{2x + 5 = x^2 - 3} \\ \hline 2(-2) + 5 & (-2)^2 - 3 \\ -4 + 5 & 4 - 3 \\ \multicolumn{2}{c}{1 = 1} \end{array}$$

- Replace x by -2.
- Simplify.
- Compare results.

The results are equal. Therefore, -2 is a solution of $2x + 5 = x^2 - 3$.

Example 1   Is $\frac{2}{3}$ of a solution of $12x - 2 = 6x + 2$?

Solution
$$\begin{array}{c|c} \multicolumn{2}{c}{12x - 2 = 6x + 2} \\ \hline 12\left(\dfrac{2}{3}\right) - 2 & 6\left(\dfrac{2}{3}\right) + 2 \\ 8 - 2 & 4 + 2 \\ \multicolumn{2}{c}{6 = 6} \end{array}$$

Yes, $\frac{2}{3}$ is a solution.

You Try It 1   Is $\frac{1}{4}$ a solution of $5 - 4x = 8x + 2$?

Your solution

Example 2   Is -4 a solution of $4 + 5x = x^2 - 2x$?

Solution
$$\begin{array}{c|c} \multicolumn{2}{c}{4 + 5x = x^2 - 2x} \\ \hline 4 + 5(-4) & (-4)^2 - 2(-4) \\ 4 + (-20) & 16 - (-8) \\ \multicolumn{2}{c}{-16 \neq 24} \end{array}$$
($\neq$ means "is not equal to")
No, -4 is not a solution.

You Try It 2   Is 5 a solution of $10x - x^2 = 3x - 10$?

Your solution

Solutions on p. A6

Objective B   *To solve an equation of the form x + a = b*

To **solve** an equation means to find a solution of the equation. The simplest equation to solve is an equation of the form *variable = constant,* because the constant is the solution.

The solution of the equation $x = 5$ is 5 because $5 = 5$ is a true equation.

The solution of the equation at the right is 7 because $7 + 2 = 9$ is a true equation.

$$x + 2 = 9$$
$$7 + 2 = 9$$

Note that if 4 is added to each side of the equation $x + 2 = 9$, the solution is still 7.

$$x + 2 = 9$$
$$x + 2 + 4 = 9 + 4$$
$$x + 6 = 13 \qquad 7 + 6 = 13$$

If -5 is added to each side of the equation, $x + 2 = 9$, the solution is still 7.

$$x + 2 = 9$$
$$x + 2 + (-5) = 9 + (-5)$$
$$x - 3 = 4$$
$$7 - 3 = 4$$

Equations that have the same solution are **equivalent equations.** The equations $x + 2 = 9$, $x + 6 = 13$, and $x - 3 = 4$ are equivalent equations; each equation has 7 as its solution. These examples suggest that adding the same number to each side of an equation produces equivalent equations. This is called the *Addition Property of Equations.*

> **Addition Property of Equations**
>
> The same number can be added to each side of an equation without changing its solution. In symbols, the equation $a = b$ has the same solution as the equation $a + c = b + c$.

This property states that the same quantity can be added to each side of an equation without changing the solution of the equation.

In solving an equation, the goal is to rewrite the given equation in the form *variable = constant.* The Addition Property of Equations is used to remove a *term* from one side of the equation by adding the opposite of that term to each side of the equation.

➡ Solve: $x - 4 = 2$ • The goal is to rewrite the equation as variable = constant.

$$x - 4 = 2$$
$$x - 4 + 4 = 2 + 4$$ • Add 4 to each side of the equation.
$$x + 0 = 6$$ • Simplify.
$$x = 6$$ • The equation is in the form variable = constant.

Check: $x - 4 = 2$

$$\frac{}{6 - 4 \,\vert\, 2}$$
$$2 = 2 \quad \text{A true equation}$$

The solution is 6.

Because subtraction is defined in terms of addition, the Addition Property of Equations also makes it possible to subtract the same number from each side of an equation without changing the solution of the equation.

➡️ Solve: $y + \dfrac{3}{4} = \dfrac{1}{2}$

$$y + \frac{3}{4} = \frac{1}{2}$$

- The goal is to rewrite the equation in the form *variable = constant*.

$$y + \frac{3}{4} - \frac{3}{4} = \frac{1}{2} - \frac{3}{4}$$

- Subtract $\dfrac{3}{4}$ from each side of the equation.

$$y + 0 = \frac{2}{4} - \frac{3}{4}$$

- Simplify.

$$y = -\frac{1}{4}$$

- The equation is in the form *variable = constant*.

The solution is $-\dfrac{1}{4}$. You should check this solution.

Example 3   Solve: $x + \dfrac{3}{4} = \dfrac{1}{3}$

Solution
$$x + \frac{3}{4} = \frac{1}{3}$$
$$x + \frac{3}{4} - \frac{3}{4} = \frac{1}{3} - \frac{3}{4}$$
$$x = -\frac{5}{12} \quad \bullet\ \frac{1}{3} - \frac{3}{4} = \frac{4}{12} - \frac{9}{12}$$

The solution is $-\dfrac{5}{12}$.

You Try It 3   Solve: $\dfrac{5}{6} = y - \dfrac{3}{8}$

Your solution
$$\frac{5}{6} = y - \frac{3}{8}$$
$$\frac{5}{6} - \frac{5}{6} = y - \frac{3}{8} - \frac{5}{6}$$ ✓
$$0 = y - \frac{9}{24} - \frac{20}{24} =$$
$$-y = -\frac{29}{24}$$

Solution on p. A6

Objective C   **To solve an equation of the form $ax = b$**

The solution of the equation at the right is 3 because $2 \cdot 3 = 6$ is a true equation.

$$2x = 6$$
$$2 \cdot 3 = 6$$

Note that if each side of $2x = 6$ is multiplied by 5, the solution is still 3.

$$2x = 6$$
$$5(2x) = 5 \cdot 6$$
$$10x = 30 \qquad 10 \cdot 3 = 30$$

If each side of $2x = 6$ is multiplied by -4, the solution is still 3.

$$2x = 6$$
$$(-4)(2x) = (-4) \cdot 6$$
$$-8x = -24 \qquad -8 \cdot 3 = -24$$

The equations $2x = 6$, $10x = 30$, and $-8x = -24$ are equivalent equations; each equation has 3 as its solution. These examples suggest that multiplying each side of an equation by the same number produces equivalent equations.

> **Multiplication Property of Equations**
>
> Each side of an equation can be multiplied by the same *nonzero* number without changing the solution of the equation. In symbols, if $c \neq 0$, then equation $a = b$ has the same solutions as equation $ac = bc$.

This property states that each side of an equation can be multiplied by the same nonzero number without changing the solution of the equation.

Recall that the goal of solving an equation is to rewrite the equation in the form *variable* = *constant*. The Multiplication Property of Equations is used to remove a coefficient by multiplying each side of the equation by the reciprocal of the coefficient.

➡ Solve: $\frac{3}{4}z = 9$

$$\frac{3}{4}z = 9$$
● The goal is to rewrite the equation in the form *variable* = *constant*.

$$\frac{4}{3} \cdot \frac{3}{4}z = \frac{4}{3} \cdot 9$$
● Multiply each side of the equation by $\frac{4}{3}$.

$$1 \cdot z = 12$$
● Simplify.

$$z = 12$$
● The equation is in the form *variable* = *constant*.

The solution is 12. You should check this solution.

Because division is defined in terms of multiplication, each side of an equation can be divided by the same non zero number without changing the solution of the equation.

➡ Solve: $6x = 14$

$$6x = 14$$
● The goal is to rewrite the equation in the form *variable* = *constant*.

$$\frac{6x}{6} = \frac{14}{6}$$
● Divide each side of the equation by 6.

$$1 \cdot x = \frac{7}{3}$$
● Simplify.

$$x = \frac{7}{3}$$
● The equation is in the form *variable* = *constant*.

The solution is $\frac{7}{3}$. You should check this solution.

LOOK CLOSELY

Remember to check the solution.

Check: $\qquad 6x = 14$

$$6\left(\frac{7}{3}\right) \quad \Big| \quad 14$$

$$14 = 14$$

When using the Multiplication Property of Equations, multiply each side of the equation by the reciprocal of the coefficient when the coefficient is a fraction. Divide each side of the equation by the coefficient when the coefficient is an integer or decimal.

Example 4 Solve: $\frac{3x}{4} = -9$

Solution

$$\frac{3x}{4} = -9$$

$$\frac{4}{3} \cdot \frac{3}{4}x = \frac{4}{3}(-9) \quad \bullet \left[\frac{3x}{4} = \frac{3}{4}x\right]$$

$$x = -12$$

The solution is -12.

You Try It 4 Solve: $-\frac{2x}{5} = 6$

Your solution

Example 5 Solve: $5x - 9x = 12$

Solution

$$5x - 9x = 12$$

$$-4x = 12 \quad \bullet \text{ Combine like terms.}$$

$$\frac{-4x}{4} = \frac{12}{-4}$$

$$x = -3$$

The solution is -3.

You Try It 5 Solve: $4x - 8x = 16$

Your solution

Solutions on p. A6

Tue Riview   Anns   march 4   mid term — 1—9

Objective D   *To solve application problems* ....

An equation that is used frequently in mathematics applications is the basic percent equation.

$$\begin{array}{r} 69.9 \\ 12\overline{)900} \\ 78 \\ \overline{120} \\ 108 \\ \overline{120} \end{array}$$

> **Basic Percent Equation**
>
> Percent · Base = Amount
>
> $P \quad \cdot \quad B \quad = \quad A$

When we translate a problem involving a percent into an equation, the word "of" translates to "multiply" and the word "is" translates to "equals." In many application problems involving percent, the base follows the word "of".

⮕ Solve: 20% of what number is 30?
Use the basic percent equation. $P = 20\% = 0.20$, $A = 30$, and B is unknown.

$P \cdot B = A$

$0.20B = 30$ • Replace P and A with their values.

$\dfrac{0.20B}{0.20} = 30$ • Solve for B.

$B = 150$

The number is 150.

⮕ Solve: 70 is what percent of 80?
Use the basic percent equation. $B = 80$, $A = 70$, and P is unknown.

LOOK CLOSELY
We have written $P(80) = 70$ because that is the form of the basic percent equation. We could have written $80P = 70$. The important point is that each side of the equation is divided by 80, the coefficient of P.

$P \cdot B = A$

$P(80) = 70$ • Replace B and A with their values.

$\dfrac{P(80)}{80} = \dfrac{70}{80}$ • Solve for P.

$P = 0.875 = 87.5\%$ • Because the question asked for a percent, the decimal 0.875 was converted to a percent.

The previous problems involved just the basics of the percent equation. The strategy for solving application problems requires identifying the percent, the base, and the amount.

⮕ Solve: Getting a grade of B on a biology exam requires earning a score between 78% and 87%. A student who took this exam received 123 points out of a possible 150 points. Did this student receive a B on the exam.?

Strategy
To determine whether the student received a B, use the basic percent equation to find what percent of 150 is 123. Then compare the percent to 78% and 87%. $B = 150$, $A = 123$, and P is unknown.

Solution

$P \cdot B = A$

$P(150) = 123$ • Replace B and A with their values.

$\dfrac{P(150)}{150} = \dfrac{123}{150}$ • Solve for P.

$P = 0.82 = 82\%$

Because 82% is between 78% and 87%, the student received a B on the exam.

In most cases, you should write the percent as a decimal before solving the basic percent equation. However, some percents are more easily written as a fraction. For example,

$$33\frac{1}{3}\% = \frac{1}{3} \qquad 66\frac{2}{3}\% = \frac{2}{3} \qquad 16\frac{2}{3}\% = \frac{1}{6} \qquad 83\frac{1}{3}\% = \frac{5}{6}$$

Example 6

12 is $33\frac{1}{3}\%$ of what number?

Solution

$$A = P \cdot B$$
$$12 = \frac{1}{3}B \quad \bullet \left(33\frac{1}{3}\% = \frac{1}{3}\right)$$
$$3 \cdot 12 = 3 \cdot \frac{1}{3}B$$
$$36 = B$$

The number is 36.

You Try It 6

Find 19% of 125.

Your solution

$P \cdot B = A$

$(.19)(125) = A$

$23.75 = A$

Example 7

A new labor contract increased an employee's hourly wage by 5%. What is the amount of increase for an employee who was making $9.60 an hour?

Strategy

To find the amount of increase, solve the basic percent equation, using $B = 9.60$ and $P = 5\% = 0.05$. The amount is unknown.

Solution

$$P \cdot B = A$$
$$(0.05)(9.60) = A$$
$$0.48 = A$$

The amount of increase is $0.48.

You Try It 7

A company was producing 2500 gal of paint each week. Because of a decrease in demand, the company reduced its weekly production by 500 gal. What percent decrease does this represent?

Your strategy

Your solution

Example 8

According to Chicago Title & Trust, the 1993 median price of a home was $115,000 in the midwest and $157,000 in the northeast. What percent of the midwest price is the northeast price?

Strategy

Use the basic percent equation. $B = 115,000$, $A = 157,000$, and P is unknown.

Solution

$$P \cdot B = A$$
$$P(115,000) = 157,000$$
$$\frac{P(115,000)}{115,000} = \frac{157,000}{115,000}$$
$$P \approx 1.37 = 137\%$$

The northeast median price of a home is 137% of the midwest price.

You Try It 8

According to the Congressional Budget Office, the federal deficit will be $223 billion in 1994 and $171 billion in 1995. What percent of the projected 1995 budget deficit is the 1994 budget deficit?

Your strategy

Your solution

Solutions on pp. A6–A7

3.1 Exercises

. . . .

Objective A

1. Is 4 a solution of
$2x = 8$?

2. Is 3 a solution of
$y + 4 = 7$?

3. Is -1 a solution of
$2b - 1 = 3$?

4. Is -2 a solution of
$3a - 4 = 10$?

5. Is 1 a solution of
$4 - 2m = 3$?

6. Is 2 a solution of
$7 - 3n = 2$?

7. Is 5 a solution of
$2x + 5 = 3x$?

8. Is 4 a solution of
$3y - 4 = 2y$?

9. Is 0 a solution of
$4a + 5 = 3a + 5$?

10. Is 3 a solution of
$z^2 + 1 = 4 + 3z$?

11. Is 2 a solution of
$2x^2 - 1 = 4x - 1$?

12. Is -1 a solution of
$y^2 - 1 = 4y + 3$?

13. Is -2 a solution of
$m^2 - 4 = m + 3$?

14. Is 5 a solution of
$x^2 + 2x + 1 = (x + 1)^2$?

15. Is -6 a solution of
$(n - 2)^2 = n^2 - 4n + 4$?

16. Is 4 a solution of
$x(x + 1) = x^2 + 5$?

17. Is 3 a solution of
$2a(a - 1) = 3a + 3$?

18. Is $-\dfrac{1}{4}$ a solution of
$8t + 1 = -1$?

19. Is $\dfrac{1}{2}$ a solution of
$4y + 1 = 3$?

20. Is $\dfrac{2}{5}$ a solution of
$5m + 1 = 10m - 3$?

21. Is $\dfrac{3}{4}$ a solution of
$8x - 1 = 12x + 3$?

Objective B

Solve and check.

22. $x + 5 = 7$

23. $y + 3 = 9$

24. $b - 4 = 11$

25. $z - 6 = 10$

26. $2 + a = 8$

27. $5 + x = 12$

28. $m + 9 = 3$

29. $t + 12 = 10$

30. $n - 5 = -2$

31. $x - 6 = -5$

32. $b + 7 = 7$

33. $y - 5 = -5$

34. $a - 3 = -5$ **35.** $x - 6 = -3$ **36.** $z + 9 = 2$ **37.** $n + 11 = 1$

38. $10 + m = 3$ **39.** $8 + x = 5$ **40.** $9 + x = -3$ **41.** $10 + y = -4$

42. $b - 5 = -3$ **43.** $t - 6 = -4$ **44.** $4 + x = 10$ **45.** $9 + a = 20$

46. $2 = x + 7$ **47.** $-8 = n + 1$ **48.** $4 = m - 11$ **49.** $-6 = y - 5$

50. $12 = 3 + w$ **51.** $-9 = 5 + x$ **52.** $4 = -10 + b$ **53.** $-7 = -2 + x$

54. $13 = -6 + a$ **55.** $m + \dfrac{2}{3} = -\dfrac{1}{3}$ **56.** $c + \dfrac{3}{4} = -\dfrac{1}{4}$ **57.** $x - \dfrac{1}{2} = \dfrac{1}{2}$

58. $x - \dfrac{2}{5} = \dfrac{3}{5}$ **59.** $\dfrac{5}{8} + y = \dfrac{1}{8}$ **60.** $\dfrac{4}{9} + a = -\dfrac{2}{9}$

61. $m + \dfrac{1}{2} = -\dfrac{1}{4}$ **62.** $b + \dfrac{1}{6} = -\dfrac{1}{3}$ **63.** $x + \dfrac{2}{3} = \dfrac{3}{4}$

64. $n + \dfrac{2}{5} = \dfrac{2}{3}$ **65.** $-\dfrac{5}{6} = x - \dfrac{1}{4}$ **66.** $-\dfrac{1}{4} = c - \dfrac{2}{3}$

67. $d + 1.3619 = 2.0148$ **68.** $w + 2.932 = 4.801$ **69.** $-0.813 + x = -1.096$

70. $-1.926 + t = -1.042$ **71.** $6.149 = -3.108 + z$ **72.** $5.237 = -2.014 + x$

Objective C

Solve and check.

73. $5x = -15$

74. $4y = -28$

75. $3b = 0$

76. $2a = 0$

77. $-3x = 6$

78. $-5m = 20$

79. $-3x = -27$

80. $-\dfrac{1}{6}n = -30$

81. $20 = \dfrac{1}{4}c$

82. $18 = 2t$

83. $-32 = 8w$

84. $-56 = 7x$

85. $0 = -5x$

86. $0 = -8a$

87. $-32 = -4y$

88. $-54 = 6c$

89. $49 = -7t$

90. $\dfrac{x}{3} = 2$

91. $\dfrac{x}{4} = 3$

92. $-\dfrac{y}{2} = 5$

93. $-\dfrac{b}{3} = 6$

94. $\dfrac{3}{4}y = 9$

95. $\dfrac{2}{5}x = 6$

96. $-\dfrac{2}{3}d = 8$

97. $-\dfrac{3}{5}m = 12$

98. $\dfrac{2n}{3} = 0$

99. $\dfrac{5x}{6} = 0$

100. $\dfrac{-3z}{8} = 9$

101. $\dfrac{-4x}{5} = -12$

102. $-6 = -\dfrac{2}{3}y$

103. $-15 = -\dfrac{1}{5}x$

104. $\dfrac{2}{5}a = 3$

105. $\dfrac{3x}{4} = 2$

106. $\dfrac{3}{4}c = \dfrac{3}{5}$

107. $\dfrac{2}{9} = \dfrac{2}{3}y$

108. $-\dfrac{6}{7} = -\dfrac{3}{4}b$

109. $\dfrac{1}{5}x = -\dfrac{1}{10}$

110. $-\dfrac{2}{3}y = -\dfrac{8}{9}$

111. $-1 = \dfrac{2n}{3}$

112. $-\dfrac{3}{4} = \dfrac{a}{8}$

Solve and check.

113. $-\dfrac{2}{5}m = -\dfrac{6}{7}$ **114.** $5x + 2x = 14$ **115.** $3n + 2n = 20$ **116.** $7d - 4d = 9$

117. $10y - 3y = 21$ **118.** $2x - 5x = 9$ **119.** $\dfrac{x}{1.46} = 3.25$ **120.** $\dfrac{z}{2.95} = -7.88$

121. $3.47a = 7.1482$ **122.** $2.31m = 2.4255$ **123.** $-3.7x = 7.881$ **124.** $\dfrac{n}{2.65} = 9.08$

_____

Objective D   P·B=A

Solve.

125. What is 35% of 80?   **126.** What percent of 8 is 0.5?   **127.** Find 1.2% of 60.

128. 8 is what percent of 5?   **129.** 125% of what is 80?   **130.** What percent of 20 is 30?

131. 12 is what percent of 50?   **132.** What percent of 125 is 50?   **133.** Find 18% of 40.

134. What is 25% of 60?   **135.** 12% of what is 48?   **136.** 45% of what is 9?

137. What is $33\dfrac{1}{3}$% of 27?   **138.** Find $16\dfrac{2}{3}$% of 30.   **139.** What percent of 12 is 3?

140. 10 is what percent of 15?   **141.** 60% of what is 3?   **142.** 75% of what is 6?

143. 12 is what percent of 6?   **144.** 20 is what percent of 16?   **145.** $5\dfrac{1}{4}$% of what is 21?

146. $37\frac{1}{2}\%$ of what is 15?

147. Find 15.4% of 50.

148. What is 18.5% of 46?

149. 1 is 0.5% of what?

150. 3 is 1.5% of what?

151. $\frac{3}{4}\%$ of what is 3?

152. $\frac{1}{2}\%$ of what is 3?

153. Find 125% of 16.

154. What is 250% of 12?

155. 16.43 is what percent of 20.45? Round to the nearest hundredth of a percent.

156. Find 18.37% of 625.43. Round to the nearest hundredth.

157. A university consists of three colleges: business, engineering, and fine arts. There are 2900 students in the business college, 1500 students in the engineering college, and 1000 students in the fine arts college. What percent of the total number of students in the university, to the nearest percent, are in the fine arts college?

158. Approximately 21% of air is oxygen. Using this estimate, determine how many liters of oxygen there are in a room containing 21,600 liters of air.

4536 L

$P · B = A$

159. The baseball playoff series were increased from 5 games to 7 games. What percent increase does this represent?

514%

160. The cost of a 30-second television commercial during the 1967 Super Bowl was $175,000 (in 1994 dollars). In 1994 a 30-second commercial for the Super Bowl cost $900,000. What percent of the 1967 commercial cost is the 1994 commercial cost?

161. A ski vacation package regularly costs $850 for one week, including lift tickets. By making an early reservation, a person receives a 15% discount. How much is the early reservation discount?

162. A football stadium increased its 60,000 seat capacity by 15%. How many seats were added to the stadium?

9,000

163. To override a presidential veto, at least $66\frac{2}{3}\%$ of the Senate must vote to override the veto. There are 100 senators in the Senate. What is the minimum number of votes needed to override a veto?

164. To receive a B− grade in a history class, a student must give 75 correct responses on a test of 90 questions. What percent of the total number of questions must a student answer correctly to receive a B− grade?

165. An airline knowingly overbooks certain flights by selling 18% more tickets than there are available seats. How many tickets would this airline sell for an airplane that has 150 seats?

166. The circle graph at the right shows the sales of full-size pickup trucks for different manufacturers in 1993. What percent of the total number of pickups sold by these companies were Ford F-series pickups?

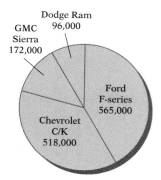

Dodge Ram 96,000
GMC Sierra 172,000
Ford F-series 565,000
Chevrolet C/K 518,000

APPLYING THE CONCEPTS

167. Solve the equation $ax = b$ for x. Is the solution you have written valid for all real numbers a and b?

168. Make up an equation of the form $x + a = b$ that has 2 as a solution.

169. Make up an equation of the form $ax = b$ that has −1 as a solution.

170. Solve.

    a. $\dfrac{3}{\frac{1}{x}} = 5$ **b.** $\dfrac{2}{\frac{1}{y}} = -2$

171. Solve.

    a. $\dfrac{3x + 2x}{3} = 2$ **b.** $\dfrac{4a - 7a}{3} = -2$

172. Write out the steps for solving the equation $\frac{1}{2}x = -3$. Identify each Property of Real Numbers or Property of Equations as you use it.

173. [W] In your own words, state the Addition Property of Equations and the Multiplication Property of Equations.

174. If a quantity increases by 100%, how many times its original value is the new value?

175. One-half of a certain number equals two-thirds of the same number. What is the number?

176. [W] To solve the equation $x + 7 = 10$, you can either add −7 to each side of the equation or subtract 7 from each side. Explain why this is possible.

177. [W] To solve the equation $2y = 10$, you can either multiply each side of the equation by $\frac{1}{2}$ or divide each side of the equation by 2. Explain why this is possible.

3.2

General Equations—Part I

Objective A   To solve an equation of the form ax + b = c

In solving an equation of the form $ax + b = c$, the goal is to rewrite the equation in the form *variable = constant*. This requires the application of both the Addition and the Multiplication Properties of Equations.

➡ Solve: $\frac{3}{4}x - 2 = -11$

The goal is to write the equation in the form *variable = constant*.

$$\frac{3}{4}x - 2 = -11$$

$$\frac{3}{4}x - 2 + 2 = -11 + 2 \qquad \bullet \text{ Add 2 to each side of the equation.}$$

LOOK CLOSELY

Check: $\frac{3}{4}x - 2 = -11$

$$\frac{3}{4}(-12) - 2 \;\Big|\; -11$$

$$-11 = -11$$

A true equation

$$\frac{3}{4}x = -9 \qquad \bullet \text{ Simplify.}$$

$$\frac{4}{3} \cdot \frac{3}{4}x = \frac{4}{3}(-9) \qquad \bullet \text{ Multiply each side of the equation by } \frac{4}{3}.$$

$$x = -12 \qquad \bullet \text{ The equation is of the form } variable = constant.$$

The solution is -12.

Example 1   Solve: $3x - 7 = -5$

Solution

$$3x - 7 = -5$$
$$3x - 7 + 7 = -5 + 7$$
$$3x = 2$$
$$\frac{3x}{3} = \frac{2}{3}$$
$$x = \frac{2}{3}$$

The solution is $\frac{2}{3}$.

You Try It 1   Solve: $5x + 7 = 10$

Your solution

$$5x + 7 = 10$$
$$3x + \frac{7}{-7} = \frac{10}{-7} + 3$$
$$\frac{5x = +3}{5} \; \frac{}{5}$$
$$x = \frac{3}{5}$$

Example 2   Solve: $5 = 9 - 2x$

Solution

$$5 = 9 - 2x$$
$$5 - 9 = 9 - 9 - 2x$$
$$-4 = -2x$$
$$\frac{-4}{-2} = \frac{-2x}{-2}$$
$$2 = x$$

The solution is 2.

You Try It 2   Solve: $2 = 11 + 3x$

Your solution

$$11 + 3x = 2$$
$$-11 \qquad\qquad -11$$
$$\frac{3x = -9}{3} \; \frac{}{3}$$
$$x = -3$$

Solutions on p. A7

Example 3

Solve: $2x + 4 - 5x = 10$

Solution

$$2x + 4 - 5x = 10 \qquad \bullet \text{ Combine like terms.}$$
$$-3x + 4 = 10$$
$$-3x + 4 - 4 = 10 - 4$$
$$-3x = 6$$
$$\frac{-3x}{-3} = \frac{6}{-3}$$
$$x = -2$$

The solution is -2.

You Try It 3

Solve: $x - 5 + 4x = 25$

Your solution

Solution on p. A4

Objective B   To solve application problems

Example 4

To determine the total cost of production, an economist uses the equation $T = U \cdot N + F$, where T is the total cost, U is the unit cost, N is the number of units made, and F is the fixed costs. Use this equation to find the number of units made during a month when the total cost was $9000, the unit cost was $25, and the fixed costs were $3000.

Strategy

To find the number of units made, replace each of the variables by their given value and solve for N.

Solution

$$T = U \cdot N + F$$
$$9000 = 25N + 3000$$
$$9000 - 3000 = 25N + 3000 - 3000$$
$$6000 = 25N$$
$$\frac{6000}{25} = \frac{25N}{25}$$
$$240 = N$$

There were 240 units made.

You Try It 4

The pressure at a certain depth in the ocean can be approximated by the equation $P = 15 + \frac{1}{2}D$, where P is the pressure in pounds per square inch and D is the depth in feet. Use this equation to find the depth when the pressure is 45 pounds per square inch.

Your strategy

Your solution

Solution on p. A4

3.2 Exercises

Objective A

Solve and check.

1. $3x + 1 = 10$ **2.** $4y + 3 = 11$ **3.** $2a - 5 = 7$ **4.** $5m - 6 = 9$

5. $5 = 4x + 9$ **6.** $2 = 5b + 12$ **7.** $2x - 5 = -11$ **8.** $3n - 7 = -19$

9. $4 - 3w = -2$ **10.** $5 - 6x = -13$ **11.** $8 - 3t = 2$ **12.** $12 - 5x = 7$

13. $4a - 20 = 0$ **14.** $3y - 9 = 0$ **15.** $6 + 2b = 0$ **16.** $10 + 5m = 0$

17. $-2x + 5 = -7$ **18.** $-5d + 3 = -12$ **19.** $-12x + 30 = -6$ **20.** $-13 = -11y + 9$

21. $2 = 7 - 5a$ **22.** $3 = 11 - 4n$ **23.** $-35 = -6b + 1$ **24.** $-8x + 3 = -29$

25. $-3m - 21 = 0$ **26.** $-5x - 30 = 0$ **27.** $-4y + 15 = 15$ **28.** $-3x + 19 = 19$

29. $9 - 4x = 6$ **30.** $3t - 2 = 0$ **31.** $9x - 4 = 0$ **32.** $7 - 8z = 0$

33. $1 - 3x = 0$ **34.** $9d + 10 = 7$ **35.** $12w + 11 = 5$ **36.** $6y - 5 = -7$

37. $8b - 3 = -9$ **38.** $5 - 6m = 2$ **39.** $7 - 9a = 4$ **40.** $9 = -12c + 5$

Solve and check.

41. $10 = -18x + 7$

42. $2y + \dfrac{1}{3} = \dfrac{7}{3}$

43. $4a + \dfrac{3}{4} = \dfrac{19}{4}$

44. $2n - \dfrac{3}{4} = \dfrac{13}{4}$

45. $3x - \dfrac{5}{6} = \dfrac{13}{6}$

46. $5y + \dfrac{3}{7} = \dfrac{3}{7}$

47. $9x + \dfrac{4}{5} = \dfrac{4}{5}$

48. $8 = 7d - 1$

49. $8 = 10x - 5$

50. $4 = 7 - 2w$

51. $7 = 9 - 5a$

52. $8t + 13 = 3$

53. $12x + 19 = 3$

54. $-6y + 5 = 13$

55. $-4x + 3 = 9$

56. $\dfrac{1}{2}a - 3 = 1$

57. $\dfrac{1}{3}m - 1 = 5$

58. $\dfrac{2}{5}y + 4 = 6$

59. $\dfrac{3}{4}n + 7 = 13$

60. $-\dfrac{2}{3}x + 1 = 7$

61. $-\dfrac{3}{8}b + 4 = 10$

62. $\dfrac{x}{4} - 6 = 1$

63. $\dfrac{y}{5} - 2 = 3$

64. $\dfrac{2x}{3} - 1 = 5$

65. $\dfrac{3c}{7} - 1 = 8$

66. $4 - \dfrac{3}{4}z = -2$

67. $3 - \dfrac{4}{5}w = -9$

68. $5 + \dfrac{2}{3}y = 3$

69. $17 + \dfrac{5}{8}x = 7$

70. $17 = 7 - \dfrac{5}{6}t$

71. $9 = 3 - \dfrac{2x}{7}$

72. $3 = \dfrac{3a}{4} + 1$

Solve and check.

73. $7 = \dfrac{2x}{5} + 4$

74. $5 - \dfrac{4c}{7} = 8$

75. $7 - \dfrac{5}{9}y = 9$

76. $6a + 3 + 2a = 11$

77. $5y + 9 + 2y = 23$

78. $7x - 4 - 2x = 6$

79. $11z - 3 - 7z = 9$

80. $2x - 6x + 1 = 9$

81. $b - 8b + 1 = -6$

82. $3 = 7x + 9 - 4x$

83. $-1 = 5m + 7 - m$

84. $8 = 4n - 6 + 3n$

85. If $2x - 3 = 7$, evaluate $3x + 4$.

86. If $3x + 5 = -4$, evaluate $2x - 5$.

87. If $4 - 5x = -1$, evaluate $x^2 - 3x + 1$.

88. If $2 - 3x = 11$, evaluate $x^2 + 2x - 3$.

89. If $5x + 3 - 2x = 12$, evaluate $4 - 5x$.

90. If $2x - 4 - 7x = 16$, evaluate $x^2 + 1$.

Objective B *Application Problems*

The distance s that an object will fall in t seconds is given by $s = 16t^2 + vt$, where v is the initial velocity of the object.

91. Find the initial velocity of an object that falls 80 ft in 2 s.

92. Find the initial velocity of an object that falls 144 ft in 3 s.

A company uses the equation $V = C - 6000t$ to determine the depreciated value V, after t years, of a milling machine that originally cost C dollars. Equations like this are used in accounting for *straight-line depreciation*.

93. A milling machine originally cost $50,000. In how many years will the depreciated value be $38,000?

$V = C - 6000 T$

$38000 = 50,000 - 6000 T$

94. A milling machine originally cost $78,000. In how many years will the depreciated value be $48,000?

Anthropologists can approximate the height of a primate by the size of its humerus (the bone from the elbow to the shoulder) by using the equation $H = 1.2L + 27.8$, where L is the length of the humerus and H is the height of the primate.

95. An anthropologist estimates the height of a primate to be 66 in. What is the approximate length of the humerus of this primate to the nearest tenth of an inch?

96. An anthropologist finds the humerus of a primate to be 28.4 in. long. Approximate the height of this primate to the nearest tenth of an inch.

A telephone company estimates that the number N of phone calls per day between two cities of population P_1 and P_2 that are d miles apart is given by the equation $N = \frac{2.51 P_1 P_2}{d^2}$.

97. Estimate the population P_2, to the nearest thousand, given that the population of one city (P_1) is 48,000, the number of phone calls is 1,100,000, and the distance between the cities is 75 mi.

98. Estimate the population P_2, to the nearest thousand, given that the population of one city (P_1) is 125,000, the number of phone calls is 2,500,000, and the distance between the cities is 50 mi.

26,000

When two lines intersect, four angles are formed as shown at the right. The angles a and b are called **vertical angles;** angles x and y are also vertical angles. The vertical angles formed by intersecting lines are equal. **Adjacent angles** are angles that share a common side. For instance, angles a and x are adjacent angles. The sum of the measures of adjacent angles formed by intersecting lines is 180°.

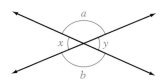

99. In Fig. 1, determine the value of y.

54°

100. In Fig. 2, determine the value of b.

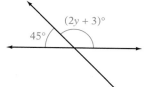

Fig. 1

APPLYING THE CONCEPTS

101. Solve $x \div 15 = 25$ remainder 10.

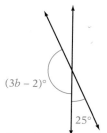

Fig. 2

102. If $2x + 3 = 4$, what is the value of $2x^2$?

103. Does the sentence "Solve $3x - 4(x - 1)$." make sense? Why or
[W] why not?

104. Explain the steps you would take to solve the equation $3x - 4 = 14$.
[W] State the Property of Real Numbers or the Property of Equations that is used at each step.

105. Explain the difference between the word *equation* and the word
[W] *expression*.

106. The following problem does not contain enough information for us
[W] to find only one solution. Supply some additional information so that the problem has exactly one solution. Then write and solve an equation.
  "The sum of two numbers is 15. Find the numbers."

107. The following problem does not contain enough information. What
[W] additional information do we need to answer the question?
  "How many hours does it take to fly from Los Angeles to New York?"

3.3 General Equations—Part II

Objective A   *To solve an equation of the form ax + b = cx + d* ....

In solving an equation of the form $ax + b = cx + d$, the goal is to rewrite the equation in the form *variable = constant*. Begin by rewriting the equation so that there is only one variable term in the equation. Then rewrite the equation so that there is only one constant term.

➡ Solve: $2x + 3 = 5x - 9$

$$2x + 3 = 5x - 9$$

$$2x - 5x + 3 = 5x - 5x - 9$$ • Subtract 5x from each side of the equation.

$$-3x + 3 = -9$$ • Simplify.

$$-3x + 3 - 3 = -9 - 3$$ • Subtract 3 from each side of the equation.

$$-3x = -12$$ • Simplify.

$$\frac{-3x}{-3} = \frac{-12}{-3}$$ • Divide each side of the equation by −3.

$$x = 4$$ • The equation is in the form *variable = constant.*

The solution is 4. You should verify this by checking this solution.

Example 1
Solve: $4x - 5 = 8x - 7$

Solution

$$4x - 5 = 8x - 7$$

$$4x - 8x - 5 = 8x - 8x - 7$$

$$-4x - 5 = -7$$

$$-4x - 5 + 5 = -7 + 5$$

$$-4x = -2$$

$$\frac{-4x}{-4} = \frac{-2}{-4}$$

$$x = \frac{1}{2}$$

The solution is $\frac{1}{2}$.

You Try It 1
Solve: $5x + 4 = 6 + 10x$

Your solution

Solution on p. A8

Example 2
Solve: $3x + 4 - 5x = 2 - 4x$

Solution

$$3x + 4 - 5x = 2 - 4x$$
$$-2x + 4 = 2 - 4x$$
$$-2x + 4x + 4 = 2 - 4x + 4x$$
$$2x + 4 = 2$$
$$2x + 4 - 4 = 2 - 4$$
$$2x = -2$$
$$\frac{2x}{2} = \frac{-2}{2}$$
$$x = -1$$

The solution is -1.

You Try It 2
Solve: $5x - 10 - 3x = 6 - 4x$

Your solution

Solution on p. A8

Objective B   *To solve an equation containing parentheses*

When an equation contains parentheses, one of the steps in solving the equation requires the use of the Distributive Property. The Distributive Property is used to remove parentheses from a variable expression.

➡ Solve: $4 + 5(2x - 3) = 3(4x - 1)$

$$4 + 5(2x - 3) = 3(4x - 1)$$

$$4 + 10x - 15 = 12x - 3$$ • Use the Distributive Property. Then simplify.

$$10x - 11 = 12x - 3$$

$$10x - 12x - 11 = 12x - 12x - 3$$ • Subtract 12x from each side of the equation.

$$-2x - 11 = -3$$ • Simplify.

$$-2x - 11 + 11 = -3 + 11$$ • Add 11 to each side of the equation.

$$-2x = 8$$ • Simplify.

$$\frac{-2x}{-2} = \frac{8}{-2}$$ • Divide each side of the equation by -2.

$$x = -4$$ • The equation is in the form *variable* = *constant*.

The solution is -4. You should verify this by checking this solution.

Example 3

Solve: $3x - 4(2 - x) = 3(x - 2) - 4$

Solution

$3x - 4(2 - x) = 3(x - 2) - 4$

$3x - 8 + 4x = 3x - 6 - 4$

$7x - 8 = 3x - 10$

$7x - 3x - 8 = 3x - 3x - 10$

$4x - 8 = -10$

$4x - 8 + 8 = -10 + 8$

$4x = -2$

$\dfrac{4x}{4} = \dfrac{-2}{4}$

$x = -\dfrac{1}{2}$

The solution is $-\dfrac{1}{2}$.

You Try It 3

Solve: $5x - 4(3 - 2x) = 2(3x - 2) + 6$

Your solution

$5x - 4(3 - 2x) = 2(3x - 2) + 6$

$5x - 12 + 8x = 6x - 4 + 6$

$-12 + 13x = 6x + 2$

$+7x = 14 \qquad x = \dfrac{14}{7} = 2$

$x = 2$

Example 4

Solve: $3[2 - 4(2x - 1)] = 4x - 10$

Solution

$3[2 - 4(2x - 1)] = 4x - 10$

$3[2 - 8x + 4] = 4x - 10$

$3[6 - 8x] = 4x - 10$

$18 - 24x = 4x - 10$

$18 - 24x - 4x = 4x - 4x - 10$

$18 - 28x = -10$

$18 - 18 - 28x = -10 - 18$

$-28x = -28$

$\dfrac{-28x}{-28} = \dfrac{-28}{-28}$

$x = 1$

The solution is 1.

You Try It 4

Solve: $-2[3x - 5(2x - 3)] = 3x - 8$

Your solution

$-2[3x - 5(2x - 3)] = 3x - 8$

$-2[3x - 10x + 15] = 3x - 8 =$

$-2[-7x + 15] = 3x - 8 =$

$14x - 30 = 3x - 8 =$

$11x = -24$

$x = \dfrac{24}{11} = 2$

Example 5

If $7x = 3x + 12$, evaluate $3x^2 - 7$.

Solution

Solve $7x = 3x + 12$ for x.

$7x - 3x = 12$

$4x = 12$

$x = 3$

Evaluate $3x^2 - 7$ for $x = 3$.

$3x^2 - 7$

$3(3)^2 - 7 = 3(9) - 7$

$= 27 - 7$

$= 20$

If $7x = 3x + 12$, the value of $3x^2 - 7$ is 20.

You Try It 5

If $2x = 5x + 6$, evaluate $-2x + 7$.

Your solution

$2x = 5x + 6$

$\dfrac{-3x}{-3} = \dfrac{+6}{+3}$

$x = +2$

$-2x + 7$

$-2 \cdot (+2) + 7 =$

$+4 + 7 =$

11

Solutions on p. A8

Objective C   To solve application problems

A lever system is shown at the right. It consists of a lever, or bar; a fulcrum; and two forces, F_1 and F_2. The distance d represents the length of the lever, x represents the distance from F_1 to the fulcrum, and $d - x$ represents the distance from F_2 to the fulcrum.

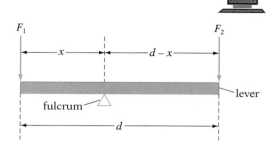

A principle of physics states that when the lever system balances, $F_1 \cdot x = F_2 \cdot (d - x)$.

Example 6
A lever is 15 ft long. A force of 50 lb is applied to one end of the lever, and a force of 100 lb is applied to the other end. Where is the fulcrum located when the system balances?

Strategy
Make a drawing.

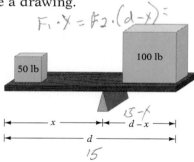

$$F_1 \cdot x = F_2 \cdot (d - x) =$$

To find the location of the fulcrum when the system balances, replace the variables F_1, F_2, and d in the lever system equation by the given values, and solve for x.

Solution

$$F_1 \cdot x = F_2 \cdot (d - x)$$
$$50x = 100(15 - x)$$
$$50x = 1500 - 100x$$
$$50x + 100x = 1500 - 100x + 100x$$
$$150x = 1500$$
$$\frac{150x}{150} = \frac{1500}{150}$$
$$x = 10$$

The fulcrum is 10 ft from the 50-lb force.

You Try It 6
A lever is 25 ft long. A force of 45 lb is applied to one end of the lever, and a force of 80 lb is applied to the other end. Where is the location of the fulcrum when the system balances?

Your strategy

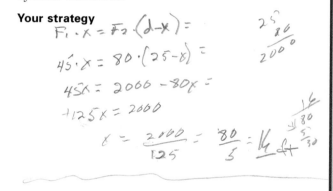

$$F_1 \cdot x = F_2 \cdot (d - x) =$$
$$45 \cdot x = 80 \cdot (25 - x) =$$
$$45x = 2000 - 80x =$$
$$+125x = 2000$$
$$x = \frac{2000}{125} = \frac{80}{5} = 16 \text{ ft}$$

Your solution

Solution on p. A8

3.3 Exercises

. . . .

Objective A

Solve and check.

1. $8x + 5 = 4x + 13$

2. $6y + 2 = y + 17$

3. $5x - 4 = 2x + 5$

4. $13b - 1 = 4b - 19$

5. $15x - 2 = 4x - 13$

6. $7a - 5 = 2a - 20$

7. $3x + 1 = 11 - 2x$

8. $n - 2 = 6 - 3n$

9. $2x - 3 = -11 - 2x$

10. $4y - 2 = -16 - 3y$

11. $2b + 3 = 5b + 12$

12. $m + 4 = 3m + 8$

13. $4y - 8 = y - 8$

14. $5a + 7 = 2a + 7$

15. $6 - 5x = 8 - 3x$

16. $10 - 4n = 16 - n$

17. $5 + 7x = 11 + 9x$

18. $3 - 2y = 15 + 4y$

19. $2x - 4 = 6x$

20. $2b - 10 = 7b$

21. $8m = 3m + 20$

22. $9y = 5y + 16$

23. $8b + 5 = 5b + 7$

24. $6y - 1 = 2y + 2$

25. $7x - 8 = x - 3$

26. $2y - 7 = -1 - 2y$

27. $2m - 1 = -6m + 5$

28. If $5x = 3x - 8$, evaluate $4x + 2$.

29. If $7x + 3 = 5x - 7$, evaluate $3x - 2$.

30. If $2 - 6a = 5 - 3a$, evaluate $4a^2 - 2a + 1$.

31. If $1 - 5c = 4 - 4c$, evaluate $3c^2 - 4c + 2$.

32. If $2y + 3 = 5 - 4y$, evaluate $6y - 7$.

33. If $3z + 1 = 1 - 5z$, evaluate $3z^2 - 7z + 8$.

Objective B

Solve and check.

34. $5x + 2(x + 1) = 23$

35. $6y + 2(2y + 3) = 16$

36. $9n - 3(2n - 1) = 15$

37. $12x - 2(4x - 6) = 28$

38. $7a - (3a - 4) = 12$

39. $9m - 4(2m - 3) = 11$

40. $5(3 - 2y) + 4y = 3$

41. $4(1 - 3x) + 7x = 9$

42. $5y - 3 = 7 + 4(y - 2)$

43. $5 + 2(3b + 1) = 3b + 5$

44. $6 - 4(3a - 2) = 2(a + 5)$

45. $7 - 3(2a - 5) = 3a + 10$

46. $2a - 5 = 4(3a + 1) - 2$

47. $5 - (9 - 6x) = 2x - 2$

48. $7 - (5 - 8x) = 4x + 3$

49. $3[2 - 4(y - 1)] = 3(2y + 8)$

50. $5[2 - (2x - 4)] = 2(5 - 3x)$

51. $3a + 2[2 + 3(a - 1)] = 2(3a + 4)$

52. $5 + 3[1 + 2(2x - 3)] = 6(x + 5)$

53. $-2[4 - (3b + 2)] = 5 - 2(3b + 6)$

54. $-4[x - 2(2x - 3)] + 1 = 2x - 3$

55. If $4 - 3a = 7 - 2(2a + 5)$, evaluate $a^2 + 7a$.

56. If $9 - 5x = 12 - (6x + 7)$, evaluate $x^2 - 3x - 2$.

57. If $2z - 5 = 3(4z + 5)$, evaluate $\dfrac{z^2}{(z - 2)}$.

58. If $3n - 7 = 5(2n + 7)$, evaluate $\dfrac{n^2}{(2n - 6)}$.

Objective C   *Application Problems*

Use the lever system equation $F_1x = F_2(d - x)$.

59. A lever 10 ft long is used to move a 100-lb rock (Fig. 1). The fulcrum is placed 2 ft from the rock. What force must be applied to the other end of the lever to move the rock?

60. A screwdriver 9 in. long is used as a lever to open a can of paint (Fig. 2). The tip of the screwdriver is placed under the lip of the can with the fulcrum 0.15 in. from the lip. A force of 30 lb is applied to the other end of the screwdriver. Find the force on the lip of the can.

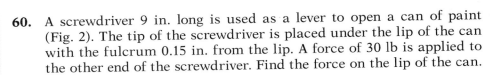

Fig. 1

61. An adult and a child are on a see-saw 14 ft long. The adult weighs 175 lb and the child weighs 70 lb. How many feet from the child must the fulcrum be placed so that the see-saw balances?

62. An 80-lb weight is applied to the left end of a see-saw 10 ft long with its fulcrum 5 ft from the 80-lb weight. A weight of 30 lb is applied to the right end of the see-saw. Is the 30-lb weight adequate to balance the see-saw?

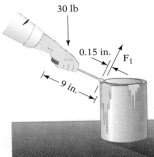

Fig. 2

63. Two children are sitting 8 ft apart on a see-saw (Fig. 3). One child weighs 60 lb and the second child weighs 50 lb. The fulcrum is 3.5 ft from the 60-lb child. Is the see-saw balanced?

64. In preparation for a stunt, two acrobats are standing on a plank 18 ft long. One acrobat weighs 128 lb and the second acrobat weighs 160 lb. How far from the 128-lb acrobat must the fulcrum be placed so that the acrobats are balanced on the plank?

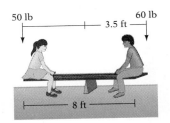

Fig. 3

$$Px = Cx + F$$

Economists use a *break-even point* to determine the number of units that must be sold so that no profit or loss occurs. The equation used is $Px = Cx + F$, where P is the selling price per unit, x is the number of units sold, C is the cost to make each unit, and F is the fixed costs.

65. An economist has determined the selling price for a television to be $250. The cost to make one television is $165 and the fixed costs are $29,750. Find the number of televisions that must be sold to break even.

66. A business analyst has determined the selling price for a cellular telephone to be $675. The cost to make one cellular telephone is $485 and the fixed costs are $36,100. Find the number of cellular telephones that must be sold to break even.

67. A manufacturing engineer determines the cost to make one compact disk to be $3.35 and the fixed costs to be $6180. The selling price for each compact disk is $8.50. Find the number of compact disks that must be sold to break even.

68. To manufacture a softball bat requires two steps. The first step is to cut a rough shape. The second step is to sand the bat to its final form. The cost to rough-shape a bat is $.45, and the cost to sand a bat to final form is $1.05. The total fixed costs for the two steps are $16,500. How many softball bats must be sold at a price of $7.00 to break even?

Two angles are **complementary angles** if the sum of the measures of the angles is 90°. Two angles are **supplementary angles** if the sum of the measures of the angles is 180°. In the diagrams at the right, angles *a* and *b* are complementary angles; angles *x* and *y* are supplementary angles.

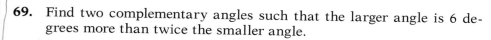

69. Find two complementary angles such that the larger angle is 6 degrees more than twice the smaller angle.

70. Find two complementary angles such that the smaller angle is 15 degrees less than one-half the larger angle.

71. Find two supplementary angles such that the larger angle is 12 degrees less than three times the smaller angle.

72. Find two supplementary angles such that the smaller angle is 10 degrees less than two-thirds the larger angle.

APPLYING THE CONCEPTS

73. Write an equation of the form $ax + b = cx + d$ that has 4 as the solution.

74. Solve the equation $x + a = b$ for *x*. Is the solution you have written valid for all real numbers *a* and *b*?

75.
[W] The equation $x = x + 1$ has no solution, whereas the solution of the equation $2x + 3 = 3$ is zero. Is there a difference between no solution and a solution of zero? Explain your answer.

Solve. If the equation has no solution, write "No solution."

76. $3(2x - 1) - (6x - 4) = -9$

77. $7(3x + 6) - 4(3 + 5x) = 13 + x$

78. $\frac{1}{5}(25 - 10a) + 4 = \frac{1}{3}(12a - 15) + 14$

79. $5[m + 2(3 - m)] = 3[2(4 - m) - 5]$

Translating Sentences into Equations

Objective A   ***To translate a number problem into an equation and solve*** ....

An equation states that two mathematical expressions are equal. Therefore, to translate a sentence into an equation requires recognition of the words or phrases that mean "equals." Some of these phrases are listed below.

$$\left.\begin{array}{l}\text{equals}\\\text{is}\\\text{is equal to}\\\text{amounts to}\\\text{represents}\end{array}\right\} \text{translate to } =$$

Once the sentence is translated into an equation, the equation can be solved by rewriting the equation in the form *variable = constant*.

➡ Translate "five less than a number is thirteen" into an equation and solve.

Assign a variable to the unknown quantity.   The unknown number: n

Find two verbal expressions for the same value.

| Five less than a number | is | thirteen |

Write a mathematical expression for each verbal expression. Write the equals sign.

$$n - 5 \quad = \quad 13$$

Solve the equation.

$$n - 5 + 5 = 13 + 5$$
$$n = 18$$

The number is 18.

Recall that the integers are the numbers $\{\ldots, -4, -3, -2, -1, 0, 1, 2, 3, 4, \ldots\}$. An **even integer** is an integer that is divisible by 2. Examples of even integers are -8, 0, and 22. An **odd integer** is an integer that is not divisible by 2. Examples of odd integers are -17, 1, and 39.

Consecutive integers are integers that follow one another in order. Examples of consecutive integers are shown at the right. (Assume that the variable n represents an integer.)

11, 12, 13
$-8, -7, -6$
$n, n + 1, n + 2$

Examples of **consecutive even integers** are shown at the right. (Assume that the variable n represents an even integer.)

24, 26, 28
$-10, -8, -6$
$n, n + 2, n + 4$

Examples of **consecutive odd integers** are shown at the right. (Assume that the variable n represents an odd integer.)

19, 21, 23
$-1, 1, 3$
$n, n + 2, n + 4$

➡ The sum of three consecutive odd integers is 45. Find the integers.

Represent three consecutive odd integers.

First odd integer: n
Second odd integer: $n + 2$
Third odd integer: $n + 4$

The sum of the three odd integers is 45.

$$n + (n + 2) + (n + 4) = 45$$
$$3n + 6 = 45$$
$$3n = 39$$
$$n = 13$$

$$n + 2 = 13 + 2 = 15$$

$$n + 4 = 13 + 4 = 17$$

The three consecutive odd integers are 13, 15, and 17.

Example 1

Translate "three more than twice a number is the number plus six" into an equation and solve.

Solution

The unknown number: n

Three more than twice a number	is	the number plus six

$$2n + 3 = n + 6$$
$$2n - n + 3 = n - n + 6$$
$$n + 3 = 6$$
$$n + 3 - 3 = 6 - 3$$
$$n = 3$$

The number is 3.

You Try It 1

Translate "four less than one-third of a number equals five minus two-thirds of the number" into an equation and solve.

Your solution

$$\frac{1}{3}x - 4 = 5 - \frac{2}{3}x$$

$$\frac{x}{3} - \frac{4}{1} = 5 - \frac{2x}{3} =$$

$$x = 9$$

Solution on p. A9

$T = x$

Example 2

The sum of two numbers is sixteen. The difference between four times the smaller number and two is two more than twice the larger number. Find the two numbers.

Solution

The smaller number: n
The larger number: $16 - n$

The difference between four times the smaller and two	is	two more than twice the larger

$$4n - 2 = 2(16 - n) + 2$$
$$4n - 2 = 32 - 2n + 2$$
$$4n - 2 = 34 - 2n$$
$$4n + 2n - 2 = 34 - 2n + 2n$$
$$6n - 2 = 34$$
$$6n - 2 + 2 = 34 + 2$$
$$6n = 36$$
$$\frac{6n}{6} = \frac{36}{6}$$
$$n = 6 \quad 16 - 6 = 10$$

The smaller number is 6.
The larger number is 10.

Example 3

Find three consecutive even integers such that three times the second is four more than the sum of the first and third.

Strategy

- First even integer: n
  Second even integer: $n + 2$
  Third even integer: $n + 4$
- Three times the second equals four more than the sum of the first and third.

Solution

$$3(n + 2) = n + (n + 4) + 4$$
$$3n + 6 = 2n + 8$$
$$n + 6 = 8$$
$$n = 2$$
$$n + 2 = 2 + 2 = 4$$
$$n + 4 = 2 + 4 = 6$$

The three even integers are 2, 4, and 6.

You Try It 2

The sum of two numbers is twelve. The total of three times the smaller number and six amounts to seven less than the product of four and the larger number. Find the two numbers.

Your solution

$x = $ one number
$12 - x = $ 2nd number

$$3x + (6) = [4(12 - x)] - 7$$
$$3x + 6 = [42 - 4x] - 7$$
$$3x + 4 = 41 - 4x$$
$$\frac{+4x}{7x} = \frac{35}{7}$$
$$x = 5$$

You Try It 3

Find three consecutive integers whose sum is -6.

Your strategy

x
$x + 1$
$x + 2$

$$3x + 3 = -6$$
$$-3$$
$$\frac{3x}{3} = \frac{-9}{3}$$
$$x = -3$$

Your solution

Solutions on p. A9

Objective B *To solve application problems*

Example 4

A Fahrenheit temperature of 68° is 32° more than $\frac{9}{5}$ the Celsius temperature. Find the Celsius temperature.

Strategy

Write and solve an equation.

Solution

68	is	32 degrees more than $\frac{9}{5}$ the Celsius temperature

$$68 = \frac{9}{5}C + 32$$

$$68 - 32 = \frac{9}{5}C + 32 - 32$$

$$36 = \frac{9}{5}C$$

$$\frac{5}{9} \cdot 36 = \frac{5}{9} \cdot \frac{9}{5}C$$

$$20 = C \qquad \text{The Celsius temperature is 20°.}$$

You Try It 4

A Celsius temperature of 20° is $\frac{5}{9}$ of the difference between the Fahrenheit temperature and 32. Find the Fahrenheit temperature.

Your strategy

$$20° = \frac{5}{9}(F - 32)$$

Your solution

$$20° = \frac{5}{9}F + \frac{5}{9} \cdot \frac{-32}{1}$$

$$20 = \frac{5}{9}F - \frac{140}{9} = 17\frac{7}{9}$$

$$20 = \frac{180}{90}$$

Example 5

A board 20 ft long is cut into two pieces. Five times the length of the smaller piece is 2 ft more than twice the length of the longer piece. Find the length of each piece.

Strategy

Make a drawing.

Write and solve an equation.

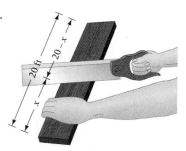

Solution

Five times the smaller piece	is	two feet more than twice the longer

$$5x = 2(20 - x) + 2$$
$$5x = 40 - 2x + 2$$
$$5x = 42 - 2x$$
$$5x + 2x = 42 - 2x + 2x$$
$$7x = 42$$
$$\frac{7x}{7} = \frac{42}{7}$$
$$x = 6 \qquad \text{The shorter piece is 6 ft.}$$
$$20 - x = 14 \qquad \text{The longer piece is 14 ft.}$$

You Try It 5

A company makes 140 televisions per day. Three times the number of black and white TV's made equals 20 less than the number of color TV's made. Find the number of color TV's made each day.

$x = Color$

Your strategy

$$T = 140$$
$$140 - x = B + W$$
$$3(140 - x) = x - 20$$
$$420 - 3x = x - 20$$
$$-4x = \frac{-440}{-4} =$$
$$x = 110$$

Your solution

Solutions on pp. A9–A10

3.4 Exercises

Objective A

Translate into an equation and solve.

1. The difference between a number and fifteen is seven. Find the number.

2. The sum of five and a number is three. Find the number.

3. The product of seven and a number is negative twenty-one. Find the number.

4. The quotient of a number and four is two. Find the number.

5. Four less than three times a number is five. Find the number.

6. The difference between five and twice a number is one. Find the number.

7. Four times the sum of twice a number and three is twelve. Find the number.

    $4(2x + 3) = 12$ $8x + 12 = 12$ $8x = 0$ $x = \frac{0}{8}$

8. Seventeen is three times the difference between four times a number and 5. Find the number.

9. Twelve is six times the difference between a number and three. Find the number.

10. Twice the difference between a number and twenty-five is three times the number. Find the number.

11. Four times a number is three times the difference between thirty-five and the number. Find the number.

12. The sum of two numbers is twenty. Three times the smaller is equal to two times the larger. Find the two numbers.

13. The sum of two numbers is fifteen. One less than three times the smaller is equal to the larger. Find the two numbers.

    $3x - 1 = 15 - x$ $4x = \frac{16}{4} = 4$ $x = sm$ $15 - x = Lg$

14. The sum of two numbers is twenty-four. Four less than three times the smaller is twelve less than twice the larger. Find the two numbers.

15. The sum of two numbers if fourteen. The difference between two times the smaller and the larger is one. Find the two numbers.

16. The sum of two numbers is eighteen. The total of three times the smaller and twice the larger is forty-four. Find the two numbers.

17. The sum of two numbers is 2. The difference between eight and twice the smaller number is two less than four times the larger. Find the two numbers.

    $x = sm$ $2 - x = Lg$

    $8 - 2x = 4(2x - 4) - 2 =$
    $8 - 2x = 6 - 4x =$
    $2x = -2 = -1$

Solve.

18. Find two consecutive even integers such that four times the first is three times the second.

19. The sum of three consecutive odd integers is 51. Find the integers.

20. Find three consecutive odd integers such that three times the middle integer is one more than the sum of the first and third.

21. Twice the smallest of three consecutive odd integers is seven more than the largest. Find the integers.

22. The sum of three consecutive integers is 60. Find the integers.

23. Find two consecutive even integers such that three times the first equals twice the second.

24. Seven times the first of two consecutive odd integers is five times the second. Find the integers.

25. Find three consecutive even integers such that three times the middle integer is four more than the sum of the first and third.

26. The sum of three consecutive even integers is 42. Find the integers.

27. Five times the first of two consecutive odd integers equals three times the second. Find the integers.

28. Find three consecutive even integers whose sum is negative eighteen.

29. The sum of three consecutive even integers is 66. Find the integers.

30. The sum of three consecutive odd integers is 75. Find the integers.

31. Find three consecutive integers whose sum is negative twenty-one.

32. Three times the smallest of three consecutive even integers is four more than twice the largest. Find the integers.

33. The sum of three consecutive integers is 48. Find the integers.

Objective B   *Application Problems*

Write an equation and solve.

34. The operating speed of a personal computer is 8 megahertz. This is one-fourth the speed of a newer model personal computer. Find the speed of the newer personal computer.

$2 = \frac{1}{4} x$ $\frac{18}{4}$ $\frac{x}{9}$

$x = SPEED$

35. One measure of computer speed is *mips*, *millions of instructions per second*. One computer has a rating of 10 mips. This is two-thirds the speed of a second computer. Find the mips rating of the second computer.

36. The score for one team in a football game was 26. The team scored twice as many field goals (3 points each) as it did touchdowns with extra points (7 points each). Find the number of field goals that this team scored.

37. A rectangular wood patio is constructed so that the length of the patio is 3 ft less than three times the width (Fig. 1). The sum of the length and width of the patio is 21 ft. Find the length of the patio.

38. A university employs a total of 600 teaching assistants and research assistants. There are three times as many teaching assistants as research assistants. Find the number of research assistants employed by the university.

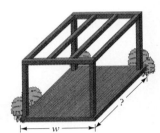

Fig. 1

39. A basketball team scored 105 points. There were as many two-point baskets as free throws (one point each). The number of three-point baskets was five less than the number of free throws. Find the number of three-point baskets.

40. Greek architects thought that a rectangle whose length was approximately 1.6 times the width yielded the most eye-pleasing proportions for the front of a building. The sum of the length and width of a certain rectangle constructed in this manner is 130 ft. Find the width and length of the rectangle.

41. The sum of the angles of a triangle, measured in degrees, is always 180°. One angle of a certain triangle is twice the smallest angle (Fig. 2). The third angle is three times the smallest angle. Find each angle.

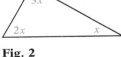

Fig. 2

42. A computer screen consists of tiny dots of light called pixels. In a certain graphics mode, there are 640 horizontal pixels. This is 40 more than three times the number of vertical pixels. Find the number of vertical pixels.

43. The cost of electricity in a certain city is $.08 for each of the first 300 kWh (kilowatt-hours) and $.13 for each kWh over 300 kWh. Find the number of kilowatt-hours used for a family with a $51.95 electric bill.

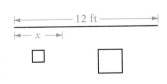

44. A wire 12 ft long is cut into two pieces (Fig. 3). Each piece is bent into the shape of a square. The perimeter of the larger square is twice the perimeter of the smaller square. Find the perimeter of the larger square.

Fig. 3

A line intersecting two other lines at two different points is called a transversal. If the lines cut by a transversal are parallel lines and the transversal is not perpendicular to the parallel lines, all four acute angles have the same measure and all four obtuse angles have the same measure. Thus for the figure at the right, $\angle a = \angle c = \angle w = \angle y$ and $\angle b = \angle d = \angle x = \angle z$.

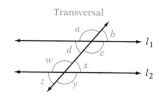

The angles on opposite sides of the transversal and between the parallel lines are called *alternate interior angles*. Angles on opposite sides and outside the parallel lines are called *alternate exterior angles*. *Corresponding angles* are two angles that are on the same side of the transversal and are both acute angles or are both obtuse angles.

45. In the figure at the right, l_1 is parallel to l_2. Find the measures of angles a, b, and c.

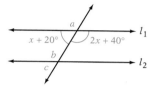

46. In the figure at the right, l_1 is parallel to l_2. Find the measures of angles a, b, and c.

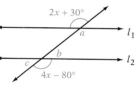

47. In the figure at the right, l_1 is parallel to l_2. Find the measures of angles a, b, and c.

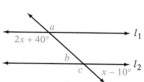

APPLYING THE CONCEPTS

48.
[W] Make up a word problem that has the solution of the equation $6x = 124$ as the answer to the problem.

49.
[W] Make up a word problem that has the solution of the equation $8x + 120 = 300$ as the answer to the problem.

50. A formula is an equation that relates variables in a known way. Find two examples of formulas that are used in your college major. Explain what each of the variables represents.

An equation that is never true is called a contradiction. For example, the equation $x = x + 1$ is a contradiction. There is no value of x that will make the equation true. An equation that is true for all real numbers is called an identity. The equation $x + x = 2x$ is an identity. This equation is true for any real number. A conditional equation is one that is true for some real numbers and false for some real numbers. The equation $2x = 4$ is a conditional equation. This equation is true when x is 2 and false for any other real number. Determine whether each of the following equations is a contradiction, an identity, or a conditional equation. If it is a conditional equation, find the solution.

51. $6x + 2 = 5 + 3(2x - 1)$

52. $3 - 2(4x + 1) = 5 + 8(1 - x)$

53. $3t - 5(t + 1) = 2(2 - t) - 9$

54. $6 + 4(2y + 1) = 5 - 8y$

55. $3y - 2 = 5y - 2(2 + y)$

56. $6x = 15x$

Project in Mathematics

Nielsen Ratings The A. C. Nielsen company surveys television viewers to determine the number of people watching particular shows. Each **rating point** indicates that 942,000 TV households are watching a particular program. For instance, if *60 Minutes* receives a rating of 22.1, then 22.1 × 942,000, or 20,818,200, TV households are watching that program.

Exercises:

1. Assume that an evening news program receives a Nielsen rating of 10.3. How many TV households are watching that program?

A rating point does *not* mean that 942,000 *people* are watching a program. A rating point refers to the number of TV sets tuned to that program. And of course, there may be more than one person watching a television set in a household.

2. During a week in which *Monday Night Football* received a rating of 17.4, there were 23.3 million people watching the program. To the nearest tenth, what is the average number of people per TV household?

The cost to advertise on a program is related to its Nielsen rating. The sponsor (the company paying for the advertisement) pays a certain number of dollars for each rating point a show receives.

3. Suppose a television network charges $15,000 per rating point for a 30-second commercial on a daytime talk show. Determine the cost for three 30-second commercials if the Nielsen rating of the show is 11.5.

Sometimes advertisers are charged on the *estimated* Nielsen rating rather than on the actual rating. This estimated rating is based on past experience for the type of program that is being televised. During the 1993 World Series, sponsors complained to the television network that the estimated rating was much higher than the actual rating. As a result, these sponsors claim they paid too much for the commercials.

4. Suppose that a network charges $18,000 per Nielsen rating point for a 30-second commercial and that the estimated rating is 19.3. A Nielsen survey after the program was televised showed that the actual rating was 18.7. What is the difference between the cost of a 30-second commercial at the two different ratings?

Chapter Summary

Key Words   An *equation* expresses the equality of two mathematical expressions.

A *solution* of an equation is a number that, when substituted for the variable, results in a true equation.

Equivalent equations have the same solution.

To *solve* an equation means to find a solution of the equation. The goal is to rewrite the equation in the form *variable = constant*.

To *translate* a sentence into an equation requires recognition of the words or phrases that mean "equals." Some of these phrases are *equals, is, is equal to, amounts to,* and *represents.*

Essential Rules   *Addition Property of Equations*

The same quantity can be added to each side of   If $a = b$, then
an equation without changing the solution of $a + c = b + c$.
the equation.

Multiplication Property of Equations

Each side of an equation can be multiplied by   If $a = b$ and $c \neq 0$,
the same non-zero number without changing   then $ac = bc$.
the solution of the equation.

Basic Percent Equation   Percent $\cdot$ Base = Amount
$$P \cdot B = A$$

Chapter Review

SECTION 3.1

1. Is 3 a solution of $5x - 2 = 4x + 5$?

2. Is -4 a solution of $x^2 - 3x = (x + 2)^2$?

3. Solve: $x + 3 = 24$

4. Solve: $a - \dfrac{1}{6} = \dfrac{2}{3}$

5. Solve: $4.6 = 2.1 + x$

6. Solve: $\dfrac{3}{5}a = 12$

7. 30 is what percent of 12?

8. $\dfrac{1}{2}\%$ of what is 8?

SECTION 3.2

9. Solve: $5x - 6 = 29$

10. Solve: $-4x - 2 = 10$

11. Solve: $32 = 9x - 4 - 3x$

12. Solve: $14x + 7x + 8 = -10$

13. Use the equation $I = Prt$, where I is the simple interest, P is the principal, r is the annual simple interest rate, and t is the time in years. If $1500 is invested at the annual simple interest rate of 0.08, in how many years will the interest be $1800?

14. Use the equation $T = c + rc$, where T is the total amount of a purchase including tax, r is the sales tax rate, and c is the cost before tax is added. Find the cost of a cookbook before tax is added if the sales tax rate is 0.07 and the total price of the book, including tax, is $29.96.

15. Use the equation $F = \dfrac{9}{5}C + 32$, where F is the temperature in degrees Fahrenheit and C is the temperature in degrees Celsius. Find the temperature in degrees Celsius of a person whose temperature is 98.6 degrees Fahrenheit.

16. Use the equation $S = R - D$, where S is the sale price, R is the regular price, and D is the discount. Find the discount on a mattress set if the sale price is $220.75 and the regular price is $300.00.

SECTION 3.3

17. Solve: $5x + 3 = 10x - 17$

18. Solve: $12y - 1 = 3y + 2$

18. Solve: $-6x + 16 = -2x$

20. Solve: $-7x - 5 = 4x + 50$

21. Solve: $6x + 3(2x - 1) = -27$

22. Solve: $5 + 2(x + 1) = 13$

23. Solve: $x + 5(3x - 20) = 10(x - 4)$

24. Solve: $7 - [4 + 2(x - 3)] = 11(x + 2)$

25. A lever is 12 ft long. At a distance of 2 ft from the fulcrum, a force of 120 lb is applied. How large a force must be applied to the other end of the lever so that the system will balance? Use the lever system equation $F_1 \cdot x = F_2 \cdot (d - x)$.

26. A lever is 8 ft long. A force of 25 lb is applied to one end of the lever and a force of 15 lb is applied to the other end. Find the location of the fulcrum when the system balances. Use the lever system equation $F_1 \cdot x = F_2 \cdot (d - x)$.

SECTION 3.4

27. Translate "four less than the product of five and a number is sixteen" into an equation and solve.

28. Four times the second of three consecutive integers equals the sum of the first and third integers. Find the integers.

29. The sum of two numbers is twenty-one. Three times the smaller number is two less than twice the larger number. Translate into an equation and then find the two numbers.

30. A piano wire is 35 in. long. A fifth chord can be produced by dividing this wire into two parts so that three times the length of the shorter piece is twice the length of the longer piece. Find the length of each piece.

31. An optical engineer's consulting fee was $600. This included $80 for supplies and $65 for each hour of consulting. Find the number of hours of consulting.

32. The Empire State building is 1472 ft tall. This is 514 ft less than twice the height of the Eiffel Tower. Find the height of the Eiffel Tower.

Chapter Test

1. Is -2 a solution of $x^2 - 3x = 2x - 6$?

2. What is 0.5% of 8?

3. Solve: $x - 3 = -8$

4. Solve: $\dfrac{3}{4}x = -9$

5. Solve: $3x - 5 = -14$

6. Solve: $7 - 4x = -13$

7. A financial manager has determined that the cost per unit for a calculator is \$15 and that the fixed costs per month are \$2000. Find the number of calculators produced during a month in which the total cost was \$5000. Use the equation $T = U \cdot N + F$, where T is the total cost, U is the cost per unit, N is the number of units produced, and F is the fixed costs.

8. Solve: $3x - 2 = 5x + 8$

9. Solve: $6 - 5x = 5x + 11$

10. Solve: $5x + 3 - 7x = 2x - 5$

11. Solve: $4 - 2(3 - 2x) = 2(5 - x)$

12. Solve: $5x - 2(4x - 3) = 6x + 9$

13. Solve: $9 - 3(2x - 5) = 12 + 5x$

14. A chemist mixes 100 g of water at 80°C with 50 g of water at 20°C. Use the equation $m_1 \cdot (T_1 - T) = m_2 \cdot (T - T_2)$ to find the final temperature of the water after mixing. m_1 is the quantity of water at the hotter temperature. T_1 is the temperature of the hotter water, m_2 is the quantity of water at the cooler temperature, T_2 is the temperature of the cooler water, and T is the final temperature of the water after mixing.

15. Translate "the difference between three times a number and fifteen is twenty-seven" into an equation and solve.

16. Find three consecutive even integers whose sum is 36.

17. The sum of two numbers is 18. The difference between four times the smaller number and seven is equal to the sum of two times the larger number and five. Find the two numbers.

18. A train travels between two cities in 26 h. This is 5 h more than the product of three and the time required for a plane to fly between the two cities. Find the number of hours required for the plane to fly between the two cities.

19. A board 18 ft long is cut into two pieces. Two feet less than the product of five and the length of the smaller piece is equal to the difference between three times the length of the longer piece and eight. Find the length of each piece.

20. The time it takes a hard disk controller in a computer to access data is 28 milliseconds (ms). This is 2 ms less than twice the time a new hard disk controller can access the data. How many milliseconds are required for the new disk controller to access the data?

Cumulative Review

1. Subtract: $-6 - (-20) - 8$

2. Multiply: $(-2)(-6)(-4)$

3. Simplify: $-\frac{5}{6} - \left(-\frac{7}{16}\right)$

4. Simplify: $-2\frac{1}{3} \div 1\frac{1}{6}$

5. Simplify: $-4^2 \cdot \left(-\frac{3}{2}\right)^3$

6. Simplify: $25 - 3\frac{(5-2)^2}{2^3+1} - (-2)$

7. Evaluate $3(a - c) - 2ab$ when $a = 2, b = 3,$ and $c = -4$.

8. Simplify: $3x - 8x + (-12x)$

9. Simplify: $2a - (-3b) - 7a - 5b$

10. Simplify: $(16x)\left(\frac{1}{8}\right)$

11. Simplify: $-4(-9y)$

12. Simplify: $-2(-x^2 - 3x + 2)$

13. Simplify: $-2(x - 3) + 2(4 - x)$

14. Simplify: $-3[2x - 4(x - 3] + 2$

15. Is -3 a solution of $x^2 + 6x + 9 = x + 3$?

16. Is $\frac{1}{2}$ a solution of $3 - 8x = 12x - 2$?

17. Find 32% of 60.

18. Solve: $\frac{3}{5}x = -15$

19. Solve: $7x - 8 = -29$

20. Solve: $13 - 9x = -14$

21. Solve: $8x - 3(4x - 5) = -2x - 11$

22. Solve: $6 - 2(5x - 8) = 3x - 4$

23. Solve: $5x - 8 = 12x + 13$

24. Solve: $11 - 4x = 2x + 8$

25. A business manager has determined that the cost per unit for a camera is $70 and that the fixed costs per month are $3500. Find the number of cameras that are produced during a month in which the total cost was $21,000. Use the equation $T = U \cdot N + F$, where T is the total cost, U is the cost per unit, N is the number of units produced, and F is the fixed cost.

26. A chemist mixes 300 g of water at 75°C with 100 g of water at 15°C. Use the equation $m_1 \cdot (T_1 - T) = m_2 \cdot (T - T_2)$ to find the final temperature of the water, where m_1 is the quantity of water at the hotter temperature, T_1 is the temperature of the hotter water, m_2 is the quantity of water at the cooler temperature, T_2 is the temperature of the cooler water, and T is the final temperature of the water after mixing.

27. Translate "the difference between twelve and the product of five and a number is negative eighteen" into an equation and solve.

28. Translate "the sum of six times a number and thirteen is five less than the product of three and the number" into an equation and solve.

29. The area of the cement foundation of a house is 2000 ft². This is 200 ft² more than three times the area of the garage. Find the area of the garage.

30. A board 16 ft long is cut into two pieces. Four feet more than the product of three and the length of the shorter piece is equal to three feet less than twice the length of the longer piece. Find the length of each piece.

Solving Equations: Applications

Objectives

Section 4.1
To solve markup problems
To solve discount problems

Section 4.2
To solve investment problems

Section 4.3
To solve value mixture problems
To solve percent mixture problems

Section 4.4
To solve uniform motion problems

Section 4.5
To solve perimeter problems
To solve problems involving the angles of a
    triangle

Word Problems

Word problems have been challenging students of mathematics for a long time. Here are two types of problems that you may have seen before:

A number added to $\frac{1}{7}$ of the number is 19. What is the number?

A dog is chasing a rabbit that has a head start of 150 feet. The dog jumps 9 feet every time the rabbit jumps 7 feet. In how many jumps will the dog catch up with the rabbit?

What is unusual about these problems is that the first one is about 4000 years old and occurred as Problem 1 in the Rhind Papyrus. The second problem is about 1500 years old and comes from a Latin algebra book written in A.D. 450.

These examples illustrate that word problems have been around for a long time. The long history of word problems also reflects the importance that each generation has placed on solving these problems. It is through word problems that the initial steps of applying mathematics are taken.

Markup and Discount

Objective A   To solve markup problems

Cost is the price that a business pays for a product. **Selling price** is the price for which a business sells a product to a customer. The difference between selling price and cost is called **markup**. Markup is added to a retailer's cost to cover the expenses of operating a business. Markup is usually expressed as a percent of the retailer's cost. This percent is called the **markup rate.**

The basic markup equations used by a business are

Selling price = cost + markup   Markup = markup rate × cost

$$S = C + M \qquad\qquad M = r \times C$$

Substituting $r \times C$ for M in the first equation, we can also write selling price as $S = C + (r \times C) = C + rC$.

➡ The manager of a clothing store buys a suit for $80 and sells the suit for $116. Find the markup rate.

$$S = C + rC$$
$$116 = 80 + 80r$$
$$\boxed{116 - 80 = 80 - 80 + 80r}$$
$$36 = 80r$$
$$\boxed{\frac{36}{80} = \frac{80r}{80}}$$
$$\frac{36}{80} = r$$
$$0.45 = r$$

- Given: C = $80; S = $116
- Use the equation $S = C + rC$.
- Do this step mentally.

- Do this step mentally.

The markup rate is 45%.

Example 1
A hardware store manager uses a markup rate of 40% on all items. The selling price of a lawn mower is $105. Find the cost.

Strategy
Given: $r = 40\% = 0.40$ $S = \$105$
Unknown: C
Use the equation $S = C + rC$.

Solution
$$S = C + rC$$
$$105 = C + 0.40C$$
$$105 = 1.40C$$
$$75 = C$$
The cost is $75.

You Try It 1
The cost to the manager of a sporting goods store for a tennis racket is $40. The selling price of the racket is $60. Find the markup rate.

Your strategy $S = C + rC$
$$60 = 40 + r \cdot 40$$
$$-40$$
$$20 = r \cdot 40$$

Your solution
$$\frac{20}{40} = \frac{}{40}$$

Solution on p. A10

Objective B **To solve discount problems** ....

Discount is the amount by which a retailer reduces the regular price of a product for a promotional sale. Discount is usually expressed as a percent of the regular price. This percent is called the **discount rate.**

The basic discount equations used by a business are

$$\begin{matrix} \text{Sale} \\ \text{price} \end{matrix} = \begin{matrix} \text{regular} \\ \text{price} \end{matrix} - \text{discount} \qquad\qquad \text{Discount} = \begin{matrix} \text{discount} \\ \text{rate} \end{matrix} \cdot \begin{matrix} \text{regular} \\ \text{price} \end{matrix}$$

$$S = R - D \qquad\qquad D = r \cdot R$$

Substituting $r \cdot R$ for D in the first equation, we can also write sale price as $S = R - (r \cdot R) = R - rR$.

➡ In a garden supply store, the regular price of a 100-foot garden hose is $32. During an "after-summer sale," the hose is being sold for $24. Find the discount rate.

$$S = R - rR$$
$$24 = 32 - 32r$$
$$\boxed{24 - 32 = 32 - 32 - 32r}$$
$$-8 = -32r$$
$$\boxed{\frac{-8}{-32} = \frac{-32r}{-32}}$$
$$\frac{1}{4} = r$$
$$0.25 = r$$

- Given: $R = \$32$; $S = \$24$
- Use the equation $S = R - rR$.
- Do this step mentally.

- Do this step mentally.

The discount rate is 25%.

Example 2
The sale price for a chemical sprayer is $23.40. This price is 35% off the regular price. Find the regular price.

Strategy

Given: $S = \$23.40$
$\qquad r = 35\% = 0.35$
Unknown: R
Use the equation $S = R - rR$.

Solution

$$S = R - rR$$
$$23.40 = R - 0.35R$$
$$23.40 = 0.65R$$
$$36 = R$$

The regular price is $36.

You Try It 2
A case of motor oil that regularly sells for $27.60 is on sale for $20.70. What is the discount rate?

Your strategy

$$S = R - rR$$
$$20.70 = 27.40 - r\,27.60$$
$$\frac{-27.40}{6.90} = \frac{27.4r}{-27.6}$$
$$-27.6$$

Your solution

$$.25 = r = 25\%$$

Solution on p. A10

4.1 Exercises

. . . .

Objective A   *Application Problems* $S = C + rC$

Solve.

1. A computer software retailer uses a markup rate of 40%. Find the selling price of a computer game that cost the retailer $25.

2. A car dealer advertises a 5% markup over cost. Find the selling price of a car that cost the dealer $12,000.

3. The pro in a golf shop purchases a one iron for $40 and fixes a selling price on the club of $75. Find the markup rate.

4. A jeweler purchases a diamond ring for $350 and fixes a selling price of $700 on the ring. Find the markup rate.

5. A leather jacket costs a clothing store manager $140. Find the selling price of the leather jacket when the markup rate is 40%.

6. The cost to a landscape architect for a 25-gallon tree is $65. Find the selling price of the tree when the markup rate used by the architect is 30%.

7. A digitally recorded compact disk costs the manager of a music store $8.50. The selling price of the disk is $11.90. Find the markup rate.

8. A grocer purchases a can of fruit juice for $.68. The selling price for the fruit juice is $.85. Find the markup rate.

9. An electronics store manager adds $50 to the cost of every 17-inch television, regardless of the cost of the television. Find the markup rate, to the nearest tenth of a percent, on a television that costs the manager $215.

10. A department store uses a markup rate of 40% on items that cost over $100 and a markup rate of 50% on items that cost less than $100. Find the selling price of a ceramic bowl that costs the department store $86.

Objective B   *Application Problems* $S = R - rR$

Solve.

11. A tennis racket that regularly sells for $55 is on sale for 25% off the regular price. Find the sale price.

12. A fax machine that regularly sells for $975 is on sale for $33\frac{1}{3}$% off the regular price. Find the sale price.

Solve.

13. A car stereo system that regularly sells for $425 is on sale for $318.75. Find the discount rate.

14. A car dealer is having a year-end clearance sale that offers $2500 off the regular price of a car. Find the discount rate for a car that regularly sells for $12,500.

15. A gold bracelet that regularly sells for $1250 is on sale for $750. Find the discount rate.

16. A pair of skis that regularly sells for $325 is on sale for $250. Find the discount rate to the nearest percent.

17. A clothing wholesaler offers a discount of 10% per shirt when 10 to 20 shirts are purchased and a discount of 15% per shirt when 21 to 50 shirts are purchased. For a shirt that regularly sells for $17, find the sale price per shirt when 35 shirts are purchased.

18. A supplier of electrical equipment offers a 10% discount for a cash purchase or a 5% discount for a purchase that is paid for within 30 days. Find the discount price of a transformer that regularly sells for $230 and is paid for 10 days after the purchase.

19. A service station offers a discount of $10 per tire when 2 tires are purchased and a discount of $25 per tire when 4 tires are purchased. Find the discount rate, to the nearest percent, when a customer purchases 4 tires that regularly sell for $95 each.

20. A department store offers a discount of $3 per dinner plate when 5 or fewer plates are purchased and a discount of $5 per plate when more than 5 plates are purchased. Find the discount rate, to the nearest percent, when a customer purchases 3 dinner plates that regularly sell for $18 each.

APPLYING THE CONCEPTS

21. A pair of shoes that now sells for $63 has been marked up 40%. Find the markup on the pair of shoes.

22. The sale price of a typewriter is 25% off the regular price. The discount is $70. Find the sale price.

23. The sale price of a television was $180. Find the regular price if the sale price was computed by taking $\frac{1}{3}$ off the regular price followed by an additional 15% discount on the reduced price.

24. A customer buys four tires, three at the regular price and one for 20% off the regular price. The four tires cost $209. What was the regular price of a tire?

25. A lamp, originally priced at under $100, was on sale for 25% off the regular price. When the regular price, a whole number of dollars, was discounted, the discounted price was also a whole number of dollars. Find the largest possible number of dollars in the regular price of the lamp.

Investment Problems

Objective A *To solve investment problems*

The annual simple interest that an investment earns is given by the equation $I = Pr$, where I is the simple interest, P is the principal, or the amount invested, and r is the simple interest rate.

➡ The annual interest rate on a $2500 investment is 8%. Find the annual simple interest earned on the investment.

$I = Pr$ • **Given:** $P = \$2500$; $r = 8\% = 0.08$
$I = 2500(0.08)$
$I = 200$

The annual simple interest is $200.

➡ An investor has a total of $10,000 to deposit into two simple interest accounts. On one account, the annual simple interest rate is 6%. On the second account, the annual simple interest rate is 10%. How much should be invested in each account so that the total annual interest earned is $900?

> **Strategy for Solving a Problem Involving Money Deposited in Two Simple Interest Accounts**
>
> 1. For each amount invested, write a numerical or variable expression for the principal, the interest rate, and the interest earned. The results can be recorded in a table.

The sum of the amounts at each interest rate is $10,000.

Amount invested at 6%: x
Amount invested at 10%: $\$10,000 - x$

	Principal, P ·	Interest Rate, r =	Interest Earned, I
Amount at 6%	x ·	0.06 =	$0.06x$
Amount at 10%	$10,000 - x$ ·	0.10 =	$0.10(10,000 - x)$

> 2. Determine how the amounts of interest earned on each amount are related. For example, the total interest earned by both accounts may be known, or it may be known that the interest earned on one account is equal to the interest earned on the other account.

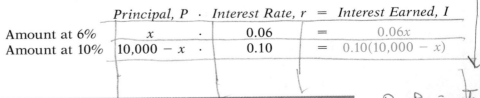

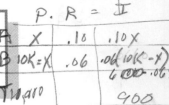

$.10x + 600 - .06x = 900$

The total annual interest earned is $900.

$$0.06x + 0.10(10,000 - x) = 900$$
$$0.06x + 1000 - 0.10x = 900$$
$$-0.04x + 1000 = 900$$
$$-0.04x = -100$$
$$x = 2500$$
$$10,000 - x = 10,000 - 2500 = 7500$$

The amount invested at 6% is $2500.
The amount invested at 10% is $7500.

Example 1

An investment counselor invested 75% of a client's money in an 8% annual simple interest money market fund. The remainder was invested in 5% annual simple interest government securities. Find the amount invested in each if the total annual interest earned is $3625.

You Try It 1

An investment of $5000 is made at an annual simple interest rate of 6%. How much additional money must be invested at 9% so that the total interest earned will be 8% of the total investment?

Strategy

• Amount invested: x
  Amount invested at 5%: $0.25x$
  Amount invested at 8%: $0.75x$

	Principal	Rate	Interest
Amount at 5%	$0.25x$	0.05	$0.0125x$
Amount at 8%	$0.75x$	0.08	$0.06x$

• The sum of the interest earned by the two investments equals the total annual interest earned ($3625).

Your strategy

Solution

$$0.0125x + 0.06x = 3625$$
$$0.0725x = 3625$$
$$x = 50,000$$

$$0.25x = 0.25(50,000) = 12,500$$

$$0.75x = 0.75(50,000) = 37,500$$

The amount invested at 5% is $12,500.
The amount invested at 8% is $37,500.

Your solution

Solution on p. A11

Content and Format © 1995 HMCo.

A
B
TOTAL

4.2 Exercises

· ·

Objective A *Application Problems*

Solve.

1. An investment of $3000 is made at an annual simple interest rate of 5%. How much additional money must be invested at an annual simple interest rate of 9% so that the total annual interest earned is 7.5% of the total investment?

2. A total of $6000 is invested into two simple interest accounts. The annual simple interest rate on one account is 9%; on the second account, the annual simple interest rate is 6%. How much should be invested in each account so that both accounts earn the same amount of annual interest?

3. An engineer invested a portion of $15,000 in a 7% annual simple interest account and the remainder in a 6.5% annual simple interest government bond. The amount of interest earned for one year was $1020. How much was invested in each account?

4. An investment club invested part of $20,000 in preferred stock that pays 8% annual simple interest and the remainder in a municipal bond that pays 7% annual simple interest. The amount of interest earned each year is $1520. How much was invested in each account?

5. A grocery checker deposited an amount of money into a high-yield mutual fund that returns a 9% annual simple interest rate. A second deposit, $2500 more than the first, was placed in a certificate of deposit that returns a 5% annual simple interest rate. The total interest earned on both investments for one year was $475. How much money was deposited in the mutual fund?

6. A deposit was made into a 7% annual simple interest account. Another deposit, $1500 less than the first deposit, was placed in a 9% annual simple interest certificate of deposit. The total interest earned on both accounts for one year was $505. How much money was deposited in the certificate of deposit?

7. A corporation gave a university $300,000 to support product safety research. The university deposited some of the money in a 10% simple interest account and the remainder in an 8.5% simple interest account. How much should be deposited in each account so that the annual interest earned in $28,500?

8. A financial consultant advises a client to invest part of $30,000 in municipal bonds that earn 6.5% annual simple interest and the remainder of the money in 8.5% corporate bonds. How much should be invested in each account so that the total annual interest earned each year is $2190?

Solve.

9. To provide for retirement income, an auto mechanic purchases a $5000 bond that earns 7.5% annual simple interest. How much money must be invested in additional bonds that have an interest rate of 8% so that the total annual interest earned from the two investments is $615?

10. The portfolio manager for an investment group invested $40,000 in a certificate of deposit that earns 7.25% annual simple interest. How much money must be invested in additional certificates that have an interest rate of 8.5% so that the total annual interest earned from the two investments is $5025?

11. A financial planner recommended that 40% of a client's cash account be placed in preferred stock that earns 9% annual simple interest. The remainder of the client's cash was placed in treasury bonds that earn 7% annual interest. The total annual interest earned from the two investments was $2496. What was the total amount invested?

12. The manager of a mutual fund placed 30% of the fund's available cash in a 6% simple interest account, 25% in 8% corporate bonds, and the remainder in a money market fund that earns 7.5% annual simple interest. The total annual interest from the investments was $35,875. What was the total amount invested?

13. The manager of a trust decided to invest 30% of a client's cash in government bonds that earn 6.5% annual simple interest. Another 30% was placed in utility stocks that earn 7% annual simple interest. The remainder of the cash was placed in an account earning 8% annual simple interest. The total annual interest earned from the investments was $5437.50. What was the total amount invested?

APPLYING THE CONCEPTS

14. A sales representative invests in a stock paying 9% dividends. A research consultant invests $5000 more than the sales representative in bonds paying 8% annual simple interest. The research consultant's income from the investment is equal to the sales representative's. Find the amount of the research consultant's investment.

15. A financial manager invested 20% of a client's money in bonds paying 9% annual simple interest, 35% in an 8% simple interest account, and the remainder in 9.5% corporate bonds. Find the amount invested in each if the total annual interest earned is $5325.

16. A plant manager invested $3000 more in stocks than in bonds. The stocks paid 8% annual simple interest, and the bonds paid 9.5% annual simple interest. Both investments yielded the same income. Find the total annual interest received on both investments.

17. A bank offers a customer a 2-year certificate of deposit (CD) that earns 8% compound annual interest. This means that the interest earned each year is added to the principal before the interest for the next year is calculated. Find the value in 2 years of a nurse's investment of $2500 in this CD.

4.3

Mixture Problems

Objective A *To solve value mixture problems*

A value mixture problem involves combining two ingredients that have different prices into a single blend. For example, a coffee merchant may blend two types of coffee into a single blend, or a candy manufacturer may combine two types of candy to sell as a "variety pack."

The solution of a value mixture problem is based on the equation $V = AC$, where V is the value of an ingredient, A is the amount of the ingredient, and C is the cost per unit of the ingredient.

➡ A coffee merchant wants to make 6 lb of a blend of coffee to sell for $5 per pound. The blend is made using a $6-per-pound grade and a $3-per-pound grade of coffee. How many pounds of each of these grades should be used?

> **Strategy for Solving a Value Mixture Problem**
>
> 1. For each ingredient in the mixture, write a numerical or variable expression for the amount of the ingredient used, the unit cost of the ingredient and the value of the amount used. For the blend, write a numerical or variable expression for the amount, the unit cost of the blend, and the value of the amount. The results can be recorded in a table.

The sum of the amounts is 6 lb.

Amount of $6 coffee: x
Amount of $3 coffee: $6 - x$

	Amount, A	·	Unit Cost, C	=	Value, V
$6 grade	x	·	$6	=	$6x$
$3 grade	$6 - x$	·	$3	=	$3(6 - x)$
$5 blend	6	·	$5	=	$5(6)$

$$6x + 3(6-x) = 5(6)$$

> 2. Determine how the values of each ingredient are related. Use the fact that the sum of the values of all the ingredients is equal to the value of the blend.

The sum of the values of the $6 grade and the $3 grade is equal to the value of the $5 blend.

$$6x + 3(6 - x) = 5(6)$$
$$6x + 18 - 3x = 30$$
$$3x + 18 = 30$$
$$3x = 12$$
$$x = 4$$

$$6 - x = 6 - 4 = 2$$

The merchant must use 4 lb of the $6 coffee and 2 lb of the $3 coffee.

Example 1

How many ounces of a silver alloy that costs $4 an ounce must be mixed with 10 oz of an alloy that costs $6 an ounce to make a mixture that costs $4.32 an ounce?

Strategy

• Ounces of $4 alloy: x

	Amount	Cost	Value
$4 alloy	x	$4	$4x$
$6 alloy	10	$6	6(10)
$4.32 mixture	10 + x	$4.32	4.32(10 + x)

• The sum of the values before mixing equals the value after mixing.

Solution

$$4x + 6(10) = 4.32(10 + x)$$
$$4x + 60 = 43.2 + 4.32x$$
$$-0.32x + 60 = 43.2$$
$$-0.32x = -16.8$$
$$x = 52.5$$

52.5 oz of the $4 silver alloy must be used.

You Try It 1

A gardener has 20 lb of a lawn fertilizer that costs $.80 per pound. How many pounds of a fertilizer that costs $.55 per pound should be mixed with this 20 lb of lawn fertilizer to produce a mixture that costs $.75 per pound?

Your strategy

Your solution

Solution on p. A11

Objective B   *To solve percent mixture problems* ....

The amount of a substance in a solution can be given as a percent of the total solution. For example, a 5% salt water solution means that 5% of the total solution is salt. The remaining 95% is water.

The solution of a percent mixture problem is based on the equation $Q = Ar$, where Q is the quantity of a substance in the solution, r is the percent of concentration, and A is the amount of solution.

⇒ A 500-milliliter bottle contains a 4% solution of hydrogen peroxide. Find the amount of hydrogen peroxide in the solution.

$Q = Ar$ • Given: $A = 500$; $r = 4\% = 0.04$
$Q = 500(0.04)$
$Q = 20$

The bottle contains 20 ml of hydrogen peroxide.

→ How many gallons of a 20% salt solution must be mixed with 6 gal of a 30% salt solution to make a 22% salt solution?

> **Strategy for Solving a Percent Mixture Problem**
>
> 1. For each solution, write a numerical or variable expression for the amount of solution, the percent of concentration, and the quantity of the substance in the solution. The results can be recorded in a table.

The unknown quantity of 20% solution: x

	Amount of Solution, A	·	Percent of Concentration, r	=	Quantity of Substance, Q
20% solution	x	·	0.20	=	$0.20x$
30% solution	6	·	0.30	=	$0.30(6)$
22% solution	$x + 6$	·	0.22	=	$0.22(x + 6)$

> 2. Determine how the quantities of the substances in each solution are related. Use the fact that the sum of the quantities of the substances being mixed is equal to the quantity of the substance after mixing.

The sum of the quantities of the substances in the 20% solution and the 30% solution is equal to the quantity of the substance in the 22% solution.

$0.20x + 0.30(6) = 0.22(x + 6)$
$0.20x + 1.80 = 0.22x + 1.32$
$-0.02x + 1.80 = 1.32$
$-0.02x = -0.48$
$x = 24$

24 gal of the 20% solution are required.

Example 2
A chemist wishes to make 2 L of an 8% acid solution by mixing a 10% acid solution and a 5% acid solution. How many liters of each solution should the chemist use?

You Try It 2
A pharmacist dilutes 5 L of a 12% solution by a 6% solution. How many liters of the 6% solution are added to make an 8% solution?

Strategy
- Liters of 10% solution: x
  Liters of 5% solution: $2 - x$

	Amount	Percent	Quantity
10%	x	0.10	$0.10x$
5%	$2 - x$	0.05	$0.05(2 - x)$
8%	2	0.08	$0.08(2)$

- The sum of the quantities before mixing is equal to the quantity after mixing.

Your strategy

Solution

$$0.10x + 0.05(2 - x) = 0.08(2)$$
$$0.10x + 0.10 - 0.05x = 0.16$$
$$0.05x + 0.10 = 0.16$$
$$0.05x = 0.06$$
$$x = 1.2$$

$$2 - x = 2 - 1.2 = 0.8$$

The chemist needs 1.2 L of the 10% solution and 0.8 L of the 5% solution.

Your solution

Solution on p. A11

4.3 Exercises

A = amount
U. = unit

Objective A   *Application Problems*

value

Solve.

1. A high-protein diet supplement that costs $6.75 per pound is mixed with a vitamin supplement that costs $3.25 per pound. How many pounds of each should be used to make 5 lb of a mixture that sells for $4.65 per pound?

2. A 20-oz alloy of platinum that costs $220 per ounce is mixed with an alloy that costs $400 per ounce. How many ounces of the $400 alloy should be mixed with the 20-oz alloy to make an alloy that costs $300 per ounce?

3. Find the selling price per pound of a coffee mixture made from 8 lb of coffee that costs $9.20 per pound and 12 lb of coffee that costs $5.50 per pound.

4. How many pounds of tea that costs $4.20 per pound must be mixed with 12 lb of tea that costs $2.25 per pound to make a mixture that costs $3.40 per pound?

5. A goldsmith combined an alloy that costs $4.30 per ounce with an alloy that costs $1.80 per ounce. How many ounces of each were used to make a mixture of 200 oz to sell for $2.50 per ounce?

6. How many liters of a solvent that costs $80 per liter must be mixed with 6 L of a solvent that costs $25 per liter to make a solvent that sells for $36 per liter?

7. Find the selling price per pound of a trail mix made from 40 pounds of raisins that cost $4.40 per pound and 100 pounds of granola that costs $2.30 per pound.

8. Find the selling price per ounce of a mixture of 200 oz of cologne that costs $5.50 per ounce and 500 oz of cologne that costs $2.00 per ounce.

9. How many kilograms of hard candy that costs $7.50 per kilogram must be mixed with 24 kg of jelly beans that cost $3.25 per kilogram to make a mixture that sells for $4.50 per kilogram?

10. A grocery store offers a cheese and fruit sampler that combines cheddar cheese that costs $8 per kilogram with kiwis that cost $3 per kilogram. How many kilograms of each were used to make a 5-kg mixture that costs $4.50 per kilogram?

Solve.

11. A ground meat mixture is formed by combining meat that costs $2.20 per pound with meat that costs $4.20 per pound. How many pounds of each were used to make a 50-lb mixture that costs $3.00 per pound?

12. A lumber company combined oak wood chips that cost $3.10 per pound with pine wood chips that cost $2.50 per pound. How many pounds of each were used to make an 80-lb mixture that costs $2.65 per pound?

13. How many kilograms of a soil supplement that costs $7.00 per kilogram must be mixed with 20 kg of aluminum nitrate that costs $3.50 per kilogram to make a fertilizer that costs $4.50 per kilogram?

14. A caterer makes an ice cream punch by combining fruit juice that costs $2.25 per gallon with ice cream that costs $3.25 per gallon. How many gallons of each should be used to make 100 gal of punch to sell for $2.50 per gallon?

15. The manager of a specialty food store combined almonds that cost $4.50 per pound with walnuts that cost $2.50 per pound. How many pounds of each were used to make a 100-lb mixture that costs $3.24 per pound?

16. Find the cost per gallon of a carbonated fruit drink made from 12 gal of fruit juice that costs $4.00 per gallon and 30 gal of carbonated water that costs $2.25 per gallon.

17. Find the cost per pound of a sugar-coated breakfast cereal made from 40 pounds of sugar that costs $1.00 per pound and 120 pounds of corn flakes that cost $.60 per pound.

18. Find the cost per ounce of a gold alloy made from 25 oz of pure gold that costs $482 per ounce and 40 oz of an alloy that costs $300 per ounce.

19. How many pounds of lima beans that cost $.90 per pound must be mixed with 16 lb of corn that costs $.50 per pound to make a mixture of vegetables that costs $.65 per pound?

20. How many liters of a blue dye that costs $1.60 per liter must be mixed with 18 L of anil that costs $2.50 per liter to make a mixture that costs $1.90 per liter?

Objective B   *Application Problems*

Solve.

21. A chemist wants to make 50 ml of a 16% acid solution. How many milliliters each of a 13% acid solution and an 18% acid solution should be mixed to produce the desired solution?

22. A blend of coffee was made by combining some coffee that was 40% java beans with 80 lb of coffee that was 30% java beans to make a mixture that is 32% java. How many pounds of the 40% java coffee were used?

23. Thirty ounces of pure silver are added to 50 oz of a silver alloy that is 20% silver. What is the percent concentration of the silver in the resulting mixture?

24. Two hundred liters of a punch that contains 35% fruit juice is mixed with 300 L of another punch. The resulting fruit punch is 20% fruit juice. Find the percent of fruit juice in the 300 liters of punch.

25. The manager of a garden shop mixes grass seed that is 60% rye grass with 70 lb of grass seed that is 80% rye grass to make a mixture that is 74% rye grass. How much of the 60% mixture is used?

26. Ten grams of sugar are added to a 40-g serving of a breakfast cereal that is 30% sugar. What is the percent concentration of sugar in the resulting mixture?

27. A dermatologist mixes 50 g of a cream that is 0.5% hydrocortizone with 150 g of another hydrocortizone cream. The resulting mixture is 0.68% hydrocortizone. Find the percent of hydrocortizone in the 150-g cream.

28. A carpet manufacturer blends two fibers, one 20% wool and the second 50% wool. How many pounds of each fiber should be woven together to produce 500 lb of a fabric that is 35% wool?

29. A hair dye is made by blending some 7% hydrogen peroxide solution with some 4% hydrogen peroxide solution. How many milliliters of each should be mixed to make a 300-ml solution that is 5% hydrogen peroxide?

30. How many grams of pure salt must be added to 40 g of a 20% salt solution to make a solution that is 36% salt?

Solve.

31. How many ounces of pure water must be added to 50 oz of a 15% saline solution to make a saline solution that is 10% salt?

32. A paint is blended by using a paint that contains 21% green dye and a paint that contains 15% green dye. How many gallons of each must be mixed to produce 60 gal of paint that is 19% green dye?

33. A goldsmith mixes 8 oz of a 30% gold alloy with 12 oz of a 25% gold alloy. What is the percent concentration of the resulting alloy?

34. A physicist mixes 40 L of oxygen with 50 L of air that contains 64% oxygen. What is the percent concentration of the resulting air?

35. A 50-oz box of cereal is 40% bran flakes. How many ounces of pure bran flakes must be added to this box to produce a mixture that is 50% bran flakes?

36. A pastry chef has 150 ml of a chocolate topping that is 50% chocolate. How many milliliters of pure chocolate must be added to this topping to make a topping that is 75% chocolate?

APPLYING THE CONCEPTS

37. Find the cost per ounce of a mixture of 30 oz of an alloy that costs $4.50 per ounce, 40 oz of an alloy that costs $3.50 per ounce, and 30 oz of an alloy that costs $3.00 per ounce.

38. A grocer combined walnuts that cost $1.60 per pound and cashews that cost $2.50 per pound with 20 lb of peanuts that cost $1.00 per pound. Find the amount of walnuts and the amount of cashews used to make a 50-lb mixture that costs $1.72 per pound.

39. How many ounces of water must be evaporated from 50 oz of a 12% salt solution to produce a 15% salt solution?

40. A chemist mixed pure acid with water to make 10 L of a 30% acid solution. How much pure acid and how much water did the chemist use?

41. How many grams of pure water must be added to 50 g of pure acid to make a solution that is 40% acid?

42. A radiator contains 15 gal of a 20% antifreeze solution. How many gallons must be drained from the radiator and replaced by pure antifreeze so that the radiator will contain 15 gal of a 40% antifreeze solution?

Uniform Motion Problems

Objective A   *To solve uniform motion problems* ·································

A train that travels constantly in a straight line at 50 mph is in *uniform motion*. **Uniform motion** means that the speed or direction of an object does not change.

The solution of a uniform motion problem is based on the equation $d = rt$, where d is the distance traveled, r is the rate of travel, and t is the time traveled.

⇒ A car leaves a town traveling at 40 mph. Two hours later, a second car leaves the same town, on the same road, traveling at 60 mph. In how many hours will the second car pass the first car?

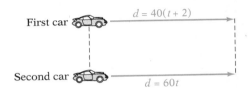

> **Strategy for Solving a Uniform Motion Problem**
> 1. For each object, write a numerical or variable expression for the distance, rate, and time. The results can be recorded in a table.

The first car traveled 2 h longer than the second car.

Unknown time for the second car: t
Time for the first car: $t + 2$

	Rate, r	·	Time, t	=	Distance, d
First car	40	·	$t + 2$	=	$40(t + 2)$
Second car	60	·	t	=	$60t$

> 2. Determine how the distances traveled by each object are related. For example, the total distance traveled by both objects may be known, or it may be known that the two objects traveled the same distance.

The two cars travel the same distance.

$$40(t + 2) = 60t$$
$$40t + 80 = 60t$$
$$80 = 20t$$
$$4 = t$$

second car will pass the first car in 4 h.

Example 1

Two cars, one traveling 10 mph faster than the other car, start at the same time from the same point and travel in opposite directions. In 3 h they are 300 mi apart. Find the rate of each car.

Strategy

• Rate of 1st car: r
  Rate of 2nd car: $r + 10$

	Rate	Time	Distance
1st car	r	3	$3r$
2nd car	$r + 10$	3	$3(r + 10)$

• The total distance traveled by the two cars is 300 mi.

Solution

$$3r + 3(r + 10) = 300$$
$$3r + 3r + 30 = 300$$
$$6r + 30 = 300$$
$$6r = 270$$
$$r = 45$$

$$r + 10 = 45 + 10 = 55$$

The first car is traveling 45 mph.
The second car is traveling 55 mph.

Example 2

How far can the members of a bicycling club ride out into the country at a speed of 12 mph and return over the same road at 8 mph if they travel a total of 10 h?

Strategy

• Time spent riding out: t
  Time spent riding back: $10 - t$

	Rate	Time	Distance
Out	12	t	$12t$
Back	8	$10 - t$	$8(10 - t)$

• The distance out equals the distance back.

Solution

$$12t = 8(10 - t)$$
$$12t = 80 - 8t$$
$$20t = 80$$
$$t = 4 \quad \text{(The time is 4 h.)}$$

The distance out $= 12t = 12(4) = 48$ mi.
The club can ride 48 mi into the country.

You Try It 1

Two trains, one traveling at twice the speed of the other, start at the same time on parallel tracks from stations that are 288 mi apart and travel toward each other. In 3 h, the trains pass each other. Find the rate of each train.

Your strategy

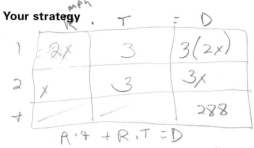

Your solution

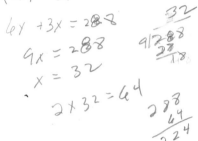

You Try It 2

A pilot flew out to a parcel of land and back in 5 h. The rate out was 150 mph and the rate returning was 100 mph. How far away was the parcel of land?

Your strategy

Your solution

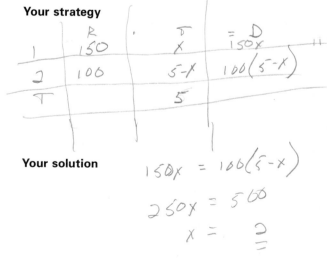

Solutions on p. A12

4.4 Exercises

· ·

Objective A   *Application Problems*

Solve.

1. A 555-mile, 5-hour plane trip was flown at two speeds. For the first part of the trip, the average speed was 105 mph. For the remainder of the trip, the average speed was 115 mph. For how long did the plane fly at each speed?

2. An executive drove from home at an average speed of 30 mph to an airport where a helicopter was waiting. The executive boarded the helicopter and flew to the corporate offices at an average speed of 60 mph. The entire distance was 150 mi. The entire trip took 3 h. Find the distance from the airport to the corporate offices.

3. After a saiboat had been on the water for 3 h, a change in wind direction reduced the average speed of the boat by 5 mph. The entire distance sailed was 57 mi. The total time spent sailing was 6 h. How far did the sailboat travel in the first 3 h?

4. A car and a bus set out at 2 P.M. from the same point headed in the same direction. The average speed of the car is 30 mph slower than twice the speed of the bus. In 2 h the car is 20 mi ahead of the bus. Find the rate of the car.

5. A passenger train leaves a train depot 2 h after a freight train leaves the same depot. The freight train is traveling 20 mph slower than the passenger train. Find the rate of each train if the passenger train overtakes the freight train in 3 h.

6. Two cyclists start at the same time from opposite ends of a course that is 45 mi long. One cyclist is riding at 14 mph and the second cyclist is riding at 16 mph. How long after they begin will they meet?

7. A cyclist and a jogger set out at 11 A.M. from the same point headed in the same direction. The average speed of the cyclist is twice the average speed of the jogger. In 1 h the cyclist is 8 mi ahead of the jogger. Find the rate of the cyclist.

8. Two cyclists start from the same point and ride in opposite directions. One cyclist rides twice as fast as the the other. In 3 h they are 72 mi apart. Find the rate of each cyclist.

9. Two small planes start from the same point and fly in opposite directions. The first plane is flying 25 mph slower than the second plane. In 2 h the planes are 430 mi apart. Find the rate of each plane.

Solve.

10. A motorboat leaves a harbor and travels at an average speed of 8 mph toward a small island. Two hours later a cabin cruiser leaves the same harbor and travels at an average speed of 16 mph toward the same island. In how many hours after the cabin cruiser leaves will it be alongside the motorboat?

11. Two joggers start at the same time from opposite ends of a 10-mile course. One jogger is running at 4 mph and the other is running at 6 mph. How long after they begin will they meet?

12. On a 195-mile trip, a car traveled at an average speed of 45 mph and then reduced its speed to an average of 30 mph for the remainder of the trip. The trip took a total of 5 h. How long did the car travel at each speed?

13. A long-distance runner started on a course running at an average speed of 6 mph. One hour later, a second runner began the same course at an average speed of 8 mph. How long after the second runner started will the second runner overtake the first runner?

14. A family drove to a resort at an average speed of 30 mph and later returned over the same road at an average speed of 50 mph. Find the distance to the resort if the total driving time was 8 h.

15. Three campers left their campsite by canoe and paddled downstream at an average rate of 8 mph. They then turned around and paddled back upstream at an average rate of 4 mph to return to their campsite. How long did it take the campers to canoe downstream if the total trip took 1 h?

16. A car traveling at 48 mph overtakes a cyclist who, riding at 12 mph, had a 3-h head start. How far from the starting point does the car overtake the cyclist?

APPLYING THE CONCEPTS

17. At 10 A.M., two campers left their campsite by canoe and paddled downstream at an average speed of 12 mph. They then turned around and paddled back upstream at an average rate of 4 mph. The total trip took 1 h. At what time did the campers turn around downstream?

18. At 7 A.M., two joggers start from opposite ends of a 8-mi course. One jogger is running at a rate of 4 mph, and the other is running at a rate of 6 mph. At what time will the joggers meet?

19. A truck leaves a depot at 11 A.M. and travels at a speed of 45 mph. At noon, a van leaves the same place and travels the same route at a speed of 65 mph. At what time does the van overtake the truck?

20. A bicyclist rides for 2 h at a speed of 10 mph and then returns at a speed of 20 mph. Find the cyclist's average speed for the trip.

21. A car travels a 1-mile track at an average speed of 30 mph. At what average speed must the car travel the next mile so that the average speed for the 2 mi is 60 mph?

Content and Format © 1995 HMCo.

4.5 Geometry Problems

Objective A   To solve perimeter problems

The **perimeter** of a geometric figure is a measure of the distance around the figure. The equations for the perimeters of a rectangle and a triangle are shown at the right.

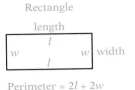

Rectangle

Perimeter = $2l + 2w$

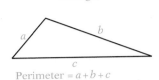

Triangle

Perimeter = $a + b + c$

➡ The perimeter of a rectangle is 26 ft. The length of the rectangle is 1 ft more than twice the width. Find the width of the rectangle.

POINT OF INTEREST
Leonardo DaVinci painted the Mona Lisa on a rectangular canvas whose height was approximately 1.6 times its width. Rectangles with these proportions, called golden rectangles, were used extensively in Renaissance art.

Strategy for Solving a Perimeter Problem

1. Let a variable represent the measure of one of the unknown sides of the figure. Express the measures of the remaining sides in terms of that variable.

Width: w
Length: $2w + 1$

2. Determine which perimeter equation to use. Use the equation for the perimeter of a rectangle.

$$2l + 2w = P$$
$$2(2w + 1) + 2w = 26$$
$$4w + 2 + 2w = 26$$
$$6w + 2 = 26$$
$$6w = 24$$
$$w = 4$$

The width is 4 ft.

Example 1
The perimeter of a triangle is 25 ft. Two sides of the triangle are equal. The third side is 2 ft less than the length of one of the equal sides. Find the measure of one of the equal sides.

Strategy
• Each equal side: x
  The third side: $x - 2$
• Use the equation for the perimeter of a triangle.

Solution
$$a + b + c = P$$
$$x + x + (x - 2) = 25$$
$$3x - 2 = 25$$
$$3x = 27$$
$$x = 9$$

Each of the equal sides measures 9 ft.

You Try It 1
The perimeter of a rectangle is 34 m. The width of the rectangle is 3 m less than the length. Find the measure of the width.

Your strategy

$$2L + 2w = P$$
$$34 = 2x + 2(x - 3)$$

Your solution

Solution on p. A12

Objective B **To solve problems involving the angles of a triangle** ....

In a triangle, the sum of the measures of all the angles is 180°.

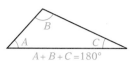

$A + B + C = 180°$

Two special types of triangles are shown at the right. A **right triangle** has one **right angle** (90°). An **isosceles triangle** has two equal angles and two equal sides.

Right triangle

Equal angles

Isosceles triangle

➡ In a certain right triangle, the measure of one angle is twice the measure of the smallest angle. Find the measure of the smallest angle.

Strategy for Solving a Problem Involving the Angles of a Triangle

1. Let a variable represent one of the unknown angles. Express the other angles in terms of that variable.

Measure of smallest angle: x
Measure of second angle: $2x$
Measure of right angle: $90°$

2. Use the equation $A + B + C = 180°$

$$x + 2x + 90 = 180$$
$$3x + 90 = 180$$
$$3x = 90$$
$$x = 30$$

The measure of the smallest angle is 30°

Example 2
In an isosceles triangle, the measure of one angle is 20° more than twice the measure of one of the equal angles. Find the measure of one of the equal angles.

Strategy
• Measure of one of the equal angles: x
  Measure of the second equal angle: x
  Measure of the third angle: $2x + 20$
• Use the equation $A + B + C = 180°$

Solution $\quad x + x + (2x + 20) = 180$
$$4x + 20 = 180$$
$$4x = 160$$
$$x = 40$$

The measure of one of the equal angles is 40°.

You Try It 2
In a triangle, the measure of one angle is twice the measure of the second angle. The measure of the third angle is 4° less than the measure of the second angle. Find the measure of each angle.

Your strategy

$2x + x + (x-4) = 180$
$9 = 2x$
$b = x$
$c = x - 4$
$x = 46$

Your solution

Solution on p. A12

4.5 Exercises

· ·

Objective A *Application Problems*

Solve.

1. In an isosceles triangle, two sides are equal. The third side is 50% of the length of one of the equal sides. Find the length of each side when the perimeter is 125 ft.

2. The width of a rectangle is 25% of the length. The perimeter is 250 cm. Find the length and width of the rectangle.

3. In an isosceles triangle, two sides are equal. The length of one of the equal sides is 3 times the length of the third side. The perimeter is 21 m. Find the length of each side.

4. The perimeter of a rectangle is 42 m. The length of the rectangle is 3 m less than twice the width. Find the length and width of the rectangle.

5. The perimeter of a rectangle is 120 ft. The length of the rectangle is twice the width. Find the length and width of the rectangle.

6. The perimeter of a triangle is 110 cm. One side is twice the second side. The third side is 30 cm more than the second side. Find the measure of each side.

7. The perimeter of a triangle is 33 ft. One side of the triangle is 1 ft longer than the second side. The third side is 2 ft longer than the second side. Find the measure of each side.

8. The perimeter of a rectangle is 50 m. The width of the rectangle is 5 m less than the length. Find the length and width of the rectangle.

9. The width of a rectangle is 30% of the length. The perimeter of the rectangle is 338 ft. Find the length and width of the rectangle.

10. The perimeter of a rectangle is 48 m. The width of the rectangle is 8 m less than the length. Find the length and width of the rectangle.

Objective B *Application Problems*

11. In an isosceles triangle, one angle is 5° less than three times the measure of one of the equal angles. Find the measure of each angle.

→12. The first angle of a triangle is twice the measure of the second angle. The third angle is 10° less than the measure of the first angle. Find the measure of each angle.

13. One angle of a triangle is twice the measure of the second angle. The third angle is three times the measure of the first angle. Find the measure of each angle.

14. In a triangle, one angle is twice the measure of the second angle. The third angle is three times the measure of the second angle. Find the measure of each angle.

15. The first angle of a triangle is three times the measure of the second angle. The third angle is 33° more than the measure of the first angle. Find the measure of each angle.

16. In an isosceles triangle, one angle is 12° more than twice the measure of one of the equal angles. Find the measure of each angle.

17. One angle of a right triangle is 3° less than twice the measure of the smallest angle. Find the measure of each angle.

18. In an equiangular triangle, all three angles are equal. Find the measures of the equal angles.

19. In a triangle, one angle is 5° more than the measure of the second angle. The third angle is 10° more than the measure of the second angle. Find the measure of each angle.

20. One angle of a triangle is three times the measure of the third angle. The second angle is 5° less than the measure of the third angle. Find the measure of each angle.

APPLYING THE CONCEPTS

21. A rectangle and an equilateral triangle have the same perimeter. The length of the rectangle is three times the width. Each side of the triangle is 8 cm. Find the length and width of the rectangle.

22. An equilateral triangle and a rectangle have the same perimeter. The length of the rectangle is 3 cm less than twice the width. Each side of the triangle is 10 cm. Find the length and width of the rectangle.

23. The length of a rectangle is 1 cm more than twice the width. If the length of the rectangle is decreased by 2 cm and the width is decreased by 1 cm, the perimeter is 20 cm. Find the length and width of the original rectangle.

24. The length of a rectangle is $14x$. The perimeter is $50x$. Find the width of the rectangle in terms of the variable x.

25. The width of a rectangle is $8x$. The perimeter is $48x$. Find the length of the rectangle in terms of the variable x.

Content and Format © 1995 HMCo.

Project in Mathematics

In some of the exercises in this text, you are asked to determine whether a statement is true or false. For instance, the statement "every real number has a reciprocal" is false because 0 is a real number and 0 does not have a reciprocal.

Finding an example, such as 0 has no reciprocal, to show that a statement is not always true is called *finding a counterexample*. A counterexample is an example that shows that a statement is not always true.

Consider the statement "the product of two numbers is greater than either factor." A counterexample to this statement is the numbers $\frac{2}{3}$ and $\frac{3}{4}$. The product of these numbers is $\frac{1}{2}$, and $\frac{1}{2}$ is *smaller* than $\frac{2}{3}$ or $\frac{3}{4}$. There are many other counterexamples to the given statement.

Here are some counterexamples to the statement that the square root of a number is smaller than the number.

$$\sqrt{\frac{1}{4}} = \frac{1}{2} \text{ but } \frac{1}{2} > \frac{1}{4} \qquad \sqrt{1} = 1 \text{ but } 1 = 1$$

Exercises:

For each of the next five statements, find at least one counterexample to show that the statement, or conjecture is false.

1. The product of two integers is always a positive number.
2. The sum of two prime numbers is never a prime number.
3. For all real numbers, $|x + y| = |x| + |y|$.
4. If x and y are nonzero real numbers and $x > y$, then $x^2 > y^2$.
5. The quotient of any two nonzero real numbers is less than either one of the numbers.

When a problem is posed, it may not be known whether the problem statement is true or false. For instance, Christian Goldbach (1690–1764) stated that every even integer greater than 2 can be written as the sum of two prime numbers. No one has been able to find a counterexample to this statement, but neither has anyone been able to prove that it is always true.

Exercises:

In the next set of problems, answer true if the statement is always true or give a counterexample if there is an instance when the statement is false.

1. The reciprocal of a positive number is always smaller, than the number.
2. If $x < 0$, then $|x| = -x$.
3. For any two real numbers x and y, $x + y > x - y$.
4. For any positive integer n, $n^2 + n + 17$ is a prime number.
5. The list of numbers $1, 11, 111, 1111, 11111, \ldots$ contains infinitely many composite numbers. *Hint:* A number is divisible by 3 if the sum of the digits of the number is divisible by 3.

Chapter Summary

Key Words   *Cost* is the price that a business pays for a product.

Selling price is the price for which a business sells a product to a customer.

Markup is the difference between selling price and cost.

Discount is the amount by which a retailer reduces the regular price of a product.

Uniform motion means that an object at a constant speed moves in a straight line.

The *perimeter* of a geometric figure is a measure of the distance around the figure.

A *right angle* is an angle whose measure is 90 degrees.

A *right triangle* has one right angle.

An *isosceles triangle* has two equal angles and two equal sides.

Essential Rules   *Basic Markup Equation*

$$\text{Selling price} = \text{cost} + \text{markup}$$
$$S = C + M$$

$$\text{Markup} = \text{markup rate} \times \text{cost}$$
$$M = r \times C$$

Basic Discount Equations

$$\text{Sale price} = \text{regular price} - \text{discount}$$
$$S = R - D$$

$$\text{Discount} = \text{discount rate} \times \text{regular price}$$
$$D = r \times R$$

Annual Simple Interest Equation

$$\text{Simple interest} = \text{principal} \times \text{simple interest rate}$$
$$I = P \times r$$

Value Mixture Equation

$$\text{Value} = \text{amount} \times \text{unit cost}$$
$$V = A \times C$$

Percent Mixture Equation

$$\text{Quantity} = \text{amount} \times \text{percent concentration}$$
$$Q = A \times r$$

Uniform Motion Equation

$$\text{Distance} = \text{rate} \times \text{time}$$
$$d = r \times t$$

Sum of the Angles in a Triangle

$$A + B + C = 180°$$

Chapter Review

. . . .

SECTION 4.1

1. A furniture store uses a markup rate of 60%. The store sells a solid oak curio cabinet for $1074. Find the cost of the curio cabinet.

2. A pair of athletic shoes, which regularly sell for $55, are on sale for 25% off the regular price. Find the sale price.

3. The sale price for a carpet sweeper is $26.56, which is 17% off the regular price. Find the regular price.

4. A ceiling fan, which regularly sells for $60, is on sale for $40. Find the discount rate.

SECTION 4.2

5. A college sports foundation deposited a total of $24,000 into two simple interest accounts. The annual simple interest rate on one account is 4%; on the second account, the annual simple interest rate is 9%. How much should be invested in each account so that the total interest earned is 7% of the total investment?

6. An investment banker decided to invest 55% of the bank's available cash in an account that earns 8.25% annual simple interest. The remainder of the cash was placed in an account that earns 10% annual simple interest. The interest earned in one year was $58,743.75. What was the total amount invested?

7. A club treasurer has $2400 to be deposited into two simple interest accounts. On one account the annual simple interest rate is 6.75%. The annual simple interest on the other account is 9.45%. How much should the treasurer deposit in each account so that the interest earned in each account is the same?

8. An engineering consultant invested $14,000 in a 5.5% annual simple interest Individual Retirement Account. How much additional money must be deposited into an account that pays 9.5% annual simple interest so that the total interest earned on both accounts is 8% of the total investment? Round to the nearest cent.

SECTION 4.3

9. A health food store combined cranberry juice that cost $1.79 per quart with apple juice that cost $1.19 per quart. How many quarts of each were used to make 10 qt of a cranapple juice mixture to sell for $1.61 per quart?

10. Find the selling price per pound of a meatloaf mixture made from three pounds of ground beef that costs $1.99 per pound and one pound of ground turkey that costs $1.39 per pound.

11. A dairy mixed five gallons of cream that is 30% butterfat with eight gallons of milk that is 4% butterfat. What is the percent of butterfat in the resulting mixture?

12. A pharmacist has 15 liters of an 80% alcohol solution. How many liters of pure water should be added to the alcohol solution to make a diluted alcohol solution that is 75% alcohol?

SECTION 4.4

13. A jet plane traveling at 600 mph overtakes a propeller-driven plane that had a 2-h head start. The propeller-driven plane is traveling at 200 mph. How far from the starting point does the jet overtake the propeller-driven plane?

14. A bus traveled on a level road for 2 h at an average speed of 20 mph faster than it traveled on a winding road. The time spent on the winding road was 3 h. Find the average speed on the winding road if the total trip was 200 mi.

SECTION 4.5

15. The perimeter of a triangle is 35 in. The second side is 4 in. longer than the first side. The third side is 1 in. shorter than twice the first side. Find the measure of each side.

16. The length of a rectangle is four times the width of the rectangle. The perimeter is 200 ft. Find the length and the width of the rectangle.

17. In an isosceles triangle, one angle is 25° less than half the measure of one of the equal angles. Find the measure of each angle.

18. One angle of a triangle is 15° more than the measure of the second angle. The third angle is 15° less than the measure of the second angle. Find the measure of each angle.

Chapter Test

1. The manager of a jewelry store uses a markup rate of 75%. The selling price for a gold ring is $612.50. Find the cost of the gold ring.

2. A television that regularly sells for $450 is on sale for $360. Find the discount rate.

3. The cost at a college bookstore for a textbook is $27. The bookstore sells the textbook for $36. Find the markup rate.

4. An airline is offering a 35% discount on round-trip air fares. Find the sale price of a round-trip ticket that normally costs $385.

5. A charity deposited a total of $54,000 into two simple interest accounts. The annual simple interest rate on one account is 6%; on the second account, the annual simple interest rate is 9%. How much should be invested in each account so that the total interest earned is 8% of the total investment?

6. A total of $5000 is deposited into two simple interest accounts. The annual simple interest rate on one account is 6%; on the second account, the simple interest rate is 9%. How much should be invested in each account so that the total annual interest earned is $345?

7. A total of $8000 is invested into two simple interest accounts. The annual simple interest rate on one account is 7%; on the second account, the annual simple interest rate is 9%. How much should be invested in each account so that both accounts earn the same amount of interest?

8. A baker wants to make a 15-lb blend of flour that cost $.60 per pound. The blend is made using a rye flour that costs $.70 per pound and a wheat flour that costs $.40 per pound. How many pounds of each flour should be used?

9. Find the selling price per pound of a trail mix made from 60 pounds of raisins that cost $5.00 per pound and 200 pounds of granola that costs $2.40 per pound.

10. How many gallons of water must be mixed with 5 gal of a 20% salt solution to make a 16% salt solution?

11. Forty ounces of pure silver are added to 120 oz of a silver alloy that is 10% silver. What is the percent concentration of the silver in the resulting mixture?

12. A recipe for a rice dish calls for 12 oz of a rice mixture that is 20% wild rice and 8 oz of pure wild rice. What is the percent concentration of wild rice in the 20-oz mixture?

13. A cross-country skier leaves a camp to explore a wilderness area. Two hours later a friend leaves the camp in a snowmobile, traveling 4 mph faster than the skier, and overtakes the skier 1 h later. Find the rate of the snowmobile.

14. As part of flight training, a student pilot was required to fly to an airport and then return. The average speed to the airport was 90 mph, and the average speed returning was 120 mph. Find the distance between the two airports if the total flying time was 7 h.

15. The perimeter of a rectangular sandbox is 20 ft. The length of the sandbox is 2 ft less than twice the width. Find the length and width of the sandbox.

16. The perimeter of a triangle is 23 ft. One side is twice the second side. The third side is 3 ft more than the second side. Find the measure of each side.

17. The width of a rectangle is 40% of the length. The perimeter of the rectangle is 112 m. Find the length and width of the rectangle.

18. In an isosceles triangle, two angles are equal. The third angle of the triangle is 30° less than one of the equal angles. Find the measure of one of the equal angles.

19. In an isosceles triangle, one angle is three times the measure of one of the equal angles. Find the measure of each angle.

20. In a triangle, the second angle is 10° less than the measure of the first angle. The third angle is 31° less than the measure of the second angle. Find the measure of each angle of the triangle.

Cumulative Review

1. Simplify: $8 \cdot (-6) - 4(-3)$

2. Simplify: $\left(-\dfrac{3}{8}\right)^2 \cdot \left(-\dfrac{4}{9}\right)$

3. Simplify: $\dfrac{5}{8} - \left(\dfrac{1}{2}\right)^2 \div \left(\dfrac{1}{3} - \dfrac{3}{4}\right)$

4. Evaluate $a^2 - (2a - b^2)$ when $a = 2$ and $b = -3$.

5. Simplify: $6a - 4b - (-2a) - 5b$

6. Simplify: $-3(3x^2 + 4x - 7)$

7. Simplify: $-2[4 - 2(2x + 1) - x]$

8. Is -1 a solution of $3 + 4x = x^2 - 2$?

9. Solve: $7 - x = 9$

10. Solve: $\dfrac{2}{3}x = -8$

11. Solve: $-2 = 6 - 4x$

12. Solve: $3x - 4 = 4 - 2(x - 1)$

13. Write 55% as a fraction.

14. Write $66\dfrac{2}{3}\%$ as a fraction.

15. The sum of two numbers is ten. The difference of four times the smaller number and eight equals ten less than the product of three and the larger number. Find the two numbers.

16. The repair bill for a washing machine was $73. This includes $28 for parts and $30 for each hour of labor. Find the number of hours of labor.

17.   Write 1.03 as a percent.

18.   Write $\dfrac{9}{20}$ as a percent.

19.   Find $16\dfrac{2}{3}\%$ of 90.

20.   25% of what number is 30?

21.   The value of an investment today is $4400. This is a 10% increase over the value of the investment 1 year ago. Find the value of the investment last year.

22.   A deposit of $2400 is made into an account that earns 5% simple interest. How much additional money must be deposited into an account that pays 8% simple interest so that the total interest earned is 7% of the total investment?

23.   A tree farm buys a 5-gal fruit tree for $3.30 and sells it for $4.62. Find the markup rate.

24.   The toner cartridge for a personal copier, which normally sells for $62.00, is on sale for 40% off the regular price. Find the sale price.

25.   How many pounds of an oat flour that costs $.80 per pound must be mixed with 40 pounds of a wheat flour that costs $.50 per pound to make a blend that costs $.60 per pound?

26.   How many grams of pure gold must be added to 100 g of a 20% gold alloy to make an alloy that is 36% gold?

27.   The perimeter of a rectangular office is 44 ft. The length of the office is 2 ft more than the width. Find the dimensions of the office.

28.   In an equilateral triangle, all angles are equal. Find the measure of one of the angles of an equilateral triangle.

29.   Four times the second of three consecutive integers equals the sum of the first and third integers. Find the integers.

30.   A sprinter ran to the end of a track at an average rate of 8 m/s and then jogged back to the starting point at an average rate of 3 m/s. The sprinter took 55 s to run to the end of the track and jog back. Find the length of the track.

Polynomials

Objectives

Section 5.1
To add polynomials
To subtract polynomials

Section 5.2
To multiply monomials
To simplify powers of monomials

Section 5.3
To multiply a polynomial by a monomial
To multiply two polynomials
To multiply two binomials
To multiply binomials that have special
   products
To solve application problems

Section 5.4
To divide monomials
To write a number in scientific notation

Section 5.5
To divide a polynomial by a monomial
To divide polynomials

Early Egyptian Arithmetic Operations

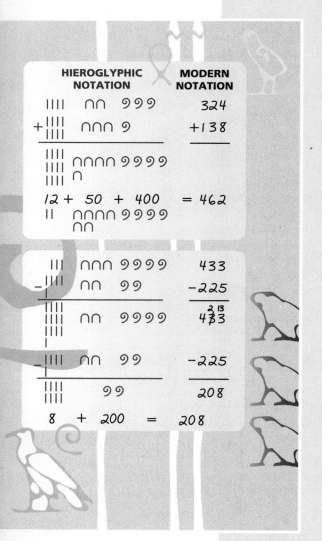

HIEROGLYPHIC NOTATION	MODERN NOTATION

The early Egyptian arithmetic processes are recorded on the early Rhind Papyrus without showing the underlying principles. Scholars of today can only guess how these early developments were discovered.

Egyptian hieroglyphics used a base-ten system of numbers in which a vertical line represented 1, a heel bone, ∩, represented 10, and a scroll, 𝒢, represented 100.

The symbols at the right represent the number 237.
There are 7 vertical lines, 3 heel bones, and 2 scrolls.
Thus the symbols at the right represent 7 + 30 + 200, or 237.

Addition in hieroglyphic notation does not require memorization of addition facts. Addition is done just by counting symbols.

Addition is a simple grouping operation.

Write down the total of each kind of symbol.

Group the 10 straight lines into one heel bone.

Subtraction in the hieroglyphic system is similar to making change. For example, what change do you get from a $1.00 bill when buying a $.55 item?

5 cannot be subtracted from 0, so a 10 is "borrowed" and 10 ones are added.

Note that no zero is provided in this number system. That place value symbol is just not used. As shown at the left, the heel bone is not used because there are no 10's necessary in 208.

Content and Format © 1995 HMCo.

5.1 Addition and Subtraction of Polynomials

Objective A   To add polynomials

A **monomial** is a number, a variable, or a product of numbers and variables. For instance,

7	b	$\dfrac{2}{3}a$	$12xy^2$
A number	A variable	A product of a number and a variable	A product of a number and variables

The expression $3\sqrt{x}$ is not a monomial because $\sqrt{x}$ cannot be written as a product of variables. The expression $\dfrac{2x}{y^2}$ is not a monomial because it is a *quotient* of variables.

A **polynomial** is a variable expression in which the terms are monomials.

A polynomial of *one* term is a **monomial.** $-7x^2$ is a monomial.
A polynomial of *two* terms is a **binomial.** $4x + 2$ is a binomial.
A polynomial of *three* terms is a **trinomial.** $7x^2 + 5x - 7$ is a trinomial.

The terms of a polynomial in one variable are usually arranged so that the exponents of the variable decrease from left to right. This is called **descending order.**

$$5x^3 - 4x^2 + 6x - 1$$
$$7z^4 + 4z^3 + z - 6$$
$$2y^4 + y^3 - 2y^2 + 4y - 1$$

The **degree** of a polynomial in one variable is its largest exponent (on a variable). The degree of $4x^3 - 5x^2 + 7x - 8$ is 3; the degree of $2y^4 + y^2 - 1$ is 4.

To add polynomials, add the coefficients of the like terms. Either a horizontal format or a vertical format can be used.

➡ Simplify $(3x^3 - 7x + 2) + (7x^2 + 2x - 7)$. Use a horizontal format.

$(3x^3 - 7x + 2) + (7x^2 + 2x - 7)$
$\quad = 3x^3 + 7x^2 + (-7x + 2x) + (2 - 7)$

• Use the Commutative and Associative Properties of Addition to rearrange and group like terms.

$\quad = 3x^3 + 7x^2 - 5x - 5$

• Then combine terms.

➡ Simplify $(-4x^2 + 6x - 9) + (12 - 8x + 2x^3)$. Use a vertical format.

$$
\begin{array}{r}
-4x^2 + 6x - 9 \\
2x^3 \qquad - 8x + 12 \\
\hline
2x^3 - 4x^2 - 2x + 3
\end{array}
$$

• Arrange the terms of each polynomial in descending order with like terms in the same column.

• Combine the terms in each column.

Example 1
Use a horizontal format to simplify
$(8x^2 - 4x - 9) + (2x^2 + 9x - 9)$.

Solution
$(8x^2 - 4x - 9) + (2x^2 + 9x - 9)$
$\quad = (8x^2 + 2x^2) + (-4x + 9x) + (-9 - 9)$
$\quad = 10x^2 + 5x - 18$

You Try It 1
Use a horizontal format to simplify
$(-4x^3 + 2x^2 - 8) + (4x^3 + 6x^2 - 7x + 5)$.

Your solution

Solution on p. A13

Example 2
Use a vertical format to simplify
$(-5x^3 + 4x^2 - 7x + 9) + (2x^3 + 5x - 11)$.

Solution
$$\begin{array}{r} -5x^3 + 4x^2 - 7x + 9 \\ 2x^3 \quad\quad + 5x - 11 \\ \hline -3x^3 + 4x^2 - 2x - 2 \end{array}$$

You Try It 2
Use a vertical format to simplify
$(6x^3 + 2x + 8) + (-9x^3 + 2x^2 - 12x - 8)$.

Your solution

Solution on p. A13

Objective B   **To subtract polynomials** ....

The **opposite** of the polynomial $(3x^2 - 7x + 8)$ is $-(3x^2 - 7x + 8)$.

To simplify the opposite of a polynomial, change the sign of each term inside the parentheses.

$$-(3x^2 - 7x + 8) = -3x^2 + 7x - 8$$

To subtract two polynomials, add the opposite of the second polynomial to the first. Either a horizontal or a vertical format can be used.

➡ Simplify $(4y^2 - 6y + 7) - (2y^3 - 5y - 4)$. Use a horizontal format.
   Add the opposite of the second polynomial to the first. Combine like terms.

$$\begin{aligned} (4y^2 - 6y + 7) - (2y^3 - 5y - 4) &= (4y^2 - 6y + 7) + (-2y^3 + 5y + 4) \\ &= -2y^3 + 4y^2 + (-6y + 5y) + (7 + 4) \\ &= -2y^3 + 4y^2 - y + 11 \end{aligned}$$

➡ Simplify $(9 + 4y + 3y^3) - (2y^2 + 4y - 21)$. Use a vertical format.
   Arrange $(9 + 4y + 3y^3)$ and the opposite of $(2y^2 + 4y - 21)$ in descending order with like terms in the same column. Then add.

$$\begin{array}{r} 3y^3 \quad\quad + 4y + 9 \\ - 2y^2 - 4y + 21 \\ \hline 3y^3 - 2y^2 \quad\quad + 30 \end{array}$$

- Write $9 + 4y + 3y^3$ in descending order.
- The opposite of $2y^2 + 4y - 21$ is $-2y^2 - 4y + 21$.
- Note that $4y - 4y = 0$ but 0 is not written.

Example 3
Use a horizontal format to simplify
$(7c^2 - 9c - 12) - (9c^2 + 5c - 8)$.

Solution
$$\begin{aligned} (7c^2 - 9c - 12) &- (9c^2 + 5c - 8) \\ = (7c^2 - 9c - 12) &- (-9c^2 - 5c + 8) \\ = -2c^2 - 14c - 4 \end{aligned}$$

You Try It 3
Use a horizontal format to simplify
$(-4w^3 + 8w - 8) - (3w^3 - 4w^2 - 2w - 1)$.

Your solution

Example 4
Use a vertical format to simplify
$(3k^2 - 4k + 1) - (k^3 + 3k^2 - 6k - 8)$.

Solution
$$\begin{array}{r} 3k^2 - 4k + 1 \\ -k^3 - 3k^2 + 6k + 8 \\ \hline -k^3 \quad\quad + 2k + 9 \end{array}$$
- Add the opposite of $(k^3 + 3k^2 - 6k - 8)$ to the first polynomial.

You Try It 4
Use a vertical format to simplify
$(13y^3 - 6y - 7) - (4y^2 - 6y - 9)$.

Your solution

Solutions on p. A13

5.1 Exercises

. . . .

Objective A

State whether the expression is a monomial.

1. 17

2. $3x^4$

3. $\dfrac{17}{\sqrt{x}}$

4. xyz

5. $\dfrac{2}{3}y$

6. $\dfrac{xy}{z}$

7. $\sqrt{5}x$

8. πx

State whether the expression is a monomial, binomial, trinomial, or none of these.

9. $3x + 5$

10. $2y - 3\sqrt{y}$

11. $9x^2 - x - 1$

12. $x^2 + y^2$

13. $\dfrac{2}{x} - 3$

14. $\dfrac{ab}{4}$

15. $6x^2 + 7x$

16. $12a^4 - 3a + 2$

Simplify. Use a vertical format.

17. $(x^2 + 7x) + (-3x^2 - 4x)$

18. $(3y^2 - 2y) + (5y^2 + 6y)$

19. $(y^2 + 4y) + (-4y - 8)$

20. $(3x^2 + 9x) + (6x - 24)$

21. $(2x^2 + 6x + 12) + (3x^2 + x + 8)$

22. $(x^2 + x + 5) + (3x^2 - 10x + 4)$

23. $(-7x + x^3 + 4) + (2x^2 + x - 10)$

24. $(y^2 + 3y^3 + 1) + (-4y^3 - 6y - 3)$

25. $(2a^3 - 7a + 1) + (1 - 4a - 3a^2)$

26. $(5r^3 - 6r^2 + 3r) + (-3 - 2r + r^2)$

Simplify. Use a horizontal format.

27. $(4x^2 + 2x) + (x^2 + 6x)$

28. $(-3y^2 + y) + (4y^2 + 6y)$

29. $(4x^2 - 5xy) + (3x^2 + 6xy - 4y^2)$

30. $(2x^2 - 4y^2) + (6x^2 - 2xy + 4y^2)$

31. $(2a^2 - 7a + 10) + (a^2 + 4a + 7)$

32. $(-6x^2 + 7x + 3) + (3x^2 + x + 3)$

33. $(7x + 5x^3 - 7) + (10x^2 - 8x + 3)$

34. $(4y + 3y^3 + 9) + (2y^2 + 4y - 21)$

35. $(7 - 5r + 2r^2) + (3r^3 - 6r)$

36. $(14 + 4y + 3y^3) + (-4y^2 + 21)$

37. $(7x + 3x^2 + 10) + (1 + 3x - 2x^3)$

38. $(4x - 1 + 7x^3) + (2 - 6x + 2x^2)$

Objective B

Simplify. Use a vertical format.

39. $(x^2 - 6x) - (x^2 - 10x)$

40. $(y^2 + 4y) - (y^2 + 10y)$

41. $(2y^2 - 4y) - (-y^2 + 2)$

42. $(-3a^2 - 2a) - (4a^2 - 4)$

43. $(x^2 - 2x + 1) - (x^2 + 5x + 8)$

44. $(3x^2 + 2x - 2) - (5x^2 - 5x + 6)$

45. $(4x^3 + 5x + 2) - (1 + 2x - 3x^2)$

46. $(5y^2 - y + 2) - (-3 + 3y - 2y^3)$

47. $(-2y + 6y^2 + 2y^3) - (4 + y^2 + y^3)$

48. $(4 - x - 2x^2) - (-2 + 3x - x^3)$

Simplify. Use a horizontal format.

49. $(y^2 - 10xy) - (2y^2 + 3xy)$

50. $(x^2 - 3xy) - (-2x^2 + xy)$

51. $(3x^2 + x - 3) - (4x + x^2 - 2)$

52. $(5y^2 - 2y + 1) - (-y - 2 - 3y^2)$

53. $(-2x^3 + x - 1) - (-x^2 + x - 3)$

54. $(2x^2 + 5x - 3) - (3x^3 + 2x - 5)$

55. $(1 - 2a + 4a^3) - (a^3 - 2a + 3)$

56. $(7 - 8b + b^2) - (4b^3 - 7b - 8)$

57. $(-1 - y + 4y^3) - (3 - 3y - 2y^2)$

58. $(-3 - 2x + 3x^2) - (4 - 2x^2 + 2x^3)$

APPLYING THE CONCEPTS

59. What polynomial must be added to $3x^2 - 6x + 9$ so that the sum is $4x^2 + 3x - 2$?

60. What polynomial must be added to $9x^2 + x - 1$ so that the sum is $-x^2 + 7x - 2$?

61. What polynomial must be subtracted from $2x^2 - x - 2$ so that the difference is $5x^2 + 3x + 1$?

62. What polynomial must be subtracted from $4x^3 - x^2 + 1$ so that the difference is $x^3 + 4x - 2$?

63. [W] In your own words, explain the terms monomial, binomial, trinomial, and polynomial. Give an example of each.

64. [W] Is it possible to subtract two polynomials, each of degree 3, and have the difference be a polynomial of degree 2? If so, give an example. If not, explain why not.

65. [W] Is it possible to add two polynomials, each of degree 3, and have the sum be a polynomial of degree 2? If so, give an example. If not, explain why not.

Multiplication of Monomials

. . . .

Objective A   To multiply monomials . . . .

Recall that in an exponential expression such as x^6, x is the base and 6 is the exponent. The exponent indicates the number of times the base occurs as a factor.

The product of exponential expressions with the *same* base can be simplified by writing each expression in factored form and writing the result with an exponent.

$$x^3 \cdot x^2 = \overbrace{(x \cdot x \cdot x)}^{3 \text{ factors}} \cdot \overbrace{(x \cdot x)}^{2 \text{ factors}}$$
$$\underbrace{}_{5 \text{ factors}}$$
$$= x^5$$

Note that adding the exponents results in the same product.

$$x^3 \cdot x^2 = x^{3+2} = x^5$$

Rule for Multiplying Exponential Expressions

If m and n are positive integers, then $x^m \cdot x^n = x^{m+n}$.

➡ Simplify: $y^4 \cdot y \cdot y^3$

$$y^4 \cdot y \cdot y^3 = \boxed{y^{4+1+3}}$$
$$= y^8$$

- The bases are the same. Add the exponents. This step is often done mentally. Recall that $y = y^1$.

➡ Simplify: $(-3a^4b^3)(2ab^4)$

$$(-3a^4b^3)(2ab^4) = (-3 \cdot 2)(a^4 \cdot a)(b^3 \cdot b^4)$$

- Use the Commutative and Associative Properties of Multiplication to rearrange and group factors.

$$= -6(a^{4+1})(b^{3+4})$$

- Multiply variables with the same base by adding their exponents.

$$= -6a^5b^7$$

- Simplify.

LOOK CLOSELY

The Rule for Multiplying Exponential Expressions requires the bases to be the same. The expression a^5b^7 cannot be simplified.

Example 1   Simplify: $(-5ab^3)(4a^5)$

Solution
$$(-5ab^3)(4a^5) = (-5 \cdot 4)(a \cdot a^5)b^3$$
$$= -20a^6b^3$$

You Try It 1   Simplify: $(8m^3n)(-3n^5)$

Your solution

Example 2   Simplify: $(6x^3y^2)(4x^4y^5)$

Solution
$$(6x^3y^2)(4x^4y^5) = (6 \cdot 4)(x^3 \cdot x^4)(y^2 \cdot y^5)$$
$$= 24x^7y^7$$

You Try It 2   Simplify: $(12p^4q^3)(-3p^5q^2)$

Your solution

Solutions on p. A13

Objective B **To simplify powers of monomials** ....

POINT OF INTEREST

One of the first symbolic representations of powers was given by Diophantus (c. A.D. 250) in his book *Arithmetica*. He used Δ^{Υ} for x^2 and κ^{Υ} for x^3. The symbol Δ^{Υ} was the first two letters of the Greek word *dunamis* meaning power; κ^{Υ} was from the Greek word *kubos* meaning cube. He also combined these symbols to denote higher powers. For instance, $\Delta\kappa^{\Upsilon}$ was the symbol for x^5.

The power of a monomial can be simplified by writing the power in factored form and then using the Rule for Multiplying Exponential Expressions.

$(x^4)^3 = x^4 \cdot x^4 \cdot x^4$ $\qquad$ $(a^2b^3)^2 = (a^2b^3)(a^2b^3)$ $\qquad$ • Write in factored form.

$\qquad = x^{4+4+4}$ $\qquad\qquad = a^{2+2}b^{3+3}$ $\qquad$ • Use the Rule for Multiplying Exponential Expressions.

$\qquad = x^{12}$ $\qquad\qquad\qquad = a^4b^6$

Note that multiplying each exponent inside the parentheses by the exponent outside the parentheses results in the same product.

$(x^4)^3 = x^{4 \cdot 3} = x^{12}$ $\qquad$ $(a^2b^3)^2 = a^{2 \cdot 2}b^{3 \cdot 2} = a^4b^6$ $\qquad$ • Multiply each exponent inside the parentheses by the exponent outside the parentheses.

Rule for Simplifying the Power of an Exponential Expression

If m and n are positive integers, then $(x^m)^n = x^{m \cdot n}$.

Rule for Simplifying Powers of Products

If m, n, and p are positive integers, then $(x^m y^n)^p = x^{m \cdot p} y^{n \cdot p}$.

➡ Simplify: $(5x^2y^3)^3$

$(5x^2y^3)^3 = 5^{1 \cdot 3}x^{2 \cdot 3}y^{3 \cdot 3}$ $\qquad$ • Multiply each exponent inside the parentheses by the exponent outside the parentheses. Note that $5 = 5^1$.

$\qquad\qquad = 5^3x^6y^9$

$\qquad\qquad = 125x^6y^9$ $\qquad$ • Evaluate 5^3.

Example 3 Simplify: $(-2p^3r)^4$

Solution $(-2p^3r)^4 = (-2)^{1 \cdot 4}p^{3 \cdot 4}r^{1 \cdot 4}$
$\qquad\qquad = (-2)^4p^{12}r^4 = 16p^{12}r^4$

You Try It 3 Simplify: $(-3a^4bc^2)^3$

Your solution

Example 4 Simplify: $(2a^2b)(2a^3b^2)^3$

Solution
$(2a^2b)(2a^3b^2)^3 = (2a^2b)(2^{1 \cdot 3}a^{3 \cdot 3}b^{2 \cdot 3})$
$\qquad\qquad = (2a^2b)(2^3a^9b^6)$
$\qquad\qquad = (2a^2b)(8a^9b^6) = 16a^{11}b^7$

You Try It 4 Simplify: $(-xy^4)(-2x^3y^2)^2$

Your solution

Solutions on p. A13

5.2 Exercises

Objective A

Simplify.

1. $(6x^2)(5x)$

2. $(-4y^3)(2y)$

3. $(7c^2)(-6c^4)$

4. $(-8z^5)(5z^8)$

5. $(-3a^3)(-3a^4)$

6. $(-5a^6)(-2a^5)$

7. $(x^2)(xy^4)$

8. $(x^2y^4)(xy^7)$

9. $(-2x^4)(5x^5y)$

10. $(-3a^3)(2a^2b^4)$

11. $(-4x^2y^4)(-3x^5y^4)$

12. $(-6a^2b^4)(-4ab^3)$

13. $(2xy)(-3x^2y^4)$

14. $(-3a^2b)(-2ab^3)$

15. $(x^2yz)(x^2y^4)$

16. $(-ab^2c)(a^2b^5)$

17. $(-a^2b^3)(-ab^2c^4)$

18. $(-x^2y^3z)(-x^3y^4)$

19. $(-5a^2b^2)(6a^3b^6)$

20. $(7xy^4)(-2xy^3)$

21. $(-6a^3)(-a^2b)$

22. $(-2a^2b^3)(-4ab^2)$

23. $(-5y^4z)(-8y^6z^5)$

24. $(3x^2y)(-4xy^2)$

25. $(x^2y)(yz)(xyz)$

26. $(xy^2z)(x^2y)(z^2y^2)$

27. $(3ab^2)(-2abc)(4ac^2)$

28. $(-2x^3y^2)(-3x^2z^2)(-5y^3z^3)$

29. $(4x^4z)(-yz^3)(-2x^3z^2)$

30. $(-a^3b^4)(-3a^4c^2)(4b^3c^4)$

31. $(-2x^2y^3)(3xy)(-5x^3y^4)$

32. $(4a^2b)(-3a^3b^4)(a^5b^2)$

33. $(3a^2b)(-6bc)(2ac^2)$

Objective B

Simplify.

34. $(z^4)^3$

35. $(x^3)^5$

36. $(y^4)^2$

37. $(x^7)^2$

38. $(-y^5)^3$

39. $(-x^2)^4$

40. $(-x^2)^3$

41. $(-y^3)^4$

42. $(-3y)^3$

43. $(-2x^2)^3$

44. $(a^3b^4)^3$

45. $(x^2y^3)^2$

46. $(2x^3y^4)^5$

47. $(3x^2y)^2$

48. $(-2ab^3)^4$

49. $(-3x^3y^2)^5$ **50.** $(3b^2)(2a^3)^4$ **51.** $(-2x)(2x^3)^2$ **52.** $(2y)(-3y^4)^3$

53. $(3x^2y)(2x^2y^2)^3$ **54.** $(a^3b)^2(ab)^3$ **55.** $(ab^2)^2(ab)^2$ **56.** $(-x^2y^3)^2(-2x^3y)^3$

57. $(-2x)^3(-2x^3y)^3$ **58.** $(-3y)(-4x^2y^3)^3$ **59.** $(-2x)(-3xy^2)^2$ **60.** $(-3y)(-2x^2y)^3$

61. $(ab^2)(-2a^2b)^3$ **62.** $(a^2b^2)(-3ab^4)^2$ **63.** $(-2a^3)(3a^2b)^3$ **64.** $(-3b^2)(2ab^2)^3$

APPLYING THE CONCEPTS

Simplify.

65. $3x^2 + (3x)^2$ **66.** $4x^2 - (4x)^2$ **67.** $2x^6y^2 + (3x^2y)^2$ **68.** $(x^2y^2)^3 + (x^3y^3)^2$

69. $(2a^3b^2)^3 - 8a^9b^6$ **70.** $4y^2z^4 - (2yz^2)^2$ **71.** $(x^2y^4)^2 + (2xy^2)^4$ **72.** $(3a^3)^2 - 4a^6 + (2a^2)^3$

For Exercises 73–76, answer true or false. If the answer is false, correct the right-hand side of the equation.

73. $(-a)^5 = -a^5$ **74.** $(-b)^8 = b^8$ **75.** $(x^2)^5 = x^{2+5} = x^7$ **76.** $x^3 + x^3 = 2x^{3+3} = 2x^6$

77. Evaluate $(2^3)^2$ and $2^{(3^2)}$. Are the results the same? If not, which expression has the larger value?

78. What is the Order of Operations for the expression x^{m^n}?

79. If n is a positive integer and $x^n = y^n$, does $x = y$? Explain your
[W] answer.

80. The distance a rock will fall in t seconds is $16t^2$ (neglecting air resistance). Find other examples of quantities that can be expressed in terms of an exponential expression, and explain where the expression is used.

81. Explain in your own words how to multiply monomials.

Multiplication of Polynomials

Objective A *To multiply a polynomial by a monomial* ....

To multiply a polynomial by a monomial, use the Distributive Property and the Rule for Multiplying Exponential Expressions.

➡ Simplify: $-3a(4a^2 - 5a + 6)$

$$-3a(4a^2 - 5a + 6) = \boxed{-3a(4a^2) - (-3a)(5a) + (-3a)(6)}$$
$$= -12a^3 + 15a^2 - 18a$$

● Use the Distributive Property. This step is frequently done mentally.

Example 1
Simplify: $(5x + 4)(-2x)$

Solution
$(5x + 4)(-2x) = -10x^2 - 8x$

Example 2
Simplify: $2a^2b(4a^2 - 2ab + b^2)$

Solution
$2a^2b(4a^2 - 2ab + b^2)$
$\quad = 8a^4b - 4a^3b^2 + 2a^2b^3$

You Try It 1
Simplify: $(-2y + 3)(-4y)$

Your solution

You Try It 2
Simplify: $-a^2(3a^2 + 2a - 7)$

Your solution

Solutions on p. A13

Objective B *To multiply two polynomials* ....

Multiplication of two polynomials requires the repeated application of the Distributive Property.

$$(y - 2)(y^2 + 3y + 1) = (y - 2)(y^2) + (y - 2)(3y) + (y - 2)(1)$$
$$= y^3 - 2y^2 + 3y^2 - 6y + y - 2$$
$$= y^3 + y^2 - 5y - 2$$

A convenient method of multiplying two polynomials is to use a vertical format similar to that used for multiplication of whole numbers.

$$
\begin{array}{r}
y^2 + 3y + 1 \\
y - 2 \\
\hline
-2y^2 - 6y - 2 \\
y^3 + 3y^2 + 1y \\
\hline
y^3 + 1y^2 - 5y - 2
\end{array}
$$

$-2y^2 - 6y - 2 = -2(y^2 + 3y + 1)$
$y^3 + 3y^2 + 1y \quad = y(y^2 + 3y + 1)$

● Multiply by -2.
● Multiply by y.
● Add the terms in each column.

➡ Simplify: $(2a^3 + a - 3)(a + 5)$

$$
\begin{array}{r}
2a^3 + a - 3 \\
a + 5 \\
\hline
10a^3 + 5a - 15 \\
2a^4 + a^2 - 3a \\
\hline
2a^4 + 10a^3 + a^2 + 2a - 15
\end{array}
$$

- Note that spaces are provided in each product so that like terms are in the same column.
- Add the terms in each column.

Example 3
Simplify: $(2b^3 - b + 1)(2b + 3)$

Solution

$$
\begin{array}{r}
2b^3 - -b + 1 \\
2b + 3 \\
\hline
6b^3 - 3b + 3 \\
4b^4 + - 2b^2 + 2b \\
\hline
4b^4 + 6b^3 - 2b^2 - b + 3
\end{array}
$$

You Try It 3
Simplify: $(2y^3 + 2y^2 - 3)(3y - 1)$

Your solution

Solution on p. A13

Objective C To multiply two binomials

It is frequently necessary to find the product of two binomials. The product can be found using a method called **FOIL**, which is based on the Distributive Property. The letters of FOIL stand for **F**irst, **O**uter, **I**nner, and **L**ast.

➡ Simplify: $(2x + 3)(x + 5)$

Multiply the **F**irst terms.	$(2x + 3)(x + 5)$	$2x \cdot x = 2x^2$
Multiply the **O**uter terms.	$(2x + 3)(x + 5)$	$2x \cdot 5 = 10x$
Multiply the **I**nner terms.	$(2x + 3)(x + 5)$	$3 \cdot x = 3x$
Multiply the **L**ast terms.	$(2x + 3)(x + 5)$	$3 \cdot 5 = 15$

$$ \text{F} \text{O} \text{I} \text{L}$$

Add the products. $(2x + 3)(x + 5) = 2x^2 + 10x + 3x + 15$

Combine like terms. $ = 2x^2 + 13x + 15$

LOOK CLOSELY
FOIL is not really a different way of multiplying. It is based on the Distributive Property.

$(2x + 3)(x + 5)$
$ = 2x(x + 5) + 3(x + 5)$
$ \text{F} \text{O} \text{I} \text{L}$
$ = 2x^2 + 10x + 3x + 15$
$ = 2x^2 + 13x + 15$

➡ Simplify: $(4x - 3)(3x - 2)$

$(4x - 3)(3x - 2) = \boxed{4x(3x) + 4x(-2) + (-3)(3x) + (-3)(-2)}$
$ = 12x^2 - 8x - 9x + 6$
$ = 12x^2 - 17x + 6$

- Do this step mentally.

➡ Simplify: $(3x - 2y)(x + 4y)$

$(3x - 2y)(x + 4y) = \boxed{3x(x) + 3x(4y) + (-2y)(x) + (-2y)(4y)}$
$ = 3x^2 + 12xy - 2xy - 8y^2$
$ = 3x^2 + 10xy - 8y^2$

- Do this step mentally.

Example 4

Simplify: $(2a - 1)(3a - 2)$

Solution

$(2a - 1)(3a - 2) = 6a^2 - 4a - 3a + 2$
$= 6a^2 - 7a + 2$

Example 5

Simplify: $(3x - 2)(4x + 3)$

Solution

$(3x - 2)(4x + 3) = 12x^2 + 9x - 8x - 6$
$= 12x^2 + x - 6$

You Try It 4

Simplify: $(4y - 5)(2y - 3)$

Your solution

You Try It 5

Simplify: $(3b + 2)(3b - 5)$

Your solution

Solutions on p. A13

Objective D   *To mulitply binomials that have special products* ....

Using FOIL, it is possible to find a pattern for the product of the sum and difference of two terms and for the square of a binomial.

The Sum and Difference of Two Terms

$$(a + b)(a - b) = a^2 - ab + ab - b^2$$
$$= a^2 - b^2$$

Square of first term ————————
Square of second term ————————

The Square of a Binomial

$$(a + b)^2 = (a + b)(a + b) = a^2 + ab + ab + b^2$$
$$= a^2 + 2ab + b^2$$

Square of first term ————————
Twice the product of the two terms —
Square of the last term ————————

➡ Simplify: $(2x + 3)(2x - 3)$

$(2x + 3)(2x - 3)$ is the sum and difference of two terms.

$(2x + 3)(2x - 3) = \boxed{(2x)^2 - 3^2}$ • Do this step mentally.
$= 4x^2 - 9$

➡ Simplify: $(3x - 2)^2$

$(3x - 2)^2$ is the square of a binomial.

$(3x - 2)^2 = \boxed{(3x)^2 + 2(3x)(-2) + (-2)^2}$ • Do this step mentally.
$= 9x^2 - 12x + 4$

Example 6
Simplify: $(4z - 2w)(4z + 2w)$

Solution
$(4z - 2w)(4z + 2w) = 16z^2 - 4w^2$

You Try It 6
Simplify: $(2a + 5c)(2a - 5c)$

Your solution

Example 7
Simplify: $(2r - 3s)^2$

Solution
$(2r - 3s)^2 = 4r^2 - 12rs + 9s^2$

You Try It 7
Simplify: $(3x + 2y)^2$

Your solution

Solutions on pp. A13–A14

Objective E   To solve application problems

Example 8
The length of a rectangle is $(x + 7)$ meters. The width is $(x - 4)$ meters. Find the area of the rectangle in terms of the variable x.

Strategy
To find the area, replace the variables l and w in the equation $A = l \cdot w$ by the given values and solve for A.

Solution
$A = l \cdot w$
$A = (x + 7)(x - 4)$
$A = x^2 - 4x + 7x - 28$
$A = x^2 + 3x - 28$

The area is $(x^2 + 3x - 28)$ m^2

You Try It 8
The radius of a circle is $(x - 4)$ft. Use the equation $A = \pi r^2$, where r is the radius, to find the area of the circle in terms of x. Use $\pi \approx 3.14$.

Your strategy

Your solution

Solutions on p. A14

5.3 Exercises

- -

_____
Objective A

Simplify.

1. $x(x - 2)$ **2.** $y(3 - y)$ **3.** $-x(x + 7)$ **4.** $-y(7 - y)$

5. $3a^2(a - 2)$ **6.** $4b^2(b + 8)$ **7.** $-5x^2(x^2 - x)$ **8.** $-6y^2(y + 2y^2)$

9. $-x^3(3x^2 - 7)$ **10.** $-y^4(2y^2 - y^6)$ **11.** $2x(6x^2 - 3x)$ **12.** $3y(4y - y^2)$

13. $(2x - 4)3x$ **14.** $(3y - 2)y$ **15.** $(3x + 4)x$ **16.** $(2x + 1)2x$

17. $-xy(x^2 - y^2)$ **18.** $-x^2y(2xy - y^2)$ **19.** $x(2x^3 - 3x + 2)$ **20.** $y(-3y^2 - 2y + 6)$

21. $-a(-2a^2 - 3a - 2)$ **22.** $-b(5b^2 + 7b - 35)$ **23.** $x^2(3x^4 - 3x^2 - 2)$

24. $y^3(-4y^3 - 6y + 7)$ **25.** $2y^2(-3y^2 - 6y + 7)$ **26.** $4x^2(3x^2 - 2x + 6)$

27. $(a^2 + 3a - 4)(-2a)$ **28.** $(b^3 - 2b + 2)(-5b)$ **29.** $-3y^2(-2y^2 + y - 2)$

30. $-5x^2(3x^2 - 3x - 7)$ **31.** $xy(x^2 - 3xy + y^2)$ **32.** $ab(2a^2 - 4ab - 6b^2)$

_____
Objective B

Simplify.

33. $(x^2 + 3x + 2)(x + 1)$ **34.** $(x^2 - 2x + 7)(x - 2)$ **35.** $(a^2 - 3a + 4)(a - 3)$

Simplify.

36. $(x^2 - 3x + 5)(2x - 3)$ **37.** $(-2b^2 - 3b + 4)(b - 5)$ **38.** $(-a^2 + 3a - 2)(2a - 1)$

39. $(-2x^2 + 7x - 2)(3x - 5)$ **40.** $(-a^2 - 2a + 3)(2a - 1)$ **41.** $(x^2 + 5)(x - 3)$

42. $(y^2 - 2y)(2y + 5)$ **43.** $(x^3 - 3x + 2)(x - 4)$ **44.** $(y^3 + 4y^2 - 8)(2y - 1)$

45. $(5y^2 + 8y - 2)(3y - 8)$ **46.** $(3y^2 + 3y - 5)(4y - 3)$ **47.** $(5a^3 - 5a + 2)(a - 4)$

48. $(3b^3 - 5b^2 + 7)(6b - 1)$ **49.** $(y^3 + 2y^2 - 3y + 1)(y + 2)$ **50.** $(2a^3 - 3a^2 + 2a - 1)(2a - 3)$

Objective C

Simplify.

51. $(x + 1)(x + 3)$ **52.** $(y + 2)(y + 5)$ **53.** $(a - 3)(a + 4)$ **54.** $(b - 6)(b + 3)$

55. $(y + 3)(y - 8)$ **56.** $(x + 10)(x - 5)$ **57.** $(y - 7)(y - 3)$ **58.** $(a - 8)(a - 9)$

59. $(2x + 1)(x + 7)$ **60.** $(y + 2)(5y + 1)$ **61.** $(3x - 1)(x + 4)$ **62.** $(7x - 2)(x + 4)$

63. $(4x - 3)(x - 7)$ **64.** $(2x - 3)(4x - 7)$ **65.** $(3y - 8)(y + 2)$ **66.** $(5y - 9)(y + 5)$

67. $(3x + 7)(3x + 11)$ **68.** $(5a + 6)(6a + 5)$ **69.** $(7a - 16)(3a - 5)$ **70.** $(5a - 12)(3a - 7)$

71. $(3a - 2b)(2a - 7b)$ **72.** $(5a - b)(7a - b)$ **73.** $(a - 9b)(2a + 7b)$

74. $(2a + 5b)(7a - 2b)$ **75.** $(10a - 3b)(10a - 7b)$ **76.** $(12a - 5b)(3a - 4b)$

77. $(5x + 12y)(3x + 4y)$ **78.** $(11x + 2y)(3x + 7y)$ **79.** $(2x - 15y)(7x + 4y)$

80. $(5x + 2y)(2x - 5y)$ **81.** $(8x - 3y)(7x - 5y)$ **82.** $(2x - 9y)(8x - 3y)$

Objective D

Simplify.

83. $(y - 5)(y + 5)$ **84.** $(y + 6)(y - 6)$ **85.** $(2x + 3)(2x - 3)$ **86.** $(4x - 7)(4x + 7)$

87. $(x + 1)^2$ **88.** $(y - 3)^2$ **89.** $(3a - 5)^2$ **90.** $(6x - 5)^2$

91. $(3x - 7)(3x + 7)$ **92.** $(9x - 2)(9x + 2)$ **93.** $(2a + b)^2$ **94.** $(x + 3y)^2$

95. $(x - 2y)^2$ **96.** $(2x - 3y)^2$ **97.** $(4 - 3y)(4 + 3y)$

98. $(4x - 9y)(4x + 9y)$ **99.** $(5x + 2y)^2$ **100.** $(2a - 9b)^2$

Objective E   *Application Problems*

Solve.

101. Let L represent the length of a rectangle. The width of the rectangle is 2 ft less than twice the length. Express the area of the rectangle in terms of the variable L.

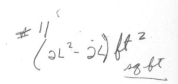

102. The side of a square is $(2x + 1)$ ft. Express the area of the square in terms of the variable x.

103.   A softball diamond has dimensions 45 ft by 45 ft. A base path border x ft wide lies on both the first-base side and third-base side of the diamond. Express the total area of the softball diamond and the base path in terms of the variable x.

104.   A field has dimensions 30 yd by 100 yd. An endzone that is w yd wide borders each end of the field. Express the total area of the field and the endzones in terms of the variable w.

105.   The radius of a circle is $(x + 3)$ centimeters. Use the equation $A = \pi r^2$, where r is the radius, to find the area of the circle in terms of the variable x. Use $\pi \approx 3.14$.

106.   The radius of a circle is $(x - 2)$ inches. Use the equation $A = \pi r^2$, where r is the radius, to find the area of the circle in terms of the variable x. Use $\pi \approx 3.14$.

APPLYING THE CONCEPTS

Simplify.

107. $(a + b)^2 - (a - b)^2$ **108.** $(a + b)^2 + (a - b)^2$ **109.** $(x^2 + x - 3)^2$

110. $(x^2 - 2x + 2)^2$ **111.** $(a + 3)^3$ **112.** $(x - 4)^3$

113.   What polynomial has quotient $3x - 4$ when divided by $4x + 5$?

114.   What polynomial has quotient $x^2 + 2x - 1$ when divided by $x + 3$?

115.   Add $x^2 + 2x - 3$ to the product of $2x - 5$ and $3x + 1$.

116.   Subtract $4x^2 - x - 5$ from the product of $x^2 + x + 3$ and $x - 4$.

117.   If a polynomial of degree 3 is multiplied by a polynomial of degree 2, what is the degree of the resulting polynomial?

118.   [W] Is it possible to multiply a polynomial of degree 2 by a polynomial of degree 2 and have the product a polynomial of degree 3? If so, give an example. If not explain why not.

Integer Exponents and Scientific Notation

Objective A   *To divide monomials*

The quotient of two exponential expressions with the same base can be simplified by writing each expression in factored form, dividing by the common factors, and then writing the result with an exponent.

$$\frac{x^5}{x^2} = \frac{\cancel{x} \cdot \cancel{x} \cdot x \cdot x \cdot x}{\cancel{x} \cdot \cancel{x}} = x^3$$

Note that subtracting the exponents gives the same result.

$$\frac{x^5}{x^2} = x^{5-2} = x^3$$

To divide two monomials with the same base, subtract the exponents of the like bases.

➡ Simplify: $\dfrac{a^7}{a^3}$

$$\frac{a^7}{a^3} = \boxed{a^{7-3}}$$
$$= a^4$$

• The bases are the same. Subtract the exponents. This step is often done mentally.

➡ Simplify: $\dfrac{r^8 t^6}{r^7 t}$

$$\frac{r^8 t^6}{r^7 t} = \boxed{r^{8-7} t^{6-1}}$$
$$= r t^5$$

• Subtract the exponents of the like bases. This step is often done mentally.

➡ Simplify: $\dfrac{p^7}{z^4}$

Because the bases are not the same, $\dfrac{p^7}{z^4}$ is already in simplest form.

Consider the expression $\dfrac{x^4}{x^4}$, $x \neq 0$. This expression can be simplified, as shown below, by subtracting exponents and dividing by common factors.

$$\frac{x^4}{x^4} = x^{4-4} = x^0 \qquad\qquad \frac{x^4}{x^4} = \frac{\cancel{x} \cdot \cancel{x} \cdot \cancel{x} \cdot \cancel{x}}{\cancel{x} \cdot \cancel{x} \cdot \cancel{x} \cdot \cancel{x}} = 1$$

The equations $\dfrac{x^4}{x^4} = x^0$ and $\dfrac{x^4}{x^4} = 1$ suggest the following definition of x^0.

Definition of Zero as an Exponent

If $x \neq 0$, then $x^0 = 1$. The expression 0^0 is not defined.

➡ Simplify: $(12a^3)^0$, $a \neq 0$

$(12a^3)^0 = 1$ • Any non-zero expression to the zero power is 1.

➡ Simplify: $-(4x^3y^7)^0$

$-(4x^3y^7)^0 = -(1) = -1$

POINT OF INTEREST

In the 15th century, the expression $12^{\overline{2m}}$ was used to mean $12x^{-2}$. The use of $\overline{m}$ reflected an Italian influence. In Italy, m was used for minus and p was used for plus. It was understood that $2\overline{m}$ referred to an unnamed variable. Issac Newton, in the 17th century, advocated the current use of a negative exponent.

Consider the expression $\dfrac{x^4}{x^6}$, $x \neq 0$. This expression can be simplified, as shown below, by subtracting exponents and dividing by common factors.

$$\frac{x^4}{x^6} = x^{4-6} = x^{-2} \qquad\qquad \frac{x^4}{x^6} = \frac{\overset{1}{\cancel{x}} \cdot \overset{1}{\cancel{x}} \cdot \overset{1}{\cancel{x}} \cdot \overset{1}{\cancel{x}}}{\underset{1}{\cancel{x}} \cdot \underset{1}{\cancel{x}} \cdot \underset{1}{\cancel{x}} \cdot \underset{1}{\cancel{x}} \cdot x \cdot x} = \frac{1}{x^2}$$

The equations $\dfrac{x^4}{x^6} = x^{-2}$ and $\dfrac{x^4}{x^6} = \dfrac{1}{x^2}$ suggest that $x^{-2} = \dfrac{1}{x^2}$. *no — exp* *— base yes*

Definition of a Negative Exponent

If $x \neq 0$ and n is a positive integer, then

$$x^{-n} = \frac{1}{x^n} \qquad \text{and} \qquad \frac{1}{x^{-n}} = x^n$$

LOOK CLOSELY

Note from the example at the right that 2^{-4} is a *positive* number. A negative exponent does not change the sign of a number.

➡ Evaluate 2^{-4}.

$2^{-4} = \dfrac{1}{2^4}$ • Use the Definition of a Negative Exponent.

$= \dfrac{1}{16}$ • Evaluate the expression.

When the Definition of a Negative Exponent is applied to fractions, we have the following result.

Rule for Negative Exponents on Fractional Expressions

If $a \neq 0$, $b \neq 0$, and n is a positive integer, then

$$\left(\frac{a}{b}\right)^{-n} = \left(\frac{b}{a}\right)^{n}$$

➡ Evaluate $\left(\dfrac{3}{4}\right)^{-2}$.

$$\left(\frac{3}{4}\right)^{-2} = \left(\frac{4}{3}\right)^2$$ • Use the Rule for Negative Exponents on Fractional Expressions.

$$= \left(\frac{4}{3}\right)\left(\frac{4}{3}\right) = \frac{16}{9}$$

> **Simplest Form of an Exponential Expression**
>
> An exponential expression is in simplest form when it is written with only positive exponents.

➡ Simplify: $3n^{-5}$

$$3n^{-5} = 3 \cdot \frac{1}{n^5} = \frac{3}{n^5}$$

● Use the Definition of Negative Exponents to rewrite the expression with a positive exponent.

➡ Simplify: $\dfrac{2}{5a^{-4}}$

$$\frac{2}{5a^{-4}} = \frac{2}{5} \cdot \frac{1}{a^{-4}} = \frac{2}{5} \cdot a^4 = \frac{2a^4}{5}$$

● Use the Definition of Negative Exponents to rewrite the expression with a positive exponent.

Now that zero as an exponent and negative exponents have been defined, a rule for dividing exponential expressions can be stated.

> **Rule for Dividing Exponential Expressions**
>
> If m and n are integers and $x \neq 0$, then $\dfrac{x^m}{x^n} = x^{m-n}$.

➡ Evaluate $\dfrac{5^{-2}}{5^{1}}$.

$$\frac{5^{-2}}{5} = 5^{-2-1} = 5^{-3}$$

● Use the Rule for Dividing Exponential Expressions.

$$= \frac{1}{5^3} = \frac{1}{125}$$

● Use the Definition of Negative Exponents to rewrite the expression with a positive exponent. Then evaluate.

➡ Simplify: $\dfrac{x^4}{x^9}$

$$\frac{x^4}{x^9} = \boxed{x^{4-9}}$$

● Use the Rule for Dividing Exponential Expressions. This step is usually done mentally.

$$= x^{-5}$$

● Subtract the exponents.

$$= \frac{1}{x^5}$$

● Use the Definition of Negative Exponents to rewrite the expression with a positive exponent.

The rules for simplifying exponential expressions and powers of exponential expressions are true for all integers. These rules are restated here along with the rules for dividing exponential expressions.

> **Rule of Exponents**
>
> If m, n, and p are integers, then
>
> $x^m \cdot x^n = x^{m+n}$ $(x^m)^n = x^{mn}$ $(x^m y^n)^p = x^{mp} y^{np}$
>
> $\dfrac{x^m}{x^n} = x^{m-n}, \; x \neq 0$ $\left(\dfrac{x^m}{y^n}\right)^p = \dfrac{x^{mp}}{y^{np}}, \; y \neq 0$ $x^{-n} = \dfrac{1}{x^n}, \; x \neq 0$
>
> $x^0 = 1, \; x \neq 0$

➡ Simplify: $(3ab^{-4})(-2a^{-3}b^7)$

$(3ab^{-4})(-2a^{-3}b^7) = \boxed{[3 \cdot (-2)](a^{1+(-3)}b^{-4+7})}$

            $= -6a^{-2}b^3$

            $= -\dfrac{6b^3}{a^2}$

- When multiplying expressions, add the exponents on like bases. Do this step mentally.

➡ Simplify: $\dfrac{4a^{-2}b^5}{6a^5b^2}$

$\dfrac{4a^{-2}b^5}{6a^5b^2} = \dfrac{\not{2} \cdot 2a^{-2}b^5}{\not{2} \cdot 3a^5b^2} = \dfrac{2a^{-2}b^5}{3a^5b^2}$

       $= \dfrac{2a^{-2-5}b^{5-2}}{3}$

       $= \dfrac{2a^{-7}b^3}{3} = \dfrac{2b^3}{3a^7}$

- Divide the coefficients by their common factor.
- Use the Rule for Dividing Exponential Expressions.
- Use the Definition of Negative Exponents to rewrite the expression with a positive exponent.

➡ Simplify: $\left[\dfrac{6m^2n^3}{8m^7n^2}\right]^{-3}$

$\left[\dfrac{6m^2n^3}{8m^7n^2}\right]^{-3} = \left[\dfrac{3m^{2-7}n^{3-2}}{4}\right]^{-3}$

       $= \left[\dfrac{3m^{-5}n}{4}\right]^{-3}$

       $= \dfrac{3^{-3}m^{15}n^{-3}}{4^{-3}}$

       $= \dfrac{4^3m^{15}}{3^3n^3} = \dfrac{64m^{15}}{27n^3}$

- Simplify inside the brackets.
- Subtract the exponents.
- Use the Rule for Simplifying Powers of Quotients.
- Use the Definition of Negative Exponents to rewrite the expression with a positive exponent. Then simplify.

Example 1   Simplify: $\dfrac{-35y^5}{25y^9}$

Solution

$$\dfrac{-35y^5}{25y^9} = \dfrac{-\cancel{5} \cdot 7y^{5-9}}{\cancel{5} \cdot 5}$$

$$= \dfrac{-7y^{-4}}{5}$$

$$= -\dfrac{7}{5y^4}$$

You Try It 1   Simplify: $\dfrac{18y^3}{-27y^7}$

Your solution

Example 2   Simplify: $(-2x)(3x^{-2})^{-3}$

Solution

$$(-2x)(3x^{-2})^{-3} = (-2x)(3^{-3}x^6)$$

$$= \dfrac{-2x^{1+6}}{3^3}$$

$$= -\dfrac{2x^7}{27}$$

You Try It 2   Simplify: $(-2x^2)(x^{-3}y^{-4})^{-2}$

Your solution

Example 3   Simplify: $\dfrac{(2r^2t^{-1})^{-3}}{(r^{-3}t^4)^2}$

Solution

$$\dfrac{(2r^2t^{-1})^{-3}}{(r^{-3}t^4)^2} = \dfrac{2^{-3}r^{-6}t^3}{r^{-6}t^8}$$

$$= 2^{-3}r^{-6-(-6)}t^{3-8}$$

$$= 2^{-3}r^0t^{-5}$$

$$= \dfrac{1}{2^3t^5}$$

$$= \dfrac{1}{8t^5}$$

You Try It 3   Simplify: $\dfrac{(6a^{-2}b^3)^{-1}}{(4a^3b^{-2})^{-2}}$

Your solution

Example 4   Simplify: $\left[\dfrac{4a^{-2}b^3}{6a^4b^{-2}}\right]^{-3}$

Solution

$$\left[\dfrac{4a^{-2}b^3}{6a^4b^{-2}}\right]^{-3} = \left[\dfrac{2a^{-6}b^5}{3}\right]^{-3}$$

$$= \dfrac{2^{-3}a^{18}b^{-15}}{3^{-3}}$$

$$= \dfrac{27a^{18}}{8b^{15}}$$

You Try It 4   Simplify: $\left[\dfrac{6r^3s^{-3}}{9r^3s^{-1}}\right]^{-2}$

Your solution

Solutions on p. A14

Objective B   *To write a number in scientific notation* ....

Very large and very small numbers abound in the natural sciences. For example, the mass of an electron is 0.00000000000000000000000000000911 kg. Numbers such as this are difficult to read, so a more convenient system called **scientific notation** is used. In scientific notation, a number is expressed as the product of two factors, one a number between 1 and 10, and the other a power of ten.

To express a number in scientific notation, write it in the form $a \times 10^n$, where a is a number between 1 and 10 and n is an integer.

For numbers greater than or equal to 10, move the decimal point to the right of the first digit. The exponent n is positive and equal to the number of places the decimal point has been moved.

$$240{,}000 = 2.4 \times 10^5$$
$$93{,}000{,}000 = 9.3 \times 10^7$$

For numbers less than 1, move the decimal point to the right of the first nonzero digit. The exponent n is negative. The absolute value of the exponent is equal to the number of places the decimal point has been moved.

$$0.0003 = 3.0 \times 10^{-4}$$
$$0.0000832 = 8.32 \times 10^{-5}$$

Changing a number written in scientific notation to decimal notation also requires moving the decimal point.

When the exponent is positive, move the decimal point to the right the same number of places as the exponent.

$$3.45 \times 10^6 = 3{,}450{,}000$$
$$2.3 \times 10^8 = 230{,}000{,}000$$

When the exponent is negative, move the decimal point to the left the same number of places as the absolute value of the exponent.

$$8.1 \times 10^{-3} = 0.0081$$
$$6.34 \times 10^{-7} = 0.000000634$$

Example 5   One picosecond is 0.000000001 of a second. Write one picosecond in scientific notation.

Solution $0.000000001 = 1.0 \times 10^{-9}$

You Try It 5   Light travels approximately 16,000,000,000,000 mi in one day. Write this number in scientific notation.

Your solution

Solution on p. A14

5.4 Exercises

· ·

___

Objective A

Evaluate.

1. 5^{-2} **2.** 3^{-3} **3.** $\dfrac{1}{8^{-2}}$ **4.** $\dfrac{1}{12^{-1}}$

5. $\dfrac{3^{-2}}{3}$ **6.** $\dfrac{5^{-3}}{5}$ **7.** $\dfrac{2^{-2}}{2^{-3}}$ **8.** $\dfrac{3^2}{3^2}$

Simplify.

9. x^{-2} **10.** y^{-10} **11.** $\dfrac{1}{a^{-6}}$ **12.** $\dfrac{1}{b^{-4}}$

13. $4x^{-7}$ **14.** $-6y^{-1}$ **15.** $\dfrac{2}{3}z^{-2}$ **16.** $\dfrac{4}{5}a^{-4}$

17. $\dfrac{5}{b^{-8}}$ **18.** $\dfrac{-3}{v^{-3}}$ **19.** $\dfrac{1}{3x^{-2}}$ **20.** $\dfrac{2}{5c^{-6}}$

21. $(ab^5)^0$ **22.** $(32x^3y^4)^0$ **23.** $-(3p^2q^5)^0$ **24.** $-\left(\dfrac{2}{3}xy\right)^0$

25. $\dfrac{y^7}{y^3}$ **26.** $\dfrac{z^9}{z^2}$ **27.** $\dfrac{a^8}{a^5}$ **28.** $\dfrac{c^{12}}{c^5}$

29. $\dfrac{p^5}{p}$ **30.** $\dfrac{w^9}{w}$ **31.** $\dfrac{4x^8}{2x^5}$ $2x$ **32.** $\dfrac{12z^7}{4z^3}$

33. $\dfrac{22k^5}{11k^4}$ **34.** $\dfrac{14m^{11}}{7m^{10}}$ **35.** $\dfrac{m^9n^7}{m^4n^5}$ **36.** $\dfrac{y^5z^6}{yz^3}$

37. $\dfrac{6r^4}{4r^2}$ **38.** $\dfrac{8x^9}{12x^6}$ **39.** $\dfrac{-16a^7}{24a^6}$ **40.** $\dfrac{-18b^5}{27b^4}$

41. $\dfrac{y^3}{y^8}$ **42.** $\dfrac{z^4}{z^6}$ **43.** $\dfrac{a^5}{a^{11}}$ **44.** $\dfrac{m}{m^7}$

Simplify.

45. $\dfrac{4x^2}{12x^5}$

46. $\dfrac{6y^8}{8y^9}$

47. $\dfrac{-12x}{-18x^6}$

48. $\dfrac{-24c^2}{-36c^{11}}$

49. $\dfrac{x^6y^5}{x^8y}$

50. $\dfrac{a^3b^2}{a^2b^3}$

51. $\dfrac{2m^6n^2}{5m^9n^{10}}$

52. $\dfrac{5r^3t^7}{6r^5t^7}$

53. $\dfrac{pq^3}{p^4q^4}$

54. $\dfrac{a^4b^5}{a^5b^6}$

55. $\dfrac{3x^4y^5}{6x^4y^8}$

56. $\dfrac{14a^3b^6}{21a^5b^6}$

57. $\dfrac{14x^4y^6z^2}{16x^3y^9z}$

58. $\dfrac{24a^2b^7c^9}{36a^7b^5c}$

59. $\dfrac{15mn^9p^3}{30m^4n^9p}$

60. $\dfrac{25x^4y^7z^2}{20x^5y^9z^{11}}$

61. $(-2xy^{-2})^3$

62. $(-3x^{-1}y^2)^2$

63. $(3x^{-1}y^{-2})^2$

64. $(5xy^{-3})^{-2}$

65. $(2x^{-1})(x^{-3})$

66. $(-2x^{-5})x^7$

67. $(-5a^2)(a^{-5})^2$

68. $(2a^{-3})(a^7b^{-1})^3$

69. $(-2ab^{-2})(4a^{-2}b)^{-2}$

70. $(3ab^{-2})(2a^{-1}b)^{-3}$

71. $(-5x^{-2}y)(-2x^{-2}y^2)$

72. $\dfrac{a^{-3}b^{-4}}{a^2b^2}$

73. $\dfrac{3x^{-2}y^2}{6xy^2}$

74. $\dfrac{2x^{-2}y}{8xy}$

75. $\dfrac{3x^{-2}y}{xy}$

76. $\dfrac{2x^{-1}y^4}{x^2y^3}$

77. $\dfrac{2x^{-1}y^{-4}}{4xy^2}$

78. $\dfrac{(x^{-1}y)^2}{xy^2}$

79. $\dfrac{(x^{-2}y)^2}{x^2y^3}$

80. $\dfrac{(x^{-3}y^{-2})^2}{x^6y^8}$

81. $\dfrac{(a^{-2}y^3)^{-3}}{a^2y}$

82. $\dfrac{12a^2b^3}{-27a^2b^2}$

83. $\dfrac{-16xy^4}{96x^4y^4}$

84. $\dfrac{-8x^2y^4}{44y^2z^5}$

Simplify.

85. $\dfrac{22a^2b^4}{-132b^3c^2}$

86. $\dfrac{-(8a^2b^4)^3}{64a^3b^8}$

87. $\dfrac{-(14ab^4)^2}{28a^4b^2}$

88. $\dfrac{(2a^{-2}b^3)^{-2}}{(4a^2b^{-4})^{-1}}$

89. $\dfrac{(3^{-1}r^4s^{-3})^{-2}}{(6r^2t^{-2}s^{-1})^2}$

90. $\left(\dfrac{6x^{-4}yz^{-1}}{14xy^{-4}z^2}\right)^{-3}$

91. $\left(\dfrac{15m^3n^{-2}p^{-1}}{25m^{-2}n^{-4}}\right)^{-3}$

92. $\left(\dfrac{18a^4b^{-2}c^4}{12ab^{-3}d^2}\right)^{-2}$

Objective B

Write in scientific notation.

93. 0.00000000324 3.24×10^{-9}

94. 0.00000012

95. 0.000000000000000003

96. 1,800,000,000

97. 32,000,000,000,000,000

98. 76,700,000,000,000

99. 0.000000000000000000122

100. 0.00137

101. 547,000,000.

5.47×10^{8}

Write in decimal notation.

102. 2.3×10^{-12}

103. 1.67×10^{-4}

104. 2×10^{15}

105. 6.8×10^{7}

106. 9×10^{-21}

107. 3.05×10^{-5}

108. 9.05×10^{11}

109. 1.02×10^{-9}

110. 7.2×10^{-3}

111. In 1994, the national debt of the United States was approximately 4.3 trillion dollars. Write the national debt in scientific notation.

112. Avogadro's number is used in chemistry, and its value is approximately 602,300,000,000,000,000,000,000. Express this number in scientific notation.

113. 5,980,000,000,000,000,000,000,000 kg is the approximate mass of the earth. Write the mass of the earth in scientific notation.

114. A parsec is a distance measurement that is used by astronomers. One parsec is 3,086,000,000,000,000,000 cm. Write this number in scientific notation.

115. The electric charge on an electron is 0.00000000000000000016 coulomb. Write this number in scientific notation.

116. The length of an ultraviolet light wave is approximately 0.0000037 m. Write this number in scientific notation.

117. The area of the earth is approximately 5.1×10^{14} m². Write this number in decimal notation.

118. Approximately 35 teragrams (35×10^{12} grams) of sulfur in the atmosphere is converted to sulfate each year. Write this number in decimal notation.

APPLYING THE CONCEPTS

119. Evaluate 2^x when $x = -2, -1, 0, 1,$ and 2.

120. Evaluate 3^x when $x = -2, -1, 0, 1,$ and 2.

121. Evaluate 2^{-x} when $x = -2, -1, 0, 1,$ and 2.

122. Evaluate 3^{-x} when $x = -2, -1, 0, 1,$ and 2.

123. Evaluate 2^{-x^2} when $x = -2, -1, 0, 1,$ and 2.

124. Evaluate 3^{-x^2} when $x = -2, -1, 0, 1,$ and 2.

Determine whether each equation for exercises 125–130 is true or false. If the equation is false, change the right-hand side of the equation to make a true equation.

125. $(2a)^{-3} = \dfrac{2}{a^3}$

126. $((a^{-1})^{-1})^{-1} = \dfrac{1}{a}$

127. $\left(\dfrac{a}{b}\right)^2 = \left(\dfrac{b}{a}\right)^{-2}$

128. $\dfrac{x^{-3}}{y^{-3}} = \left(\dfrac{x}{y}\right)^{-3}$

129. $(2 + 3)^{-1} = 2^{-1} + 3^{-1}$

130. If $x \neq \dfrac{1}{3}$, then $(3x - 1)^0 = (1 - 3x)^0$.

131. Simplify: $\left(\dfrac{6x^4yz^3}{2x^2y^3}\right)\left(\dfrac{2x^2z^3}{4y^2z}\right) \div \left(\dfrac{6x^2y^3}{x^4y^2z}\right)$

132. Simplify: $\left(\dfrac{x^3yz^2}{3x^3y^2z^2}\right)\left(\dfrac{2x^3y^2z^2}{2y^2z}\right) \div \left(\dfrac{6y^2z^3}{x^3y^4z}\right)$

133.
[W] If x is a non-zero real number, is x^{-2} always positive, always negative, or positive or negative depending on whether x is positive or negative? Explain your answer.

134.
[W] If x is a nonzero real number, is x^{-3} always positive, always negative, or positive or negative depending on whether x is positive or negative? Explain your answer.

135.
[W] Why is the condition $x \neq \dfrac{1}{3}$ given in Exercise 130?

5.5 Division of Polynomials

Objective A   To divide a polynomial by a monomial ....

To divide a polynomial by a monomial, divide each term in the numerator by the denominator and write the sum of the quotients.

➡ Simplify: $\dfrac{6x^3 - 3x^2 + 9x}{3x}$

$$\dfrac{6x^3 - 3x^2 + 9x}{3x} = \dfrac{6x^3}{3x} - \dfrac{3x^2}{3x} + \dfrac{9x}{3x}$$

  • Divide each term of the polynomial by the monomial.

$$= 2x^2 - x + 3$$

  • Simplify each expression.

Example 1   Simplify: $\dfrac{12x^2y - 6xy + 4x^2}{2xy}$

You Try It 1   Simplify: $\dfrac{24x^2y^2 - 18xy + 6y}{6xy}$

Solution

$$\dfrac{12x^2y - 6xy + 4x^2}{2xy} = \dfrac{12x^2y}{2xy} - \dfrac{6xy}{2xy} + \dfrac{4x^2}{2xy} = 6x - 3 + \dfrac{2x}{y}$$

Your solution

Solution on p. A14

Objective B   To divide polynomials ....

The procedure for dividing two polynomials is similar to the one for dividing whole numbers.

➡ Simplify: $(x^2 - 5x + 8) \div (x - 3)$

$$
\begin{array}{r}
x \\
x - 3 \overline{)\,x^2 - 5x + 8} \\
\underline{x^2 - 3x} \\
-2x + 8
\end{array}
$$

  • Think: $x)\overline{x^2} = \dfrac{x^2}{x} = x$
  • Multiply: $x(x - 3) = x^2 - 3x$
  • Subtract: $(x^2 - 5x) - (x^2 - 3x) = -2x$
    Bring down the 8.

$$
\begin{array}{r}
x - 2 \\
x - 3 \overline{)\,x^2 - 5x + 8} \\
\underline{x^2 - 3x} \\
-2x + 8 \\
\underline{-2x + 6} \\
2
\end{array}
$$

  • Think: $x)\overline{-2x} = \dfrac{-2x}{x} = -2$
  • Multiply: $-2(x - 3) = -2x + 6$
  • Subtract: $(-2x + 8) - (-2x + 6) = 2$
  • The remainder is 2.

$$(x^2 - 5x - 8) \div (x - 3) = x - 2 + \dfrac{2}{x - 3}$$

The same equation used to check division of whole numbers is used to check division of polynomials.

(Quotient × divisor) + remainder = dividend

Check: $(x - 2)(x - 3) + 2 = x^2 - 5x + 6 + 2 = x^2 - 5x + 8$

If a power of the dividend is missing from a polynomial, a zero can be inserted for that term. This helps keep like terms in the same column.

➡ Simplify: $(6x + 26 + 2x^3) \div (2 + x)$

$(2x^3 + 6x + 26) \div (x + 2)$

$$
\begin{array}{r}
2x^2 - 4x + 14 \\
x + 2 \overline{)2x^3 + 0 + 6x + 26} \\
\underline{2x^3 + 4x^2} \\
-4x^2 + 6x \\
\underline{-4x^2 - 8x} \\
14x + 26 \\
\underline{14x + 28} \\
-2
\end{array}
$$

- Arrange the terms of each polynomial in descending order.

- There is no x^2 term in $2x^3 + 6x + 26$. Insert a zero for the missing term.

$(2x^3 + 6x + 26) \div (x + 2) = 2x^2 - 4x + 14 - \dfrac{2}{x + 2}$

Check:

$(2x^2 - 4x + 14)(x + 2) + (-2) = (2x^3 + 6x + 28) + (-2) = 2x^3 + 6x + 26$

Example 2
Simplify: $(8x^2 + 4x^3 + x - 4) \div (2x + 3)$

Solution

$$
\begin{array}{r}
2x^2 + x - 1 \\
2x + 3 \overline{)4x^3 + 8x^2 + x - 4} \\
\underline{4x^3 + 6x^2} \\
2x^2 + x \\
\underline{2x^2 + 3x} \\
-2x - 4 \\
\underline{-2x - 3} \\
-1
\end{array}
$$

- Write the dividend in descending powers of x.

$(4x^3 + 8x^2 + x - 4) \div (2x + 3)$
$= 2x^2 + x - 1 - \dfrac{1}{2x + 3}$

You Try It 2
Simplify: $(2x^3 + x^2 - 8x - 3) \div (2x - 3)$

Your solution

Example 3
Simplify: $(x^2 - 1) \div (x + 1)$

Solution

$$
\begin{array}{r}
x - 1 \\
x + 1 \overline{)x^2 + 0 - 1} \\
\underline{x^2 + x} \\
-x - 1 \\
\underline{-x - 1} \\
0
\end{array}
$$

- Insert a zero for the missing term.

$(x^2 - 1) \div (x + 1) = x - 1$

You Try It 3
Simplify: $(x^3 - 2x + 1) \div (x - 1)$

Your solution

Solutions on p. A14

5.5 Exercises

· ·

Objective A

Simplify.

1. $\dfrac{10a - 25}{5}$

2. $\dfrac{16b - 40}{8}$

3. $\dfrac{6y^2 + 4y}{y}$

4. $\dfrac{4b^3 - 3b}{b}$

5. $\dfrac{3x^2 - 6x}{3x}$

6. $\dfrac{10y^2 - 6y}{2y}$

7. $\dfrac{5x^2 - 10x}{-5x}$

8. $\dfrac{3y^2 - 27y}{-3y}$

9. $\dfrac{x^3 + 3x^2 - 5x}{x}$

10. $\dfrac{a^3 - 5a^2 + 7a}{a}$

11. $\dfrac{x^6 - 3x^4 - x^2}{x^2}$

12. $\dfrac{a^8 - 5a^5 - 3a^3}{a^2}$

13. $\dfrac{5x^2y^2 + 10xy}{5xy}$

14. $\dfrac{8x^2y^2 - 24xy}{8xy}$

15. $\dfrac{9y^6 - 15y^3}{-3y^3}$

16. $\dfrac{4x^4 - 6x^2}{-2x^2}$

17. $\dfrac{3x^2 - 2x + 1}{x}$

18. $\dfrac{8y^2 + 2y - 3}{y}$

19. $\dfrac{-3x^2 + 7x - 6}{x}$

20. $\dfrac{2y^2 - 6y + 9}{y}$

21. $\dfrac{16a^2b - 20ab + 24ab^2}{4ab}$

22. $\dfrac{22a^2b - 11ab - 33ab^2}{11ab}$

23. $\dfrac{9x^2y + 6xy - 3xy^2}{xy}$

24. $\dfrac{5a^2b - 15ab + 30ab^2}{5ab}$

Objective B

Simplify.

25. $(b^2 - 14b + 49) \div (b - 7)$

26. $(x^2 - x - 6) \div (x - 3)$

27. $(y^2 + 2y - 35) \div (y + 7)$

28. $(2x^2 + 5x + 2) \div (x + 2)$

29. $(2y^2 - 13y + 21) \div (y - 3)$

30. $(4x^2 - 16) \div (2x + 4)$

31. $(2y^2 + 7) \div (y - 3)$ **32.** $(x^2 + 1) \div (x - 1)$ **33.** $(x^2 + 4) \div (x + 2)$

34. $(6x^2 - 7x) \div (3x - 2)$ **35.** $(6y^2 + 2y) \div (2y + 4)$ **36.** $(5x^2 + 7x) \div (x - 1)$

37. $(6x^2 - 5) \div (x + 2)$ **38.** $(a^2 + 5a + 10) \div (a + 2)$ **39.** $(b^2 - 8b - 9) \div (b - 3)$

40. $(2y^2 - 9y + 8) \div (2y + 3)$ **41.** $(3x^2 + 5x - 4) \div (x - 4)$

42. $(8x + 3 + 4x^2) \div (2x - 1)$ **43.** $(10 + 21y + 10y^2) \div (2y + 3)$

44. $(15a^2 - 8a - 8) \div (3a + 2)$ **45.** $(12a^2 - 25a - 7) \div (3a - 7)$

46. $(5 - 23x + 12x^2) \div (4x - 1)$ **47.** $(24 + 6a^2 + 25a) \div (3a - 1)$

48. $(x^3 + 3x^2 + 5x + 3) \div (x + 1)$ **49.** $(x^3 - 6x^2 + 7x - 2) \div (x - 1)$

50. $(x^4 - x^2 - 6) \div (x^2 + 2)$ **51.** $(x^4 + 3x^2 - 10) \div (x^2 - 2)$

APPLYING THE CONCEPTS

52. In your own words, explain how to divide exponential expressions.
[W]

53. The product of a monomial and $4b$ is $12ab^2$. Find the monomial.

54. The quotient of a polynomial and $x - 3$ is $x^2 - x + 8 + \dfrac{22}{x - 3}$. Find the polynomial.

Project in Mathematics

Intensity of Illumination

The rate at which light falls upon a one-square-unit area of surface is called the **intensity of illumination.** Intensity of illumination is measured in **lumens** (lm). A lumen is defined in the following illustration.

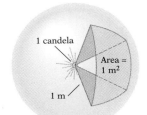

Picture a source of light equal to one candela (1 cd) positioned at the center of a hollow sphere that has a radius of 1 m. The rate at which light falls upon 1 m² of the inner surface of the sphere is equal to one lumen. If a light source equal to 4 cd is positioned at the center of the sphere, each square meter of the inner surface receives four times as much illumination, or 4 lm.

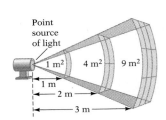

Light rays diverge as they leave a light source. The light that falls upon an area of 1 m² at a distance of 1 m from the source of light spreads out over an area of 4 m² when it is 2 m from the source. The same light spreads out over an area of 9 m² when it is 3 m from the light source and over an area of 16 m² when it is 4 m from the light source. Therefore, as a surface moves farther away from the source of light, the intensity of illumination on the surface decreases from its value at 1 m to $\left(\frac{1}{2}\right)^2$, or $\frac{1}{4}$, that value at 2 m; to $\left(\frac{1}{3}\right)^2$, or $\frac{1}{9}$, that value at 3 m; and to $\left(\frac{1}{4}\right)^2$, or $\frac{1}{16}$, that value at 4 m. The formula for the intensity of illumination is

$$I = \frac{s}{r^2}$$

where I is the intensity of illumination, s is the strength of the light source in candelas, and r is the distance in meters between the light source and the illuminated surface.

A 30-s lamp is 0.5 m above a desk. Find the illumination on the desk.

$$I = \frac{s}{r^2} = \frac{30}{(0.5)^2} = 120$$

The illumination on the desk is 120 lm.

Exercises:

Solve.

1. A 100-cd light is hanging 5 m above a floor. What is the intensity of illumination on the floor beneath it?
2. A 25-cd source of light is 2 m above a desk. Find the intensity of illumination on the desk.
3. How strong a light source is needed to cast 80 lm of light on a surface 5 m from the source?
4. Two lights cast the same intensity of illumination on a wall. One light is 6 m from the wall and has a rating of 36 cd. The second light is 8 m from the wall. Find the candela rating of the second light.

Chapter Summary

Key Words   A *monomial* is a number, a variable, or a product of numbers and variables.

A *polynomial* is a variable expression in which the terms are monomials.

A *monomial* is a polynomial of *one* term.

A *binomial* is a polynomial of *two* terms.

A *trinomial* is a polynomial of *three* terms.

The *degree of a polynomial* in one variable is the largest exponent on a variable.

Essential Rules   *Rule for Multiplying Exponential Expressions*
If m and n are integers, then $x^m \cdot x^n = x^{m+n}$.

Rule for Simplifying Powers of Exponential Expressions
If m and n are integers, then $(x^m)^n = x^{m \cdot n}$.

Rule For Simplifying Powers of Products
If m, n, and p are integers, then $(x^m \cdot y^n)^p = x^{m \cdot p} y^{n \cdot p}$.

Rule for Zero as an Exponent
If $x \neq 0$, then $x^0 = 1$.

Rule for Negative Exponents
If n is a positive integer and $x \neq 0$, then $x^{-n} = \dfrac{1}{x^n}$.

Rule for Simplifying Powers of Quotients
If m, n, and p are integers, then $\left(\dfrac{x^m}{y^n}\right)^p = \dfrac{x^{m \cdot p}}{y^{n \cdot p}}$, $y \neq 0$.

Rule for Dividing Exponential Expressions
If m and n are integers and $x \neq 0$, then $\dfrac{x^m}{x^n} = x^{m-n}$.

The Sum and Difference of Two Terms
$(a + b)(a - b) = a^2 - b^2$

The Square of a Binomial
$(a + b)^2 = a^2 + 2ab + b^2$
$(a - b)^2 = a^2 - 2ab + b^2$

Dividend = (quotient × divisor) + remainder

Chapter Review

. . . .

SECTION 5.1

1. Simplify: $(12y^2 + 17y - 4) + (9y^2 - 13y + 3)$

$9y^2 - 13y + 3$

$21y^2 + 4y - 1$

2. Simplify: $(2x^3 + 7x^2 + x) + (2x^2 - 4x - 12)$

3. Simplify: $(5a^2 + 6a - 11) + (5a^2 + 6a - 11)$

4. Simplify: $(5x^2 - 2x - 1) - (3x^2 - 5x + 7)$

5. Simplify: $(13y^3 - 7y - 2) - (12y^2 - 2y - 1)$

6. Simplify: $(6y^2 + 2y + 7) - (8y^2 + y + 12)$

SECTION 5.2

7. Simplify: $(5xy^2)(-4x^2y^3)$

8. Simplify: $(xy^5z^3)(x^3y^3z)$

9. Simplify: $(2a^{12}b^3)(-9b^2c^6)(3ac)$

10. Simplify: $(2^3)^2$

11. Simplify: $(-3x^2y^3)^2$

12. Simplify: $(5a^7b^6)^2(4ab)$

SECTION 5.3

13. Simplify: $-2x(4x^2 + 7x - 9)$

14. Simplify: $2ab^3(4a^2 - 2ab + 3b^2)$

15. Simplify: $(3y^2 + 4y - 7)(2y + 3)$

16. Simplify: $(6b^3 - 2b^2 - 5)(2b^2 - 1)$

17. Simplify: $(5a - 7)(2a + 9)$

18. Simplify: $(2b - 3)(4b + 5)$

19. Simplify: $(a + 7)(a - 7)$

20. Simplify: $(5y - 7)^2$

21. The length of a ping-pong table is 1 ft less than twice the width of the table. Let w represent the width of the ping-pong table. Express the area of the ping-pong table in terms of the variable w.

22. The side of a square checkerboard is $(3x - 2)$ inches. Express the area of the checkboard in terms of the variable x.

SECTION 5.4

23. Simplify: $\dfrac{8x^{12}}{12x^9}$

$= \dfrac{2}{3} \dfrac{x^{12}}{x^9} = \dfrac{2}{3} x^{12-4} =$

$\dfrac{2}{3} x^3$

24. Simplify: $\dfrac{-18a^6b}{27a^3b^4}$

$\dfrac{-18a^6b}{27a^3b^4} \quad \dfrac{2}{3} a^{6-3} \dfrac{b^{-4}}{b^4}$

$\dfrac{2a^3b^{-4}}{3}$

$\dfrac{2}{3} a^3 \dfrac{1}{b^4}$

25. Simplify: $(-3x^{-2}y^{-3})^{-2}$

26. Simplify: $\dfrac{(4a^{-2}b^{-3})^2}{(2a^{-1}b^{-2})^4}$

27. Write 0.000000127 in scientific notation.

28. Write 3.2×10^{-12} in decimal notation.

SECTION 5.5

29. Simplify: $\dfrac{16y^2 - 32y}{-4y}$

30. Simplify: $\dfrac{12b^7 + 36b^5 - 3b^3}{3b^3}$

31. Simplify: $(6y^2 - 35y + 36) \div (3y - 4)$

32. Simplify: $(b^3 - 2b^2 - 33b - 7) \div (b - 7)$

Chapter Test

1. Simplify:
   $(3x^3 - 2x^2 - 4) + (8x^2 - 8x + 7)$

2. Simplify:
   $(3a^2 - 2a - 7) - (5a^3 + 2a - 10)$

3. Simplify: $(ab^2)(a^3b^5)$

4. Simplify: $(-2xy^2)(3x^2y^4)$

5. Write 0.00000000302 in scientific notation.

6. Simplify: $(-2a^2b)^3$

7. Simplify: $2x(2x^2 - 3x)$

8. Simplify: $-3y^2(-2y^2 + 3y - 6)$

9. Simplify: $(x - 3)(x^2 - 4x + 5)$

10. Simplify: $(-2x^3 + x^2 - 7)(2x - 3)$

11. Simplify: $(a - 2b)(a + 5b)$

12. Simplify: $(2x - 7y)(5x - 4y)$

13. Simplify: $(4y - 3)(4y + 3)$

14. Simplify: $(2x - 5)^2$

15. The radius of a circle is $(x - 5)$ meters. Use the equation $A = \pi r^2$, where r is the radius, to find the area of the circle in terms of the variable x. Use $\pi \approx 3.14$.

16. Simplify: $\dfrac{12x^2}{-3x^8}$

17. Simplify: $\dfrac{-(2x^2y)^3}{4x^3y^3}$

18. Simplify: $\dfrac{2a^{-1}b}{2^{-2}a^{-2}b^{-3}}$

19. Simplify: $\dfrac{(3x^{-2}y^3)^3}{3x^4y^{-1}}$

20. Simplify: $\dfrac{20a - 35}{5}$

21. Simplify: $\dfrac{16x^5 - 8x^3 + 20x}{4x}$

22. Simplify: $\dfrac{12x^3 - 3x^2 + 9}{3x^2}$

23. Simplify: $(x^2 + 6x - 7) \div (x - 1)$

24. Simplify: $(x^2 + 1) \div (x + 1)$

25. Simplify: $(4x^2 - 7) \div (2x - 3)$

Cumulative Review

1. Simplify: $\dfrac{3}{16} - \left(-\dfrac{5}{8}\right) - \dfrac{7}{9}$

2. Evaluate $-3^2 \cdot \left(\dfrac{2}{3}\right)^3 \cdot \left(-\dfrac{5}{8}\right)$.

3. Simplify: $\left(-\dfrac{1}{2}\right)^3 \div \left(\dfrac{3}{8} - \dfrac{5}{6}\right) + 2$

4. Evaluate $\dfrac{b - (a - b)^2}{b^2}$ when $a = -2$ and $b = 3$.

5. Simplify: $-2x - (-xy) + 7x - 4xy$

6. Simplify: $(12x)\left(-\dfrac{3}{4}\right)$

7. Simplify: $-2[3x - 2(4 - 3x) + 2]$

8. Solve: $12 = -\dfrac{3}{4}x$

9. Solve: $2x - 9 = 3x + 7$

10. Solve: $2 - 3(4 - x) = 2x + 5$

11. 35.2 is what percent of 160?

12. Simplify: $(4b^3 - 7b^2 - 7) + (3b^2 - 8b + 3)$

13. Simplify: $(3y^3 - 5y + 8) - (-2y^2 + 5y + 8)$

14. Simplify: $(a^3b^5)^3$

15. Simplify: $(4xy^3)(-2x^2y^3)$

16. Simplify: $-2y^2(-3y^2 - 4y + 8)$

17. Simplify: $(2a - 7)(5a^2 - 2a + 3)$

18. Simplify: $(3b - 2)(5b - 7)$

19. Simplify: $(3b + 2)^2$

20. Simplify: $\dfrac{(-2a^2b^3)^2}{8a^4b^8}$

21. Write 6.09×10^{-5} in decimal notation.

22. Simplify: $\dfrac{-18a^3 + 12a^2 - 6}{-3a^2}$

23. Simplify: $(a^2 - 4a - 21) \div (a + 3)$

24. Translate "the difference between eight times a number and twice a number is eighteen" into an equation and solve.

25. A calculator costs a retailer $24. Find the selling price when the markup rate is 80%.

26. Fifty ounces of pure orange juice are added to 200 oz of a fruit punch that is 10% orange juice. What is the percent concentration of orange juice in the resulting mixture?

27. A car traveling at 50 mph overtakes a cyclist who, riding at 10 mph, has had a 2-h head start. How far from the starting point does the car overtake the cyclist?

28. The width of a rectangle is 40% of the length. The perimeter of the rectangle is 42 m. Find the length and width of the rectangle.

29. Five times the first of two consecutive odd integers is thirteen more than four times the second. Find the two integers.

30. The length of a side of a square is $2x + 3$. Use the equation $A = s^2$, where s is the length of the side of a square, to find the area of the square in terms of the variable x.

6

Factoring

Objectives

Section 6.1
To factor a monomial from a polynomial
To factor by grouping

Section 6.2
To factor a trinomial of the form $x^2 + bx + c$
To factor completely

Section 6.3
To factor a trinomial of the form $ax^2 + bx + c$
    by using trial factors
To factor a trinomial of the form $ax^2 + bx + c$
    by grouping

Section 6.4
To factor the difference of two squares and
    perfect-square trinomials
To factor completely

Section 6.5
To solve equations by factoring
To solve application problems

Algebra from Geometry

The early Babylonians made substantial progress in both algebra and geometry. Often the progress they made in algebra was based on geometric concepts.

Here are some geometric proofs of algebraic identities the Babylonians understood.

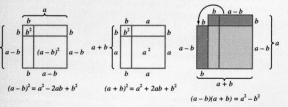

$$(a-b)^2 = a^2 - 2ab + b^2$$

$$(a+b)^2 = a^2 + 2ab + b^2$$

$$(a-b)(a+b) = a^2 - b^2$$

Common Factors

Objective A   *To factor a monomial from a polynomial* ....

The **greatest common factor (GCF)** of two or more monomials is the product of the GCF of the coefficients and the common variable factors.

$$6x^3y = 2 \cdot 3 \cdot x \cdot x \cdot x \cdot y$$
$$8x^2y^2 = 2 \cdot 2 \cdot 2 \cdot x \cdot x \cdot y \cdot y$$
$$GCF = 2 \cdot x \cdot x \cdot y = 2x^2y$$

Note that the exponent of each variable in the GCF is the same as the *smallest* exponent of that variable in either of the monomials.

The GCF of $6x^3y$ and $8x^2y^2$ is $2x^2y$

➡ Find the GCF of $12a^4b$ and $18a^2b^2c$.

The common variable factors are a^2 and b. c is not a common variable factor.

$$12a^4b = 2 \cdot 2 \cdot 3 \cdot a^4 \cdot b$$
$$18a^2b^2c = 2 \cdot 3 \cdot 3 \cdot a^2 \cdot b^2 \cdot c$$
$$GCF = 2 \cdot 3 \cdot a^2 \cdot b = 6a^2b$$

To **factor** a polynomial means to write the polynomial as a product of other polynomials.

Multiply ←

Polynomial   =   **Factors**
$2x^2 + 10x$ $2x(x + 5)$

→ Factor

In the example above, $2x$ is the GCF of the terms $2x^2$ and $10x$. It is a **common factor** of the terms.

➡ Factor: $5x^3 - 35x^2 + 10x$
Find the GCF of the terms of the polynomial.

$$5x^3 = 5 \cdot x^3 \qquad 35x^2 = 5 \cdot 7 \cdot x^2 \qquad 10x = 2 \cdot 5 \cdot x$$
The GCF is $5x$.

Rewrite the polynomial, expressing each term as a product with the GCF as one of the factors.

$$5x^3 - 35x^2 + 10x = 5x(x^2) + 5x(-7x) + 5x(2)$$
$$= 5x(x^2 - 7x + 2)$$

● Use the Distributive Property to write the polynomial as a product of factors.

LOOK CLOSELY
The factors in color are determined by dividing each term of the trinomial by the GCF, $5x$. For instance,

$$\left(\frac{5x^3}{5x}\right) = x^2,$$

$$\left(\frac{-35x^2}{5x}\right) = -7x, \text{ and}$$

$$\left(\frac{10x}{5x}\right) = 2$$

Content and Format © 1995 HMCo.

Example 1
Factor: $8x^2 + 2xy$

You Try It 1
Factor: $14a^2 - 21a^4b$

Solution
The GCF is $2x$.

$8x^2 + 2xy = 2x(4x) + 2x(y) = 2x(4x + y)$

Your solution

Example 2
Factor: $n^3 - 5n^2 + 2n$

You Try It 2
Factor: $27b^2 + 18b + 9$

Solution
The GCF is n.

$n^3 - 5n^2 + 2n = n(n^2) + n(-5n) + n(2)$
$\qquad\qquad\qquad = n(n^2 - 5n + 2)$

Your solution

Example 3
Factor: $16x^2y + 8x^4y^2 - 12x^4y^5$

You Try It 3
Factor: $6x^4y^2 - 9x^3y^2 + 12x^2y^4$

Solution
The GCF is $4x^2y$.

$16x^2y + 8x^4y^2 - 12x^4y^5$
$\quad = 4x^2y(4) + 4x^2y(2x^2y) + 4x^2y(-3x^2y^4)$
$\quad = 4x^2y(4 + 2x^2y - 3x^2y^4)$

Your solution

Solutions on p. A15

Objective B   *To factor by grouping* ....

In the examples at the right, the binomials in parentheses are called **binomial factors.**

$$2a(a + b)^2$$
$$3xy(x - y)$$

The Distributive Property is used to factor a common binomial factor from an expression.

① The common binomial factor of the expression $6x(x - 3) + y^2(x - 3)$ is $(x - 3)$. To factor that expression, use the Distributive Property to write the expression as a product of factors.

$$6x(x - 3) + y^2(x - 3) = (x - 3)(6x + y^2)$$

② Consider the following simplification of $-(a - b)$.

$$-(a - b) = -1(a - b) = -a + b = b - a$$

Thus, $$b - a = -(a - b)$$

This equation is sometimes used to factor a common binomial from an expression.

➡ Factor: $2x(x - y) + 5(y - x)$

$$2x(x - y) + 5(y - x) = 2x(x - y) - 5(x - y)$$
$$= (x - y)(2x - 5)$$

• $5(y - x) = 5[(-1)(x - y)]$
    $= -5(x - y)$

Some polynomials can be factored by grouping terms in such a way that a common binomial factor is found.

➡ Factor: $ax + by - ay - by$

$$ax + bx - ay - by = (ax + bx) - (ay + by)$$

• Group the first two terms and the last two terms. Note that $-ay - by = -(ay + by)$.

$$= x(a + b) - y(a + b)$$
$$= (a + b)(x - y)$$

• Factor the GCF, $(a + b)$, from each group.

➡ Factor: $6x^2 - 9x - 4xy + 6y$

$$6x^2 - 9x - 4xy + 6y = (6x^2 - 9x) - (4xy - 6y)$$

• Group the first two terms and the last two terms. Note that $-4xy + 6y = -(4xy - 6y)$.

$$= 3x(2x - 3) - 2y(2x - 3)$$
$$= (2x - 3)(3x - 2y)$$

• Factor the GCF, $(2x - 3)$, from each group.

Example 4
Factor: $4x(3x - 2) - 7(3x - 2)$

Solution
$4x(3x - 2) - 7(3x - 2)$
$\quad = (3x - 2)(4x - 7)$

You Try It 4
Factor: $2y(5x - 2) - 3(2 - 5x)$

Your solution

Example 5
Factor: $9x^2 - 15x - 6xy + 10y$

Solution
$9x^2 - 15x - 6xy + 10y$
$\quad = (9x^2 - 15x) - (6xy - 10y)$
$\quad = 3x(3x - 5) - 2y(3x - 5)$
$\quad = (3x - 5)(3x - 2y)$

You Try It 5
Factor: $a^2 - 3a + 2ab - 6b$

Your solution

Example 6
Factor: $3x^2y - 4x - 15xy + 20$

Solution
$3x^2y - 4x - 15xy + 20$
$\quad = (3x^2y - 4x) - (15xy - 20)$
$\quad = x(3xy - 4) - 5(3xy - 4)$
$\quad = (3xy - 4)(x - 5)$

You Try It 6
Factor: $2mn^2 - n + 8mn - 4$

Your solution

Solutions on p. A15

6.1 Exercises

Objective A

Factor.

1. $5a + 5$

2. $7b - 7$

3. $16 - 8a^2$

4. $12 + 12y^2$

5. $8x + 12$

6. $16a - 24$

7. $30a - 6$

8. $20b + 5$

9. $7x^2 - 3x$

10. $12y^2 - 5y$

11. $3a^2 + 5a^5$

12. $9x - 5x^2$

13. $14y^2 + 11y$

14. $6b^3 - 5b^2$

15. $2x^4 - 4x$

16. $3y^4 - 9y$

17. $10x^4 - 12x^2$

18. $12a^5 - 32a^2$

19. $8a^8 - 4a^5$

20. $16y^4 - 8y^7$

21. $x^2y^2 - xy$

22. $a^2b^2 + ab$

23. $3x^2y^4 - 6xy$

24. $12a^2b^5 - 9ab$

25. $x^2y - xy^3$

26. $3x^3 + 6x^2 + 9x$

27. $5y^3 - 20y^2 + 10y$

28. $2x^4 - 4x^3 + 6x^2$

29. $3y^4 - 9y^3 - 6y^2$

30. $2x^3 + 6x^2 - 14x$

31. $3y^3 - 9y^2 + 24y$

32. $2y^5 - 3y^4 + 7y^3$

33. $6a^5 - 3a^3 - 2a^2$

34. $x^3y - 3x^2y^2 + 7xy^3$

35. $2a^2b - 5a^2b^2 + 7ab^2$

36. $5y^3 + 10y^2 - 25y$

37. $4b^5 + 6b^3 - 12b$

38. $3a^2b^2 - 9ab^2 + 15b^2$

39. $8x^2y^2 - 4x^2y + x^2$

Objective B

Factor.

40. $x(b + 4) + 3(b + 4)$

41. $y(a + z) + 7(a + z)$

42. $a(y - x) - b(y - x)$

43. $3r(a - b) + s(a - b)$

44. $x(x - 2) + y(2 - x)$

45. $t(m - 7) + 7(7 - m)$

Factor.

46. $2x(7 + b) - y(b + 7)$

47. $2y(4a - b) - (b - 4a)$

48. $8c(2m - 3n) + (3n - 2m)$

49. $x^2 + 2x + 2xy + 4y$

50. $x^2 - 3x + 4ax - 12a$

51. $p^2 - 2p - 3rp + 6r$

52. $t^2 + 4t - st - 4s$

53. $ab + 6b - 4a - 24$

54. $xy - 5y - 2x + 10$

55. $2z^2 - z + 2yz - y$

56. $2y^2 - 10y + 7xy - 35x$

57. $8v^2 - 12vy + 14v - 21y$

58. $21x^2 + 6xy - 49x - 14y$

59. $2x^2 - 5x - 6xy + 15y$

60. $4a^2 + 5ab - 10b - 8a$

61. $3y^2 - 6y - ay + 2a$

62. $2ra + a^2 - 2r - a$

63. $3xy - y^2 - y + 3x$

64. $2ab - 3b^2 - 3b + 2a$

65. $3st + t^2 - 2t - 6s$

66. $4x^2 + 3xy - 12y - 16x$

APPLYING THE CONCEPTS

67. Factor: **a.** $2x^2 + 6x + 5x + 15$ **b.** $2x^2 + 5x + 6x + 15$

68. Look at parts a and b of Exercise 67. Do different groupings of the terms in a polynomial affect the binomial factors?

A whole number is a perfect number if it is the sum of all of its factors less than itself. For example, 6 is a perfect number because all the factors of 6 that are less than 6 are 1, 2, and 3, and $1 + 2 + 3 = 6$.

69. Find the one perfect number between 20 and 30.

70. Find the one perfect number between 490 and 500.

71. In the equation $P = 2L + 2W$, what is the effect on P when the quantity $L + W$ doubles?

72. Write an expression in factored form for each of the shaded portions in the following diagrams. (Use the equation for the area of a rectangle $A = bh$ and the equation for the area of a circle $A = \pi r^2$.)

a.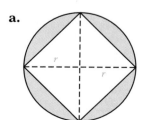

b.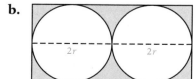

c.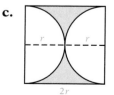

Factoring Polynomials of the Form $x^2 + bx + c$

Objective A   *To factor a trinomial of the form $x^2 + bx + c$* ....

Trinomials of the form $x^2 + bx + c$, where b and c are integers, are shown at the right.

$x^2 + 8x + 12$; $b = 8$, $c = 12$
$x^2 - 7x + 12$; $b = -7$, $c = 12$
$x^2 - 2x - 15$; $b = -2$, $c = -15$

To factor a trinomial of this form means to express the trinomial as the product of two binomials.

Trinomials expressed as the product of binomials are shown at the right.

$x^2 + 8x + 12 = (x + 6)(x + 2)$
$x^2 - 7x + 12 = (x - 3)(x - 4)$
$x^2 - 2x - 15 = (x + 3)(x - 5)$

The method by which factors of a trinomial are found is based on FOIL. Consider the following binomial products, noting the relationship between the constant terms of the binomials and the terms of the trinomials.

Signs in the binomials are the same.

$(x + 6)(x + 2) = x^2 + 2x + 6x + (6)(2) \quad = x^2 + 8x + 12$
sum of 6 and 2
product of 6 and 2

$(x - 3)(x - 4) = x^2 - 4x - 3x + (-3)(-4) = x^2 - 7x + 12$
sum of -3 and -4
product of -3 and -4

Signs in the binomials are opposite.

$(x + 3)(x - 5) = x^2 - 5x + 3x + (3)(-5) \quad = x^2 - 2x - 15$
sum of 3 and -5
product of 3 and -5

$(x - 4)(x + 6) = x^2 + 6x - 4x + (-4)(6) \quad = x^2 + 2x - 24$
sum of -4 and 6
product of -4 and 6

IMPORTANT RELATIONSHIPS

1. When the constant term of the trinomial is positive, the constant terms of the binomials have the same sign. They are both positive when the coefficient of the x term in the trinomial is positive. They are both negative when the coefficient of the x term in the trinomial is negative.

2. When the constant term of the trinomial is negative, the constant terms of the binomials have opposite signs.

3. In the trinomial, the coefficient of x is the sum of the constant terms of the binomials.

4. In the trinomial, the constant term is the product of the constant terms of the binomials.

➡ Factor: $x^2 - 7x + 10$

Because the constant term is positive and the coefficient of x is negative, the binomial constants will be negative. Find two negative factors of 10 whose sum is -7. The results can be recorded in a table.

Negative Factors of 10	Sum
$-1, -10$	-11
$-2, -5$	-7

• These are the correct factors.

$x^2 - 7x + 10 = (x - 2)(x - 5)$ • Write the trinomial as a product of its factors.

You can check the proposed factorization by multiplying the two binomials.

LOOK CLOSELY

Always check your proposed factorization to ensure accuracy.

Check: $(x - 2)(x - 5) = x^2 - 5x - 2x + 10$
$= x^2 - 7x + 10$

➡ Factor: $x^2 - 9x - 36$

The constant term is negative. The binomial constants will have opposite signs. Find two factors of -36 whose sum is -9.

Factors of -36	Sum
$+1, -36$	-35
$-1, +36$	35
$+2, -18$	-16
$-2, +18$	16
$+3, -12$	-9

• Once the correct factors are found, it is not necessary to try the remaining factors.

$x^2 - 9x - 36 = (x + 3)(x - 12)$ • Write the trinomials as a product of its factors.

➡ Factor: $x^2 + 7x + 8$

Because the constant term is positive and the coefficient of x is positive, the binomial constants will be positive. Find two positive factors of 8 whose sum is 7.

Positive Factors of 8	Sum
$1, 8$	8
$2, 4$	6

There are no positive integer factors of 8 whose sum is 7. The trinomial $x^2 + 7x + 8$ is said to be **nonfactorable over the integers.** Just as 17 is a prime number, $x^2 + 7x + 8$ is a **prime polynomial.** Binomials of the form $x - a$ and $x + a$ are also prime polynomials.

Example 1
Factor: $x^2 - 8x + 15$

Solution
Find two negative factors of 15 whose sum is -8.

Factors	Sum
$-1, -15$	-16
$-3, -5$	-8

$x^2 - 8x + 15 = (x - 3)(x - 5)$

You Try It 1
Factor: $x^2 + 9x + 20$

Your solution

Solution on p. A15

Example 2
Factor: $x^2 + 6x - 27$

Solution
Find two factors of
−27 whose sum is 6.

Factors	Sum
+1, −27	−26
−1, +27	26
+3, −9	−6
−3, +9	6

$x^2 + 6x - 27 = (x - 3)(x + 9)$

You Try It 2
Factor: $x^2 + 7x - 18$

Your solution

Solution on p. A15

Objective B   To factor completely ....

LOOK CLOSELY
The first step in *any* factoring problem is to determine whether the terms of the polynomials have a **common factor.** If they do, factor it out first.

A polynomial is factored completely when it is written as a product of factors that are nonfactorable over the integers.

➡ Factor: $4y^3 - 4y^2 - 24y$

$4y^3 - 4y^2 - 24y = \boxed{4y(y^2) - 4y(y) - 4y(6)}$

$= 4y(y^2 - y - 6)$

$= 4y(y + 2)(y - 3)$

- The GCF is 4y. Do this step mentally.
- Use the Distributive Property to factor out the GCF.
- Factor $y^2 - y - 6$. The two factors of −6 whose sum is −1 are 2 and −3.

It is always possible to check the proposed factorization by multiplying the polynomials. Here is the check for the last example.

Check: $4y(y + 2)(y - 3) = 4y(y^2 - 3y + 2y - 6)$
$= 4y(y^2 - y - 6)$
$= 4y^3 - 4y^2 - 24y$

- This is the original polynomial.

➡ Factor: $5x^2 + 60xy + 100y^2$

$5x^2 + 60xy + 100y^2 = \boxed{5(x^2) + 5(12xy) + 5(20)y^2}$

$= 5(x^2 + 12xy + 20y^2)$

$= 5(x + 2y)(x + 10y)$

- The GCF is 5. Do this step mentally.
- Use the Distributive Property to factor out the GCF.
- Factor $x^2 + 12xy + 20y^2$. The two factors of 20 whose sum is 12 are 2 and 10.

LOOK CLOSELY
2y and 10y are placed in the binomials. This is necessary so that the middle term contains *xy* and the last term contains y^2.

Note that 2y and 10y were placed in the binomials. The following check shows that was necessary.

Check: $5(x + 2y)(x + 10y) = 5(x^2 + 10xy + 2xy + 20y^2)$
$= 5(x^2 + 12xy + 20y^2)$
$= 5x^2 + 60xy + 100y^2$

- The original polynomial

➡ Factor: $15 - 2x - x^2$

Because the coefficient of x^2 is -1, factor -1 from the trinomial and then write the resulting trinomial in descending order.

$$15 - 2x - x^2 = -(x^2 + 2x - 15)$$

$$= -(x + 5)(x - 3)$$

- $15 - 2x - x^2 = -1(-15 + 2x + x^2)$
  $$= -(x^2 + 2x - 15)$$
- Factor $x^2 + 2x - 15$. The two factors of -15 whose sum is 2 are 5 and -3.

Check: $-(x + 5)(x-3) = -(x^2 + 2x - 15)$
$$= -x^2 - 2x + 15$$
$$= 15 - 2x - x^2$$

- The original polynomial

Example 3
Factor: $-3x^3 + 9x^2 + 12x$

Solution
The GCF is $-3x$.
$-3x^3 + 9x^2 + 12x = -3x(x^2 - 3x - 4)$
Factor the trinomial $x^2 - 3x - 4$. Find two factors of -4 whose sum is -3.

Factors	Sum
$-2, +2$	0
$+1, -4$	-3

$-3x^3 + 9x^2 + 12x = -3x(x + 1)(x - 4)$

You Try It 3
Factor: $-2x^3 + 14x^2 - 12x$

Your solution

Example 4
Factor: $4x^2 - 40xy + 84y^2$

Solution
The GCF is 4.
$4x^2 - 40xy + 84y^2 = 4(x^2 - 10xy + 21y^2)$
Factor the trinomial $x^2 - 10xy + 21y^2$. Find two negative factors of 21 whose sum is -10.

Factors	Sum
$-1, -21$	-22
$-3, -7$	-10

$4x^2 - 40xy + 84y^2 = 4(x - 3y)(x - 7y)$

You Try It 4
Factor: $3x^2 - 9xy - 12y^2$

Your solution

Solutions on p. A15

6.2 Exercises

Objective A

Factor.

1. $x^2 + 3x + 2$ **2.** $x^2 + 5x + 6$ **3.** $x^2 - x - 2$ **4.** $x^2 + x - 6$

5. $a^2 + a - 12$ **6.** $a^2 - 2a - 35$ **7.** $a^2 - 3a + 2$ **8.** $a^2 - 5a + 4$

9. $a^2 + a - 2$ **10.** $a^2 - 2a - 3$ **11.** $b^2 - 6b + 9$ **12.** $b^2 + 8b + 16$

13. $b^2 + 7b - 8$ **14.** $y^2 - y - 6$ **15.** $y^2 + 6y - 55$ **16.** $z^2 - 4z - 45$

17. $y^2 - 5y + 6$ **18.** $y^2 - 8y + 15$ **19.** $z^2 - 14z + 45$ **20.** $z^2 - 14z + 49$

21. $z^2 - 12z - 160$ **22.** $p^2 + 2p - 35$ **23.** $p^2 + 12p + 27$ **24.** $p^2 - 6p + 8$

25. $x^2 + 20x + 100$ **26.** $x^2 + 18x + 81$ **27.** $b^2 + 9b + 20$ **28.** $b^2 + 13b + 40$

29. $x^2 - 11x - 42$ **30.** $x^2 + 9x - 70$ **31.** $b^2 - b - 20$ **32.** $b^2 + 3b - 40$

33. $y^2 - 14y - 51$ **34.** $y^2 - y - 72$ **35.** $p^2 - 4p - 21$ **36.** $p^2 + 16p + 39$

37. $y^2 - 8y + 32$ **38.** $y^2 - 9y + 81$ **39.** $x^2 - 20x + 75$

Factor.

40. $p^2 + 24p + 63$ **41.** $x^2 - 15x + 56$ **42.** $x^2 + 21x + 38$

43. $x^2 + x - 56$ **44.** $x^2 + 5x - 36$ **45.** $a^2 - 21a - 72$

46. $a^2 - 7a - 44$ **47.** $a^2 - 15a + 36$ **48.** $a^2 - 21a + 54$

49. $z^2 - 9z - 136$ **50.** $z^2 + 14z - 147$ **51.** $c^2 - c - 90$

52. $c^2 - 3c - 180$ **53.** $z^2 + 15z + 44$ **54.** $p^2 + 24p + 135$

55. $c^2 + 19c + 34$ **56.** $c^2 + 11c + 18$ **57.** $x^2 - 4x - 96$

58. $x^2 + 10x - 75$ **59.** $x^2 - 22x + 112$ **60.** $x^2 + 21x - 100$

61. $b^2 + 8b - 105$ **62.** $b^2 - 22b + 72$ **63.** $a^2 - 9a - 36$

64. $a^2 + 42a - 135$ **65.** $b^2 - 23b + 102$ **66.** $b^2 - 25b + 126$

67. $a^2 + 27a + 72$ **68.** $z^2 + 24z + 144$ **69.** $x^2 + 25x + 156$

70. $x^2 - 29x + 100$ **71.** $x^2 - 10x - 96$ **72.** $x^2 + 9x - 112$

Objective B

Factor.

73. $2x^2 + 6x + 4$ **74.** $3x^2 + 15x + 18$ **75.** $18 + 7x - x^2$

76. $12 - 4x - x^2$ **77.** $ab^2 + 2ab - 15a$ **78.** $ab^2 + 7ab - 8a$

79. $xy^2 - 5xy + 6x$ **80.** $xy^2 + 8xy + 15x$ **81.** $z^3 - 7z^2 + 12z$

82. $-2a^3 - 6a^2 - 4a$ **83.** $-3y^3 + 15y^2 - 18y$ **84.** $4y^3 + 12y^2 - 72y$

85. $3x^2 + 3x - 36$ **86.** $2x^3 - 2x^2 + 4x$ **87.** $5z^2 - 15z - 140$

88. $6z^2 + 12z - 90$ **89.** $2a^3 + 8a^2 - 64a$ **90.** $3a^3 - 9a^2 - 54a$

91. $x^2 - 5xy + 6y^2$ **92.** $x^2 + 4xy - 21y^2$ **93.** $a^2 - 9ab + 20b^2$

94. $a^2 - 15ab + 50b^2$ **95.** $x^2 - 3xy - 28y^2$ **96.** $s^2 + 2st - 48t^2$

97. $y^2 - 15yz - 41z^2$ **98.** $y^2 + 85yz + 36z^2$ **99.** $z^4 - 12z^3 + 35z^2$

100. $z^4 + 2z^3 - 80z^2$ **101.** $b^4 - 22b^3 + 120b^2$ **102.** $b^4 - 3b^3 - 10b^2$

103. $2y^4 - 26y^3 - 96y^2$ **104.** $3y^4 + 54y^3 + 135y^2$ **105.** $-x^4 - 7x^3 + 8x^2$

106. $-x^4 + 11x^3 + 12x^2$ **107.** $4x^2y + 20xy - 56y$ **108.** $3x^2y - 6xy - 45y$

Factor.

109. $c^3 + 18c^2 - 40c$

110. $-3x^3 + 36x^2 - 81x$

111. $-4x^3 - 4x^2 + 24x$

112. $x^2 - 8xy + 15y^2$

113. $y^2 - 7xy - 8x^2$

114. $a^2 - 13ab + 42b^2$

115. $y^2 + 4yz - 21z^2$

116. $y^2 + 8yz + 7z^2$

117. $y^2 - 16yz + 15z^2$

118. $3x^2y + 60xy - 63y$

119. $4x^2y - 68xy - 72y$

120. $3x^3 + 3x^2 - 36x$

121. $4x^3 + 12x^2 - 160x$

122. $4z^3 + 32z^2 - 132z$

123. $5z^3 - 50z^2 - 120z$

124. $4x^3 + 8x^2 - 12x$

125. $5x^3 + 30x^2 + 40x$

126. $5p^2 + 25p - 420$

127. $4p^2 - 28p - 480$

128. $p^4 + 9p^3 - 36p^2$

129. $p^4 + p^3 - 56p^2$

130. $t^2 - 12ts + 35s^2$

131. $a^2 - 10ab + 25b^2$

132. $a^2 - 8ab - 33b^2$

133. $x^2 + 4xy - 60y^2$

134. $5x^4 - 30x^3 + 40x^2$

135. $6x^3 - 6x^2 - 120x$

APPLYING THE CONCEPTS

Factor.

136. $2 + c^2 + 9c$

137. $x^2y - 54y - 3xy$

138. $45a^2 + a^2b^2 - 14a^2b$

Find all integers k such that the trinomial can be factored over the integers.

139. $x^2 + kx + 35$

140. $x^2 + kx + 18$

141. $x^2 + kx + 21$

Determine the positive integer values of k for which the following polynomials are factorable over the integers.

142. $y^2 + 4y + k$

143. $z^2 + 7z + k$

144. $a^2 - 6a + k$

145. $c^2 - 7c + k$

146. $x^2 - 3x + k$

147. $y^2 + 5y + k$

148. In Exercise 142–145, there was the stated requirement that $k > 0$. If k is allowed to be any integer, how many different values of k are possible for each polynomial?

Factoring Polynomials of the Form $ax^2 + bx + c$

Objective A   *To factor a trinomial of the form $ax^2 + bx + c$ by using trial factors* ....

Trinomials of the form $ax^2 + bx + c$, where a, b, and c are integers, are shown at the right.

$3x^2 - x + 4;\ a = 3,\ b = -1,\ c = 4$
$6x^2 + 2x - 3;\ a = 6,\ b = 2,\ c = -3$

These trinomials differ from those in the previous section in that the coefficient of x^2 is not 1. There are various methods of factoring these trinomials. The method described in this objective is factoring polynomials using trial factors.

To reduce the number of trial factors that must be considered, remember the following:

1. Use the signs of the constant term and the coefficient of x in the trinomial to determine the signs of the binomial factors. If the constant term is positive, the signs of the binomial factors will be the same as the sign of the coefficient of x in the trinomial. If the sign of the constant term is negative, the constant terms in the binomials have opposite signs.

2. If the terms of the trinomial do not have a common factor, then the terms of neither of the binomial factors will have a common factor.

➡ Factor: $2x^2 - 7x + 3$

The terms have no common factor. The constant term is positive. The coefficient of x is negative. The binomial constants will be negative.

Positive *Factors of 2* (coefficient of x^2)	*Negative* *Factors of 3* (constant term)
1, 2	−1, −3

Write trial factors. Use the **O**uter and **I**nner products of FOIL to determine the middle term, $-7x$, of the trinomial.

Trial Factors	*Middle Term*
$(x - 1)(2x - 3)$	$-3x - 2x = -5x$
$(x - 3)(2x - 1)$	$-x - 6x = -7x$

Write the factors of the trinomial. $2x^2 - 7x + 3 = (x - 3)(2x - 1)$

➡ Factor: $6x^2 - 13x + 6$

The terms have no common factor. The constant term is positive. The coefficient of x is negative. The binomial constants will be negative.

Positive *Factors of 6* (coefficient of x^2)	*Negative* *Factors of 6* (constant term)
2, 3	−2, −3
1, 6	−1, −6

Write trial factors. Use the **O**uter and **I**nner products of FOIL to determine the middle term, $-13x$, of the trinomial.

Trial Factors	*Middle Term*
$(2x - 3)(3x - 2)$	$-4x - 9x = -13x$
$(x - 6)(6x - 1)$	$-x - 36x = -37x$

Write the factors of the trinomial. $6x^2 - 13x + 6 = (2x - 3)(3x - 2)$

➡ Factor $6x^3 + 14x^2 - 12x$.

Factor the GCF, $2x$, from the terms.	$6x^3 + 14x^2 - 12x = 2x(3x^2 + 7x - 6)$

Factor the trinomial. The constant term is negative. The binomial constants will have opposite signs.

Positive Factors of 3	*Factors of −6*
1, 3	−1, 6
	1, −6
	−2, 3
	2, −3

Write trial factors. Use the **O**uter and **I**nner products of FOIL to determine the middle term, $7x$, of the trinomial.

It is not necessary to test trial factors that have a common factor.

Trial Factors	*Middle Term*
$(x - 1)(3x + 6)$	Common factor
$(x + 6)(3x - 1)$	$-x + 18x = 17x$
$(x + 1)(3x - 6)$	Common factor
$(x - 6)(3x + 1)$	$x - 18x = -17x$
$(x - 2)(3x + 3)$	Common factor
$(x + 3)(3x - 2)$	$-2x + 9x = 7x$
$(x + 2)(3x - 3)$	Common factor
$(x - 3)(3x + 2)$	$2x - 9x = -7x$

Write the factors of the trinomial. $6x^3 + 14x^2 - 12x = 2x(x + 3)(3x - 2)$

For this example, all the trial factors were listed. Once the correct factors have been found, the remaining trial factors can be omitted. For the examples and solutions in this text, all trial factors except those that have a common factor will be listed.

Example 1
Factor: $3x^2 + x - 2$

Solution

Positive	Factors of −2: 1, −2
factors of 3: 1, 3	−1, 2

Trial Factors	*Middle Term*
$(1x + 1)(3x - 2)$	$-2x + 3x = x$
$(1x - 2)(3x + 1)$	$x - 6x = -5x$
$(1x - 1)(3x + 2)$	$2x - 3x = -x$
$(1x + 2)(3x - 1)$	$-x + 6x = 5x$

$3x^2 + x - 2 = (x + 1)(3x - 2)$

You Try It 1
Factor $2x^2 - x - 3$.

Your solution

Example 2
Factor: $-12x^3 - 32x^2 + 12x$

Solution
The GCF is $-4x$.
$-12x^3 - 32x^2 + 12x = -4x(3x^2 + 8x - 3)$
Factor the trinomial.

Positive	Factors of −3: 1, −3
factors of 3: 1, 3	−1, 3

Trial Factors	*Middle Term*
$(x - 3)(3x + 1)$	$x - 9x = -8x$
$(x + 3)(3x - 1)$	$-x + 9x = 8x$

$-12x^3 - 32x^2 + 12x = -4x(x + 3)(3x - 1)$

You Try It 2
Factor $-45y^3 + 12y^2 + 12y$.

Your solution

Solutions on p. A16

Objective B   *To factor a trinomial of the form $ax^2 + bx + c$ by grouping* ....

In the previous objective, trinomials of the form $ax^2 + bx + c$ were factored by using trial factors. In this objective, these trinomials will be factored by grouping.

To factor $ax^2 + bx + c$, first find two factors of $a \cdot c$ whose sum is b. Then use factoring by grouping to write the factorization of the trinomial.

➡ Factor: $2x^2 + 13x + 15$

Find two positive factors of 30 $(2 \cdot 15)$ whose sum is 13.

Positive Factors of 30	Sum
1, 30	31
2, 15	17
3, 10	13

     • **When the required sum has been found, the remaining factors need not be checked.**

$2x^2 + 13x + 15 = 2x^2 + 3x + 10x + 15$ • **Use the factors of 30 whose sum is 13 to write $13x$ as $3x + 10x$.**

$ = (2x^2 + 3x) + (10x + 15)$ • **Factor by grouping.**
$ = x(2x + 3) + 5(2x + 3)$
$ = (2x + 3)(x + 5)$

Check your answer. $(2x + 3)(x + 5) = 2x^2 + 10x + 3x + 15$
$ = 2x^2 + 13x + 15$

➡ Factor: $6x^2 - 11x - 10$

Find two factors of -60 $[6 \cdot (-10)]$ whose sum is -11.

Factors of -60	Sum
1, -60	-59
-1, 60	59
2, -30	-28
-2, 30	28
3, -20	-17
-3, 20	17
4, -15	-11

$6x^2 - 11x - 10 = 6x^2 + 4x - 15x - 10$ • **Use the factors of -60 whose sum is -11 to write $-11x$ as $4x - 15x$.**

$ = (6x^2 + 4x) - (15x + 10)$ • **Factor by grouping. Recall $-15x - 10 = -(15x + 10)$.**
$ = 2x(3x + 2) - 5(3x + 2)$
$ = (3x + 2)(2x - 5)$

➡ Factor; $3x^2 - 2x - 4$

Find two factors of 12 $[3 \cdot (-4)]$ whose sum is -2.

Factors of -12	Sum
1, -12	-11
-1, 12	11
2, -6	-4
-2, 6	4
3, -4	-1
-3, 4	1

Because no integer factors of -12 have a sum of -2, $3x^2 - 2x - 4$ is nonfactorable over the integers. $3x^2 - 2x - 4$ is a **prime polynomial.**

Example 3
Factor: $2x^2 + 19x - 10$

Solution

Factors of -20 [2(-10)]	Sum
-1, 20	19

$$2x^2 + 19x - 10 = 2x^2 - x + 20x - 10$$
$$= (2x^2 - x) + (20x - 10)$$
$$= x(2x - 1) + 10(2x - 1)$$
$$= (2x - 1)(x + 10)$$

You Try It 3
Factor: $2a^2 + 13a - 7$

Your solution

Example 4
Factor: $24x^2y - 76xy + 40y$

Solution
The GCF is $4y$.

$$24x^2y - 76xy + 40y = 4y(6x^2 - 19x + 10)$$

Negative Factors of 60 [6(10)]	Sum
-1, -60	-61
-2, -30	-32
-3, -20	-23
-4, -15	-19

$$6x^2 - 19x + 10 = 6x^2 - 4x - 15x + 10$$
$$= (6x^2 - 4x) - (15x - 10)$$
$$= 2x(3x - 2) - 5(3x - 2)$$
$$= (3x - 2)(2x - 5)$$

$$24x^2y - 76xy + 40y = 4y(6x^2 - 19x + 10)$$
$$= 4y(3x - 2)(2x - 5)$$

You Try It 4
Factor: $15x^3 + 40x^2 - 80x$

Your solution

Solutions on p. A16

6.3 Exercises

Objective A

Factor by using trial factors.

1. $2x^2 + 3x + 1$ **2.** $5x^2 + 6x + 1$ **3.** $2y^2 + 7y + 3$ **4.** $3y^2 + 7y + 2$

5. $2a^2 - 3a + 1$ **6.** $3a^2 - 4a + 1$ **7.** $2b^2 - 11b + 5$ **8.** $3b^2 - 13b + 4$

9. $2x^2 + x - 1$ **10.** $4x^2 - 3x - 1$ **11.** $2x^2 - 5x - 3$ **12.** $3x^2 + 5x - 2$

13. $2t^2 - t - 10$ **14.** $2t^2 + 5t - 12$ **15.** $3p^2 - 16p + 5$ **16.** $6p^2 + 5p + 1$

17. $12y^2 - 7y + 1$ **18.** $6y^2 - 5y + 1$ **19.** $6z^2 - 7z + 3$ **20.** $9z^2 + 3z + 2$

21. $6t^2 - 11t + 4$ **22.** $10t^2 + 11t + 3$ **23.** $8x^2 + 33x + 4$ **24.** $7x^2 + 50x + 7$

25. $5x^2 - 62x - 7$ **26.** $9x^2 - 13x - 4$ **27.** $12y^2 + 19y + 5$ **28.** $5y^2 - 22y + 8$

29. $7a^2 + 47a - 14$ **30.** $11a^2 - 54a - 5$ **31.** $3b^2 - 16b + 16$ **32.** $6b^2 - 19b + 15$

33. $2z^2 - 27z - 14$ **34.** $4z^2 + 5z - 6$ **35.** $3p^2 + 22p - 16$ **36.** $7p^2 + 19p + 10$

Factor by using trial factors.

37. $4x^2 + 6x + 2$ **38.** $12x^2 + 33x - 9$ **39.** $15y^2 - 50y + 35$ **40.** $30y^2 + 10y - 20$

41. $2x^3 - 11x^2 + 5x$ **42.** $2x^3 - 3x^2 - 5x$ **43.** $3a^2b - 16ab + 16b$ **44.** $2a^2b - ab - 21b$

45. $3z^2 + 95z + 10$ **46.** $8z^2 - 36z + 1$ **47.** $36x - 3x^2 - 3x^3$ **48.** $-2x^3 + 2x^2 + 4x$

49. $80y^2 - 36y + 4$ **50.** $24y^2 - 24y - 18$ **51.** $8z^3 + 14z^2 + 3z$ **52.** $6z^3 - 23z^2 + 20z$

53. $6x^2y - 11xy - 10y$ **54.** $8x^2y - 27xy + 9y$ **55.** $10t^2 - 5t - 50$

56. $16t^2 + 40t - 96$ **57.** $3p^3 - 16p^2 + 5p$ **58.** $6p^3 + 5p^2 + p$

59. $26z^2 + 98z - 24$ **60.** $30z^2 - 87z + 30$ **61.** $10y^3 - 44y^2 + 16y$

62. $14y^3 + 94y^2 - 28y$ **63.** $4yz^3 + 5yz^2 - 6yz$ **64.** $12a^3 + 14a^2 - 48a$

65. $42a^3 + 45a^2 - 27a$ **66.** $36p^2 - 9p^3 - p^4$ **67.** $9x^2y - 30xy^2 + 25y^3$

68. $8x^2y - 38xy^2 + 35y^3$ **69.** $9x^3y - 24x^2y^2 + 16xy^3$ **70.** $9x^3y + 12x^2y + 4xy$

___

Objective B

Factor by grouping.

71. $6x^2 - 17x + 12$ **72.** $15x^2 - 19x + 6$ **73.** $5b^2 + 33b - 14$ **74.** $8x^2 - 30x + 25$

75. $6a^2 + 7a - 24$ **76.** $14a^2 + 15a - 9$ **77.** $4z^2 + 11z + 6$ **78.** $6z^2 - 25z + 14$

79. $22p^2 + 51p - 10$ **80.** $14p^2 - 41p + 15$ **81.** $8y^2 + 17y + 9$ **82.** $12y^2 - 145y + 12$

83. $18t^2 - 9t - 5$ **84.** $12t^2 + 28t - 5$ **85.** $6b^2 + 71b - 12$ **86.** $8b^2 + 65b + 8$

87. $9x^2 + 12x + 4$ **88.** $25x^2 - 30x + 9$ **89.** $6b^2 - 13b + 6$ **90.** $20b^2 + 37b + 15$

91. $33b^2 + 34b - 35$ **92.** $15b^2 - 43b + 22$ **93.** $18y^2 - 39y + 20$ **94.** $24y^2 + 41y + 12$

95. $15a^2 + 26a - 21$ **96.** $6a^2 + 23a + 21$ **97.** $8y^2 - 26y + 15$ **98.** $18y^2 - 27y + 4$

99. $8z^2 + 2z - 15$ **100.** $10z^2 + 3z - 4$ **101.** $15x^2 - 82x + 24$ **102.** $13z^2 + 49z - 8$

103. $10z^2 - 29z + 10$ **104.** $15z^2 - 44z + 32$ **105.** $36z^2 + 72z + 35$ **106.** $16z^2 + 8z - 35$

107. $3x^2 + xy - 2y^2$ **108.** $6x^2 + 10xy + 4y^2$ **109.** $3a^2 + 5ab - 2b^2$ **110.** $2a^2 - 9ab + 9b^2$

Factor by grouping.

111. $4y^2 - 11yz + 6z^2$ **112.** $2y^2 + 7yz + 5z^2$ **113.** $28 + 3z - z^2$ **114.** $15 - 2z - z^2$

115. $8 - 7x - x^2$ **116.** $12 + 11x - x^2$ **117.** $9x^2 + 33x - 60$ **118.** $16x^2 - 16x - 12$

119. $24x^2 - 52x + 24$ **120.** $60x^2 + 95x + 20$ **121.** $35a^4 + 9a^3 - 2a^2$

122. $15a^4 + 26a^3 + 7a^2$ **123.** $15b^2 - 115b + 70$ **124.** $25b^2 + 35b - 30$

125. $3x^2 - 26xy + 35y^2$ **126.** $4x^2 + 16xy + 15y^2$ **127.** $216y^2 - 3y - 3$

128. $360y^2 + 4y - 4$ **129.** $21 - 20x - x^2$ **130.** $18 + 17x - x^2$

131. $15a^2 + 11ab - 14b^2$ **132.** $15a^2 - 31ab + 10b^2$ **133.** $33z - 8z^2 - z^3$

APPLYING THE CONCEPTS

134.
[W]   In your own words, explain how the signs of the last terms of the two binomial factors of a trinomial are determined.

Factor.

135. $(x + 1)^2 - (x + 1) - 6$ **136.** $(x - 2)^2 + 3(x - 2) + 2$ **137.** $(y + 3)^2 - 5(y + 3) + 6$

138. $2(y + 2)^2 - (y + 2) - 3$ **139.** $3(a + 2)^2 - (a + 2) - 4$ **140.** $4(y - 1)^2 - 7(y - 1) - 2$

Find all integers k such that the trinomial can be factored over the integers.

141. $2x^2 + kx + 3$ **142.** $2x^2 + kx - 3$ **143.** $3x^2 + kx + 2$

144. $3x^2 + kx - 2$ **145.** $2x^2 + kx + 5$ **146.** $2x^2 + kx - 5$

Special Factoring

Objective A   *To factor the difference of two squares and perfect-square trinomials* ....

Recall that the product of the sum and difference of the same terms equals the square of the first term minus the square of the second term.

Sum and difference of two terms   Difference of two squares

$$(a + b)(a - b) \quad\quad = \quad\quad a^2 - b^2$$

This suggests that the difference of two squares can be factored as follows:

$$a^2 - b^2 = (a + b)(a - b)$$

Note that the polynomial $x^2 + y^2$ is the *sum* of two squares. The sum of two squares is nonfactorable over the integers.

➡ Factor: $x^2 - 16$

$$
\begin{aligned}
x^2 - 16 &= (x)^2 - (4)^2 \\
&= (x - 4)(x + 4)
\end{aligned}
$$
  • Write $x^2 - 16$ as the difference of two squares. Then factor.

Check: $(x - 4)(x + 4) = x^2 + 4x - 4x - 16$
$$= x^2 - 16$$

➡ Factor: $8x^3 - 18x$

$$
\begin{aligned}
8x^3 - 18x &= 2x(4x^2 - 9) \\
&= 2x[(2x)^2 - 3^2] \\
&= 2x(2x - 3)(2x + 3)
\end{aligned}
$$
  • The GCF is $2x$.
  • Factor the difference of two squares.

You should check the factorization.

➡ Factor: $x^2 - 10$

Because 10 cannot be written as the square of an integer, $x^2 - 10$ is nonfactorable over the integers.

Recall from an earlier discussion the relationship between the terms of the trinomial and the terms of the binomial.

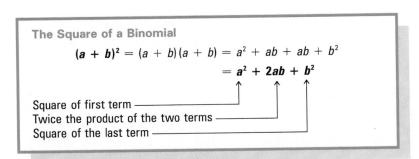

The Square of a Binomial
$$(a + b)^2 = (a + b)(a + b) = a^2 + ab + ab + b^2$$
$$= a^2 + 2ab + b^2$$

Square of first term ————
Twice the product of the two terms ————
Square of the last term ————

This relationship is used to factor a **perfect-square trinomial.**

➡ Factor: $4x^2 - 20x + 25$

Because the first and last terms are squares $[(2x)^2 = 4x^2; 5^2 = 25]$, try to factor as the square of a binomial. Check the factorization.

$$4x^2 - 20x + 25 = (2x - 5)^2$$

Check: $(2x - 5)^2 = (2x)^2 + 2(2x)(-5) + 5^2$
$$= 4x^2 - 20x + 25$$
$$4x^2 - 20x + 25 = (2x - 5)^2$$

● **The factorization is correct.**

➡ Factor: $4x^2 + 37x + 9$

Because the first and last terms are squares $[(2x)^2 = 4x^2; 3^2 = 9]$, try to factor as the square of a binomial. Check the proposed factorization.

$$4x^2 + 37x + 9 = (2x + 3)^2$$

Check: $(2x + 3)^2 = (2x)^2 + 2(2x)(3) + 3^2$
$$= 4x^2 + 12x + 9$$

Because $4x^2 + 12x + 9 \neq 4x^2 + 37x + 9$, the proposed factorization is not correct. In this case, the polynomial is not a perfect-square trinomial. It may, however, still factor. In fact, $4x^2 + 37x + 9 = (4x + 1)(x + 9)$.

Example 1
Factor: $16x^2 - y^2$

Solution
$16x^2 - y^2 = (4x)^2 - y^2 = (4x + y)(4x - y)$

Example 2
Factor: $z^4 - 16$

Solution
$z^4 - 16 = (z^2)^2 - 4^2 = (z^2 + 4)(z^2 - 4)$
$\quad\quad\;\; = (z^2 + 4)(z - 2)(z + 2)$

Example 3
Factor: $9x^2 - 30x + 25$

Solution
Because $9x^2 = (3x)^2$, $25 = (-5)^2$, and $-30x = 2(3x)(-5)$, the trinomial is a perfect-square trinomial.
$9x^2 - 30x + 25 = (3x - 5)^2$

You Try It 1
Factor: $25a^2 - b^2$

Your solution

You Try It 2
Factor: $n^4 - 81$

Your solution

You Try It 3
Factor: $16y^2 + 8y + 1$

Your solution

Solutions on p. A17

Example 4
Factor: $9x^2 + 40x + 16$

Solution
Because $9x^2 = (3x)^2$, $16 = 4^2$, and
$40x \neq 2(3x)(4)$, the trinomial is not
a perfect-square trinomial.

Try to factor by another method.

$9x^2 + 40x + 16 = (9x + 4)(x + 4)$

You Try It 4
Factor: $x^2 + 15x + 36$

Your solution

Example 5
Factor: $(r + 2)^2 - 4$

Solution
$$(r + 2)^2 - 4 = (r + 2)^2 - 2^2$$
$$= (r + 2 - 2)(r + 2 + 2)$$
$$= r(r + 4)$$

You Try It 5
Factor: $(x^2 - 6x + 9) - y^2$

Your solution

Solutions on p. A17

Objective B   *To factor completely* ....

When factoring a polynomial completely, ask the following questions about the polynomial.

1. Is there a common factor? If so, factor out the common factor.
2. Is the polynomial the difference of two perfect squares? If so, factor.
3. Is the polynomial a perfect-square trinomial? If so, factor.
4. Is the polynomial a trinomial that is the product of two binomials? If so, factor.
5. Does the polynomial contain four terms? If so, try factoring by grouping.
6. Is each binomial factor a prime polynomial over the integers? If not, factor.

➡ Factor: $z^3 + 4z^2 - 9z - 36$

$z^3 + 4z^2 - 9z - 36 = (z^3 + 4z^2) - (9z + 36)$ • Factor by grouping. Recall
$-9z - 36 = -(9z + 36)$.

$= z^2(z + 4) - 9(z + 4)$ • $z^3 + 4z^2 = z^2(z + 4)$
$9z + 36 = 9(z + 4)$

$= (z + 4)(z^2 - 9)$ • Factor the common binomial factor $(z + 4)$.

$= (z + 4)(z + 3)(z - 3)$ • Factor the difference of squares.

Example 6
Factor: $3x^2 - 48$

Solution
The GCF is 3.

$3x^2 - 48 = 3(x^2 - 16)$
$\qquad\qquad = 3(x + 4)(x - 4)$

$3x^2 - 48 = 3(x + 4)(x - 4)$

You Try It 6
Factor: $12x^3 - 75x$

Your solution

Example 7
Factor: $x^3 - 3x^2 - 4x + 12$

Solution
Factor by grouping.

$x^3 - 3x^2 - 4x + 12 = (x^3 - 3x^2) - (4x - 12)$
$\qquad\qquad\qquad\qquad = x^2(x - 3) - 4(x - 3)$
$\qquad\qquad\qquad\qquad = (x - 3)(x^2 - 4)$
$\qquad\qquad\qquad\qquad = (x - 3)(x + 2)(x - 2)$

$x^3 - 3x^2 - 4x + 12 = (x - 3)(x + 2)(x - 2)$

You Try It 7
Factor: $a^2b - 7a^2 - b + 7$

Your solution

Example 8
Factor: $4x^2y^2 + 12xy^2 + 9y^2$

Solution
The GCF is y^2.

$4x^2y^2 + 12xy^2 + 9y^2 = y^2(4x^2 + 12x + 9)$
$\qquad\qquad\qquad\qquad = y^2(2x + 3)^2$

$4x^2y^2 + 12xy^2 + 9y^2 = y^2(2x + 3)^2$

You Try It 8
Factor: $4x^3 + 28x^2 - 120x$

Your solution

Solutions on p. A17

6.4 Exercises

Objective A

Factor.

1. $x^2 - 4$

$(x + 2)(x - 2)$

2. $x^2 - 9$

3. $a^2 - 81$

4. $a^2 - 49$

5. $y^2 + 2y + 1$

6. $y^2 + 14y + 49$

7. $a^2 - 2a + 1$

8. $x^2 - 12x + 36$

9. $4x^2 - 1$

10. $9x^2 - 16$

11. $x^6 - 9$

12. $y^{12} - 4$

13. $x^2 + 8x - 16$

14. $z^2 - 18z - 81$

15. $x^2 + 2xy + y^2$

16. $x^2 + 6xy + 9y^2$
( )( )

17. $4a^2 + 4a + 1$

18. $25x^2 + 10x + 1$

19. $9x^2 - 1$

20. $1 - 49x^2$

21. $1 - 64x^2$

22. $t^2 + 36$

23. $x^2 + 64$

24. $64a^2 - 16a + 1$

25. $9a^2 + 6a + 1$

26. $x^4 - y^2$

27. $b^4 - 16a^2$

28. $16b^2 + 8b + 1$

29. $4a^2 - 20a + 25$

30. $4b^2 + 28b + 49$

31. $9a^2 - 42a + 49$

32. $9x^2 - 16y^2$

33. $25z^2 - y^2$

34. $x^2y^2 - 4$

35. $a^2b^2 - 25$

36. $16 - x^2y^2$

Factor.

37. $25x^2 - 1$

38. $25a^2 + 30ab + 9b^2$

39. $4a^2 - 12ab + 9b^2$

40. $49x^2 + 28xy + 4y^2$

41. $4y^2 - 36yz + 81z^2$

42. $64y^2 - 48yz + 9z^2$

43. $\left[\dfrac{1}{x^2} - 4\right]$

44. $\left[\dfrac{9}{a^2} - 16\right]$

45. $9a^2b^2 - 6ab + 1$

46. $16x^2y^2 - 24xy + 9$

Objective B

Factor.

47. $8y^2 - 2$

48. $12n^2 - 48$

49. $3a^3 + 6a^2 + 3a$

50. $4rs^2 - 4rs + r$

51. $m^4 - 256$

52. $81 - t^4$

53. $9x^2 + 13x + 4$

54. $x^2 + 10x + 16$

55. $16y^4 + 48y^3 + 36y^2$

56. $36c^4 - 48c^3 + 16c^2$

57. $y^8 - 81$

58. $32s^4 - 2$

59. $25 - 20p + 4p^2$

60. $9 + 24a + 16a^2$

61. $(4x - 3)^2 - y^2$

62. $(2x + 5)^2 - 25$

63. $(x^2 - 4x + 4) - y^2$

64. $(4x^2 + 12x + 9) - 4y^2$

Factor.

65. $5x^2 - 5$

66. $2x^2 - 18$

67. $x^3 + 4x^2 + 4x$

68. $y^3 - 10y^2 + 25y$

69. $x^4 + 2x^3 - 35x^2$

70. $a^4 - 11a^3 + 24a^2$

71. $5b^2 + 75b + 180$

72. $6y^2 - 48y + 72$

73. $3a^2 + 36a + 10$

74. $5a^2 - 30a + 4$

75. $2x^2y + 16xy - 66y$

76. $3a^2b + 21ab - 54b$

77. $x^3 - 6x^2 - 5x$

78. $b^3 - 8b^2 - 7b$

79. $3y^2 - 36$

80. $3y^2 - 147$

81. $20a^2 + 12a + 1$

82. $12a^2 - 36a + 27$

83. $x^2y^2 - 7xy^2 - 8y^2$

84. $a^2b^2 + 3a^2b - 88a^2$

85. $10a^2 - 5ab - 15b^2$

86. $16x^2 - 32xy + 12y^2$

87. $50 - 2x^2$

88. $72 - 2x^2$

89. $a^2b^2 - 10ab^2 + 25b^2$

90. $a^2b^2 + 6ab^2 + 9b^2$

91. $12a^3b - a^2b^2 - ab^3$

92. $2x^3y - 7x^2y^2 + 6xy^3$

93. $12a^3 - 12a^2 + 3a$

94. $18a^3 + 24a^2 + 8a$

95. $243 + 3a^2$

96. $75 + 27y^2$

97. $12a^3 - 46a^2 + 40a$

98. $24x^3 - 66x^2 + 15x$

99. $4a^3 + 20a^2 + 25a$

100. $2a^3 - 8a^2b + 8ab^2$

Factor.

101. $27a^2b - 18ab + 3b$

102. $a^2b^2 - 6ab^2 + 9b^2$

103. $48 - 12x - 6x^2$

104. $21x^2 - 11x^3 - 2x^4$

105. $x^4 - x^2y^2$

106. $b^4 - a^2b^2$

107. $18a^3 + 24a^2 + 8a$

108. $32xy^2 - 48xy + 18x$

109. $2b + ab - 6a^2b$

110. $15y^2 - 2xy^2 - x^2y^2$

111. $4x^4 - 38x^3 + 48x^2$

112. $3x^2 - 27y^2$

113. $x^4 - 25x^2$

114. $y^3 - 9y$

115. $a^4 - 16$

116. $15x^4y^2 - 13x^3y^3 - 20x^2y^4$

117. $45y^2 - 42y^3 - 24y^4$

118. $a(2x - 2) + b(2x - 2)$

119. $4a(x - 3) - 2b(x - 3)$

120. $x^2(x - 2) - (x - 2)$

121. $y^2(a - b) - (a - b)$

122. $a(x^2 - 4) + b(x^2 - 4)$

123. $x(a^2 - b^2) - y(a^2 - b^2)$

124. $4(x - 5) - x^2(x - 5)$

APPLYING THE CONCEPTS

Find all integers k such that the trinomial is a perfect-square trinomial.

125. $4x^2 - kx + 9$

126. $x^2 + 6x + k$

127. $64x^2 + kxy + y^2$

128. $x^2 - 2x + k$

129. $25x^2 - kx + 1$

130. $x^2 + 10x + k$

Both the sum of two cubes and the difference of two cubes can always be factored.

Sum of Two Cubes	Difference of Two Cubes
$a^3 + b^3 = (a + b)(a^2 - ab + b^2)$	$a^3 - b^3 = (a - b)(a^2 + ab + b^2)$

Factor.

131. $x^3 + 8$

132. $8y^3 + 27$

133. $x^3 + 64$

134. $a^3 - 64$

135. $27y^3 - 1$

136. $8x^3 - 27y^3$

137.
[W] Select any odd integer greater than 1, square it, and then subtract 1. Is the result evenly divisible by 8? Prove that this procedure always produces a number divisible by 8. (*Suggestion:* Any odd integer greater than 1 can be expressed as $2n + 1$, where n is a natural number.)

6.5 Solving Equations

Objective A   *To solve equations by factoring*

Recall that the Multiplication Property of Zero states that the product of a number and zero is zero. This property is restated below.

If a is a real number, then $a \cdot 0 = 0 \cdot a = 0$.

Now consider $x \cdot y = 0$. For this to be a true equation, then either $x = 0$ or $y = 0$.

Principle of Zero Products

If the product of two factors is zero, then at least one of the factors must be zero.

If $a \cdot b = 0$, then $a = 0$ or $b = 0$.

The Principle of Zero Products is used to solve some equations.

➡ Solve: $(x - 2)(x - 3) = 0$

$(x - 2)(x - 3) = 0$
$x - 2 = 0 \qquad x - 3 = 0$
• Let each factor equal zero (the Principle of Zero Products)

$x = 2 \qquad\qquad x = 3$
• Rewrite each equation in the form *variable = constant.*

Check:

$$(x - 2)(x - 3) = 0 \qquad\qquad (x - 2)(x - 3) = 0$$
$$\frac{(2 - 2)(2 - 3)}{0(-1)} \,\Big|\, 0 \qquad\qquad \frac{(3 - 2)(3 - 3)}{(1)(0)} \,\Big|\, 0$$
$$0 = 0 \qquad\qquad\qquad\qquad 0 = 0$$

A true equation   A true equation

The solutions are 2 and 3.

An equation of the form $ax^2 + bx + c = 0, a \ne 0$, is a **quadratic equation.** A quadratic equation is in **standard form** when the polynomial is in descending order and equal to zero. The quadratic equations at the right are in standard form.

$3x^2 + 2x + 1 = 0$

$4x^2 - 3x + 2 = 0$

➡ Solve: $2x^2 + x = 6$

$$2x^2 + x = 6$$
$$2x^2 + x - 6 = 0$$ • Write the equation in standard form.
$$(2x - 3)(x + 2) = 0$$ • Factor.
$$2x - 3 = 0 \qquad x + 2 = 0$$ • Use the Principle of Zero Products.
$$2x = 3 \qquad\qquad x = -2$$ • Rewrite each equation in the form *variable = constant.*
$$x = \frac{3}{2}$$

$\frac{3}{2}$ and -2 check as solutions. The solutions are $\frac{3}{2}$ and -2.

Example 1
Solve: $x(x - 3) = 0$

Solution
$x(x - 3) = 0$

$x = 0 \qquad\qquad x - 3 = 0$
$\qquad\qquad\qquad x = 3$

The solutions are 0 and 3.

You Try It 1
Solve: $2x(x + 7) = 0$

Your solution

Example 2
Solve: $2x^2 - 50 = 0$

Solution
$2x^2 - 50 = 0$
$2(x^2 - 25) = 0$
$2(x + 5)(x - 5) = 0$
$x + 5 = 0 \qquad x - 5 = 0$
$\quad x = -5 \qquad\quad x = 5$

The solutions are -5 and 5.

You Try It 2
Solve: $4x^2 - 9 = 0$

Your solution

Example 3
Solve: $(x - 3)(x - 10) = -10$

Solution
$(x - 3)(x - 10) = -10$
$x^2 - 13x + 30 = -10$ • Multiply $(x - 3)(x - 10)$.
$x^2 - 13x + 40 = 0$ • Add 10 to each side of the equation. The equation is now in standard form.
$(x - 8)(x - 5) = 0$
$x - 8 = 0 \quad x - 5 = 0$
$\quad x = 8 \qquad\quad x = 5$

The solutions are 8 and 5.

You Try It 3
Solve: $(x + 2)(x - 7) = 52$

Your solution

Solutions on p. A17

Objective B   To solve application problems ....

Example 4
The sum of the squares of two consecutive positive even integers is equal to 100. Find the two integers.

You Try It 4
The sum of the squares of two consecutive positive integers is 61. Find the two integers.

Strategy
First positive even integer: n
Second positive even integer: $n + 2$

The sum of the square of the first positive even integer and the square of the second positive even integer is 100.

Your strategy

Solution
$n^2 + (n + 2)^2 = 100$
$n^2 + n^2 + 4n + 4 = 100$
$2n^2 + 4n + 4 = 100$
$2n^2 + 4n - 96 = 0$
$2(n^2 + 2n - 48) = 0$
$2(n - 6)(n + 8) = 0$

$n - 6 = 0 \qquad n + 8 = 0$
$\qquad n = 6 \qquad\qquad n = -8$

Because -8 is not a positive even integer, it is not a solution.
$n = 6$
$n + 2 = 6 + 2 = 8$

The two integers are 6 and 8.

Your solution

Solution on p. A18

Example 5

A stone is thrown into a well with an initial speed of 4 ft/s. The well is 420 ft deep. How many seconds later will the stone hit the bottom of the well? Use the equation $d = vt + 16t^2$, where d is the distance in feet, v is the initial speed, and t is the time in seconds.

Strategy

To find the time for the stone to drop to the bottom of the well, replace the variables d and v by their given values and solve for t.

Solution

$$d = vt + 16t^2$$
$$420 = 4t + 16t^2$$
$$0 = -420 + 4t + 16t^2$$
$$16t^2 + 4t - 420 = 0$$
$$4(4t^2 + t - 105) = 0$$
$$4(4t + 21)(t - 5) = 0$$

$$4t + 21 = 0 \qquad t - 5 = 0$$
$$4t = -21 \qquad\quad t = 5$$
$$t = -\frac{21}{4}$$

Because the time cannot be a negative number, $-\frac{21}{4}$ is not a solution.

The time is 5 s.

You Try It 5

The length of a rectangle is 4 in. longer than twice the width. The area of the rectangle is 96 in.². Find the length and width of the rectangle.

Your strategy

Your solution

Solution on p. A18

6.5 Exercises

· ·

Objective A

Solve.

1. $(y + 3)(y + 2) = 0$ **2.** $(y - 3)(y - 5) = 0$ **3.** $(z - 7)(z - 3) = 0$ **4.** $(z + 8)(z - 9) = 0$

5. $x(x - 5) = 0$ **6.** $x(x + 2) = 0$ **7.** $a(a - 9) = 0$ **8.** $a(a + 12) = 0$

9. $y(2y + 3) = 0$ **10.** $t(4t - 7) = 0$ **11.** $2a(3a - 2) = 0$ **12.** $4b(2b + 5) = 0$

13. $(b + 2)(b - 5) = 0$ **14.** $(b - 8)(b + 3) = 0$ **15.** $x^2 - 81 = 0$ **16.** $x^2 - 121 = 0$

17. $4x^2 - 49 = 0$ **18.** $16x^2 - 1 = 0$ **19.** $9x^2 - 1 = 0$ **20.** $16x^2 - 49 = 0$

21. $x^2 + 6x + 8 = 0$ **22.** $x^2 - 8x + 15 = 0$ **23.** $z^2 + 5z - 14 = 0$ **24.** $z^2 + z - 72 = 0$

25. $x^2 - 5x + 6 = 0$ **26.** $x^2 - 3x - 10 = 0$ **27.** $y^2 + 4y - 21 = 0$ **28.** $2y^2 - y - 1 = 0$

29. $2a^2 - 9a - 5 = 0$ **30.** $3a^2 + 14a + 8 = 0$ **31.** $6z^2 + 5z + 1 = 0$ **32.** $6y^2 - 19y + 15 = 0$

33. $x^2 - 3x = 0$ **34.** $a^2 - 5a = 0$ **35.** $x^2 - 7x = 0$ **36.** $2a^2 - 8a = 0$

37. $a^2 + 5a = -4$ **38.** $a^2 - 5a = 24$ **39.** $y^2 - 5y = -6$ **40.** $y^2 - 7y = 8$

41. $2t^2 + 7t = 4$ **42.** $3t^2 + t = 10$ **43.** $3t^2 - 13t = -4$ **44.** $5t^2 - 16t = -12$

45. $x(x - 12) = -27$ **46.** $x(x - 11) = 12$ **47.** $y(y - 7) = 18$ **48.** $y(y + 8) = -15$

Solve.

49. $p(p + 3) = -2$ **50.** $p(p - 1) = 20$ **51.** $y(y + 4) = 45$ **52.** $y(y - 8) = -15$

53. $x(x + 3) = 28$ **54.** $p(p - 14) = 15$ **55.** $(x + 8)(x - 3) = -30$ **56.** $(x + 4)(x - 1) = 14$

57. $(z - 5)(z + 4) = 52$ **58.** $(z - 8)(z + 4) = -35$ **59.** $(z - 6)(z + 1) = -10$

60. $(a + 3)(a + 4) = 72$ **61.** $(a - 4)(a + 7) = -18$ **62.** $(2x + 5)(x + 1) = -1$

63. $(z + 3)(z - 10) = -42$ **64.** $(y + 3)(2y + 3) = 5$ **65.** $(y + 5)(3y - 2) = -14$

Objective B   *Application Problems*

Solve.

66. The square of a positive number is six more than five times the positive number. Find the number.

67. The square of a negative number is sixteen more than six times the negative number. Find the number.

68. The sum of two numbers is six. The sum of the squares of the two numbers is twenty. Find the two numbers. $2 + 4$

69. The sum of two numbers is eight. The sum of the squares of the two numbers is thirty-four. Find the two numbers. $3 + 5$

70. The sum of the squares of two consecutive positive integers is eighty-five. Find the two integers.

71. The sum of the squares of two consecutive positive even integers is one hundred. Find the two integers.

72. The sum of two numbers is ten. The product of the two numbers is twenty-one. Find the two numbers. $3 + 7$

73. The sum of two numbers is twenty-three. The product of the two numbers is one hundred twenty. Find the two numbers. $8 + 15$

Solve.

The formula $S = \dfrac{n^2 + n}{2}$ gives the sum S of the first n natural numbers. Use this formula for Problems 74 and 75.

74. How many consecutive natural numbers beginning with 1 will give a sum of 78?

75. How many consecutive natural numbers beginning with 1 will give a sum of 120?

The formula $N = \dfrac{t^2 - t}{2}$ gives the number N of football games that must be scheduled in a league with t teams if each team is to play every other team once. Use this formula for Problems 76 and 77.

76. How many teams are in a league that schedules 28 games in such a way that each team plays every other team once?

77. How many teams are in a league that schedules 45 games in such a way that each team plays every other team once.

The distance s that an object will fall (neglecting air resistance) in t seconds is given by $s = vt + 16t^2$, where v is the initial velocity of the object. Use this formula for Problems 78 and 79.

78. An object is released from a plane at an altitude of 1600 ft. The initial velocity is 0 ft/s, and air resistance is neglected. How many seconds later will the object hit the ground?

79. An object is released from the top of a building 320 ft high. The initial velocity is 16 ft/s, and air resistance is neglected. How many seconds later will the object hit the ground?

The height h an object will attain (neglecting air resistance) in t seconds is given by $h = vt - 16t^2$, where v is the initial velocity of the object. Use this formula for Problems 80 and 81.

80. A foul ball leaves a bat and travels straight up with an initial velocity of 64 ft/s. How many seconds later will the ball be 64 ft above the ground?

81. A golf ball is thrown onto a cement surface and rebounds straight up. The initial velocity of the rebound is 96 ft/s. How many seconds later will the golf ball return to the ground?

82. The length of a rectangle is 5 in. more than twice the width. The area is 75 in^2. Find the length and width of the rectangle.

Solve.

83. The width of a rectangle is 5 ft less than the length. The area of the rectangle is 176 ft². Find the length and width of the rectangle.

84. The length of each side of a square is extended 2 in. The area of the resulting square is 144 in². Find the length of a side of the original square.

85. The length of each side of a square is extended 5 in. The area of the resulting square is 64 in². Find the length of a side of the original square.

86. The page of a book measures 6 in. by 9 in. A uniform border around the page leaves 28 in² for type. What are the dimensions of the type area?

87. A small garden measures 8 ft by 10 ft. A uniform border around the garden increases the total area to 168 ft². What is the width of the border?

88. The radius of a circle is increased by 3 in., increasing the area by 100 in². Find the radius of the original circle. Use $\pi \approx 3.14$. Round to hundredth.

89. A circle has a radius of 10 in. Find the increase in area that occurs when the radius is increased by 2 in. Use $\pi \approx 3.14$.

APPLYING THE CONCEPTS

90. In your own words what is the Principle of Zero Products.

91. Explain the error made in solving the equation at the right. Solve the equation correctly.

$$(x + 2)(x - 3) = 6$$
$$x + 2 = 6 \quad x - 3 = 6$$
$$x = 4 \quad\quad x = 9$$

92. Explain the error made in solving the equation at the right. Solve the equation correctly.

$$x^2 = x$$
$$\frac{x^2}{x} = \frac{x}{x}$$
$$x = 1$$

93. Find $3n^2$ if $n(n + 5) = -4$.

94. Find $2n^2$ if $n(n + 3) = 4$.

Solve.

95. $2y(y + 4) = -5(y + 3)$

96. $(b + 5)^2 = 16$

97. $p^3 = 9p^2$

98. $(x + 3)(2x - 1) = (3 - x)(5 - 3x)$

Projects in Mathematics

. . . .

Evaluating Polynomials Using a Calculator

One way to evaluate a polynomial is first to express the polynomial in a form that suggests a sequence of steps on the calculator. To illustrate this method, consider the polynomial $4x^2 - 5x + 2$. First the polynomial is rewritten as

$$4x^2 - 5x + 2 = (4x - 5)x + 2$$

To evaluate the polynomial, work through the rewritten expression from left to right, substituting the appropriate value for x.

Here are some examples:

Evaluate $5x^2 - 2x + 4$ when $x = 3$.

Rewrite the polynomial. $\qquad\qquad\qquad\qquad\qquad 5x^2 - 2x + 4 = (5x - 2)x + 4$

Replace x in the rewritten expression by the given value. $\quad (5 \cdot 3 - 2) \cdot 3 + 4$

Work through the expression from left to right.

The result in the display should be 43.

Evaluate $2x^3 - 4x^2 + 7x - 12$ when $x = 4$.

Rewrite the polynomial. $\qquad\qquad 2x^3 - 4x^2 + 7x - 12 = [(2x - 4)x + 7]x - 12$

Replace x in the rewritten expression by the $\qquad [(2 \cdot 4 - 4) \cdot 4 + 7] \cdot 4 - 12$
given value.

Work through the expression from left to right.

$$\boxed{(}\ \boxed{(}\ 2\ \boxed{\times}\ 4\ \boxed{-}\ 4\ \boxed{)}\ \boxed{\times}\ 4\ \boxed{+}\ 7\ \boxed{)}\ \boxed{\times}\ 4\ \boxed{-}\ 12\ \boxed{=}$$

The result in the display should be 80.

Evaluate $4x^2 - 3x + 5$ when $x = -2$.

Rewrite the polynomial. $\qquad\qquad\qquad\qquad\qquad 4x^2 - 3x + 5 = (4x - 3)x + 5$

Replace x in the given expression by the given value. $\quad [4 \cdot (-2) - 3] \cdot (-2) + 5$

Work through the expression from left to right.

$$\boxed{(}\ 4\ \boxed{\times}\ 2\ \boxed{+/-}\ \boxed{-}\ 3\ \boxed{)}\ \boxed{\times}\ 2\ \boxed{+/-}\ \boxed{+}\ 5\ \boxed{=}$$

The result in the display should be 27.

Exercises

Here are some practice exercises. Evaluate for the given value.

1. $2x^2 - 3x + 7; x = 4$ $\qquad\qquad$ **2.** $3x^2 + 7x - 12; x = -3$

3. $3x^3 - 2x^2 + 6x - 8; x = 3$ $\qquad$ **4.** $2x^3 + 4x^2 - x - 2; x = -2$

5. $x^4 - 3x^3 + 6x^2 + 5x - 1;$ $\qquad$ **6.** $2x^3 - 4x + 8; x = 2$

$\qquad x = 2$ $\qquad\qquad\qquad\qquad\qquad$ *Hint:* $2x^3 - 4x + 8 = 2x^3 + 0x^2 - 4x + 8$

Chapter Summary

Key Words   The *greatest common factor* (GCF) of two or more integers is the greatest integer that is a factor of all the integers.

To *factor* a polynomial means to write the polynomial as a product of other polynomials.

To *factor* a trinomial of the form $ax^2 + bx + c$ means to express the trinomial as the product of two binomials.

A polynomial that does not factor using only integers is *nonfactorable over the integers*.

A product of a term and itself is a *perfect square*.

An equation of the form $ax^2 + bx + c = 0$ is a *quadratic equation*.

A quadratic equation is in *standard form* when the polynomial is in descending order and equal to zero. The quadratic equation $ax^2 + bx + c = 0$ is in standard form.

Essential Rules   *Principle of Zero Products*

If the product of two factors is zero, then at least one of the factors must be zero.   If $a \cdot b = 0$, then $a = 0$ or $b = 0$.

General Factoring Strategy

1. Is there a common factor? If so, factor out the common factor.
2. Is the polynomial the difference of two perfect squares? If so, factor.
3. Is the polynomial a perfect-square trinomial? If so, factor.
4. Is the polynomial a trinomial that is the product of two binomials? If so, factor.
5. Does the polynomial contain four terms? If so, try factoring by grouping.
6. Is each binomial factor a prime polynomial over the integers? If not, factor.

Chapter Review

SECTION 6.1

1. Factor: $5x^3 + 10x^2 + 35x$

2. Factor: $12a^2b + 3ab^2$

3. Factor: $14y^9 - 49y^6 + 7y^3$

4. Factor: $4x(x - 3) - 5(3 - x)$

5. Factor: $10x^2 + 25x + 4xy + 10y$

6. Factor: $21ax - 35bx - 10by + 6ay$

SECTION 6.2

7. Factor: $b^2 - 13b + 30$

8. Factor: $c^2 + 8c + 12$

9. Factor: $y^2 + 5y - 36$

10. Factor completely $3a^2 - 15a - 42$.

11. Factor completely $4x^3 - 20x^2 - 24x$.

12. Factor completely $n^4 - 2n^3 - 3n^2$.

SECTION 6.3

13. Factor $6x^2 - 29x + 28$ by using trial factors.

14. Factor $12y^2 + 16y - 3$ by using trial factors.

15. Factor $2x^2 - 5x + 6$ by using trial factors.

16. Factor $3x^2 - 17x + 10$ by grouping.

17. Factor $2a^2 - 19a - 60$ by grouping.

18. Factor $18a^2 - 3a - 10$ by grouping.

SECTION 6.4

19. Factor: $a^6 - 100$

20. Factor: $9y^4 - 25z^2$

21. Factor: $a^2b^2 - 1$

22. Factor: $5x^2 - 5x - 30$

23. Factor: $3x^2 + 36x + 108$

24. Factor: $12b^3 - 58b^2 + 56b$

SECTION 6.5

25. Solve: $4x^2 + 27x = 7$

26. Solve: $(x + 1)(x - 5) = 16$

27. The length of a hockey field is 20 yd less than twice the width of the hockey field. The area of the hockey field is 6000 yd². Find the length and width of the hockey field.

28. The size (S) of an image from a slide projector depends on the distance (d) of the screen from the projector and is given by $S = d^2$. Find the distance between the projector and the screen when the size of the picture is 400 ft².

29. A rectangular photograph has dimensions 15 in. by 12 in. A picture frame around the photograph increases the total area to 270 in². What is the width of the frame?

30. The length of each side of a square garden plot is extended 4 ft. The area of the resulting square is 576 ft². Find the length of a side of the original garden plot.

Chapter Test

· ·

1. Factor: $6x^3 - 8x^2 + 10x$

2. Factor: $ab + 6a - 3b - 18$

3. Factor: $p^2 + 5p + 6$

4. Factor: $a^2 - 19a + 48$

5. Factor: $x^2 + 2x - 15$

6. Factor: $x^2 - 9x - 36$

7. Factor: $5x^2 - 45x - 15$

8. Factor: $2y^4 - 14y^3 - 16y^2$

9. Factor $2x^2 + 4x - 5$ by using trial factors.

10. Factor $6x^2 + 19x + 8$ by using trial factors.

11. Factor $8x^2 + 20x - 48$ by grouping.

12. Factor $6x^2y^2 + 9xy^2 + 3y^2$ by grouping.

13. Factor: $a(x - 2) + b(x - 2)$

14. Factor: $x(p + 1) - (p + 1)$

15.   Factor: $b^2 - 16$

16.   Factor: $4x^2 - 49y^2$

17.   Factor: $p^2 + 12p + 36$

18.   Factor: $4a^2 - 12ab + 9b^2$

19.   Factor: $3a^2 - 75$

20.   Factor: $3x^2 + 12xy + 12y^2$

21.   Solve: $(2a - 3)(a + 7) = 0$

22.   Solve: $4x^2 - 1 = 0$

23.   Solve: $x(x - 8) = -15$

24.   The sum of two numbers is 10. The sum of the squares of the two numbers is 58. Find the two numbers.

25.   The length of a rectangle is 3 cm longer than twice the width. The area of the rectangle is 90 cm². Find the length and width of the rectangle.

Cumulative Review

1. Subtract: $-2 - (-3) - 5 - (-11)$

2. Simplify: $(3 - 7)^2 \div (-2) - 3 \cdot (-4)$

3. Evaluate $-2a^2 \div (2b) - c$ when $a = -4$, $b = 2$, and $c = -1$.

4. Simplify: $-\dfrac{3}{4}(-20x^2)$

5. Simplify: $-2[4x - 2(3 - 2x) - 8x]$

6. Solve: $-\dfrac{5}{7}x = -\dfrac{10}{21}$

7. Solve: $3x - 2 = 12 - 5x$

8. Solve:
   $-2 + 4[3x - 2(4 - x) - 3] = 4x + 2$

9. 120% of what number is 54?

10. Simplify: $(-3a^3b^2)^2$

11. Simplify: $(x + 2)(x^2 - 5x + 4)$

12. Simplify: $(8x^2 + 4x - 3) \div (2x - 3)$

13. Simplify: $(x^{-4}y^3)^2$

14. Factor: $3a - 3b - ax + bx$

15. Factor: $15xy^2 - 20xy^4$

16. Factor: $x^2 - 5xy - 14y^2$

17. Factor: $p^2 - 9p - 10$

18. Factor: $18a^3 + 57a^2 + 30a$

19. Factor: $36a^2 - 49b^2$

20. Factor: $4x^2 + 28xy + 49y^2$

21. Factor: $9x^2 + 15x - 14$

22. Factor: $18x^2 - 48xy + 32y^2$

23. Factor: $3y(x - 3) - 2(x - 3)$

24. Solve: $3x^2 + 19x - 14 = 0$

25. A board 10 ft long is cut into two pieces. Four times the length of the shorter piece is 2 ft less than three times the length of the longer piece. Find the length of each piece.

26. A stereo that regularly sells for $165 is on sale for $99. Find the discount rate.

27. An investment of $4000 is made at an annual simple interest rate of 8%. How much more money must be invested at an annual simple interest rate of 11% so that the total interest earned is $1035?

28. A family drove to a resort at an average speed of 42 mph and later returned over the same road at an average speed of 56 mph. Find the distance to the resort if the total driving time was 7 h.

29. Find three consecutive even integers such that five times the middle integer is twelve more than twice the sum of the first and third.

30. The length of the base of a triangle is three times the height. The area of the triangle is 24 in². Find the length of the base of the triangle.

7

Rational Expressions

Objectives

Section 7.1
To simplify a rational expression
To multiply rational expressions
To divide rational expressions

Section 7.2
To find the least common multiple (LCM) of two or more polynomials
To express two fractions in terms of the LCM of their denominators

Section 7.3
To add or subtract rational expressions with like denominators
To add or subtract rational expressions with unlike denominators

Section 7.4
To simplify a complex fraction

Section 7.5
To solve an equation containing fractions

Section 7.6
To solve a proportion
To solve application problems

Section 7.7
To solve a literal equation for one of the variables

Section 7.8
To solve work problems
To solve uniform motion problems

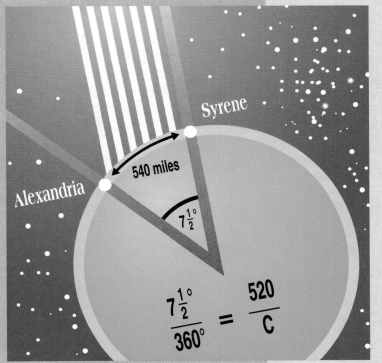

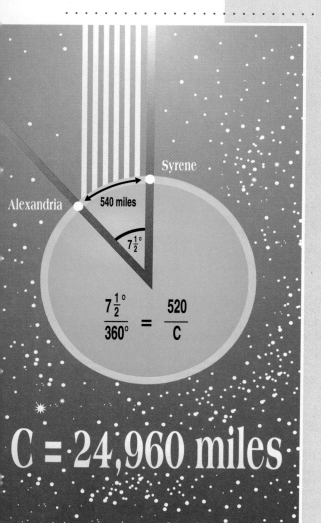

Measurement of the Circumference of the Earth

Distances on the earth, the circumference of the earth, and the distance to the moon and stars are known to great precision. Eratosthenes, the fifth librarian of Alexandria (230 B.C.), laid the foundation of scientific geography with his determination of the circumference of the earth.

Eratosthenes was familiar with certain astronomical data that enabled him to calculate the circumference of the earth by using a proportion statement.

Eratosthenes knew that on a mid-summer day, the sun was directly overhead at Syrene, as shown in the diagram. At the same time, at Alexandria the sun was at a $7\frac{1}{2}°$ angle from the zenith. The distance from Syrene to Alexandria was 5000 stadia (about 520 mi).

Knowing that the ratio of the $7\frac{1}{2}°$ angle to one revolution (360°) is equal to the ratio of the arc length (520 mi) to the circumference, Eratosthenes was able to write and solve a proportion.

This result, calculated over 2000 years ago is very close to the accepted value of 24,800 miles.

Content and Format © 1995 HMCo.

Multiplication and Division of Rational Expressions

Objective A *To simplify a rational expression* ....

A fraction in which the numerator or denominator is a polynomial is called a **rational expression.** Examples of rational expressions are shown at the right.

$$\frac{5}{z}, \quad \frac{x^2 + 1}{2x - 1}, \quad \frac{y^2 + y - 1}{4y^2 + 1}$$

Care must be exercised with a rational expression to ensure that when the variables are replaced with numbers, the resulting denominator is not zero.

Consider the rational expression at the right. The value of x cannot be 3, because the denominator would then be zero.

$$\frac{4x^2 - 9}{2x - 6}$$

$$\frac{4(3)^2 - 9}{2(3) - 6} = \frac{27}{0} \quad \begin{array}{l} \text{Not a real} \\ \text{number} \end{array}$$

A rational expression is in simplest form when the numerator and denominator have no common factors. The Multiplication Property of One is used to write a rational expression in simplest form.

➡ Simplify: $\dfrac{x^2 - 4}{x^2 - 2x - 8}$

$$\frac{x^2 - 4}{x^2 - 2x - 8} = \frac{(x - 2)(x + 2)}{(x - 4)(x + 2)}$$

- Factor the numerator and denominator.

$$= \frac{x - 2}{x - 4} \cdot \boxed{\frac{x + 2}{x + 2}} = \frac{x - 2}{x - 4} \cdot 1$$

$$= \frac{x - 2}{x - 4}, \quad x \neq -2, 4$$

- The restrictions, $x \neq -2, 4$, are necessary to prevent division by zero.

This simplification is usually shown with slashes through the common factors. The last simplification would be shown as follows.

$$\frac{x^2 - 4}{x^2 - 2x - 8} = \frac{(x - 2)\overset{1}{\cancel{(x + 2)}}}{(x - 4)\underset{1}{\cancel{(x + 2)}}}$$

- Factor the numerator and denominator.

$$= \frac{x - 2}{x - 4}, x \neq -2, 4$$

- Divide by the common factors. The restrictions, $x \neq -2, 4$, are necessary to prevent division by zero.

➡ Simplify: $\dfrac{10 + 3x - x^2}{x^2 - 4x - 5}$

$$\frac{10 + 3x - x^2}{x^2 - 4x - 5} = \frac{(5 - x)(2 + x)}{(x - 5)(x + 1)}$$

- Factor the numerator and denominator.

$$= \frac{\overset{-1}{\cancel{(5 - x)}}(2 + x)}{\underset{1}{\cancel{(x - 5)}}(x + 1)}$$

- Recall that $5 - x = -(x - 5)$. Therefore

$$\frac{5 - x}{x - 5} = \frac{-(x - 5)}{x - 5} = \frac{-1}{1} = -1$$

$$= -\frac{x + 2}{x + 1}, x \neq -1, 5$$

For the remaining examples, we will omit the restrictions on the variables that prevent division by zero and assume the values of the variables are such that division by zero is not possible.

Example 1

Simplify: $\dfrac{4x^3y^4}{6x^4y}$

Solution

$\dfrac{4x^3y^4}{6x^4y} = \dfrac{2y^3}{3x}$ • Use rules of exponents.

You Try It 1

Simplify: $\dfrac{6x^5y}{12x^2y^3}$

Your solution

Example 2

Simplify: $\dfrac{9 - x^2}{x^2 + x - 12}$

Solution

$\dfrac{9 - x^2}{x^2 + x - 12} = \dfrac{\overset{-1}{\cancel{(3 - x)}}(3 + x)}{\underset{1}{\cancel{(x - 3)}}(x + 4)} = -\dfrac{x + 3}{x + 4}$

You Try It 2

Simplify: $\dfrac{x^2 + 2x - 24}{16 - x^2}$

Your solution

Example 3

Simplify: $\dfrac{x^2 + 2x - 15}{x^2 - 7x + 12}$

Solution

$\dfrac{x^2 + 2x - 15}{x^2 - 7x + 12} = \dfrac{(x + 5)\overset{1}{\cancel{(x - 3)}}}{\underset{1}{\cancel{(x - 3)}}(x - 4)} = \dfrac{x + 5}{x - 4}$

You Try It 3

Simplify: $\dfrac{x^2 + 4x - 12}{x^2 - 3x + 2}$

Your solution

Solutions on p. A18

Objective B *To multiply rational expressions* ....

The product of two fractions is a fraction whose numerator is the product of the numerators of the two fractions and whose denominator is the product of the denominators of the two fractions.

> If $\dfrac{a}{b}$ and $\dfrac{c}{d}$ are rational numbers, then $\dfrac{a}{b} \cdot \dfrac{c}{d} = \dfrac{ac}{bd}$.

$\dfrac{2}{3} \cdot \dfrac{4}{5} = \dfrac{8}{15}$ $\dfrac{3x}{y} \cdot \dfrac{2}{z} = \dfrac{6x}{yz}$ $\dfrac{x + 2}{x} \cdot \dfrac{3}{x - 2} = \dfrac{3x + 6}{x^2 - 2x}$

➡️ Simplify: $\dfrac{x^2 + 3x}{x^2 - 3x - 4} \cdot \dfrac{x^2 - 5x + 4}{x^2 + 2x - 3}$

$$\dfrac{x^2 + 3x}{x^2 - 3x - 4} \cdot \dfrac{x^2 - 5x + 4}{x^2 + 2x - 3}$$

$$= \dfrac{x(x + 3)}{(x - 4)(x + 1)} \cdot \dfrac{(x - 4)(x - 1)}{(x + 3)(x - 1)}$$

• Factor the numerator and denominator of each fraction.

$$= \dfrac{x(x + 3)\overset{1}{\cancel{(x + 3)}}\overset{1}{\cancel{(x - 4)}}\overset{1}{\cancel{(x - 1)}}}{\underset{1}{\cancel{(x - 4)}}(x + 1)\underset{1}{\cancel{(x + 3)}}\underset{1}{\cancel{(x - 1)}}}$$

• Multiply.

$$= \dfrac{x}{x + 1}$$

• Write the answer in simplest form.

Example 4

Simplify: $\dfrac{10x^2 - 15x}{12x - 8} \cdot \dfrac{3x - 2}{20x - 25}$

Solution

$$\dfrac{10x^2 - 15x}{12x - 8} \cdot \dfrac{3x - 2}{20x - 25}$$

$$= \dfrac{5x(2x - 3)}{4(3x - 2)} \cdot \dfrac{(3x - 2)}{5(4x - 5)}$$

$$= \dfrac{\overset{1}{\cancel{5}}x(2x - 3)\overset{1}{\cancel{(3x - 2)}}}{4\underset{1}{\cancel{(3x - 2)}}\underset{1}{\cancel{5}}(4x - 5)} = \dfrac{x(2x - 3)}{4(4x - 5)}$$

You Try It 4

Simplify: $\dfrac{12x^2 + 3x}{10x - 15} \cdot \dfrac{8x - 12}{9x + 18}$

Your solution

Example 5

Simplify: $\dfrac{x^2 + x - 6}{x^2 + 7x + 12} \cdot \dfrac{x^2 + 3x - 4}{4 - x^2}$

Solution

$$\dfrac{x^2 + x - 6}{x^2 + 7x + 12} \cdot \dfrac{x^2 + 3x - 4}{4 - x^2}$$

$$= \dfrac{(x + 3)(x - 2)}{(x + 3)(x + 4)} \cdot \dfrac{(x + 4)(x - 1)}{(2 - x)(2 + x)}$$

$$= \dfrac{\overset{1}{\cancel{(x + 3)}}\overset{-1}{\cancel{(x - 2)}}\overset{1}{\cancel{(x + 4)}}(x - 1)}{\underset{1}{\cancel{(x + 3)}}\underset{1}{\cancel{(x + 4)}}\underset{1}{\cancel{(2 - x)}}(2 + x)} = -\dfrac{x - 1}{x + 2}$$

You Try It 5

Simplify: $\dfrac{x^2 + 2x - 15}{9 - x^2} \cdot \dfrac{x^2 - 3x - 18}{x^2 - 7x + 6}$

Your solution

Solutions on p. A18

Objective C   To divide rational expressions

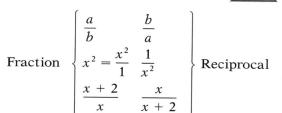

The **reciprocal** of a fraction is a fraction with the numerator and denominator interchanged.

$$\text{Fraction} \begin{cases} \dfrac{a}{b} & \dfrac{b}{a} \\[2mm] x^2 = \dfrac{x^2}{1} & \dfrac{1}{x^2} \\[2mm] \dfrac{x+2}{x} & \dfrac{x}{x+2} \end{cases} \text{Reciprocal}$$

> To divide two fractions multiply by the reciprocal of the divisor.
>
> $$\frac{a}{b} \div \frac{c}{d} = \frac{a}{b} \cdot \frac{d}{c} = \frac{ad}{bc}$$

$$\frac{4}{x} \div \frac{y}{5} = \frac{4}{x} \cdot \frac{5}{y} = \frac{20}{xy} \qquad\qquad \frac{x+4}{x} \div \frac{x-2}{4} = \frac{x+4}{x} \cdot \frac{4}{x-2} = \frac{4(x+4)}{x(x-2)}$$

The basis for the division rule is shown at the right.

$$\frac{a}{b} \div \frac{c}{d} = \frac{\dfrac{a}{b}}{\dfrac{c}{d}} = \frac{\dfrac{a}{b} \cdot \dfrac{d}{c}}{\dfrac{c}{d} \cdot \dfrac{d}{c}} = \frac{\dfrac{a}{b} \cdot \dfrac{d}{c}}{1} = \frac{a}{b} \cdot \frac{d}{c}$$

Example 6

Simplify: $\dfrac{xy^2 - 3x^2y}{z^2} \div \dfrac{6x^2 - 2xy}{z^3}$

Solution

$$\frac{xy^2 - 3x^2y}{z^2} \div \frac{6x^2 - 2xy}{z^3}$$

$$= \frac{xy^2 - 3x^2y}{z^2} \cdot \frac{z^3}{6x^2 - 2xy}$$

$$= \frac{xy(y - 3x) \cdot z^3}{z^2 \cdot 2x(3x - y)} = -\frac{yz}{2}$$

You Try It 6

Simplify: $\dfrac{a^2}{4bc^2 - 2b^2c} \div \dfrac{a}{6bc - 3b^2}$

Your solution

Example 7

Simplify: $\dfrac{2x^2 + 5x + 2}{2x^2 + 3x - 2} \div \dfrac{3x^2 + 13x + 4}{2x^2 + 7x - 4}$

Solution

$$\frac{2x^2 + 5x + 2}{2x^2 + 3x - 2} \div \frac{3x^2 + 13x + 4}{2x^2 + 7x - 4}$$

$$= \frac{2x^2 + 5x + 2}{2x^2 + 3x - 2} \cdot \frac{2x^2 + 7x - 4}{3x^2 + 13x + 4}$$

$$= \frac{(2x + 1)(x + 2) \cdot (2x - 1)(x + 4)}{(2x - 1)(x + 2) \cdot (3x + 1)(x + 4)} = \frac{2x + 1}{3x + 1}$$

You Try It 7

Simplify: $\dfrac{3x^2 + 26x + 16}{3x^2 - 7x - 6} \div \dfrac{2x^2 + 9x - 5}{x^2 + 2x - 15}$

Your solution

$$\frac{x+8}{2y-1}$$

Solutions on p. A19

7.1 Exercises

Objective A

Simplify.

1. $\dfrac{9x^3}{12x^4}$

2. $\dfrac{16x^2y}{24xy^3}$

3. $\dfrac{(x+3)^2}{(x+3)^3}$

4. $\dfrac{(2x-1)^5}{(2x-1)^4}$

5. $\dfrac{3n-4}{4-3n}$

6. $\dfrac{5-2x}{2x-5}$

7. $\dfrac{6y(y+2)}{9y^2(y+2)}$

8. $\dfrac{12x^2(3-x)}{18x(3-x)}$

9. $\dfrac{6x(x-5)}{8x^2(5-x)}$

10. $\dfrac{14x^3(7-3x)}{21x(3x-7)}$

11. $\dfrac{a^2+4a}{ab+4b}$

12. $\dfrac{x^2-3x}{2x-6}$

13. $\dfrac{4-6x}{3x^2-2x}$

14. $\dfrac{5xy-3y}{9-15x}$

15. $\dfrac{y^2-3y+2}{y^2-4y+3}$

16. $\dfrac{x^2+5x+6}{x^2+8x+15}$

17. $\dfrac{x^2+3x-10}{x^2+2x-8}$

18. $\dfrac{a^2+7a-8}{a^2+6a-7}$

19. $\dfrac{x^2+x-12}{x^2-6x+9}$

20. $\dfrac{x^2+8x+16}{x^2-2x-24}$

21. $\dfrac{x^2-3x-10}{25-x^2}$

22. $\dfrac{4-y^2}{y^2-3y-10}$

23. $\dfrac{2x^3+2x^2-4x}{x^3+2x^2-3x}$

24. $\dfrac{3x^3-12x}{6x^3-24x^2+24x}$

25. $\dfrac{6x^2-7x+2}{6x^2+5x-6}$

26. $\dfrac{2n^2-9n+4}{2n^2-5n-12}$

27. $\dfrac{x^2+3x-28}{24-2x-x^2}$

Objective B

Simplify.

28. $\dfrac{8x^2}{9y^3} \cdot \dfrac{3y^2}{4x^3}$

29. $\dfrac{14a^2b^3}{15x^5y^2} \cdot \dfrac{25x^3y}{16ab}$

30. $\dfrac{12x^3y^4}{7a^2b^3} \cdot \dfrac{14a^3b^4}{9x^2y^2}$

31. $\dfrac{18a^4b^2}{25x^2y^3} \cdot \dfrac{50x^5y^6}{27a^6b^2}$

32. $\dfrac{3x - 6}{5x - 20} \cdot \dfrac{10x - 40}{27x - 54}$

33. $\dfrac{8x - 12}{14x + 7} \cdot \dfrac{42x + 21}{32x - 48}$

34. $\dfrac{3x^2 + 2x}{2xy - 3y} \cdot \dfrac{2xy^3 - 3y^3}{3x^3 + 2x^2}$

35. $\dfrac{4a^2x - 3a^2}{2by + 5b} \cdot \dfrac{2b^3y + 5b^3}{4ax - 3a}$

36. $\dfrac{x^2 + 5x + 4}{x^3y^2} \cdot \dfrac{x^2y^3}{x^2 + 2x + 1}$

37. $\dfrac{x^2 + x - 2}{xy^2} \cdot \dfrac{x^3y}{x^2 + 5x + 6}$

38. $\dfrac{x^4y^2}{x^2 + 3x - 28} \cdot \dfrac{x^2 - 49}{xy^4}$

39. $\dfrac{x^5y^3}{x^2 + 13x + 30} \cdot \dfrac{x^2 + 2x - 3}{x^7y^2}$

40. $\dfrac{2x^2 - 5x}{2xy + y} \cdot \dfrac{2xy^2 + y^2}{5x^2 - 2x^3}$

41. $\dfrac{3a^3 + 4a^2}{5ab - 3b} \cdot \dfrac{3b^3 - 5ab^3}{3a^2 + 4a}$

42. $\dfrac{x^2 - 2x - 24}{x^2 - 5x - 6} \cdot \dfrac{x^2 + 5x + 6}{x^2 + 6x + 8}$

43. $\dfrac{x^2 - 8x + 7}{x^2 + 3x - 4} \cdot \dfrac{x^2 + 3x - 10}{x^2 - 9x + 14}$

44. $\dfrac{x^2 + 2x - 35}{x^2 + 4x - 21} \cdot \dfrac{x^2 + 3x - 18}{x^2 + 9x + 18}$

45. $\dfrac{y^2 + y - 20}{y^2 + 2y - 15} \cdot \dfrac{y^2 + 4y - 21}{y^2 + 3y - 28}$

Simplify.

46. $\dfrac{x^2 - 3x - 4}{x^2 + 6x + 5} \cdot \dfrac{x^2 + 5x + 6}{8 + 2x - x^2}$

47. $\dfrac{25 - n^2}{n^2 - 2n - 35} \cdot \dfrac{n^2 - 8n - 20}{n^2 - 3n - 10}$

48. $\dfrac{12x^2 - 6x}{x^2 + 6x + 5} \cdot \dfrac{2x^4 + 10x^3}{4x^2 - 1}$

49. $\dfrac{8x^3 + 4x^2}{x^2 - 3x + 2} \cdot \dfrac{x^2 - 4}{16x^2 + 8x}$

50. $\dfrac{16 + 6x - x^2}{x^2 - 10x - 24} \cdot \dfrac{x^2 - 6x - 27}{x^2 - 17x + 72}$

51. $\dfrac{x^2 - 11x + 28}{x^2 - 13x + 42} \cdot \dfrac{x^2 + 7x + 10}{20 - x - x^2}$

52. $\dfrac{2x^2 + 5x + 2}{2x^2 + 7x + 3} \cdot \dfrac{x^2 - 7x - 30}{x^2 - 6x - 40}$

53. $\dfrac{x^2 - 4x - 32}{x^2 - 8x - 48} \cdot \dfrac{3x^2 + 17x + 10}{3x^2 - 22x - 16}$

54. $\dfrac{2x^2 + x - 3}{2x^2 - x - 6} \cdot \dfrac{2x^2 - 9x + 10}{2x^2 - 3x + 1}$

55. $\dfrac{3y^2 + 14y + 8}{2y^2 + 7y - 4} \cdot \dfrac{2y^2 + 9y - 5}{3y^2 + 16y + 5}$

Objective C

Simplify.

56. $\dfrac{4x^2y^3}{15a^2b^3} \div \dfrac{6xy}{5a^3b^5}$

57. $\dfrac{9x^3y^4}{16a^4b^2} \div \dfrac{45x^4y^2}{14a^7b}$

58. $\dfrac{6x - 12}{8x + 32} \div \dfrac{18x - 36}{10x + 40}$

59. $\dfrac{28x + 14}{45x - 30} \div \dfrac{14x + 7}{30x - 20}$

60. $\dfrac{6x^3 + 7x^2}{12x - 3} \div \dfrac{6x^2 + 7x}{36x - 9}$

61. $\dfrac{5a^2y + 3a^2}{2x^3 + 5x^2} \div \dfrac{10ay + 6a}{6x^3 + 15x^2}$

62. $\dfrac{x^2 + 4x + 3}{x^2y} \div \dfrac{x^2 + 2x + 1}{xy^2}$

63. $\dfrac{x^3y^2}{x^2 - 3x - 10} \div \dfrac{xy^4}{x^2 - x - 20}$

64. $\dfrac{x^2 - 49}{x^4y^3} \div \dfrac{x^2 - 14x + 49}{x^4y^3}$

65. $\dfrac{x^2y^5}{x^2 - 11x + 30} \div \dfrac{xy^6}{x^2 - 7x + 10}$

Simplify.

66. $\dfrac{4ax - 8a}{c^2} \div \dfrac{2y - xy}{c^3}$

67. $\dfrac{3x^2y - 9xy}{a^2b} \div \dfrac{3x^2 - x^3}{ab^2}$

68. $\dfrac{x^2 - 5x + 6}{x^2 - 9x + 18} \div \dfrac{x^2 - 6x + 8}{x^2 - 9x + 20}$

69. $\dfrac{x^2 + 3x - 40}{x^2 + 2x - 35} \div \dfrac{x^2 + 2x - 48}{x^2 + 3x - 18}$

70. $\dfrac{x^2 + 2x - 15}{x^2 - 4x - 45} \div \dfrac{x^2 + x - 12}{x^2 - 5x - 36}$

71. $\dfrac{y^2 - y - 56}{y^2 + 8y + 7} \div \dfrac{y^2 - 13y + 40}{y^2 - 4y - 5}$

72. $\dfrac{8 + 2x - x^2}{x^2 + 7x + 10} \div \dfrac{x^2 - 11x + 28}{x^2 - x - 42}$

73. $\dfrac{x^2 - x - 2}{x^2 - 7x + 10} \div \dfrac{x^2 - 3x - 4}{40 - 3x - x^2}$

74. $\dfrac{2x^2 - 3x - 20}{2x^2 - 7x - 30} \div \dfrac{2x^2 - 5x - 12}{4x^2 + 12x + 9}$

75. $\dfrac{6n^2 + 13n + 6}{4n^2 - 9} \div \dfrac{6n^2 + n - 2}{4n^2 - 1}$

76. $\dfrac{9x^2 - 16}{6x^2 - 11x + 4} \div \dfrac{6x^2 + 11x + 4}{8x^2 + 10x + 3}$

77. $\dfrac{15 - 14x - 8x^2}{4x^2 + 4x - 15} \div \dfrac{4x^2 + 13x - 12}{3x^2 + 13x + 4}$

APPLYING THE CONCEPTS

78. Given the expression $\dfrac{9}{x^2 + 1}$, choose some values of x and evaluate the expression for those values. Is it possible to choose a value of x for which the value of the expression is greater than 10? If so, what is that value of x? If not, explain why it is not possible.

79. Given the expression $\dfrac{1}{y - 3}$, choose some values of y and evaluate the expression for those values. Is it possible to choose a value of y for which the value of the expression is greater than 10,000,000? If so, what is that value of y? If not, explain why it is not possible.

For what values of x is the algebraic fraction undefined?

80. $\dfrac{x}{(x - 2)(x + 5)}$

81. $\dfrac{7}{x^2 - 25}$

82. $\dfrac{3x - 8}{3x^2 - 10x - 8}$

Simplify.

83. $\dfrac{xy}{3} \cdot \dfrac{x}{y^2} \div \dfrac{x}{4}$

84. $\left(\dfrac{y}{3}\right) \div \left(\dfrac{y}{2} \cdot \dfrac{y}{4}\right)$

85. $\left(\dfrac{x - 4}{y^2}\right)^3 \cdot \left(\dfrac{y}{4 - x}\right)^3$

86. $\dfrac{x - 2}{x + 5} \div \dfrac{x - 3}{x + 5} \cdot \dfrac{x - 3}{x - 2}$

87. $\dfrac{b + 4}{b - 1} \div \dfrac{b + 4}{b + 2} \cdot \dfrac{b - 1}{b - 5}$

7.2 Expressing Fractions in Terms of the Least Common Multiple (LCM)

Objective A *To find the least common multiple (LCM) of two or more polynomials*

The **least common multiple (LCM)** of two or more numbers is the smallest number that contains the prime factorization of each number.

The LCM of 12 and 18 is 36 because 36 contains the prime factors of 12 and the prime factors of 18.

$$12 = 2 \cdot 2 \cdot 3$$
$$18 = 2 \cdot 3 \cdot 3$$

$$\text{Factors of 12}$$
$$\text{LCM} = 36 = \overbrace{2 \cdot 2 \cdot 3 \cdot 3}$$
$$\underbrace{\qquad\qquad}_{\text{Factors of 18}}$$

The least common multiple of two or more polynomials is the polynomial of least degree that contains the factors of each polynomial.

To find the LCM of two or more polynomials, first factor each polynomial completely. The LCM is the product of each factor the greatest number of times it occurs in any one factorization.

➡ Find the LCM of $4x^2 + 4x$ and $x^2 + 2x + 1$.

The LCM of the polynomials is the product of the LCM of the numerical coefficients and each variable factor the greatest number of times it occurs in any one factorization.

$$4x^2 + 4x = 4x(x + 1) = 2 \cdot 2 \cdot x(x + 1)$$
$$x^2 + 2x + 1 = (x + 1)(x + 1)$$

$$\text{Factors of } 4x^2 + 4x$$
$$\text{LCM} = \overbrace{2 \cdot 2 \cdot x(x + 1)}(x + 1) = 4x(x + 1)(x + 1)$$
$$\underbrace{\qquad\qquad}_{\text{Factors of } x^2 + 2x + 1}$$

LOOK CLOSELY

The LCM must contain the factors of each polynomial. As shown with the braces at the right, the LCM contains the factors of $4x^2 + 4x$ and the factors of $x^2 + 2x + 1$.

Example 1
Find the LCM of $4x^2y$ and $6xy^2$.

Solution
$4x^2y = 2 \cdot 2 \cdot x \cdot x \cdot y$
$6xy^2 = 2 \cdot 3 \cdot x \cdot y \cdot y$
$\text{LCM} = 2 \cdot 2 \cdot 3 \cdot x \cdot x \cdot y \cdot y = 12x^2y^2$

You Try It 1
Find the LCM of $8uv^2$ and $12uw$.

Your solution $2 \cdot 2 \cdot 2$

Example 2
Find the LCM of $x^2 - x - 6$ and $9 - x^2$.

Solution
$x^2 - x - 6 = (x - 3)(x + 2)$
$9 - x^2 = -(x^2 - 9) = -(x + 3)(x - 3)$
$\text{LCM} = (x - 3)(x + 2)(x + 3)$

You Try It 2
Find the LCM of $m^2 - 6m + 9$ and $m^2 - 2m - 3$.

Your solution

Solutions on p. A19

Objective B **To express two fractions in terms of the LCM of their denominators** ....

When adding and subtracting fractions, it is frequently necessary to express two or more fractions in terms of a common denominator. This common denominator is the LCM of the denominators of the fractions.

➡ Write the fractions $\dfrac{x+1}{4x^2}$ and $\dfrac{x-3}{6x^2-12x}$ in terms of the LCM of the denominators.

Find the LCM of the denominators.

The LCM is $12x^2(x-2)$.

For each fraction, multiply the numerator and denominator by the factors whose product with the denominator is the LCM.

$$\frac{x+1}{4x^2} = \frac{x+1}{4x^2} \cdot \frac{3(x-2)}{3(x-2)} = \frac{3x^2-3x-6}{12x^2(x-2)} \longleftarrow$$

$$\frac{x-3}{6x^2-12x} = \frac{x-3}{6x(x-2)} \cdot \frac{2x}{2x} = \frac{2x^2-6x}{12x^2(x-2)} \longleftarrow$$

LCM

Example 3

Write the fractions $\dfrac{x+2}{3x^2}$ and $\dfrac{x-1}{8xy}$ in terms of the LCM of the denominators.

Solution
The LCM is $24x^2y$.

$$\frac{x+2}{3x^2} = \frac{x+2}{3x^2} \cdot \frac{8y}{8y} = \frac{8xy+16y}{24x^2y}$$

$$\frac{x-1}{8xy} = \frac{x-1}{8xy} \cdot \frac{3x}{3x} = \frac{3x^2-3x}{24x^2y}$$

You Try It 3

Write the fractions $\dfrac{x-3}{4xy^2}$ and $\dfrac{2x+1}{9y^2z}$ in terms of the LCM of the denominators.

Your solution

Example 4

Write the fractions $\dfrac{2x-1}{2x-x^2}$ and $\dfrac{x}{x^2+x-6}$ in terms of the LCM of the denominators.

Solution

$$\frac{2x-1}{2x-x^2} = \frac{2x-1}{-(x^2-2x)} = -\frac{2x-1}{x^2-2x}$$

The LCM is $x(x-2)(x+3)$.

$$\frac{2x-1}{2x-x^2} = -\frac{2x-1}{x(x-2)} \cdot \frac{x+3}{x+3} = -\frac{2x^2+5x-3}{x(x-2)(x+3)}$$

$$\frac{x}{x^2+x-6} = \frac{x}{(x-2)(x+3)} \cdot \frac{x}{x} = \frac{x^2}{x(x-2)(x+3)}$$

You Try It 4

Write the fractions $\dfrac{x+4}{x^2-3x-10}$ and $\dfrac{2x}{25-x^2}$ in terms of the LCM of the denominators.

Your solution

Solutions on p. A19

7.2 Exercises

. . . .

Objective A

Find the LCM of the expressions.

1. $8x^3y$
$12xy^2$

2. $6ab^2$
$18ab^3$

3. $10x^4y^2$
$15x^3y$

4. $12a^2b$
$18ab^3$

5. $8x^2$
$4x^2 + 8x$

6. $6y^2$
$4y + 12$

7. $2x^2y$
$3x^2 + 12x$

8. $4xy^2$
$6xy^2 + 12y^2$

9. $9x(x + 2)$
$12(x + 2)^2$

10. $8x^2(x - 1)^2$
$10x^3(x - 1)$

11. $3x + 3$
$2x^2 + 4x + 2$

12. $4x - 12$
$2x^2 - 12x + 18$

13. $(x - 1)(x + 2)$
$(x - 1)(x + 3)$

14. $(2x - 1)(x + 4)$
$(2x + 1)(x + 4)$

15. $(2x + 3)^2$
$(2x + 3)(x - 5)$

16. $(x - 7)(x + 2)$
$(x - 7)^2$

17. $(x - 1)$
$(x - 2)$
$(x - 1)(x - 2)$

18. $(x + 4)(x - 3)$
$x + 4$
$x - 3$

19. $x^2 - x - 6$
$x^2 + x - 12$

20. $x^2 + 3x - 10$
$x^2 + 5x - 14$

21. $x^2 + 5x + 4$
$x^2 - 3x - 28$

22. $x^2 - 10x + 21$
$x^2 - 8x + 15$

23. $x^2 - 2x - 24$
$x^2 - 36$

24. $x^2 + 7x + 10$
$x^2 - 25$

25. $x^2 - 7x - 30$
$x^2 - 5x - 24$

26. $2x^2 - 7x + 3$
$2x^2 + x - 1$

27. $3x^2 - 11x + 6$
$3x^2 + 4x - 4$

28. $2x^2 - 9x + 10$
$2x^2 + x - 15$

29. $6 + x - x^2$
$x + 2$
$x - 3$

30. $15 + 2x - x^2$
$x - 5$
$x + 3$

31. $5 + 4x - x^2$
$x - 5$
$x + 1$

32. $x^2 + 3x - 18$
$3 - x$
$x + 6$

33. $x^2 - 5x + 6$
$1 - x$
$x - 6$

Objective B

Write each fraction in terms of the LCM of the denominators.

34. $\dfrac{4}{x}, \dfrac{3}{x^2}$

35. $\dfrac{5}{ab^2}, \dfrac{6}{ab}$

36. $\dfrac{x}{3y^2}, \dfrac{z}{4y}$

37. $\dfrac{5y}{6x^2}, \dfrac{7}{9xy}$

38. $\dfrac{y}{x(x-3)}, \dfrac{6}{x^2}$

39. $\dfrac{a}{y^2}, \dfrac{6}{y(y+5)}$

40. $\dfrac{9}{(x-1)^2}, \dfrac{6}{x(x-1)}$

41. $\dfrac{a^2}{y(y+7)}, \dfrac{a}{(y+7)^2}$

42. $\dfrac{3}{x-3}, \dfrac{5}{x(3-x)}$

43. $\dfrac{b}{y(y-4)}, \dfrac{b^2}{4-y}$

44. $\dfrac{3}{(x-5)^2}, \dfrac{2}{5-x}$

45. $\dfrac{3}{7-y}, \dfrac{2}{(y-7)^2}$

46. $\dfrac{3}{x^2+2x}, \dfrac{4}{x^2}$

47. $\dfrac{2}{y-3}, \dfrac{3}{y^3-3y^2}$

48. $\dfrac{x-2}{x+3}, \dfrac{x}{x-4}$

49. $\dfrac{x^2}{2x-1}, \dfrac{x+1}{x+4}$

50. $\dfrac{3}{x^2+x-2}, \dfrac{x}{x+2}$

51. $\dfrac{3x}{x-5}, \dfrac{4}{x^2-25}$

52. $\dfrac{5}{2x^2-9x+10}, \dfrac{x-1}{2x-5}$

53. $\dfrac{x-3}{3x^2+4x-4}, \dfrac{2}{x+2}$

54. $\dfrac{x}{x^2+x-6}, \dfrac{2x}{x^2-9}$

55. $\dfrac{x-1}{x^2+2x-15}, \dfrac{x}{x^2+6x+5}$

APPLYING THE CONCEPTS

56. When is the LCM of two expressions equal to their product?

Write each expression in terms of the LCM of the denominators.

57. $\dfrac{8}{10^3}, \dfrac{9}{10^5}$

58. $3, \dfrac{2}{n}$

59. $x, \dfrac{x}{x^2-1}$

60. $\dfrac{x^2+1}{(x-1)^3}, \dfrac{x+1}{(x-1)^2}, \dfrac{1}{x-1}$

61. $\dfrac{c}{6c^2+7cd+d^2}, \dfrac{d}{3c^2-3d^2}$

62. $\dfrac{1}{ab+3a-3b-b^2}, \dfrac{1}{ab+3a+3b+b^2}$

7.3 Addition and Subtraction of Rational Expressions

Objective A   *To add or subtract rational expressions with like denominators*

When adding rational expressions in which the denominators are the same, add the numerators. The denominator of the sum is the common denominator.

$$\frac{a}{b} + \frac{c}{b} = \frac{a+c}{b}$$

$$\frac{5x}{18} + \frac{7x}{18} = \frac{5x + 7x}{18} = \frac{12x}{18} = \frac{2x}{3}$$

$$\frac{x}{x^2 - 1} + \frac{1}{x^2 - 1} = \frac{x + 1}{x^2 - 1} = \frac{\overset{1}{\cancel{(x+1)}}}{(x-1)\underset{1}{\cancel{(x+1)}}} = \frac{1}{x - 1}$$

Note that the sum is written in simplest form

When subtracting rational expressions with like denominators, subtract the numerators. The denominator of the difference is the common denominator. Write the answer in simplest form.

$$\frac{2x}{x-2} - \frac{4}{x-2} = \frac{2x - 4}{x - 2} = \frac{2\overset{1}{\cancel{(x-2)}}}{\underset{1}{\cancel{x-2}}} = 2$$

$$\frac{3x - 1}{x^2 - 5x + 4} - \frac{2x + 3}{x^2 - 5x + 4} = \frac{(3x - 1) - (2x + 3)}{x^2 - 5x + 4} = \frac{3x - 1 - 2x - 3}{x^2 - 5x + 4}$$

$$= \frac{x - 4}{x^2 - 5x + 4} = \frac{\overset{1}{\cancel{(x-4)}}}{\underset{1}{\cancel{(x-4)}}(x - 1)} = \frac{1}{x - 1}$$

Example 1

Simplify: $\dfrac{7}{x^2} + \dfrac{9}{x^2}$

Solution

$$\frac{7}{x^2} + \frac{9}{x^2} = \frac{7 + 9}{x^2} = \frac{16}{x^2}$$

Example 2

Simplify: $\dfrac{3x^2}{x^2 - 1} - \dfrac{x + 4}{x^2 - 1}$

Solution

$$\frac{3x^2}{x^2 - 1} - \frac{x + 4}{x^2 - 1} = \frac{3x^2 - (x + 4)}{x^2 - 1}$$

$$= \frac{3x^2 - x - 4}{x^2 - 1}$$

$$= \frac{(3x - 4)\overset{1}{\cancel{(x + 1)}}}{(x - 1)\underset{1}{\cancel{(x + 1)}}} = \frac{3x - 4}{x - 1}$$

You Try It 1

Simplify: $\dfrac{3}{xy} + \dfrac{12}{xy}$ $\dfrac{15}{xy}$

Your solution

You Try It 2

Simplify: $\dfrac{2x^2}{x^2 - x - 12} - \dfrac{7x + 4}{x^2 - x - 12}$

Your solution

$$\frac{2x^2 - (7x + 4)}{x^2 - x - 12}$$

$$(x - 4)(x + 3)$$

$$2x^2 - 7x - 4$$

$$2x^2 - 8x + x - 4$$

$$2x(x - 4) + 1(x - 4)$$

$$\frac{2x + 1}{x + 3}$$

Solutions on p. A19

Example 3
Simplify:
$$\frac{2x^2 + 5}{x^2 + 2x - 3} - \frac{x^2 - 3x}{x^2 + 2x - 3} + \frac{x - 2}{x^2 + 2x - 3}$$

Solution
$$\frac{2x^2 + 5}{x^2 + 2x - 3} - \frac{x^2 - 3x}{x^2 + 2x - 3} + \frac{x - 2}{x^2 + 2x - 3}$$

$$= \frac{(2x^2 + 5) - (x^2 - 3x) + (x - 2)}{x^2 + 2x - 3}$$

$$= \frac{2x^2 + 5 - x^2 + 3x + x - 2}{x^2 + 2x - 3}$$

$$= \frac{x^2 + 4x + 3}{x^2 + 2x - 3} = \frac{\overset{1}{\cancel{(x + 3)}}(x + 1)}{\underset{1}{\cancel{(x + 3)}}(x - 1)} = \frac{x + 1}{x - 1}$$

You Try It 3
Simplify:
$$\frac{x^2 - 1}{x^2 - 8x + 12} - \frac{2x + 1}{x^2 - 8x + 12} + \frac{x}{x^2 - 8x + 12}$$

Your solution

Solution on p. A19

Objective B   To add or subtract rational expressions with unlike denominators

Before two fractions with unlike denominators can be added or subtracted, each fraction must be expressed in terms of a common denominator. This common denominator is the LCM of the denominators of the fractions.

➡ Simplify: $\dfrac{x - 3}{x^2 - 2x} + \dfrac{6}{x^2 - 4}$

Find the LCM of the denominators.   The LCM is $x(x - 2)(x + 2)$.

$$\frac{x - 3}{x^2 - 2x} + \frac{6}{x^2 - 4} = \frac{x - 3}{x(x - 2)} \cdot \frac{x + 2}{x + 2} + \frac{6}{(x - 2)(x + 2)} \cdot \frac{x}{x}$$

• Write each fraction in terms of the LCM.

$$= \frac{x^2 - x - 6}{x(x - 2)(x + 2)} + \frac{6x}{x(x - 2)(x + 2)}$$

• Multiply the factors in the numerator and then add the fractions.

$$= \frac{(x^2 - x - 6) + 6x}{x(x - 2)(x + 2)}$$

$$= \frac{x^2 + 5x - 6}{x(x - 2)(x + 2)}$$

$$= \frac{(x + 6)(x - 1)}{x(x - 2)(x + 2)}$$

The last step is to factor the numerator to determine whether there are common factors in the numerator and denominator. For this example there are no common factors so the answer is in simplest form.

Example 4

Simplify: $\dfrac{y}{x} - \dfrac{4y}{3x} + \dfrac{3y}{4x}$

Solution

The LCM of the denominators is $12x$.

$\dfrac{y}{x} - \dfrac{4y}{3x} + \dfrac{3y}{4x} = \dfrac{y}{x} \cdot \dfrac{12}{12} - \dfrac{4y}{3x} \cdot \dfrac{4}{4} + \dfrac{3y}{4x} \cdot \dfrac{3}{3}$

$\qquad = \dfrac{12y}{12x} - \dfrac{16y}{12x} + \dfrac{9y}{12x}$

$\qquad = \dfrac{12y - 16y + 9y}{12} = \dfrac{5y}{12x}$

You Try It 4

Simplify: $\dfrac{z}{8y} - \dfrac{4z}{3y} + \dfrac{5z}{4y}$

Your solution

$\dfrac{z}{8y} = \dfrac{3z}{24y}$

$- \dfrac{4z}{3y} = \dfrac{32z}{24y}$

$+ \dfrac{5z}{4y} = \dfrac{30z}{24y}$

$\dfrac{z}{24y}$

Example 5

Simplify: $\dfrac{2x}{x - 3} - \dfrac{5}{3 - x}$

Solution

Remember: $3 - x = -(x - 3)$.

Therefore, $\dfrac{5}{3 - x} = \dfrac{5}{-(x - 3)} = \dfrac{-5}{x - 3}$.

$\dfrac{2x}{x - 3} - \dfrac{5}{3 - x} = \dfrac{2x}{x - 3} - \dfrac{-5}{x - 3}$

$\qquad = \dfrac{2x - (-5)}{x - 3} = \dfrac{2x + 5}{x - 3}$

You Try It 5

Simplify: $\dfrac{5x}{x - 2} - \dfrac{3}{2 - x}$

Your solution

$\dfrac{5x}{x - 2} =$

$- \dfrac{3}{2 - x} = \dfrac{5x}{(x - 2)}$

$\dfrac{-1(x - 2)}{-1}$

$\dfrac{3}{(x - 2)}$

Example 6

Simplify: $\dfrac{2x}{2x - 3} - \dfrac{1}{x + 1}$

Solution

The LCM is $(2x - 3)(x + 1)$.

$\dfrac{2x}{2x - 3} - \dfrac{1}{x + 1}$

$= \dfrac{2x}{2x - 3} \cdot \dfrac{x + 1}{x + 1} - \dfrac{1}{x + 1} \cdot \dfrac{2x - 3}{2x - 3}$

$= \dfrac{2x^2 + 2x}{(2x - 3)(x + 1)} - \dfrac{2x - 3}{(2x - 3)(x + 1)}$

$= \dfrac{(2x^2 + 2x) - (2x - 3)}{(2x - 3)(x + 1)} = \dfrac{2x^2 + 3}{(2x - 3)(x + 1)}$

You Try It 6

Simplify: $\dfrac{4x}{3x - 1} - \dfrac{9}{x + 4}$

Your solution

$\dfrac{4x}{3x - 1} = \dfrac{}{(3x - 1)(x + 4)}$

$- \dfrac{9}{x + 4} = \dfrac{}{(3x - 1)(x + 4)}$

$\dfrac{(3x - 1)(x + 4)}{(x + 4)}$

Solutions on pp. A19–A20

Example 7

Simplify: $1 + \dfrac{3}{x^2}$

Solution

The LCM is x^2.

$$1 + \frac{3}{x^2} = 1 \cdot \frac{x^2}{x^2} + \frac{3}{x^2}$$

$$= \frac{x^2}{x^2} + \frac{3}{x^2}$$

$$= \frac{x^2 + 3}{x^2}$$

You Try It 7

Simplify: $2 - \dfrac{1}{x - 3}$

Your solution

Example 8

Simplify: $\dfrac{x + 3}{x^2 - 2x - 8} + \dfrac{3}{4 - x}$

Solution

The LCM is $(x - 4)(x + 2)$.

Recall: $\dfrac{3}{4 - x} = \dfrac{-3}{x - 4}$

$$\frac{x + 3}{x^2 - 2x - 8} + \frac{3}{4 - x}$$

$$= \frac{x + 3}{(x - 4)(x + 2)} + \frac{(-3)}{x - 4}$$

$$= \frac{x + 3}{(x - 4)(x + 2)} + \frac{(-3)}{x - 4} \cdot \frac{x + 2}{x + 2}$$

$$= \frac{x + 3}{(x - 4)(x + 2)} + \frac{(-3)(x + 2)}{(x - 4)(x + 2)}$$

$$= \frac{(x + 3) + (-3)(x + 2)}{(x - 4)(x + 2)}$$

$$= \frac{x + 3 - 3x - 6}{(x - 4)(x + 2)} = \frac{-2x - 3}{(x - 4)(x + 2)}$$

You Try It 8

Simplify: $\dfrac{2x - 1}{x^2 - 25} + \dfrac{2}{5 - x}$

Your solution

$$\frac{2x-1}{x^2 - 25}$$

$$\frac{2x-1}{(x-5)(x+5)} \quad 2 - \frac{2}{(5+x)}$$

Example 9

Simplify: $\dfrac{3x + 2}{2x^2 - x - 1} - \dfrac{3}{2x + 1} + \dfrac{4}{x - 1}$

Solution

The LCM is $(2x + 1)(x - 1)$.

$$\frac{3x + 2}{2x^2 - x - 1} - \frac{3}{2x + 1} + \frac{4}{x - 1}$$

$$= \frac{3x + 2}{(2x + 1)(x - 1)} - \frac{3}{2x + 1} \cdot \frac{x - 1}{x - 1} + \frac{4}{x - 1} \cdot \frac{2x + 1}{2x + 1}$$

$$= \frac{3x + 2}{(2x + 1)(x - 1)} - \frac{3x - 3}{(2x + 1)(x - 1)} + \frac{8x + 4}{(2x + 1)(x - 1)}$$

$$= \frac{(3x + 2) - (3x - 3) + (8x + 4)}{(2x + 1)(x - 1)}$$

$$= \frac{3x + 2 - 3x + 3 + 8x + 4}{(2x + 1)(x - 1)} = \frac{8x + 9}{(2x + 1)(x - 1)}$$

You Try It 9

Simplify: $\dfrac{2x - 3}{3x^2 - x - 2} + \dfrac{5}{3x + 2} - \dfrac{1}{x - 1}$

Your solution

Solutions on p. A20

7.3 Exercises

Objective A

Simplify.

1. $\dfrac{3}{y^2} + \dfrac{8}{y^2}$

2. $\dfrac{6}{ab} - \dfrac{2}{ab}$

3. $\dfrac{3}{x+4} - \dfrac{10}{x+4}$

4. $\dfrac{x}{x+6} - \dfrac{2}{x+6}$

5. $\dfrac{3x}{2x+3} + \dfrac{5x}{2x+3}$

6. $\dfrac{6y}{4y+1} - \dfrac{11y}{4y+1}$

7. $\dfrac{2x+1}{x-3} + \dfrac{3x+6}{x-3}$

8. $\dfrac{4x+3}{2x-7} + \dfrac{3x-8}{2x-7}$

9. $\dfrac{5x-1}{x+9} - \dfrac{3x+4}{x+9}$

10. $\dfrac{6x-5}{x-10} - \dfrac{3x-4}{x-10}$

11. $\dfrac{x-7}{2x+7} - \dfrac{4x-3}{2x+7}$

12. $\dfrac{2n}{3n+4} - \dfrac{5n-3}{3n+4}$

13. $\dfrac{x}{x^2+2x-15} - \dfrac{3}{x^2+2x-15}$

14. $\dfrac{3x}{x^2+3x-10} - \dfrac{6}{x^2+3x-10}$

15. $\dfrac{2x+3}{x^2-x-30} - \dfrac{x-2}{x^2-x-30}$

16. $\dfrac{3x-1}{x^2+5x-6} - \dfrac{2x-7}{x^2+5x-6}$

17. $\dfrac{4y+7}{2y^2+7y-4} - \dfrac{y-5}{2y^2+7y-4}$

18. $\dfrac{x+1}{2x^2-5x-12} + \dfrac{x+2}{2x^2-5x-12}$

19. $\dfrac{2x^2+3x}{x^2-9x+20} + \dfrac{2x^2-3}{x^2-9x+20} - \dfrac{4x^2+2x+1}{x^2-9x+20}$

20. $\dfrac{2x^2+3x}{x^2-2x-63} - \dfrac{x^2-3x+21}{x^2-2x-63} - \dfrac{x-7}{x^2-2x-63}$

Objective B

Simplify.

21. $\dfrac{4}{x} + \dfrac{5}{y}$

22. $\dfrac{7}{a} + \dfrac{5}{b}$

23. $\dfrac{12}{x} - \dfrac{5}{2x}$

24. $\dfrac{5}{3a} - \dfrac{3}{4a}$

25. $\dfrac{1}{2x} - \dfrac{5}{4x} + \dfrac{7}{6x}$

26. $\dfrac{7}{4y} + \dfrac{11}{6y} - \dfrac{8}{3y}$

27. $\dfrac{5}{3x} - \dfrac{2}{x^2} + \dfrac{3}{2x}$

28. $\dfrac{6}{y^2} + \dfrac{3}{4y} - \dfrac{2}{5y}$

29. $\dfrac{2}{x} - \dfrac{3}{2y} + \dfrac{3}{5x} - \dfrac{1}{4y}$

30. $\dfrac{5}{2a} + \dfrac{7}{3b} - \dfrac{2}{b} - \dfrac{3}{4a}$

31. $\dfrac{2x + 1}{3x} + \dfrac{x - 1}{5x}$

32. $\dfrac{4x - 3}{6x} + \dfrac{2x + 3}{4x}$

33. $\dfrac{x - 3}{6x} + \dfrac{x + 4}{8x}$

34. $\dfrac{2x - 3}{2x} + \dfrac{x + 3}{3x}$

35. $\dfrac{2x + 9}{9x} - \dfrac{x - 5}{5x}$

36. $\dfrac{3y - 2}{12y} - \dfrac{y - 3}{18y}$

37. $\dfrac{x + 4}{2x} - \dfrac{x - 1}{x^2}$

38. $\dfrac{x - 2}{3x^2} - \dfrac{x + 4}{x}$

39. $\dfrac{x - 10}{4x^2} + \dfrac{x + 1}{2x}$

40. $\dfrac{x + 5}{3x^2} + \dfrac{2x + 1}{2x}$

41. $\dfrac{4}{x + 4} - x$

42. $2x + \dfrac{1}{x}$

43. $5 - \dfrac{x - 2}{x + 1}$

44. $3 + \dfrac{x - 1}{x + 1}$

Simplify.

45. $\dfrac{x+3}{6x} - \dfrac{x-3}{8x^2}$

46. $\dfrac{x+2}{xy} - \dfrac{3x-2}{x^2y}$

47. $\dfrac{3x-1}{xy^2} - \dfrac{2x+3}{xy}$

48. $\dfrac{4x-3}{3x^2y} + \dfrac{2x+1}{4xy^2}$

49. $\dfrac{5x+7}{6xy^2} - \dfrac{4x-3}{8x^2y}$

50. $\dfrac{x-2}{8x^2} - \dfrac{x+7}{12xy}$

51. $\dfrac{3x-1}{6y^2} - \dfrac{x+5}{9xy}$

52. $\dfrac{4}{x-2} + \dfrac{5}{x+3}$

53. $\dfrac{2}{x-3} + \dfrac{5}{x-4}$

54. $\dfrac{6}{x-7} - \dfrac{4}{x+3}$

55. $\dfrac{3}{y+6} - \dfrac{4}{y-3}$

56. $\dfrac{2x}{x+1} + \dfrac{1}{x-3}$

57. $\dfrac{3x}{x-4} + \dfrac{2}{x+6}$

58. $\dfrac{4x}{2x-1} - \dfrac{5}{x-6}$

59. $\dfrac{6x}{x+5} - \dfrac{3}{2x+3}$

60. $\dfrac{2a}{a-7} + \dfrac{5}{7-a}$

61. $\dfrac{4x}{6-x} + \dfrac{5}{x-6}$

62. $\dfrac{x}{x^2-9} + \dfrac{3}{x-3}$

63. $\dfrac{y}{y^2-16} + \dfrac{1}{y-4}$

64. $\dfrac{2x}{x^2-x-6} - \dfrac{3}{x+2}$

65. $\dfrac{(x-1)^2}{(x+1)^2} - 1$

66. $1 - \dfrac{(y-2)^2}{(y+2)^2}$

67. $\dfrac{x}{1-x^2} - 1 + \dfrac{x}{1+x}$

68. $\dfrac{y}{x-y} + 2 - \dfrac{x}{y-x}$

Simplify.

69. $\dfrac{3x - 1}{x^2 - 10x + 25} - \dfrac{3}{x - 5}$

70. $\dfrac{2a + 3}{a^2 - 7a + 12} - \dfrac{2}{a - 3}$

71. $\dfrac{x + 4}{x^2 - x - 42} + \dfrac{3}{7 - x}$

72. $\dfrac{x + 3}{x^2 - 3x - 10} + \dfrac{2}{5 - x}$

73. $\dfrac{1}{x + 1} + \dfrac{x}{x - 6} - \dfrac{5x - 2}{x^2 - 5x - 6}$

74. $\dfrac{x}{x - 4} + \dfrac{5}{x + 5} - \dfrac{11x - 8}{x^2 + x - 20}$

75. $\dfrac{3x + 1}{x - 1} - \dfrac{x - 1}{x - 3} + \dfrac{x + 1}{x^2 - 4x + 3}$

76. $\dfrac{4x + 1}{x - 8} - \dfrac{3x + 2}{x + 4} - \dfrac{49x + 4}{x^2 - 4x - 32}$

77. $\dfrac{2x + 9}{3 - x} + \dfrac{x + 5}{x + 7} - \dfrac{2x^2 + 3x - 3}{x^2 + 4x - 21}$

78. $\dfrac{3x + 5}{x + 5} - \dfrac{x + 1}{2 - x} - \dfrac{4x^2 - 3x - 1}{x^2 + 3x - 10}$

APPLYING THE CONCEPTS

79. Find the sum of the following:
$$\dfrac{1}{1 \cdot 2} + \dfrac{1}{2 \cdot 3}$$
$$\dfrac{1}{1 \cdot 2} + \dfrac{1}{2 \cdot 3} + \dfrac{1}{3 \cdot 4}$$
$$\dfrac{1}{1 \cdot 2} + \dfrac{1}{2 \cdot 3} + \dfrac{1}{3 \cdot 4} + \dfrac{1}{4 \cdot 5}$$

Note the pattern in these sums, and find the sum of 50 terms, of 100 terms, and of 1000 terms.

80. In your own words, explain the procedure for adding rational expres-
[W] sions with different denominators.

Simplify.

81. $\dfrac{x^2 + x - 6}{x^2 + 2x - 8} \cdot \dfrac{x^2 + 5x + 4}{x^2 + 2x - 3} - \dfrac{2}{x - 1}$

82. $\dfrac{x^2 + 9x + 20}{x^2 + 4x - 5} - \dfrac{x^2 - 49}{x^2 + 6x - 7} \div \dfrac{x}{x - 7}$

83. $\dfrac{x^2 - 25}{x^2 + 10x + 25} \cdot \dfrac{x^2 - 7x + 10}{x^2 - x - 2} + \dfrac{1}{x + 1}$

7.4 Complex Fractions

Objective A To simplify a complex fraction

A **complex fraction** is a fraction whose numerator or denominator contains one or more fractions. Examples of complex fractions are shown at the right.

$$\frac{3}{2 - \frac{1}{2}}, \quad \frac{4 + \frac{1}{x}}{3 + \frac{2}{x}}, \quad \frac{\frac{1}{x-1} + x + 3}{x - 3 + \frac{1}{x+4}}$$

➡ Simplify: $\dfrac{1 - \dfrac{4}{x^2}}{1 + \dfrac{2}{x}}$

Find the LCM of the denominators of the fractions in the numerator and denominator. The LCM of x and x^2 is x^2.

$$\frac{1 - \dfrac{4}{x^2}}{1 + \dfrac{2}{x}} = \frac{1 - \dfrac{4}{x^2}}{1 + \dfrac{2}{x}} \cdot \frac{x^2}{x^2}$$

• Multiply numerator and denominator by the LCM.

$$= \frac{1 \cdot x^2 - \dfrac{4}{x^2} \cdot x^2}{1 \cdot x^2 + \dfrac{2}{x} \cdot x^2}$$

• Simplify.

$$= \frac{x^2 - 4}{x^2 + 2x} = \frac{(x - 2)\,\overset{1}{\cancel{(x+2)}}}{x\underset{1}{\cancel{(x+2)}}}$$

$$= \frac{x - 2}{x}$$

Example 1

Simplify: $\dfrac{\dfrac{1}{x} + \dfrac{1}{2}}{\dfrac{1}{x^2} - \dfrac{1}{4}}$

Solution

The LCM of x, 2, x^2, and 4 is $4x^2$.

$$\frac{\dfrac{1}{x} + \dfrac{1}{2}}{\dfrac{1}{x^2} - \dfrac{1}{4}} = \frac{\dfrac{1}{x} + \dfrac{1}{2}}{\dfrac{1}{x^2} - \dfrac{1}{4}} \cdot \frac{4x^2}{4x^2} = \frac{\dfrac{1}{x} \cdot 4x^2 + \dfrac{1}{2} \cdot 4x^2}{\dfrac{1}{x^2} \cdot 4x^2 - \dfrac{1}{4} \cdot 4x^2}$$

$$= \frac{4x + 2x^2}{4 - x^2} = \frac{2x\,\overset{1}{\cancel{(2 + x)}}}{(2 - x)\underset{1}{\cancel{(2 + x)}}} = \frac{2x}{2 - x}$$

You Try It 1

Simplify: $\dfrac{\dfrac{1}{3} - \dfrac{1}{x}}{\dfrac{1}{9} - \dfrac{1}{x^2}}$

Your solution

Solution on p. A20

Example 2

Simplify: $\dfrac{1 - \dfrac{2}{x} - \dfrac{15}{x^2}}{1 - \dfrac{11}{x} + \dfrac{30}{x^2}}$

You Try It 2

Simplify: $\dfrac{1 + \dfrac{4}{x} + \dfrac{3}{x^2}}{1 + \dfrac{10}{x} + \dfrac{21}{x^2}}$

Solution

The LCM of x and x^2 is x^2.

$$\dfrac{1 - \dfrac{2}{x} - \dfrac{15}{x^2}}{1 - \dfrac{11}{x} + \dfrac{30}{x^2}} = \dfrac{1 - \dfrac{2}{x} - \dfrac{15}{x^2}}{1 - \dfrac{11}{x} + \dfrac{30}{x^2}} \cdot \dfrac{x^2}{x^2}$$

$$= \dfrac{1 \cdot x^2 - \dfrac{2}{x} \cdot x^2 - \dfrac{15}{x^2} \cdot x^2}{1 \cdot x^2 - \dfrac{11}{x} \cdot x^2 + \dfrac{30}{x^2} \cdot x^2}$$

$$= \dfrac{x^2 - 2x - 15}{x^2 - 11x + 30}$$

$$= \dfrac{\overset{1}{\cancel{(x - 5)}}(x + 3)}{\underset{1}{\cancel{(x - 5)}}(x - 6)} = \dfrac{x + 3}{x - 6}$$

You solution

Example 3

Simplify: $\dfrac{x - 8 + \dfrac{20}{x + 4}}{x - 10 + \dfrac{24}{x + 4}}$

You Try It 3

Simplify: $\dfrac{x + 3 - \dfrac{20}{x - 5}}{x + 8 + \dfrac{30}{x - 5}}$

Solution

The LCM is $x + 4$.

$$\dfrac{x - 8 + \dfrac{20}{x + 4}}{x - 10 + \dfrac{24}{x + 4}}$$

$$= \dfrac{x - 8 + \dfrac{20}{x + 4}}{x - 10 + \dfrac{24}{x + 4}} \cdot \dfrac{x + 4}{x + 4}$$

$$= \dfrac{(x - 8)(x + 4) + \dfrac{20}{x + 4} \cdot (x + 4)}{(x - 10)(x + 4) + \dfrac{24}{x + 4} \cdot (x + 4)}$$

$$= \dfrac{x^2 - 4x - 32 + 20}{x^2 - 6x - 40 + 24} = \dfrac{x^2 - 4x - 12}{x^2 - 6x - 16}$$

$$= \dfrac{(x - 6)\overset{1}{\cancel{(x + 2)}}}{(x - 8)\underset{1}{\cancel{(x + 2)}}} = \dfrac{x - 6}{x - 8}$$

Your solution

Solutions on p. A21

7.4 Exercises

Objective A

Simplify.

1. $\dfrac{1 + \dfrac{3}{x}}{1 - \dfrac{9}{x^2}}$

2. $\dfrac{1 + \dfrac{4}{x}}{1 - \dfrac{16}{x^2}}$

3. $\dfrac{2 - \dfrac{8}{x + 4}}{3 - \dfrac{12}{x + 4}}$

4. $\dfrac{5 - \dfrac{25}{x + 5}}{1 - \dfrac{3}{x + 5}}$

5. $\dfrac{1 + \dfrac{5}{y - 2}}{1 - \dfrac{2}{y - 2}}$

6. $\dfrac{2 - \dfrac{11}{2x - 1}}{3 - \dfrac{17}{2x - 1}}$

7. $\dfrac{4 - \dfrac{2}{x + 7}}{5 + \dfrac{1}{x + 7}}$

8. $\dfrac{5 + \dfrac{3}{x - 8}}{2 - \dfrac{1}{x - 8}}$

9. $\dfrac{1 - \dfrac{1}{x} - \dfrac{6}{x^2}}{1 - \dfrac{9}{x^2}}$

10. $\dfrac{1 + \dfrac{4}{x} + \dfrac{4}{x^2}}{1 - \dfrac{2}{x} - \dfrac{8}{x^2}}$

11. $\dfrac{1 - \dfrac{5}{x} - \dfrac{6}{x^2}}{1 + \dfrac{6}{x} + \dfrac{5}{x^2}}$

12. $\dfrac{1 - \dfrac{7}{a} + \dfrac{12}{a^2}}{1 + \dfrac{1}{a} - \dfrac{20}{a^2}}$

13. $\dfrac{1 - \dfrac{6}{x} + \dfrac{8}{x^2}}{\dfrac{4}{x^2} + \dfrac{3}{x} - 1}$

14. $\dfrac{1 + \dfrac{3}{x} - \dfrac{18}{x^2}}{\dfrac{21}{x^2} - \dfrac{4}{x} - 1}$

15. $\dfrac{x - \dfrac{4}{x + 3}}{1 + \dfrac{1}{x + 3}}$

16. $\dfrac{y + \dfrac{1}{y - 2}}{1 + \dfrac{1}{y - 2}}$

17. $\dfrac{1 - \dfrac{x}{2x + 1}}{x - \dfrac{1}{2x + 1}}$

18. $\dfrac{1 - \dfrac{2x - 2}{3x - 1}}{x - \dfrac{4}{3x - 1}}$

Simplify.

19. $\dfrac{x - 5 + \dfrac{14}{x + 4}}{x + 3 - \dfrac{2}{x + 4}}$

20. $\dfrac{a + 4 + \dfrac{5}{a - 2}}{a + 6 + \dfrac{15}{a - 2}}$

21. $\dfrac{x + 3 - \dfrac{10}{x - 6}}{x + 2 - \dfrac{20}{x - 6}}$

22. $\dfrac{x - 7 + \dfrac{5}{x - 1}}{x - 3 + \dfrac{1}{x - 1}}$

23. $\dfrac{y - 6 + \dfrac{22}{2y + 3}}{y - 5 + \dfrac{11}{2y + 3}}$

24. $\dfrac{x + 2 - \dfrac{12}{2x - 1}}{x + 1 - \dfrac{9}{2x - 1}}$

25. $\dfrac{x - \dfrac{2}{2x - 3}}{2x - 1 - \dfrac{8}{2x - 3}}$

26. $\dfrac{x + 3 - \dfrac{18}{2x + 1}}{x - \dfrac{6}{2x + 1}}$

27. $\dfrac{\dfrac{1}{x} - \dfrac{2}{x - 1}}{\dfrac{3}{x} + \dfrac{1}{x - 1}}$

28. $\dfrac{\dfrac{3}{n + 1} + \dfrac{1}{n}}{\dfrac{2}{n + 1} + \dfrac{3}{n}}$

29. $\dfrac{\dfrac{3}{2x - 1} - \dfrac{1}{x}}{\dfrac{4}{x} + \dfrac{2}{2x - 1}}$

30. $\dfrac{\dfrac{4}{3x + 1} + \dfrac{3}{x}}{\dfrac{6}{x} - \dfrac{2}{3x + 1}}$

APPLYING THE CONCEPTS

Simplify.

31. $1 + \dfrac{1}{1 + \dfrac{1}{2}}$

32. $1 + \dfrac{1}{1 + \dfrac{1}{1 + \dfrac{1}{2}}}$

33. $1 - \dfrac{1}{1 - \dfrac{1}{x}}$

34. $\dfrac{a^{-1} - b^{-1}}{a^{-2} - b^{-2}}$

35. $\left(\dfrac{y}{4} - \dfrac{4}{y}\right) \div \left(\dfrac{4}{y} - 3 + \dfrac{y}{2}\right)$

36. $\left(\dfrac{b}{8} - \dfrac{8}{b}\right) \div \left(\dfrac{8}{b} - 5 + \dfrac{b}{2}\right)$

37. $\dfrac{1 + x^{-1}}{1 - x^{-1}}$

38. $\dfrac{x + x^{-1}}{x - x^{-1}}$

39. $\dfrac{x^{-1}}{y^{-1}} + \dfrac{x}{y}$

7.5 Solving Equations Containing Fractions

Objective A *To solve an equation containing fractions*

To solve an equation containing fractions, **clear denominators** by multiplying each side of the equation by the LCM of the denominators. Then solve for the variable.

➡ Solve: $\dfrac{3x-1}{4} + \dfrac{2}{3} = \dfrac{7}{6}$

$$\dfrac{3x-1}{4} + \dfrac{2}{3} = \dfrac{7}{6}$$

$$12\left(\dfrac{3x-1}{4} + \dfrac{2}{3}\right) = 12 \cdot \dfrac{7}{6}$$

- The LCM is 12. Multiply each side of the equation by the LCM.

$$12\left(\dfrac{3x-1}{4}\right) + 12 \cdot \dfrac{2}{3} = 12 \cdot \dfrac{7}{6}$$

- Simplify using the Distributive Property and the Properties of Fractions.

$$\dfrac{\overset{3}{\cancel{12}}}{1}\left(\dfrac{3x-1}{\cancel{4}}\right) + \dfrac{\overset{4}{\cancel{12}}}{1} \cdot \dfrac{2}{\underset{1}{\cancel{3}}} = \dfrac{\overset{2}{\cancel{12}}}{1} \cdot \dfrac{7}{\underset{1}{\cancel{6}}}$$

$$9x - 3 + 8 = 14$$

- Solve for x.

$$9x + 5 = 14$$
$$9x = 9$$
$$x = 1$$

1 checks as a solution. The solution is 1.

Occasionally, a value of the variable that appears to be a solution of an equation will make one of the denominators zero. In this case, the equation has no solution for that value of the variable.

➡ Solve: $\dfrac{2x}{x-2} = 1 + \dfrac{4}{x-2}$

$$\dfrac{2x}{x-2} = 1 + \dfrac{4}{x-2}$$

$$(x-2)\dfrac{2x}{x-2} = (x-2)\left(1 + \dfrac{4}{x-2}\right)$$

- The LCM is $x - 2$. Multiply each side of the equation by the LCM.

$$(x-2)\dfrac{2x}{x-2} = (x-2) \cdot 1 + (x-2)\dfrac{4}{x-2}$$

- Simplify using the Distributive Property and the Properties of Fractions.

$$\dfrac{\overset{1}{\cancel{(x-2)}}}{1}\dfrac{2x}{\cancel{x-2}} = (x-2) \cdot 1 + \dfrac{\overset{1}{\cancel{(x-2)}}}{1}\dfrac{4}{\cancel{x-2}}$$

$$2x = x - 2 + 4$$

- Solve for x.

$$2x = x + 2$$
$$x = 2$$

When x is replaced by 2, the denominators of $\dfrac{2x}{x-2}$ and $\dfrac{4}{x-2}$ are zero. Therefore, the equation has no solution.

Example 1

Solve: $\dfrac{x}{x+4} = \dfrac{2}{x}$

Solution

The LCM is $x(x+4)$.

$$\frac{x}{x+4} = \frac{2}{x}$$

$$x(x+4)\left(\frac{x}{x+4}\right) = x(x+4)\left(\frac{2}{x}\right)$$

$$\frac{x(x+4)}{1}\cdot\frac{x}{x+4} = \frac{x(x+4)}{1}\cdot\frac{2}{x}$$

$$x^2 = (x+4)2$$
$$x^2 = 2x+8$$

Solve the quadratic equation by factoring.

$$x^2 - 2x - 8 = 0$$
$$(x-4)(x+2) = 0$$
$$x-4=0 \qquad x+2=0$$
$$x=4 \qquad\quad x=-2$$

Both 4 and −2 check as solutions.
The solutions are 4 and −2.

Example 2

Solve: $\dfrac{3x}{x-4} = 5 + \dfrac{12}{x-4}$

Solution

The LCM is $x-4$.

$$\frac{3x}{x-4} = 5 + \frac{12}{x-4}$$

$$(x-4)\left(\frac{3x}{x-4}\right) = (x-4)\left(5 + \frac{12}{x-4}\right)$$

$$\frac{(x-4)}{1}\cdot\frac{3}{x-4} = \frac{(x-4)}{1}\cdot 5 + \frac{(x-4)}{1}\cdot\frac{12}{x-4}$$

$$3x = (x-4)5 + 12$$
$$3x = 5x - 20 + 12$$
$$3x = 5x - 8$$
$$-2x = -8$$
$$x = 4$$

4 does not check as a solution.
The equation has no solution.

You Try It 1

Solve: $\dfrac{x}{x+6} = \dfrac{3}{x}$

Your solution

You Try It 2

Solve: $\dfrac{5x}{x+2} = 3 - \dfrac{10}{x+2}$

Your solution

Solutions on p. A21

7.5 Exercises

· ·

Objective A

Solve.

1. $\dfrac{2x}{3} - \dfrac{5}{2} = -\dfrac{1}{2}$

2. $\dfrac{x}{3} - \dfrac{1}{4} = \dfrac{1}{12}$

3. $\dfrac{x}{3} - \dfrac{1}{4} = \dfrac{x}{4} - \dfrac{1}{6}$

4. $\dfrac{2y}{9} - \dfrac{1}{6} = \dfrac{y}{9} + \dfrac{1}{6}$

5. $\dfrac{2x - 5}{8} + \dfrac{1}{4} = \dfrac{x}{8} + \dfrac{3}{4}$

6. $\dfrac{3x + 4}{12} - \dfrac{1}{3} = \dfrac{5x + 2}{12} - \dfrac{1}{2}$

7. $\dfrac{6}{2a + 1} = 2$

8. $\dfrac{12}{3x - 2} = 3$

9. $\dfrac{9}{2x - 5} = -2$

10. $\dfrac{6}{4 - 3x} = 3$

11. $2 + \dfrac{5}{x} = 7$

12. $3 + \dfrac{8}{n} = 5$

13. $1 - \dfrac{9}{x} = 4$

14. $3 - \dfrac{12}{x} = 7$

15. $\dfrac{2}{y} + 5 = 9$

16. $\dfrac{6}{x} + 3 = 11$

17. $\dfrac{3}{x - 2} = \dfrac{4}{x}$

18. $\dfrac{5}{x + 3} = \dfrac{3}{x - 1}$

Solve.

19. $\dfrac{2}{3x - 1} = \dfrac{3}{4x + 1}$

20. $\dfrac{5}{3x - 4} = \dfrac{-3}{1 - 2x}$

21. $\dfrac{-3}{2x + 5} = \dfrac{2}{x - 1}$

22. $\dfrac{4}{5y - 1} - \dfrac{2}{2y - 1}$

23. $\dfrac{4x}{x - 4} + 5 = \dfrac{5x}{x - 4}$

24. $\dfrac{2x}{x + 2} - 5 = \dfrac{7x}{x + 2}$

25. $2 + \dfrac{3}{a - 3} = \dfrac{a}{a - 3}$

26. $\dfrac{x}{x + 4} = 3 - \dfrac{4}{x + 4}$

27. $\dfrac{x}{x - 1} = \dfrac{8}{x + 2}$

28. $\dfrac{x}{x + 12} = \dfrac{1}{x + 5}$

29. $\dfrac{2x}{x + 4} = \dfrac{3}{x - 1}$

30. $\dfrac{5}{3n - 8} = \dfrac{n}{n + 2}$

31. $x + \dfrac{6}{x - 2} = \dfrac{3x}{x - 2}$

32. $x - \dfrac{6}{x - 3} = \dfrac{2x}{x - 3}$

33. $\dfrac{8}{y} = \dfrac{2}{y - 2} + 1$

APPLYING THE CONCEPTS

34. Explain the procedure for solving an equation containing fractions.
[W]   Include in your discussion how the LCM is used to eliminate fractions in the equation.

Solve.

35. $\dfrac{3}{5}y - \dfrac{1}{3}(1 - y) = \dfrac{2y - 5}{15}$

36. $\dfrac{3}{4}a = \dfrac{1}{2}(3 - a) + \dfrac{a - 2}{4}$

37. $\dfrac{b + 2}{5} = \dfrac{1}{4}b - \dfrac{3}{10}(b - 1)$

38. $\dfrac{x}{2x^2 - x - 1} = \dfrac{3}{x^2 - 1} + \dfrac{3}{2x + 1}$

39. $\dfrac{x + 1}{x^2 + x - 2} = \dfrac{x + 2}{x^2 - 1} + \dfrac{3}{x + 2}$

40. $\dfrac{y + 2}{y^2 - y - 2} + \dfrac{y + 1}{y^2 - 4} = \dfrac{1}{y + 1}$

7.6 Ratio and Proportion

Objective A *To solve a proportion* ....

Quantities such as 4 meters, 15 seconds, and 8 gallons are number quantities written with units. In these examples the units are meters, seconds, and gallons.

A **ratio** is the quotient of two quantities that have the same unit.

The length of a living room is 16 ft and the width is 12 ft. The ratio of the length to the width is written

$$\frac{16 \text{ ft}}{12 \text{ ft}} = \frac{16}{12} = \frac{4}{3}$$ A ratio is in simplest form when the two numbers do not have a common factor. Note that the units are not written.

A **rate** is the quotient of two quantities that have different units.

There are 2 lb of salt in 8 gal of water. The salt-to-water rate is

$$\frac{2 \text{ lb}}{8 \text{ gal}} = \frac{1 \text{ lb}}{4 \text{ gal}}$$ A rate is in simplest form when the two numbers do not have a common factor. The units are written as part of the rate.

A **proportion** is an equation that states the equality of two ratios or rates. Examples of proportions are shown at the right.

$$\frac{30 \text{ mi}}{4 \text{ h}} = \frac{15 \text{ mi}}{2 \text{ h}}$$

$$\frac{4}{6} = \frac{8}{12}$$

$$\frac{3}{4} = \frac{x}{8}$$

➡ Solve the proportion $\dfrac{4}{x} = \dfrac{2}{3}$.

$$\frac{4}{x} = \frac{2}{3}$$

$$3x\left(\frac{4}{x}\right) = 3x\left(\frac{2}{3}\right)$$ • The LCM is 3x. Multiply each side of the proportion by the LCM.

$$12 = 2x$$ • Solve the equation.

$$6 = x$$

The solution is 6.

Example 1

Solve the proportion $\dfrac{8}{x + 3} = \dfrac{4}{x}$.

Solution

$$\frac{8}{x + 3} = \frac{4}{x}$$

$$x(x + 3)\frac{8}{x + 3} = x(x + 3)\frac{4}{x}$$

$$8x = 4(x + 3)$$
$$8x = 4x + 12$$
$$4x = 12$$
$$x = 3$$

The solution is 3.

You Try It 1

Solve the proportion $\dfrac{2}{x + 3} = \dfrac{6}{5x + 5}$.

Your solution

Solution on p. A22

Objective B *To solve application problems* ....

Example 2

The monthly loan payment for a car is $28.35 for each $1000 borrowed. At this rate, find the monthly payment for a $6000 car loan.

Strategy

To find the monthly payment, write and solve a proportion, using P to represent the monthly car payment.

Solution

$$\frac{28.35}{1000} = \frac{P}{6000}$$

$$6000\left(\frac{28.35}{1000}\right) = 6000\left(\frac{P}{6000}\right)$$

$$170.10 = P$$

The monthly payment is $170.10.

You Try It 2

Sixteen ceramic tiles are needed to tile a 9-ft^2 area. At this rate how many square feet can be tiled using 256 ceramic tiles?

Your strategy

Your solution

Example 3

An investment of $500 earns $60 each year. At the same rate, how much additional money must be invested to earn $90 each year?

Strategy

To find the additional amount of money that must be invested, write and solve a proportion, using x to represent the additional money. Then $500 + x$ is the total amount invested.

Solution

$$\frac{60}{500} = \frac{90}{500 + x}$$

$$\frac{3}{25} = \frac{90}{500 + x}$$

$$25(500 + x)\left(\frac{3}{25}\right) = 25(500 + x)\left(\frac{90}{500 + x}\right)$$

$$(500 + x)3 = 25(90)$$

$$1500 + 3x = 2250$$

$$3x = 750$$

$$x = 250$$

An additional $250 must be invested.

You Try It 3

Three ounces of a certain medication are required for a 150-lb adult. At the same rate, how many additional ounces of this medication are required for a 200-lb adult?

Your strategy

Your solution

Solutions on p. A22

7.6 Exercises

. . . .

Objective A

Solve.

1. $\dfrac{x}{12} = \dfrac{3}{4}$

2. $\dfrac{6}{x} = \dfrac{2}{3}$

3. $\dfrac{4}{9} = \dfrac{x}{27}$

4. $\dfrac{16}{9} = \dfrac{64}{x}$

5. $\dfrac{x+3}{12} = \dfrac{5}{6}$

6. $\dfrac{3}{5} = \dfrac{x-4}{10}$

7. $\dfrac{18}{x+4} = \dfrac{9}{5}$

8. $\dfrac{2}{11} = \dfrac{20}{x-3}$

9. $\dfrac{2}{x} = \dfrac{4}{x+1}$

10. $\dfrac{16}{x-2} = \dfrac{8}{x}$

11. $\dfrac{x+3}{4} = \dfrac{x}{8}$

12. $\dfrac{x-6}{3} = \dfrac{x}{5}$

13. $\dfrac{2}{x-1} = \dfrac{6}{2x+1}$

14. $\dfrac{9}{x+2} = \dfrac{3}{x-2}$

15. $\dfrac{2x}{7} = \dfrac{x-2}{14}$

Objective B   Application Problems

Solve.

16. Simple syrup used in making some desserts requires 2 cups of sugar for every 2/3 cup of boiling water. At this rate, how many cups of sugar are required for 2 cups of boiling water?

17. An exit poll survey showed that 4 out of every 7 voters cast a ballot in favor of an amendment to a city charter. At this rate, how many voters voted in favor of the amendment if 35,000 people voted?

18. A quality control inspector found 3 defective transistors in a shipment of 500 transistors. At this rate, how many defective transistors are there in a shipment of 2000 transistors?

19. An air conditioning specialist recommends 2 air vents for each 300 ft^2 of floor space. At this rate, how many air vents are required for a 21,000-ft^2 office building?

Solve.

20. A company decides to accept a large shipment of 10,000 computer chips if there are 2 or fewer defects in a sample of 100 randomly chosen chips. Assuming that there are 300 defective chips in the shipment and that the rate of defective chips in the sample is the same as the rate in the shipment, will the shipment be accepted?

21. A company decides to accept a large shipment of 20,000 precision bearings if there are 3 or fewer defects in a sample of 100 randomly chosen bearings. Assuming that there are 400 defective bearings in the shipment and that the rate of defective bearings in the sample is the same as the rate in the shipment, will the shipment be accepted?

22. The lighting for some billboards is provided by using solar energy. If 3 small solar energy panels can generate 10 watts of power, how many panels are necessary to provide 600 watts of power?

23. A laser printer is rated by the number of pages per minute it can print. An inexpensive laser printer can print 5 pages every 2 minutes. At this rate, how long would it take to print a document 45 pages long?

24. As part of a conservation effort for a lake, 40 fish are caught, tagged, and then released. Later 80 fish are caught. Four of the 80 fish are found to have tags. Estimate the number of fish in the lake.

25. In a wildlife preserve, 10 elk are captured, tagged, and then released. Later 15 elk are captured and 2 are found to have tags. Estimate the number of elk in the preserve.

26. A painter estimates that 5 gal of paint will cover 1200 ft² of wall space. At this rate, how many additional gallons will be necessary to cover 1680 ft²?

APPLYING THE CONCEPTS

27. Three people put their money together to buy lottery tickets. The first person put in $25, the second person put in $30, and the third person put in $35. One of their tickets was a winning ticket. If they won $4.5 million, what was the first person's share of the winnings?

28. No one belongs to both the Math Club and the Photography Club, but the two clubs join to hold a car wash. Ten members of the Math Club and 6 members of the Photography Club participate. The profits from the car wash are $120. If each club's profits are proportional to the number of members participating, what share of the profits does the Math Club receive?

29. A basketball player has made 5 out of every 6 foul shots attempted in one year of play. If 42 foul shots were missed that year, how many shots did the basketball player make?

7.7 Literal Equations

Objective A   *To solve a literal equation for one of the variables*

A **literal equation** is an equation that contains more than one variable. Examples of literal equations are shown at the right.

$$2x + 3y = 6$$
$$4w - 2x + z = 0$$

Formulas are used to express a relationship among physical quantities. A **formula** is a literal equation that states rules about measurements. Examples of formulas are shown at the right.

$$\frac{1}{R_1} + \frac{1}{R_2} = \frac{1}{R} \qquad \text{(Physics)}$$
$$s = a + (n - 1)d \qquad \text{(Mathematics)}$$
$$A = P + Prt \qquad \text{(Business)}$$

The Addition and Multiplication Properties can be used to solve a literal equation for one of the variables. The goal is to rewrite the equation so that the variable being solved for is alone on one side of the equation and all the other numbers and variables are on the other side.

➡ Solve $A = P(1 + i)$ for i.

The goal is to rewrite the equation so that i is on one side of the equation and all other variables are on the other side.

$$A = P(1 + i)$$
$$A = P + Pi \qquad \bullet \text{ Use the Distributive Property to remove parentheses.}$$
$$A - P = P - P + Pi \qquad \bullet \text{ Subtract } P \text{ from each side of the equation.}$$
$$A - P = Pi$$
$$\frac{A - P}{P} = \frac{Pi}{P} \qquad \bullet \text{ Divide each side of the equation by } P.$$
$$\frac{A - P}{P} = i$$

Example 1
Solve $3x - 4y = 12$ for y.

Solution
$$3x - 4y = 12$$
$$3x - 3x - 4y = -3x + 12$$
$$-4y = -3x + 12$$
$$\frac{-4y}{-4} = \frac{-3x + 12}{-4}$$
$$y = \frac{3}{4}x - 3$$

You Try It 1
Solve $5x - 2y = 10$ for y.

Your solution

Solution on p. A23

Example 2

Solve $I = \dfrac{E}{R + r}$ for R.

Solution

$$I = \dfrac{E}{R + r}$$

$$(R + r)I = (R + r)\dfrac{E}{R + r}$$

$$RI + rI = E$$

$$RI + rI - rI = E - rI$$

$$RI = E - rI$$

$$\dfrac{RI}{I} = \dfrac{E - rI}{I}$$

$$R = \dfrac{E - rI}{I}$$

You Try It 2

Solve $s = \dfrac{A + L}{2}$ for L.

Your solution

Example 3

Solve $L = a(1 + ct)$ for c.

Solution

$$L = a(1 + ct)$$

$$L = a + act$$

$$L - a = a - a + act$$

$$L - a = act$$

$$\dfrac{L - a}{at} = \dfrac{act}{at}$$

$$\dfrac{L - a}{at} = c$$

You Try It 3

Solve $S = a + (n - 1)d$ for n.

Your solution

Example 4

Solve $S = C - rC$ for C.

Solution

$$S = C - rC$$

$$S = (1 - r)C$$

$$\dfrac{S}{1 - r} = \dfrac{(1 - r)C}{1 - r}$$

$$\dfrac{S}{1 - r} = C$$

You Try It 4

Solve $S = C + rC$ for C.

Your solution

Solutions on p. A23

7.7 Exercises

Objective A

Solve for y.

1. $3x + y = 10$

2. $2x + y = 5$

3. $4x - y = 3$

4. $5x - y = 7$

5. $3x + 2y = 6$

6. $2x + 3y = 9$

7. $2x - 5y = 10$

8. $5x - 2y = 4$

9. $2x + 7y = 14$

10. $6x - 5y = 10$

11. $x + 3y = 6$

12. $x + 2y = 8$

13. $2x - 9y - 18 = 0$

14. $3x - y + 7 = 0$

15. $2x - y + 5 = 0$

Solve for x.

16. $x + 3y = 6$

17. $x + 6y = 10$

18. $3x - y = 3$

19. $2x - y = 6$

20. $2x + 5y = 10$

21. $4x + 3y = 12$

22. $x - 2y + 1 = 0$

23. $x - 4y - 3 = 0$

24. $5x + 4y + 20 = 0$

Solve the formula for the given variable.

25. $d = rt;\ t$ (Physics)

26. $E = IR;\ R$ (Physics)

27. $PV = nRT;\ T$ (Chemistry)

28. $A = bh;\ h$ (Geometry)

29. $P = 2l + 2w;\ l$ (Geometry)

30. $F = \dfrac{9}{5}C + 32;\ C$ (Temperature conversion)

31. $A = \dfrac{1}{2}h(b_1 + b_2);\ b_1$ (Geometry)

32. $C = \dfrac{5}{9}(F - 32);\ F$ (Temperature conversion)

Solve the formula for the given variable.

33. $V = \dfrac{1}{3}Ah; h$ (Geometry)

34. $P = R - C; C$ (Business)

35. $R = \dfrac{C - S}{t}; S$ (Business)

36. $P = \dfrac{R - C}{n}; R$ (Business)

37. $A = P + Prt; P$ (Business)

38. $T = fm - gm; m$ (Engineering)

39. $A = Sw + w; w$ (Physics)

40. $a = S - Sr; S$ (Mathematics)

APPLYING THE CONCEPTS

Break-even analysis is a method used to determine the sales volume required for a company to break even, or experience neither a profit nor a loss on the sale of a product. The break-even point represents the number of units that must be made and sold for income from sales to equal the cost of the product. The break-even point can be calculated using the formula $B = \dfrac{F}{S - V}$, where F is the fixed costs, S is the selling price per unit, and V is the variable costs per unit.

41. **a.** Solve the formula $B = \dfrac{F}{S - V}$ for S.

    b. Use your answer to part **a** to find the required selling price per unit for a company to break even. The fixed costs are \$20,000, the variable costs per unit are \$80, and the company plans to make and sell 200 desks.

    c. Use your answer to part **a** to find the required selling price per unit for a company to break even. The fixed costs are \$15,000, the variable costs per unit are \$50, and the company plans to make and sell 600 cameras.

Resistors are used to control the flow of current. The total resistance of two resistors in a circuit can be given by the formula $R = \dfrac{1}{\dfrac{1}{R_1} + \dfrac{1}{R_2}}$, where R_1 and R_2 are the two resistors in the circuit. Resistance is measured in ohms.

42. **a.** Solve the formula $R = \dfrac{1}{\dfrac{1}{R_1} + \dfrac{1}{R_2}}$ for R_1.

    b. Use your answer to part **a** to find the resistance in R_1 if the resistance in R_2 is 30 ohms and the total resistance is 12 ohms.

    c. Use your answer to part **a** to find the resistance in R_1 if the resistance in R_2 is 15 ohms and the total resistance is 6 ohms.

7.8

Application Problems

Objective A   *To solve work problems*

If a painter can paint a room in 4 h, then in 1 h the painter can paint $\frac{1}{4}$ of the room. The painter's rate of work is $\frac{1}{4}$ of the room each hour. The **rate of work** is the part of a task that is completed in one unit of time.

A pipe can fill a tank in 30 min. This pipe can fill $\frac{1}{30}$ of the tank in 1 min. The rate of work is $\frac{1}{30}$ of the tank each minute. If a second pipe can fill the tank in x min, the rate of work for the second pipe is $\frac{1}{x}$ of the tank each minute.

In solving a work problem, the goal is to determine the time it takes to complete a task. The basic equation that is used to solve work problems is

Rate of work × time worked = part of task completed

For example, if a faucet can fill a sink in 6 min, then in 5 min the faucet will fill $\frac{1}{6} \times 5 = \frac{5}{6}$ of the sink. In 5 min the faucet completes $\frac{5}{6}$ of the task.

➡ A painter can paint a wall in 20 min. The painter's apprentice can paint the same wall in 30 min. How long will it take to paint the wall when they work together?

> **Strategy for Solving a Work Problem**
>
> 1. For each person or machine, write a numerical or variable expression for the rate of work, the time worked, and the part of the task completed. The results can be recorded in a table.

Unknown time to paint the wall working together: t

	Rate of Work	·	Time Worked	=	Part of Task Completed
Painter	$\frac{1}{20}$	·	t	=	$\frac{t}{20}$
Apprentice	$\frac{1}{30}$	·	t	=	$\frac{t}{30}$

> 2. Determine how the parts of the task completed are related. Use the fact that the sum of the parts of the task completed must equal 1: the complete task.

The sum of the part of the task completed by the painter and the part of the task completed by the apprentice is 1.

$$\frac{t}{20} + \frac{t}{30} = 1$$

$$60\left(\frac{t}{20} + \frac{t}{30}\right) = 60 \cdot 1$$

• Multiply by the LCM.

$$3t + 2t = 60$$
$$5t = 60$$
$$t = 12$$

Working together, they will paint the wall in 12 min.

Example 1

A small water pipe takes three times longer to fill a tank than does a large water pipe. With both pipes open it takes 4 h to fill the tank. Find the time it would take the small pipe, working alone, to fill the tank.

You Try It 1

Two computer printers that work at the same rate are working together to print the payroll checks for a large corporation. After they work together for 2 h, one of the printers quits. The second requires 3 more hours to complete the payroll checks. Find the time it would take one printer, working alone, to print the payroll.

Strategy

- Time for large pipe to fill the tank: t
  Time for small pipe to fill the take: $3t$

	Rate	Time	Part
Small pipe	$\frac{1}{3t}$	4	$\frac{4}{3t}$
Large pipe	$\frac{1}{t}$	4	$\frac{4}{t}$

- The sum of the parts of the task completed by each pipe must equal 1.

Your strategy

Solution

$$\frac{4}{3t} + \frac{4}{t} = 1$$

$$3t\left(\frac{4}{3t} + \frac{4}{t}\right) = 3t \cdot 1$$

$$4 + 12 = 3t$$

$$16 = 3t$$

$$\frac{16}{3} = t$$

$$3t = 3\left(\frac{16}{3}\right) = 16$$

The small pipe working alone takes 16 h to fill the tank.

Your solution

Solution on p. A23

Content and Format © 1995 HMCo.

Objective B*   *To solve uniform motion problems ....

A car that travels constantly in a straight line at 30 mph is in uniform motion. **Uniform motion** means that the speed or direction of an object does not change.

The basic equation used to solve uniform motion problems is

$$\textbf{Distance = rate} \times \textbf{time}$$

An alternative form of this equation can be written by solving the equation for time.

$$\frac{\textbf{Distance}}{\textbf{Rate}} = \textbf{time}$$

This form of the equation is useful when the total time of travel for two objects or the time of travel between two points is known.

➡ The speed of a boat in still water is 20 mph. The boat traveled 75 mi down a river in the same amount of time it took to travel 45 mi up the river. Find the rate of the river's current.

Strategy for Solving a Uniform Motion Problem

1. For each object, write a numerical or variable expression for the distance, rate, and time. The results can be recorded in a table.

The unknown rate of the river's current: r

	Distance	÷	*Rate*	=	*Time*
Down river	75	÷	$20 + r$	=	$\dfrac{75}{20 + r}$
Up river	45	÷	$20 - r$	=	$\dfrac{45}{20 - r}$

---

2. Determine how the times traveled by each object are related. For example, it may be known that the times are equal, or the total time may be known.

$$\frac{75}{20 + r} = \frac{45}{20 - r}$$

• The time down the river is equal to the time up the river.

$$(20 + r)(20 - r)\frac{75}{20 + r} = (20 + r)(20 - r)\frac{45}{20 - r}$$

• Multiply by the LCM.

$$(20 - r)75 = (20 + r)45$$
$$1500 - 75r = 900 + 45r$$
$$-120r = -600$$
$$r = 5$$

The rate of the river's current is 5 mph.

Example 2

A cyclist rode the first 20 mi of a trip at a constant rate. For the next 16 mi, the cyclist reduced the speed by 2 mph. The total time for the 36 mi was 4 h. Find the rate of the cyclist for each leg of the trip.

Strategy

• Rate for the first 20 mi: r
  Rate for the next 16 mi: $r - 2$

	Distance	Rate	Time
First 20 mi	20	r	$\dfrac{20}{r}$
Next 16 mi	16	$r - 2$	$\dfrac{16}{r - 2}$

• The total time for the trip was 4 h.

Solution

$$\frac{20}{r} + \frac{16}{r - 2} = 4$$

$$r(r - 2)\left[\frac{20}{r} + \frac{16}{r - 2}\right] = r(r - 2) \cdot 4$$

$$(r - 2)20 + 16r = 4r^2 - 8r$$

$$20r - 40 + 16r = 4r^2 - 8r$$

$$36r - 40 = 4r^2 - 8r$$

Solve the quadratic equation by factoring.

$$0 = 4r^2 - 44r + 40$$

$$0 = 4(r^2 - 11r + 10)$$

$$0 = 4(r - 10)(r - 1)$$

$$r - 10 = 0 \qquad r - 1 = 0$$
$$r = 10 \qquad\quad r = 1$$

The solution $r = 1$ mph is not possible, because the rate on the last 16 mi would then be -1 mph.

10 mph was the rate for the first 20 mi.
8 mph was the rate for the next 16 mi.

You Try It 2

The total time it took for a sailboat to sail back and forth across a lake 6 km wide was 2 h. The rate sailing back was three times the rate sailing across. Find the rate sailing out across the lake.

Your strategy

Your solution

Solution on p. A23

7.8 Exercises

Objective A   *Application Problems*

Solve.

1. An experienced painter can paint a garage twice as fast as an apprentice. Working together, the painters require 4 h to paint the garage. How long would it take the experienced painter, working alone, to paint the garage?

2. One grocery clerk can stock a shelf in 20 min, whereas a second clerk requires 30 min to stock the same shelf. How long would it take to stock the shelf if the two clerks worked together?

3. One person with a skiploader requires 12 h to remove a large quantity of earth. A second, larger skiploader can remove the same amount of earth in 4 h. How long would it take to remove the earth with both skiploaders working together?

4. One worker can dig the trenches for a sprinkler system in 3 h, whereas a second worker requires 6 h to do the same task. How long would it take to dig the trenches with both people working together?

5. One computer can solve a complex prime factorization problem in 75 h. A second computer can solve the same problem in 50 h. How long would it take both computers, working together, to solve the problem?

6. A new machine can make 10,000 aluminum cans three times faster than an older machine. With both machines working, 10,000 cans can be made in 9 h. How long would it take the new machine, working alone, to make the 10,000 cans?

7. A small air conditioner will cool a room 2° in 15 min. A larger air conditioner will cool the room 2° in 10 min. How long would it take to cool the room 2° with both air conditioners operating?

8. One printing press can print the first edition of a book in 55 min, whereas a second printing press requires 66 min to print the same number of copies. How long would it take to print the first edition with both presses operating?

9. Two welders working together can complete a job in 6 h. One of the welders, working alone, can complete the task in 10 h. How long would it take the second welder, working alone, to complete the task?

10. Two oil pipelines can fill a small tank in 30 min. Using one of the pipelines would require 45 min to fill the tank. How long would it take the second pipeline alone to fill the tank?

Solve.

11. With two harvesters, a plot of land can be harvested in 1 h. One harvester, working alone, requires 1.5 h to harvest the field. How long would it take the second harvester, working alone, to harvest the field?

12. Working together, two dock workers can load a crate in 6 min. One dock worker, working alone, can load the crate in 15 min. How long would it take the second dock worker, working alone, to load the crate?

13. A cement mason can build a barbeque in 8 h, whereas it takes a second mason 12 h to do the same task. After working alone for 4 h, the first mason quits. How long will it take the second mason to complete the task?

14. A mechanic requires 2 h to repair a transmission, whereas an apprentice requires 6 h to make the same repairs. The mechanic worked alone for 1 h and then stopped. How long will it take the apprentice, working alone, to complete the repairs?

15. One computer technician can wire a modem in 4 h, whereas it takes 6 h for a second technician to do the same job. After working alone for 2 h, the first technician quit. How long will it take the second technician to complete the wiring?

16. A wallpaper hanger requires 2 h to hang the wallpaper on one wall of a room. A second wallpaper hanger requires 4 h to hang the same amount of paper. The first wallpaper hanger worked alone for 1 hour and then quit. How long will it take the second wallpaper hanger, working alone, to complete the wall?

17. Two welders who work at the same rate are welding the girders of a building. After they work together for 10 h, one of the welders quits. The second welder requires 20 more hours to complete the welds. Find the time it would have taken one of the welders, working alone, to complete the welds.

18. A large and a small heating unit are being used to heat the water of a pool. The larger unit, working alone, requires 8 h to heat the pool. After both units have been operating 2 h, the larger unit is turned off. The small unit requires 9 more hours to heat the pool. How long would it take the small unit, working alone, to heat the pool?

19. Two machines that fill cereal boxes work at the same rate. After they work together for 7 h, one machine breaks down. The second machine requires 14 more hours to finish filling the boxes. How long would it have taken one of the machines, working alone, to fill the boxes?

20. A large and a small drain are opened to drain a pool. The large drain can empty the pool in 6 h. After both drains have been open for 1 hour, the large drain becomes clogged and is closed. The smaller drain remains open and requires 9 more hours to empty the pool. How long would it have taken the small drain, working alone, to empty the pool?

Objective B   *Application Problems*

Solve.

21. A camper drove 90 mi to a recreational area and then hiked 5 mi into the wilderness. The rate of the camper while driving in the car was nine times the rate hiking. The time spent hiking and driving was 3 h. Find the rate at which the camper hiked.

22. The president of a company traveled 1800 mi by jet and 300 mi on a prop plane. The rate of the jet was four times the rate of the prop plane. The entire trip took a total of 5 h. Find the rate of the jet plane.

23. As part of a conditioning program, a jogger ran 8 mi in the same time a cyclist rode 20 mi. The rate of the cyclist was 12 mph faster than the rate of the jogger. Find the rate of the jogger and that of the cyclist.

24. An express train travels 600 mi in the same amount of time it takes a freight train to travel 360 mi. The rate of the express train is 20 mph faster than that of the freight train. Find the rate of each train.

25. To assess the damage done by a fire, a forest ranger traveled 1080 mi by jet and then an additional 180 mi by helicopter. The rate of the jet was 4 times the rate of the helicopter. The entire trip took a total of 5 h. Find the rate of the jet.

26. A twin-engine plane can fly 800 mi in the same time that it takes a single-engine plane to fly 600 mi. The rate of the twin-engine plane is 50 mph faster than that of the single-engine plane. Find the rate of the twin-engine plane.

27. Two planes leave an airport and head for another airport 900 mi away. The rate of the first plane is twice that of the second plane. The second plane arrives at the airport 3 h after the first plane. Find the rate of the second plane.

28. A car and a bus leave a town at 1 P.M. and head for a town 300 mi away. The rate of the car is twice the rate of the bus. The car arrives 5 h ahead of the bus. Find the rate of the car.

29. A car is traveling at a rate that is 36 mph faster than the rate of a cyclist. The car travels 384 mi in the same time it takes the cyclist to travel 96 mi. Find the rate of the car.

30. An engineer traveled 165 mi by car and then an additional 660 mi by plane. The rate of the plane was 4 times the rate of the car, and the total trip took 6 h. Find the rate of the car.

Solve.

31. A backpacker hiking into a wilderness area walked 9 mi at a constant rate and then reduced this rate by 1 mph. Another 4 mi was hiked at this reduced rate. The time required to hike the 4 mi was 1 h less than the time required to walk the 9 mi. Find the rate at which the hiker walked the first 9 mi.

32. A sailboat sailed 15 mi on the first leg of a trip before changing direction and sailing an additional 7 mi. Because of the wind, the change caused the sailboat to increase its speed by 2 mph. The total sailing time was 4 h. Find the rate of sailing for the first leg of the trip.

33. A small motor on a fishing boat can move the boat 6 mph in calm water. When trolling in a river, the amount of time it takes to travel 12 mi against the river's current is the same as the time it takes to travel 24 mi with the current. Find the rate of the current.

34. A commercial jet can fly 550 mph in calm air. Traveling with the jet stream, the plane flew 2400 mi in the same amount of time it takes to fly 2000 mi against the jet stream. Find the rate of the jet stream.

35. A cruise ship can sail at 28 mph in calm water. Sailing with the gulf current, the ship can sail 170 mi in the same amount of time that it can sail 110 mi against the gulf current. Find the rate of the gulf current.

36. Paddling in calm water, a canoeist can paddle at a rate of 8 mph. Traveling with the current, the canoeist went 30 mi in the same amount of time it took to travel 18 mi against the current. Find the rate of the current.

37. On a recent trip, a trucker traveled 330 mi at a constant rate. Because of road construction, the trucker then had to reduce speed by 25 mph. An additional 30 mi was traveled at the reduced rate. The total time for the entire trip was 7 h. Find the rate of the trucker for the first 330 mi.

APPLYING THE CONCEPTS

38. One pipe can fill a tank in 2 h, a second pipe can fill the tank in 4 h, and a third pipe can fill the tank in 5 h. How long will it take to fill the tank with all three pipes working?

39. A mason can construct a retaining wall in 10 h. The mason's more experienced apprentice can do the same job in 15 h. How long would it take the mason's less experienced apprentice to do the job if, working together, all three can complete the job in 5 h?

40. A surveyor traveled 32 mi by canoe and then hiked 4 mi. The rate of speed by boat was four times the rate on foot. If the time spent walking was 1 h less than the time spent canoeing, find the amount of time spent traveling by canoe.

41. Because of bad weather, a bus driver reduced the usual speed along a 150-mi bus route by 10 mph. The bus arrived only 30 min later than its usual arrival time. How fast does the bus usually travel?

Project in Mathematics

Continued Fractions   The following complex fraction is called a **continued fraction.**

$$1 + \cfrac{1}{1 + \cfrac{1}{1 + \cfrac{1}{1 + \cfrac{1}{1 + \cdots}}}}$$

The dots indicate that the pattern continues to repeat forever.

A **convergent** for a continued fraction is an approximation of the repeated pattern. For instance,

$$c_2 = 1 + \cfrac{1}{1 + \cfrac{1}{1 + 1}} \qquad c_3 = 1 + \cfrac{1}{1 + \cfrac{1}{1 + \cfrac{1}{1 + 1}}} \qquad c_4 = 1 + \cfrac{1}{1 + \cfrac{1}{1 + \cfrac{1}{1 + \cfrac{1}{1 + 1}}}}$$

Exercises:

1. Calculate c_5 for the continued fraction above.

This particular continued fraction is related to the golden rectangle, which has been used in architectural designs as diverse as the Parthenon in Athens, built around 440 B.C., and the United Nations building. A golden rectangle is one for which

$$\frac{\text{length}}{\text{width}} = \frac{\text{length} + \text{width}}{\text{length}}$$

An example of a golden rectangle is shown at the right.

Here is another continued fraction that was discovered by Leonard Euler (1707–1793). Calculating the convergents of this continued fraction yields approximations that are closer and closer to π.

$$\pi = 3 + \cfrac{1^2}{6 + \cfrac{3^2}{6 + \cfrac{5^2}{6 + \cfrac{7^2}{6 + \cdots}}}}$$

2. Calculate $c_5 = 3 + \cfrac{1^2}{6 + \cfrac{3^2}{6 + \cfrac{5^2}{6 + \cfrac{7^2}{6 + \cfrac{9^2}{6 + 11^2}}}}}$.

Chapter Summary

Key Words   A *rational expression* is a fraction in which the numerator or denominator is a polynomial.

A *rational expression* is in *simplest form* when the numerator and denominator have no common factors.

The *reciprocal* of a fraction is a fraction with the numerator and denominator interchanged.

The *least common multiple* (LCM) of two or more numbers is the smallest number that contains the prime factorization of each number.

A *complex fraction* is a fraction whose numerator or denominator contains one or more fractions.

A *ratio* is the quotient of two quantities that have the same unit.

A *rate* is the quotient of two quantities that have different units.

A *proportion* is an equation that states the equality of two ratios or rates.

A *literal equation* is an equation that contains more than one variable.

A *formula* is a literal equation that states rules about measurements.

Essential Rules   *To multiply fractions:*

$$\frac{a}{b} \cdot \frac{c}{d} = \frac{ac}{bd}$$

To divide fractions:

$$\frac{a}{b} \div \frac{c}{d} = \frac{a}{b} \cdot \frac{d}{c} = \frac{ad}{bc}$$

To add fractions:

$$\frac{a}{c} + \frac{b}{c} = \frac{a + b}{c}$$

To subtract fractions:

$$\frac{a}{c} - \frac{b}{c} = \frac{a - b}{c}$$

Equation for Work Problems:

$$\text{Rate of work} \times \text{time worked} = \text{part of task completed}$$

Uniform Motion Equation:

$$\text{Distance} = \text{rate} \times \text{time}$$

Chapter Review

. . . .

SECTION 7.1

1. Simplify: $\dfrac{16x^5y^3}{24xy^{10}}$

2. Simplify: $\dfrac{x^2 + x - 30}{15 + 2x - x^2}$

3. Simplify: $\dfrac{8ab^2}{15x^3y} \cdot \dfrac{5xy^4}{16a^2b}$

4. Simplify: $\dfrac{3x^3 + 10x^2}{10x - 2} \cdot \dfrac{20x - 4}{6x^4 + 20x^3}$

5. Simplify: $\dfrac{24x^2 - 94x + 15}{12x^2 - 49x + 15} \cdot \dfrac{24x^2 + 7x - 5}{4 - 27x + 18x^2}$

6. Simplify: $\dfrac{6a^2b^7}{25x^3y} \div \dfrac{12a^3b^4}{5x^2y^2}$

7. Simplify: $\dfrac{20x^2 - 45x}{6x^3 + 4x^2} \div \dfrac{40x^3 - 90x^2}{12x^2 + 8x}$

8. Simplify: $\dfrac{10 - 23y + 12y^2}{6y^2 - y - 5} \div \dfrac{4y^2 - 13y + 10}{18y^2 + 3y - 10}$

SECTION 7.2

9. Find the LCM of $10x^2 - 11x + 3$ and $20x^2 - 17x + 3$.

10. Write each fraction in terms of the LCM of the denominators.

$$\dfrac{x}{12x^2 + 16x - 3}, \dfrac{4x^2}{6x^2 + 7x - 3}$$

SECTION 7.3

11. Simplify: $\dfrac{5x + 3}{2x^2 + 5x - 3} - \dfrac{3x + 4}{2x^2 + 5x - 3}$

12. Simplify: $\dfrac{x + 7}{15x} + \dfrac{x - 2}{20x}$

13. Simplify: $\dfrac{2y}{5y - 7} + \dfrac{3}{7 - 5y}$

14. Simplify: $\dfrac{x - 1}{x + 2} + \dfrac{3x - 2}{5 - x} + \dfrac{5x^2 + 15x - 11}{x^2 - 3x - 10}$

SECTION 7.4

15. Simplify: $\dfrac{1 - \dfrac{1}{x}}{1 - \dfrac{8x - 7}{x^2}}$

16. Simplify: $\dfrac{x + \dfrac{6}{x - 5}}{1 + \dfrac{2}{x - 5}}$

17. Simplify: $\dfrac{x - \dfrac{16}{5x - 2}}{3x - 4 - \dfrac{88}{5x - 2}}$

SECTION 7.5

18. Solve: $\dfrac{5}{7} + \dfrac{x}{2} = 2 - \dfrac{x}{7}$

19. Solve: $\dfrac{x + 8}{x + 4} = 1 + \dfrac{5}{x + 4}$

20. Solve: $\dfrac{20}{2x + 3} = \dfrac{17x}{2x + 3} - 5$

SECTION 7.6

21. Solve: $\dfrac{3}{20} = \dfrac{x}{80}$

22. Solve: $\dfrac{20}{x + 2} = \dfrac{5}{16}$

23. Solve: $\dfrac{6}{x - 7} = \dfrac{8}{x - 6}$

24. A pitcher's ERA, or "earned run average," is the average number of runs allowed in 9 innings of pitching. If a pitcher allows 15 runs in 100 innings, find the pitcher's ERA.

SECTION 7.7

25. Solve $4x + 9y = 18$ for y.

26. Solve $i = \dfrac{100m}{c}$ for c.

27. Solve $T = 2(ab + bc + ca)$ for a.

SECTION 7.8

28. One hose can fill a pool in 15 h. A second hose can fill the pool in 10 h. How long would it take to fill the pool using both hoses?

29. A car travels 315 mi in the same amount of time that a bus travels 245 mi. The rate of the car is 10 mph faster than that of the bus. Find the rate of the car.

30. The rate of a jet is 400 mph in calm air. Traveling with the wind, the jet can fly 2100 mi in the same amount of time it takes to fly 1900 mi against the wind. Find the rate of the wind.

Chapter Test

1. Simplify: $\dfrac{16x^5y}{24x^2y^4}$

2. Simplify: $\dfrac{x^2 + 4x - 5}{1 - x^2}$

3. Simplify: $\dfrac{x^3y^4}{x^2 - 4x + 4} \cdot \dfrac{x^2 - x - 2}{x^6y^4}$

4. Simplify: $\dfrac{x^2 + 2x - 3}{x^2 + 6x + 9} \cdot \dfrac{2x^2 - 11x + 5}{2x^2 + 3x - 5}$

5. Simplify: $\dfrac{x^2 + 3x + 2}{x^2 + 5x + 4} \div \dfrac{x^2 - x - 6}{x^2 + 2x - 15}$

6. Find the LCM of $6x - 3$ and $2x^2 + x - 1$.

7. Write each fraction in terms of the LCM of the denominators.

   $\dfrac{3}{x^2 - 2x}, \dfrac{x}{x^2 - 4}$

8. Simplify: $\dfrac{2x}{x^2 + 3x - 10} - \dfrac{4}{x^2 + 3x - 10}$

9. Simplify: $\dfrac{2}{2x - 1} - \dfrac{3}{3x + 1}$

10. Simplify: $\dfrac{x}{x + 3} - \dfrac{2x - 5}{x^2 + x - 6}$

11. Simplify: $\dfrac{1 + \dfrac{1}{x} - \dfrac{12}{x^2}}{1 + \dfrac{2}{x} - \dfrac{8}{x^2}}$

12. Solve: $\dfrac{6}{x} - 2 = 1$

13. Solve: $\dfrac{2x}{x + 1} - 3 = \dfrac{-2}{x + 1}$

14. Solve the proportion. $\dfrac{3}{x + 4} = \dfrac{5}{x + 6}$

15. A salt water solution is formed by mixing 4 lb of salt with 10 gal of water. At this rate, how many additional pounds of salt are required for 15 gal of water?

16. A landscape architect uses three sprinklers for each 200 ft² of lawn. At this rate, how many sprinklers are needed for a 3600-ft² lawn?

17. Solve $3x - 8y = 16$ for y.

18. Solve $d = s + rt$ for t.

19. A pool can be filled with one pipe in 6 h, whereas a second pipe requires 12 h to fill the pool. How long would it take to fill the pool with both pipes turned on?

20. A small plane can fly at 110 mph in calm air. Flying with the wind, the plane can fly 260 mi in the same amount of time it takes to fly 180 mi against the wind. Find the rate of the wind.

Cumulative Review

. . . .

1. Simplify: $\left(\dfrac{2}{3}\right)^2 \div \left(\dfrac{3}{2} - \dfrac{2}{3}\right) + \dfrac{1}{2}$

2. Evaluate $-a^2 + (a - b)^2$ when $a = -2$ and $b = 3$.

3. Simplify: $-2x - (-3y) + 7x - 5y$

4. Simplify: $2[3x - 7(x - 3) - 8]$

5. Solve: $4 - \dfrac{2}{3}x = 7$

6. Solve: $3[x - 2(x - 3)] = 2(3 - 2x)$

7. Find $16\dfrac{2}{3}\%$ of 60.

8. Simplify: $(a^2 b^5)(ab^2)$

9. Simplify: $(a - 3b)(a + 4b)$

10. Simplify: $\dfrac{15b^4 - 5b^2 + 10b}{5b}$

11. Simplify: $(x^3 - 8) \div (x - 2)$

12. Factor: $12x^2 - x - 1$

13. Factor: $y^2 - 7y + 6$

14. Factor: $2a^3 + 7a^2 - 15a$

15. Factor: $4b^2 - 100$

16. Solve: $(x + 3)(2x - 5) = 0$

17. Simplify: $\dfrac{12x^4 y^2}{18xy^7}$

18. Simplify: $\dfrac{x^2 - 7x + 10}{25 - x^2}$

19. Simplify: $\dfrac{x^2 - x - 56}{x^2 + 8x + 7} \div \dfrac{x^2 - 13x + 40}{x^2 - 4x - 5}$

20. Simplify: $\dfrac{2}{2x - 1} - \dfrac{1}{x + 1}$

21. Simplify: $\dfrac{1 - \dfrac{2}{x} - \dfrac{15}{x^2}}{1 - \dfrac{25}{x^2}}$

22. Solve: $\dfrac{3x}{x - 3} - 2 = \dfrac{10}{x - 3}$

23. Solve the proportion. $\dfrac{2}{x - 2} = \dfrac{12}{x + 3}$

24. Solve $f = v + at$ for t.

25. Translate "the difference between five times a number and thirteen is the opposite of eight" into an equation and solve.

26. A silversmith mixes 60 g of an alloy that is 40% silver with 120 g of another silver alloy. The resulting alloy is 60% silver. Find the percent of silver in the 120-g alloy.

27. The length of the base of a triangle is 2 in. less than twice the height. The area of the triangle is 30 in². Find the base and height of the triangle.

28. A life insurance policy costs $16 for every $1000 of coverage. At this rate, how much money would a policy of $5000 cost?

29. One water pipe can fill a tank in 9 min, whereas a second pipe requires 18 min to fill the tank. How long would it take both pipes, working together, to fill the tank?

30. The rower of a boat can row at a rate of 5 mph in calm water. Rowing with the current, the boat travels 14 mi in the same amount of time it takes to travel 6 mi against the current. Find the rate of the current.

Linear Equations in Two Variables

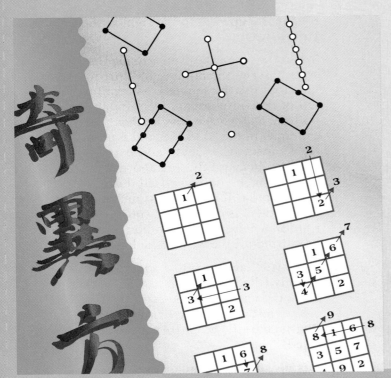

Objectives

Section 8.1
To graph points in a rectangular coordinate system

To determine ordered-pair solutions of an equation in two variables

To determine whether a set of ordered pairs is a function

To evaluate a function written in functional notation

Section 8.2
To graph an equation of the form $y = mx + b$

To graph an equation of the form $Ax + By = C$

To solve application problems

Section 8.3
To find the x- and y-intercepts of a straight line

To find the slope of a straight line

To graph a line using the slope and the y-intercept

Section 8.4
To find the equation of a line given a point and the slope

To find the equation of a line given two points

To solve application problems

Magic Squares

A magic square is a square array of distinct integers so arranged that the numbers along any row, column, or main diagonal have the same sum. An example of a magic square is shown at the right.

8	3	4
1	5	9
6	7	2

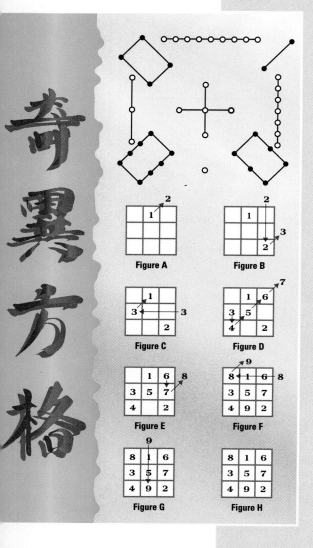

Figure A

Figure B

Figure C

Figure D

Figure E

Figure F

Figure G

Figure H

The oldest known example of a magic square comes from China. Estimates are that this magic square is over 4000 years old. It is shown at the left.

There is a simple way to produce a magic square with an odd number of cells. Start by writing a 1 in the top middle cell. The rule then is to proceed diagonally upward to the right with the successive integers.

When the rule takes you outside the square, write the number by shifting either across the square from right to left or down the square from top to bottom, as the case may be. For example, in Fig. B the second number (2) is outside the square above a column. Because the 2 is above a column, it should be shifted down to the bottom cell in that column. In Fig. C, the 3 is outside the square to the right of a column and should therefore be shifted all the way to the left.

If the rule takes you to a square that is already filled (as shown in Fig. D), then write the number in the cell directly below the last number written. Continue until the entire square is filled.

It is possible to begin a magic square with any integer and proceed by using the above rule and consecutive integers.

For an odd magic square beginning with 1, the sum of a row, column or diagonal is $\frac{n(n^2 + 1)}{2}$, where n is the number of rows.

The Rectangular Coordinate System

Objective A   *To graph points in a rectangular coordinate system* ....

Before the 15th century, geometry and algebra were considered separate branches of mathematics. That all changed when René Descartes, a French mathematician who lived from 1596 to 1650, founded **analytic geometry.** In this geometry, a *coordinate system* is used to study relationships between variables.

A **rectangular coordinate system** is formed by two number lines, one horizontal and one vertical, that intersect at the zero point of each line. The point of intersection is called the **origin**. The two lines are called **coordinate axes,** or simply **axes**.

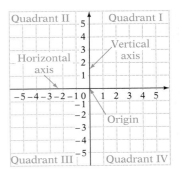

The axes determine a **plane**, which can be thought of as a large, flat sheet of paper. The two axes divide the plane into four regions called **quadrants**, which are numbered counterclockwise from I to IV.

Each point in the plane can be identified by a pair of numbers called an **ordered pair.** The first number of the pair measures a horizontal distance and is called the **abscissa**. The second number of the pair measures a vertical distance and is called the **ordinate**. The **coordinates** of the point are the numbers in the ordered pair associated with the point. The abscissa is also called the **first coordinate** of the ordered pair, and the ordinate is also called the **second coordinate** of the ordered pair.

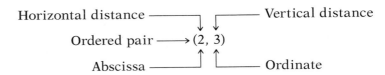

To **graph** or **plot** a point in the plane, place a dot at the location given by the ordered pair. The **graph of an ordered pair** is the dot drawn at the coordinates of the point in the plane. The points whose coordinates are $(3, 4)$ and $(-2.5, -3)$ are graphed in the figures below.

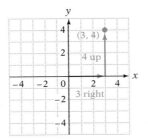

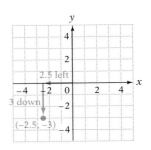

LOOK CLOSELY

This is very important. An **ordered pair** is a *pair* of coordinates, and the *order* in which the coordinates appear is important.

The points whose coordinates are $(3, -1)$ and $(-1, 3)$ are graphed at the right. Note that the graphed points are in different locations. *The order of the coordinates of an ordered pair is important.*

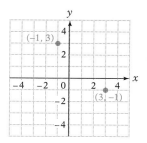

Each point in the plane is associated with an ordered pair, and each ordered pair is associated with a point in the plane. Although only the labels for integers are given on a coordinate grid, the graph of any ordered pair can be approximated. For example, the point whose coordinates are $(-2.3, 4.1)$ and $(\pi, 1)$ are shown on the graph at the right.

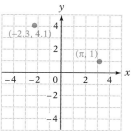

Example 1   Graph the ordered pairs $(-2, -3)$, $(3, -2)$, $(0, -2)$, and $(3, 0)$.

Solution

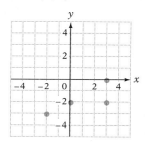

You Try It 1   Graph the ordered pairs $(-4, 1)$, $(3, -3)$, $(0, 4)$, and $(-3, 0)$.

Your solution

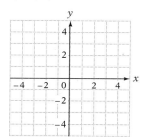

Example 2   Give the coordinates of the points labeled A and B. Give the abscissa of Point C and the ordinate of point D.

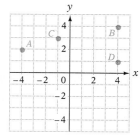

You Try It 2   Give the coordinates of the points labeled A and B. Give the abscissa of point C and the ordinate of point D.

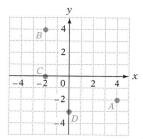

Solution   The coordinates of A are $(-4, 2)$.
The coordinates of B are $(4, 4)$.
The abscissa of C is -1.
The ordinate of D is 1.

Your solution

Solutions on p. A24

Objective B   ***To determine ordered-pair solutions of an equation in two variables*** ....

When drawing a rectangular coordinate system, we often label the horizontal axis x and the vertical axis y. In this case, the coordinate system is called an **xy-coordinate system.** The coordinates of the points are given by ordered pairs (x, y), where the abscissa is called the **x-coordinate** and the ordinate is called the **y-coordinate.**

A coordinate system is used to study the relationship between two variables. Frequently this relationship is given by an equation. Examples of equations in two variables include

$$y = 2x - 3 \qquad 3x + 2y = 6 \qquad x^2 - y = 0$$

A **solution of an equation in two variables** is an ordered pair (x, y) whose coordinates make the equation a true statement.

➡ Is $(-3, 7)$ a solution of $y = -2x + 1$?

$$y = -2x + 1$$

$$\begin{array}{c|l} 7 & -2(-3) + 1 \\ & 6 + 1 \end{array}$$ • Replace x by -3; replace y by 7.

$7 = 7$ • The results are equal.

$(-3, 7)$ is a solution of the equation $y = -2x + 1$.

Besides $(-3, 7)$, there are many other ordered-pair solutions of $y = -2x + 1$. For example, $(0, 1)$, $\left(-\dfrac{3}{2}, 4\right)$, and $(4, -7)$ are also solutions. In general, an equation in two variables has an infinite number of solutions. By choosing any value of x and substituting that value into the equation, we can calculate a corresponding value of y.

➡ Find the ordered-pair solution of $y = \dfrac{2}{3}x - 3$ that corresponds to $x = 6$.

$$y = \dfrac{2}{3}x - 3$$

$$= \dfrac{2}{3}(6) - 3$$ • Replace x by 6.

$$= 4 - 3$$ • Solve for y.

$$= 1$$

The ordered-pair solution is $(6, 1)$.

The solutions of an equation in two variables can be graphed in an xy-coordinate system.

➡ Graph the ordered-pair solutions of $y = -2x + 1$ when $x = -2, -1, 0, 1,$ and 2.

Use the values of x to determine ordered-pair solutions of the equation. It is convenient to record these in a table.

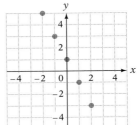

x	$y = -2x + 1$	y	(x, y)
-2	$-2(-2) + 1$	5	$(-2, 5)$
-1	$-2(-1) + 1$	3	$(-1, 3)$
0	$-2(0) + 1$	1	$(0, 1)$
1	$-2(1) + 1$	-1	$(1, -1)$
2	$-2(2) + 1$	-3	$(2, -3)$

Example 3
Is $(3, -2)$ a solution of $3x - 4y = 15$?

You Try It 3
Is $(-2, 4)$ a solution of $x - 3y = -14$?

Solution

$$3x - 4y = 15$$

$$\begin{array}{c|c} 3(3) - 4(-2) & 15 \\ 9 + 8 & \\ & 17 \neq 15 \end{array}$$

• **Replace x by 3 and y by -2.**

No, $(3, -2)$ is not a solution of $3x - 4y = 15$.

Your solution

Example 4
Graph the ordered-pair solutions of $2x - 3y = 6$ when $x = -3, 0, 3, 6$.

You Try It 4
Graph the ordered-pair solutions of $x + 2y = 4$ when $x = -4, -2, 0, 2$.

Solution   Solve $2x - 3y = 6$ for y.

$$2x - 3y = 6$$
$$-3y = -2x + 6$$
$$y = \frac{2}{3}x - 2$$

Replace x in $y = \frac{2}{3}x - 2$ by $-3, 0, 3$, and 6. For each value of x, determine the value of y.

Your solution

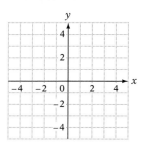

x	$y = \dfrac{2}{3}x - 2$	y	(x, y)
-3	$\dfrac{2}{3}(-3) - 2$	-4	$(-3, -4)$
0	$\dfrac{2}{3}(0) - 2$	-2	$(0, -2)$
3	$\dfrac{2}{3}(3) - 2$	0	$(3, 0)$
6	$\dfrac{2}{3}(6) - 2$	2	$(6, 2)$

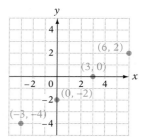

Solutions on p. A24

Objective C **To determine whether a set of ordered pairs is a function** ....

Discovering a relationship between two variables is an important task in the application of mathematics. Here are some examples.

- Botanists study the relationship between the number of bushels of wheat yielded per acre and the amount of watering per acre.
- Environmental scientists study the relationship between the incidents of skin cancer and the amount of ozone in the atmosphere.
- Business analysts study the relationship between the price of a product and the number of products that are sold at that price.

Each of these relationships can be described by a set of ordered pairs.

> **Definition of a Relation**
>
> A **relation** is any set of ordered pairs.

The following table shows the number of hours that each of 10 students spent studying for a midterm exam and the grade that each of these 10 students received.

Hours	3	3.5	2.75	2	4	4.5	3	2.5	5
Grade	78	75	70	65	85	85	80	75	90

This information can be written as the relation

$$\{(3, 78), (3.5, 75), (2.75, 70), (2, 65), (4, 85), (4.5, 85), (3, 80), (2.5, 75), (5, 90)\}$$

where the first coordinate of the ordered pair is the hours spent studying and the second coordinate is the score on the midterm.

The **domain** of a relation is the set of first coordinates of the ordered pairs; the **range** is the set of second coordinates. For the relation above,

$$\text{Domain} = \{2, 2.5, 2.75, 3, 3.5, 4, 4.5, 5\} \qquad \text{Range} = \{65, 70, 75, 78, 80, 85, 90\}$$

The **graph of a relation** is the graph of the ordered pairs that belong to the relation. The graph of the relation given above is shown at the right. The horizontal axis represents the hours spent studying (the domain); the vertical axis represents the test score (the range). The axes could be labeled H for hours studied and S for test score.

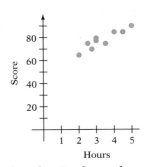

A *function* is a special type of relation for which no two ordered pairs have the same first coordinate.

> **Definition of a Function**
>
> A **function** is a relation in which no two ordered pairs that have the same first coordinate have different second coordinates.

The table at the right is the grading scale for a 50-point test. This table defines a relationship between the *score* on the test and a *letter grade*. Some of the ordered pairs of this function are (38, C), (47, A), (44, B), and (42, B).

Score	Grade
46–50	A
40–45	B
33–39	C
28–32	D
0–27	F

The grading-scale table defines a function, because no two ordered pairs can have the *same* first coordinate and *different* second coordinates. For instance, it is not possible to have the ordered pairs (32, C) and (32, B)—same first coordinate (test score) but different second coordinate (test grade). The domain of this function is $\{0, 1, 2, \ldots, 49, 50\}$. The range is $\{A, B, C, D, F\}$

The example of hours spent studying and test score given earlier is *not* a function, because (3, 78) and (3, 80) are ordered pairs of the relation that have the *same* first coordinate but *different* second coordinates.

Consider, again, the grading-scale example. Note that (44, B), and (42, B) are ordered pairs of the function. Ordered pairs of a function may have the same *second* coordinates but not the same first coordinates.

Although relations and function are given by tables, they are frequently given by an equation in two variables.

The equation $y = 2x$ expresses the relationship between a number, x, and twice the number, y. For instance, if $x = 3$, then $y = 6$, which is twice 3. To indicate exactly which ordered pairs are determined by the equation, the domain (values of x) is specified. If $x \in \{-2, -1, 0, 1, 2, 3\}$, then the ordered pairs determined by the equation are $\{(-2, -4), (-1, -2), (0, 0), (1, 2), (2, 4), (3, 6)\}$. This relation is a function because no two ordered pairs have the same first coordinate.

The graph of the function is shown at the right. The horizontal axis (domain) is labeled x; the vertical axis (range) is labeled y.

The domain $\{-2, -1, 0, 1, 2\}$ was chosen arbitrarily. Other domains could have been selected. The type of application usually influences the choice of the domain.

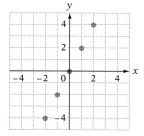

For the equation $y = 2x$, we say that "y is a function of x" because the set of ordered pairs is a function.

Not all equations, however, define a function. For instance, the equation $|y| = x + 2$ does not define y as a function of x. The ordered pairs (2, 4) and (2, −4) both satisfy the equation. Thus there are two ordered pairs with the same first coordinate but different second coordinates.

Example 5

To test a heart medicine, Dr. Lisa Nabib measures the heart rate of 5 patients before and after taking the medicine. The results are given in the table below.

Before medicine	85	80	85	75	90
After medicine	75	70	70	80	80

Write a relation where the first coordinate is the heart rate before taking the medicine and the second coordinate is the heart rate after taking the medicine. Is the relation a function?

You Try It 5

Professor Kaplan records the midterm and final exam scores for 5 graduate students. The results are given in the table below.

Midterm	82	78	81	87	81
Final Exam	91	86	96	79	87

Write a relation where the first coordinate is the score on the midterm and the second coordinate is the score on the final exam. Is the relation a function?

Solution

$\{(85, 75), (80, 70), (85, 70), (75, 80), (90, 80)\}$

No, the relation is not a function. The two ordered pairs $(85, 75)$ and $(85, 70)$ have the same first coordinate but different second coordinates.

Your solution

Example 6

Does $y = x^2 + 3$, where $x \in \{-2, -1, 1, 3\}$, define y as a function of x?

You Try It 6

Does $y = \frac{1}{2}x + 1$, where $x \in \{-4, 0, 2\}$, define y as a function of x?

Solution

Determine the ordered pairs defined by the equation. Replace x in $y = x^2 + 3$ by the given values and solve for y.

$\{(-2, 7), (-1, 4), (1, 4), (3, 12)\}$

No two ordered pairs have the same first coordinate. Therefore, the relation is a function and the equation $y = x^2 + 3$ defines y as a function of x.

Note that $(-1, 4)$ and $(1, 4)$ are ordered pairs that belong to this function. Ordered pairs of a function may have the same *second* coordinates, but not the same *first* coordinates.

Your solution

Solutions on p. A24

Objective D *To evaluate a function written in functional notation*

When an equation defines y as a function of x, **functional notation** is frequently used to emphasize that the relation is a function. In this case, it is common to use the notation $f(x)$, where

$$f(x) \text{ is read "} f \text{ of } x \text{" or "the value of } f \text{ at } x\text{."}$$

For instance, the equation $y = x^2 + 3$ from Example 6 defined y as a function of x. The equation can also be written in functional notation as

$$f(x) = x^2 + 3$$

where y has been replaced by $f(x)$.

The symbol $f(x)$ is called the **value of the function** at x because it is the result of evaluating a variable expression. For instance, $f(4)$ means to replace x by 4 and then simplify the resulting numerical expression.

$$f(x) = x^2 + 3$$
$$f(4) = 4^2 + 3 \qquad \bullet \text{ Replace } x \text{ by 4.}$$
$$= 16 + 3 = 19$$

This process is called **evaluating the function.**

➡ Given $f(x) = x^2 + x - 3$, find $f(-2)$.

$$f(x) = x^2 + x - 3$$
$$f(-2) = (-2)^2 + (-2) - 3 \qquad \bullet \text{ Replace } x \text{ by } -2.$$
$$= 4 - 2 - 3 = -1$$
$$f(-2) = -1$$

In this example, $f(-2)$ is the second coordinate of an ordered pair of the function; the first coordinate is -1. Therefore, an ordered pair of this function is $(-2, f(-2))$, which simplifies to $(-2, -1)$.

For the function given by $y = f(x) = x^2 + x - 3$, y is called the **dependent variable** because its value depends on the value of x. The **independent variable** is x.

Functions can be written using other letters or even combinations of letters. For instance, some calculators use $ABS(x)$ for the absolute-value function. Thus the equation $y = |x|$ would be written $ABS(x) = |x|$, where $ABS(x)$ replaces y.

Example 7

Given $G(t) = \dfrac{3t}{t + 4}$, find $G(1)$.

Solution

$$G(t) = \frac{3t}{t + 4}$$

$$G(1) = \frac{3(1)}{1 + 4} \qquad \bullet \text{ Replace } t \text{ by 1. Then simplify.}$$

$$G(1) = \frac{3}{5}$$

You Try It 7

Given $H(x) = \dfrac{x}{x - 4}$, find $H(8)$.

Your solution

Solution on p. A24

8.1 Exercises

Objective A

1. Graph: $(-2, 1)$, $(3, -5)$, $(-2, 4)$, and $(0, 3)$.

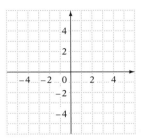

2. Graph: $(5, -1)$, $(-3, -3)$, $(-1, 0)$, and $(1, -1)$.

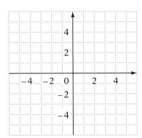

3. Graph: $(0, 0)$, $(0, -5)$, $(-3, 0)$, and $(0, 2)$.

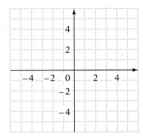

4. Graph the ordered pairs $(-4, 5)$, $(-3, 1)$, $(3, -4)$, and $(5, 0)$.

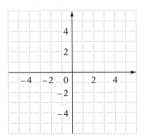

5. Graph the ordered pairs $(-1, 4)$, $(-2, -3)$, $(0, 2)$, and $(4, 0)$.

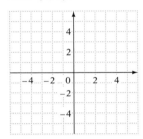

6. Graph the ordered pairs $(5, 2)$, $(-4, -1)$, $(0, 0)$, and $(0, 3)$.

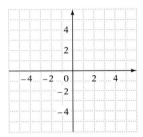

7. Find the coordinates of each of the points.

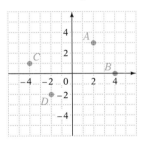

8. Find the coordinates of each of the points.

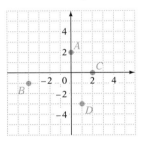

9. Find the coordinates of each of the points.

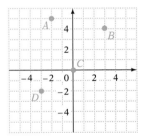

10. Find the coordinates of each of the points.

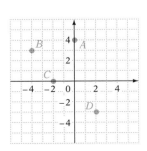

11. **a.** Name the abscissas of points A and C.
b. Name the ordinates of points B and D.

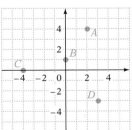

12. **a.** Name the abscissas of points A and C.
b. Name the ordinates of points B and D.

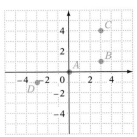

Objective B

13. Is $(3, 4)$ a solution of $y = -x + 7$?

14. Is $(2, -3)$ a solution of $y = x + 5$?

15. Is $(-1, 2)$ a solution of $y = \frac{1}{2}x - 1$?

16. Is $(1, -3)$ a solution of $y = -2x - 1$?

17. Is $(4, 1)$ a solution of $2x - 5y = 4$?

18. Is $(-5, 3)$ a solution of $3x - 2y = 9$?

19. Is $(0, 4)$ a solution of $3x - 4y = -4$?

20. Is $(-2, 0)$ a solution of $x + 2y = -1$?

Graph the ordered pair solutions of each equation for the given values of x.

21. $y = 2x;\ x = -2, -1, 0, 2$ **22.** $y = -2x;\ x = -2, -1, 0, 2$ **23.** $y = x + 2;\ x = -4, -2, 0, 3$

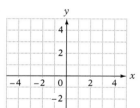
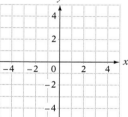
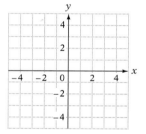

24. $y = \frac{1}{2}x - 1;$
$x = -2, 0, 2, 4$

25. $y = \frac{2}{3}x + 1;\ x = -3, 0, 3$

26. $y = -\frac{1}{3}x - 2;\ x = -3, 0, 3$

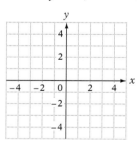
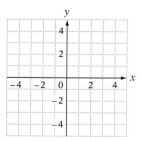
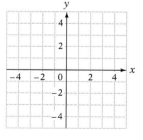

27. $2x + 3y = 6;\ x = -3, 0, 3$ **28.** $x - 2y = 4;\ x = -2, 0, 2$ **29.** $2x + y = 3;\ x = -1, 0, 1, 2$

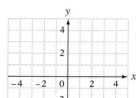

30. $3x - 4y = 8;\ x = -4, 0, 4$ **31.** $5x + 2y = 0;\ x = -2, 0, 2$ **32.** $-x - 2y = 0;\ x = -2, 0, 2$

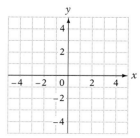
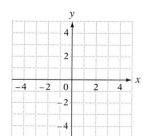
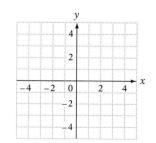

Objective C

33. The runs scored by a baseball team and whether the team won, W, or lost, L, its game are recorded in the following table. Write a relation where the first component is the runs scored and the second component is a win or a loss. Is the relation a function?

Runs scored	4	6	4	2	1	6
Win or loss	L	W	W	W	L	W

34. The number of aces served by a tennis player during a set and the number of games that player won in the set are given in the following table. Write a relation where the first component is the aces served and the second component is the number of games won. Is the relation a function?

Aces served	6	8	4	10	6
Games won	6	5	3	6	4

35. The number of concerts given in one year by five musical groups and the number of hit records that group had that year are given in the following table. Write a relation where the first component is the number of hit records and the second component is the number of concerts. Is the relation a function?

Hit records	11	9	7	12	12
Number of concerts	200	150	200	175	250

36. The monthly salary (in thousands) of an employee and the number of years of college for that employee are given in the following table. Write a relation where the first component is the years of college and the second component is the monthly salary. Is the relation a function?

Years of college	4	4	5	4	6	8
Salary	3	3.5	4	2.5	5	5.5

37. Does $y = -2x - 3$, where $x \in \{-2, -1, 0, 3\}$, define y as a function of x?

38. Does $y = 2x + 3$, where $x \in \{-2, -1, 1, 4\}$, define y as a function of x?

39. Does $|y| = x - 1$, where $x \in \{1, 2, 3, 4\}$, define y as a function of x?

40. Does $|y| = x + 2$, where $x \in \{-2, -1, 0, 3\}$, define y as a function of x?

41. Does $y = x^2$, where $x \in \{-2, -1, 0, 1, 2\}$, define y as a function of x?

42. Does $y = x^2 - 1$, where $x \in \{-2, -1, 0, 1, 2\}$, define y as a function of x?

Objective D

43. Given $f(x) = 3x - 4$, find $f(4)$.

44. Given $f(x) = 5x + 1$, find $f(2)$.

45. Given $f(x) = x^2$, find $f(3)$.

46. Given $f(x) = x^2 - 1$, find $f(1)$.

47. Given $G(x) = x^2 + x$, find $G(-2)$.

48. Given $H(x) = x^2 - x$, find $H(-2)$.

49. Given $s(t) = \dfrac{3}{t - 1}$, find $s(-2)$.

50. Given $P(x) = \dfrac{4}{2x + 1}$, find $P(-2)$.

51. Given $h(x) = 3x^2 - 2x + 1$, find $h(3)$.

52. Given $Q(r) = 4r^2 - r - 3$, find $Q(2)$.

53. Given $f(x) = \dfrac{x}{x + 5}$, find $f(-3)$.

54. Given $v(t) = \dfrac{2t}{2t + 1}$, find $v(3)$.

55. Given $g(x) = x^3 - x^2 + 2x - 7$, find $g(0)$.

56. Given $F(z) = \dfrac{z}{z^2 + 1}$, find $F(0)$.

APPLYING THE CONCEPTS

57. Suppose you are helping a student who is having trouble graphing
[W] ordered pairs. The work of the student is at the right. What can you
say to this student to correct the error that is being made?

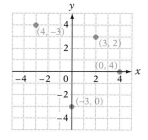

58. Write a few sentences that describe the similarities and differences
[W] between relations and functions.

59. The graph of $y^2 = x$, where $x \in \{0, 1, 4, 9\}$, is shown at the right. Is this
[W] the graph of a function? Explain your answer.

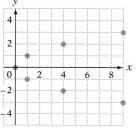

60. Is it possible to evaluate $f(x) = \dfrac{5}{x - 1}$ when $x = 1$? If so, what is $f(1)$? If
[W] not, explain why not.

Linear Equations in Two Variables

. . . .

Objective A   *To graph an equation of the form y = mx + b* ....

The **graph of an equation in two variables** is a graph of the ordered-pair solutions of the equation.

Consider $y = 2x + 1$. Choosing $x = -2$, $-1, 0, 1$, and 2 and determining the corresponding values of y produces some of the ordered pairs of the equation. These are recorded in the table at the right. See the graph of the ordered pairs in Fig. 1.

x	$y = 2x + 1$	y	(x, y)
-2	$2(-2) + 1$	-3	$(-2, -3)$
-1	$2(-1) + 1$	-1	$(-1, -1)$
0	$2(0) + 1$	1	$(0, 1)$
1	$2(1) + 1$	3	$(1, 3)$
2	$2(2) + 1$	5	$(2, 5)$

Choosing values of x that are not integers produces more ordered pairs to graph, such as $\left(-\frac{5}{2}, -4\right)$ and $\left(\frac{3}{2}, 4\right)$, as shown in Fig. 2. Choosing still other values of x would result in more and more ordered pairs being graphed. The result would be so many dots that the graph would appear as the straight line shown in Fig. 3, which is the graph of $y = 2x + 1$.

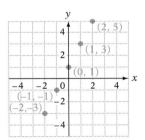

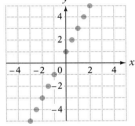

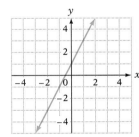

Fig. 1   **Fig. 2**   **Fig. 3**

Equations in two variables have characteristic graphs. The equation $y = 2x + 1$ is an example of a *linear equation*, or *linear function*, because its graph is a straight line.

> **Linear Equation in Two Variables**
>
> Any equation of the form $y = mx + b$, where m and b are constants, is a **linear equation in two variables.** The graph of a linear equation in two variables is a straight line.

Examples of linear equations are shown at the right. These equations represent linear functions because there is only one possible y for each x. Note from $y = 3 - 2x$ that m is the coefficient of x and b is the constant.

$y = 2x + 1$ $(m = 2, b = 1)$

$y = x - 4$ $(m = 1, b = -4)$

$y = -\dfrac{3}{4}x$ $\left(m = -\dfrac{3}{4}, b = 0\right)$

$y = 3 - 2x$ $(m = -2, b = 3)$

The equation $y = x^2 + 4x + 3$ is not a linear equation in two variables because there is a term with a variable squared. The equation $y = \dfrac{3}{x - 4}$ is not a linear equation because a variable occurs in the denominator of a fraction.

POINT OF INTEREST
The Project in Mathematics at the end of this chapter contains information on using calculators to graph an equation.

To graph a linear equation, choose some values of x and then find the corresponding values of y. Because a straight line is determined by two points, it is sufficient to find only two ordered-pair solutions. However, it is recommended that at least three ordered-pair solutions be used to ensure accuracy.

➡ Graph $y = -\dfrac{3}{2}x + 2$.

This is a linear equation with $m = -\dfrac{3}{2}$ and $b = 2$. Find at least three solutions. Because m is a fraction, choose values of x that will simplify the calculations. We have chosen $-2, 0,$ and 4 for x. (Any values of x could have been selected.)

x	$y = -\dfrac{3}{2}x + 2$	y	(x, y)
-2	$-\dfrac{3}{2}(-2) + 2$	5	$(-2, 5)$
0	$-\dfrac{3}{2}(0) + 2$	2	$(0, 2)$
4	$-\dfrac{3}{2}(4) + 2$	-4	$(4, -4)$

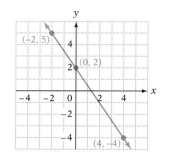

The graph of $y = -\dfrac{3}{2}x + 2$ is shown at the right.

Remember that a graph is a drawing of the ordered-pair solutions of the equation. Therefore, every point on the graph is a solution of the equation and every solution of the equation is a point on the graph.

The graph at the right is the graph of $y = x + 2$. Note that $(-4, -2)$ and $(1, 3)$ are points on the graph and that these points are solutions of $y = x + 2$. The point whose coordinates are $(4, 1)$ is not a point on the graph and is not a solution of the equation.

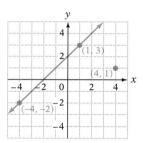

Example 1   Graph $y = 3x - 2$.

Solution

x	y
0	-2
-1	-5
2	4

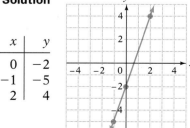

You Try It 1   Graph $y = 3x + 1$.

Your solution

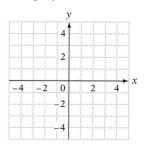

Solution on p. A24

Example 2   Graph $y = 2x$.

Solution

x	y
0	0
2	4
-2	-4

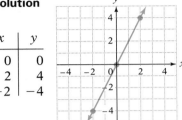

You Try It 2   Graph $y = -2x$.

Your solution

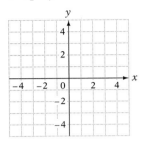

Example 3   Graph $y = \dfrac{1}{2}x - 1$.

Solution

x	y
0	-1
2	0
-2	-2

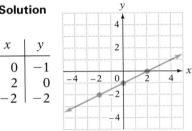

You Try It 3   Graph $y = \dfrac{1}{3}x - 3$.

Your solution

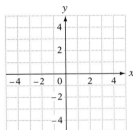

Solutions on pp. A24–A25

Objective B   To graph an equation of the form Ax + By = C ....

The equation $Ax + By = C$, where A, B, and C are constants, is also a linear equation. Examples of these equations are shown at the right.

$2x + 3y = 6$ $(A = 2, B = 3, C = 6)$
$x - 2y = -4$ $(A = 1, B = -2, C = -4)$
$2x + y = 0$ $(A = 2, B = 1, C = 0)$
$4x - 5y = 2$ $(A = 4, B = -5, C = 2)$

To graph an equation of the form $Ax + By = C$, first solve the equation for y. Then follow the same procedure used for graphing $y = mx + b$.

⟹ Graph $3x + 4y = 12$.

$$3x + 4y = 12$$
- Solve for y.

$$4y = -3x + 12$$
- Subtract $3x$ from each side of the equation.

$$y = -\dfrac{3}{4}x + 3$$
- Divide each side of the equation by 4.

x	y
0	3
4	0
-4	6

- Find 3 ordered-pair solutions of the equation.

- Graph the ordered pairs and then draw a line through the points.

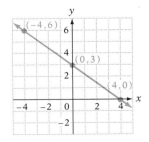

The graph of an equation with one of the variables missing is either a horizontal or a vertical line.

The equation $y = 2$ could be written $0 \cdot x + y = 2$. Because $0 \cdot x = 0$ for any value of x, the value of y is always 2 no matter what value of x is chosen. For instance, replace x by -4, -1, 0, or 3. In each case, $y = 2$.

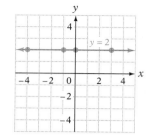

$$0x + y = 2$$
$$0(-4) + y = 2 \qquad (-4, 2) \text{ a solution.}$$
$$0(-1) + y = 2 \qquad (-1, 2) \text{ a solution.}$$
$$0(0) + y = 2 \qquad (0, 2) \text{ a solution.}$$
$$0(3) + y = 2 \qquad (3, 2) \text{ a solution.}$$

The solutions are plotted in the graph above, and a line is drawn through the plotted points. Note that the line is horizontal.

Graph of a Horizontal Line

The graph of $y = b$ is a horizontal line passing through $(0, b)$.

The equation $x = -2$ could be written $x + 0 \cdot y = -2$. Because $0 \cdot y = 0$ for any value of y, the value of x is always -2 no matter what value of y is chosen. For instance, replace y by -2, 0, 2, or 3. In each case, $x = -2$.

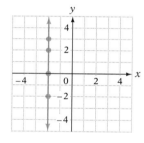

$$x + 0y = -2$$
$$x + 0(-2) = -2 \qquad (-2, -2) \text{ a solution.}$$
$$x + 0(0) = -2 \qquad (-2, 0) \text{ a solution.}$$
$$x + 0(2) = -2 \qquad (-2, 2) \text{ a solution.}$$
$$x + 0(3) = -2 \qquad (-2, 3) \text{ a solution.}$$

The solutions are plotted in the graph at the right, and a line is drawn through the plotted points. Note that the line is vertical.

Graph of a Vertical Line

The graph of $x = a$ is a vertical line passing through $(a, 0)$.

➡ Graph $x = -3$ and $y = 2$ in the same coordinate grid.

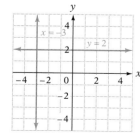

- The graph of $x = -3$ is a vertical line passing through $(-3, 0)$.
- The graph of $y = 2$ is a horizontal line passing through $(0, 2)$.

Example 4   Graph $2x - 5y = 10$.

Solution

$2x - 5y = 10$
$-5y = -2x + 10$
$y = \dfrac{2}{5}x - 2$

x	y
-0	-2
5	0
-5	-4

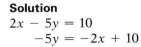

You Try It 4   Graph $5x - 2y = 10$.

Your solution

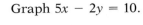

Example 5   Graph $x + 2y = 6$.

Solution

$x + 2y = 6$
$2y = -x + 6$
$y = -\dfrac{1}{2}x + 3$

x	y
0	3
-2	4
4	1

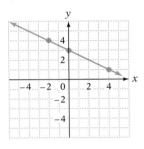

You Try It 5   Graph $x - 3y = 9$.

Your solution

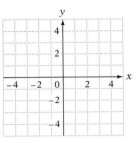

Example 6   Graph $y = -2$.

Solution   The graph of an equation of the form $y = b$ is a horizontal line passing through the point $(0, b)$.

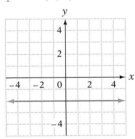

You Try It 6   Graph $y = 3$.

Your solution

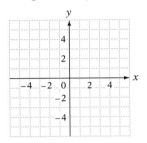

Example 7   Graph $x = 3$.

Solution   The graph of an equation of the form $x = a$ is a vertical line passing through the point $(a, 0)$.

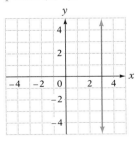

You Try It 7   Graph $x = -4$.

Your solution

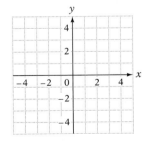

Solutions on p. A25

Objective C   *To solve application problems*

There are a variety of applications of linear functions.

⇒ Solve: An installer of marble kitchen countertops charges $250 plus $180 per foot of countertop. The equation that describes the total cost, C, to have x feet of countertop installed is $C = 180x + 250$.

a. Graph this equation for $0 \le x \le 25$. (*Note:* In many applications, the domain of the variable is given so that the equation makes sense. For instance, it would not be sensible to have values of x that are less than 0. This would mean negative countertop! The choice of 25 is somewhat arbitrary, but most kitchens have less than 25 feet of counter space.)

b. The point whose coordinates are (8, 1690) is on the graph. Write a sentence that describes this ordered pair.

Solution

a.

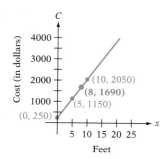

• Choosing $x = 0, 5$, and 10, you find that the corresponding ordered pairs are (0, 250), (5, 1150), and (10, 2050). Plot these points and draw a line through them.

b. The point whose coordinates are (8, 1690) means that 8 feet of countertop costs $1690 to install.

Example 8
The value, y, of an investment of $2500 at an annual simple interest rate of 6% is given by the equation $y = 150x + 2500$, where x is the number of years the investment is held. Graph this equation for $0 \le x \le 10$. The point whose coordinates are (5, 3250) is on the graph. Write a sentence that describes this ordered pair.

Solution

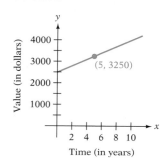

The ordered pair (5, 3250) means that in 5 years the value of the investment will be $3250.

You Try It 8
A car is traveling at a uniform speed of 40 miles per hour. The distance, d, the car travels in t hours is given by $d = 40t$. Graph this equation for $0 \le t \le 5$. The point whose coordinates are (3, 120) is on the graph. Write a sentence that describes this ordered pair.

Your solution

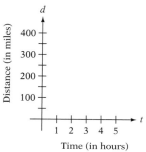

Solution on p. A25

8.2 Exercises

Objective A

Graph.

1. $y = 2x - 3$

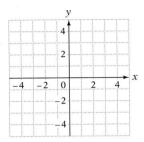

2. $y = -2x + 2$

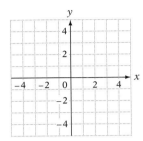

3. $y = \dfrac{1}{3}x$

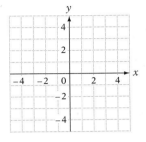

4. $y = -3x$

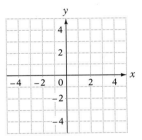

5. $y = \dfrac{2}{3}x - 1$

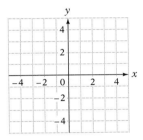

6. $y = \dfrac{3}{4}x + 2$

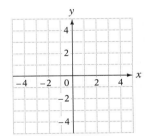

7. $y = -\dfrac{1}{4}x + 2$

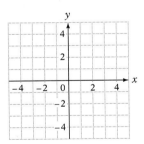

8. $y = -\dfrac{1}{3}x + 1$

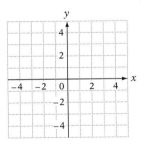

9. $y = -\dfrac{2}{5}x + 1$

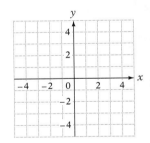

10. $y = -\dfrac{1}{2}x + 3$

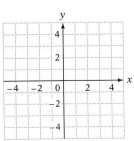

11. $y = 2x - 4$

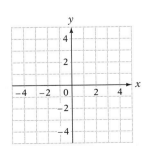

12. $y = 3x - 4$

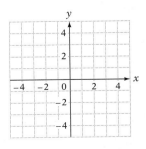

13. $y = x - 3$

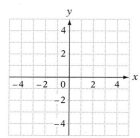

14. $y = x + 2$

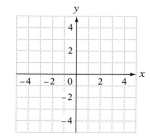

15. $y = -x + 2$

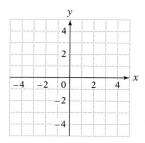

16. $y = -x - 1$

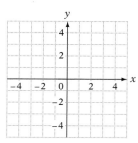

17. $y = -\dfrac{2}{3}x + 1$

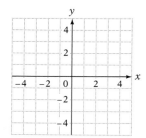

18. $y = 5x - 4$

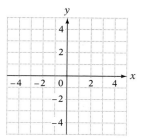

Objective B

Graph.

19. $3x + y = 3$

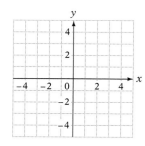

20. $2x + y = 4$

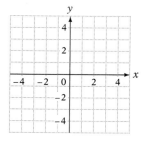

21. $2x + 3y = 6$

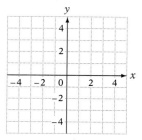

22. $3x + 2y = 4$

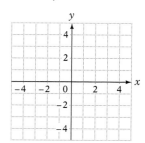

23. $x - 2y = 4$

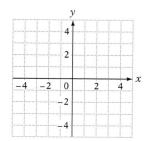

24. $x - 3y = 6$

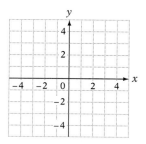

Graph.

25. $2x - 3y = 6$

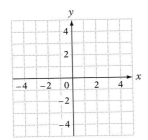

26. $3x - 2y = 8$

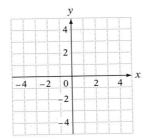

27. $2x + 5y = 10$

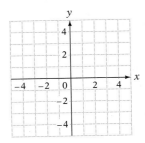

28. $3x + 4y = 12$

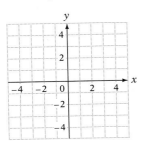

29. $x = 3$

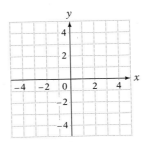

30. $y = -4$

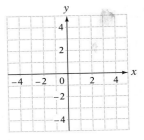

31. $x + 4y = 4$

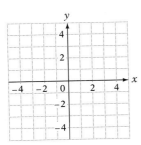

32. $4x - 3y = 12$

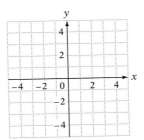

33. $y = 4$

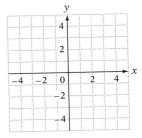

34. $x = -2$

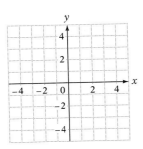

35. $\dfrac{x}{5} + \dfrac{y}{4} = 1$

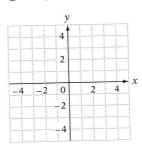

36. $\dfrac{x}{4} - \dfrac{y}{3} = 1$

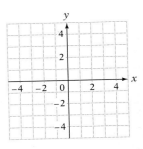

Objective C   *Application Problems*

Solve.

37. Depreciation is the declining value of an asset. For instance, a com-
[W]   pany that purchases a truck for $20,000 has an asset worth $20,000.
In five years, however, the value of the truck will have declined and
it may be worth only $4000. An equation that represents this decline
is $V = 20{,}000 - 3200x$, where V is the value of the truck after x
years. Graph this equation for $0 \le x \le 5$. The point $(3, 10{,}400)$ is on
the graph (Fig. 1). Write a sentence that describes the meaning of this
ordered pair.

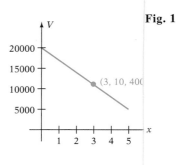

Fig. 1

38. A company uses the equation $V = 30{,}000 - 5000x$ to estimate the de-
[W]   preciated value (see Exercise 37) of a computer. Graph this equation
for $0 \le x \le 4$. The point $(1, 25{,}000)$ is on the graph (Fig. 2). Write a
sentence that describes the meaning of this ordered pair.

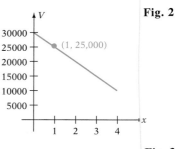

Fig. 2

39. An architect charges a fee of $500 plus $2.65 per square foot to
[W]   design a house. The equation that represents the architect's fee is
given by $F = 2.65x + 500$, where F is the fee and x is the number of
square feet in the house. Graph this equation for $0 \le x \le 5000$. The
point $(3500, 9775)$ is on the graph (Fig. 3). Write a sentence that describes the meaning of this ordered pair.

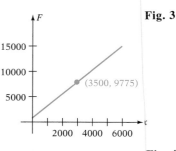

Fig. 3

40. A rental car company charges a "drop-off" fee of $50 to return a car
[W]   to a location different from that from which it was rented. In addition, a fee of $0.18 per mile the car is driven is charged. An equation
that represents the total cost to rent a car with a drop-off fee is
$C = 0.18x + 50$, where C is the total cost and x is the number of miles
the car is driven. Graph this equation for $0 \le x \le 1000$. The point
$(500, 140)$ is on the graph (Fig. 4). Write a sentence that describes the
meaning of this ordered pair.

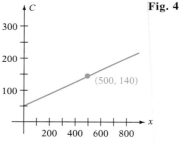

Fig. 4

APPLYING THE CONCEPTS

41. For the equation $y = 3x + 2$, when the value of x changes from 1 to 2,
does the value of y increase or decrease? What is the change in y?
Suppose that the value of x changes from 13 to 14. What is the change
in y?

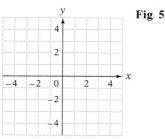

Fig. 5

42. For the equation $y = -2x + 1$, when the value of x changes from 1 to 2,
does the value of y increase or decrease? What is the change in y?
Suppose the value of x changes from 13 to 14. What is the change
in y?

43. Graph $y = 2x - 2$, $y = 2x$, and $y = 2x + 3$ in the same coordinate
[W]   system in Fig. 5. What observation can you make about the graphs?

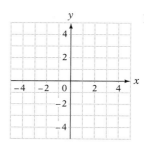

Fig. 6

44. Graph $y = x + 3$, $y = 2x + 3$, and $y = -\frac{1}{2}x + 3$ in the same coordi-
[W]   nate system in Fig. 6. What observation can you make about the
graphs?

8.3 Intercepts and Slopes of Straight Lines

Objective A   To find the x- and y-intercepts of a straight line

The graph of the equation $2x + 3y = 6$ is shown at the right. The graph crosses the x-axis at the point $(3, 0)$. This point is called the **x-intercept.** The graph also crosses the y-axis at the point $(0, 2)$. This point is called the **y-intercept.**

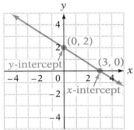

➡ Find the x-intercept and the y-intercept of the graph of the equation $2x - 3y = 12$.

LOOK CLOSELY

To find the x-intercept, let $y = 0$. To find the y-intercept, let $x = 0$.

To find the x-intercept, let $y = 0$. (Any point on the x-axis has y-coordinate 0.)

$$2x - 3y = 12$$
$$2x - 3(0) = 12$$
$$2x = 12$$
$$x = 6$$

The x-intercept is $(6, 0)$.

To find the y-intercept, let $x = 0$. (Any point on the y-axis has x-coordinate 0.)

$$2x - 3y = 12$$
$$2(0) - 3y = 12$$
$$-3y = 12$$
$$y = -4$$

The y-intercept is $(0, -4)$.

➡ Find the y-intercept of $y = 3x + 4$.

$$y = 3x + 4 = 3(0) + 4 = 4 \qquad \bullet \text{ Let } x = 0.$$

The y-intercept is $(0, 4)$.

For any equation of the form $y = mx + b$, the y-intercept is $(0, b)$.

Some linear equations can be graphed by finding the x- and y-intercepts and then drawing a line through these two points.

Example 1   Find the x- and y-intercepts for $x - 2y = 4$. Graph the line.

Solution

x-intercept:
$$x - 2y = 4$$
$$x - 2(0) = 4$$
$$x = 4$$
$$(4, 0)$$

y-intercept:
$$x - 2y = 4$$
$$0 - 2y = 4$$
$$-2y = 4$$
$$y = -2$$
$$(0, -2)$$

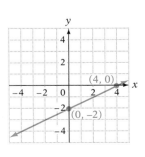

You Try It 1   Find the x- and y-intercepts for $y = 2x - 4$. Graph the line.

Solution

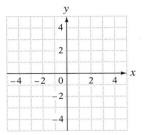

Solution on p. A25

Objective B   *To find the slope of a straight line* ....

The graphs of $y = \frac{2}{3}x + 1$ and $y = 2x + 1$ are shown in Fig. 1. Each graph crosses the y-axis at the point $(0, 1)$, but the graphs have different slants. The **slope** of a line is a measure of the slant of a line. The symbol for slope is m.

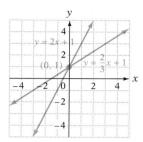

Fig. 1

The slope of a line containing two points is the ratio of the change in the y values of the two points to the change in the x values. The line containing the points $(-2, -3)$ and $(6, 1)$ is graphed in Fig. 2. The change in the y values is the difference between the two ordinates.

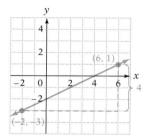

Fig. 2

Change in $y = 1 - (-3) = 4$

The change in the x values is the difference between the two abscissas (Fig. 3).

Change in $x = 6 - (-2) = 8$

$$\text{Slope} = m = \frac{\text{change in } y}{\text{change in } x} = \frac{4}{8} = \frac{1}{2}$$

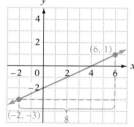

Fig. 3

> **Slope Formula**
>
> If $P_1(x_1, y_1)$ and $P_2(x_2, y_2)$ are two points on a line and $x_1 \neq x_2$, then $m = \dfrac{y_2 - y_1}{x_2 - x_1}$ (Fig. 4). If $x_1 = x_2$, the slope is undefined.

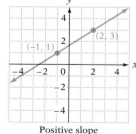

Fig. 4

➡ Find the slope of the line containing the points $(-1, 1)$ and $(2, 3)$.
Let P_1 be $(-1, 1)$ and P_2 be $(2, 3)$. Then, $x_1 = -1, y_1 = 1; x_2 = 2, y_2 = 3$.

$$m = \frac{y_2 - y_1}{x_2 - x_1} = \frac{3 - 1}{2 - (-1)} = \frac{2}{3}$$

The slope is $\frac{2}{3}$.

LOOK CLOSELY
Positive slope means that the value of y increases as the value of x increases.

A line that slants upward to the right always has a **positive slope.**

Positive slope

Note that you obtain the same results if the points are named oppositely. If P_1 is $(2, 3)$ and P_2 is $(-1, 1)$, then $x_1 = 2, y_1 = 3; x_2 = -1, y_2 = 1$.

$$m = \frac{y_2 - y_1}{x_2 - x_1} = \frac{1 - 3}{-1 - 2} = \frac{-2}{-3} = \frac{2}{3}$$

The slope is $\frac{2}{3}$.

Therefore, it does not matter which point is named P_1 and which P_2; the slope remains the same.

Content and Format © 1995 HMCo.

LOOK CLOSELY

Negative slope means that the value of *y* decreases as *x* increases. Compare this to positive slope.

➡ Find the slope of the line containing the points $(-3, 4)$ and $(2, -2)$.

Let P_1 be $(-3, 4)$ and P_2 be $(2, -2)$.

$$m = \frac{y_2 - y_1}{x_2 - x_1} = \frac{-2 - 4}{2 - (-3)} = \frac{-6}{5} = -\frac{6}{5}$$

The slope is $-\frac{6}{5}$.

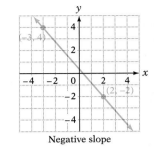

Negative slope

A line that slants downward to the right always has a **negative slope.**

➡ Find the slope of the line containing the points $(-1, 3)$ and $(4, 3)$.

Let P_1 be $(-1, 3)$ and P_2 be $(4, 3)$.

$$m = \frac{y_2 - y_1}{x_2 - x_1} = \frac{3 - 3}{4 - (-1)} = \frac{0}{5} = 0$$

The slope is 0.

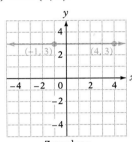

Zero slope

A horizontal line has **zero slope.**

➡ Find the slope of a line containing the points $(2, -2)$ and $(2, 4)$.

Let P_1 be $(2, -2)$ and P_2 be $(2, 4)$.

$$m = \frac{y_2 - y_1}{x_2 - x_1} = \frac{4 - (-2)}{2 - 2} = \frac{6}{0} \quad \text{Not a real number}$$

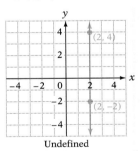

Undefined

The slope of a vertical line is undefined.

The are many applications of the concept of slope. Here are two possibilities. In 1988, when Florence Griffith-Joyner set the world record for the 100-meter dash, her average rate of speed was approximately 9.5 meters per second. The graph at the right shows the distance she ran during her record-setting run. From the graph, note that after 4 seconds she had traveled 38 meters and that after 6 seconds she had traveled 57 meters. The slope of the line between these two points is

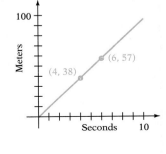

$$m = \frac{57 - 38}{6 - 4} = \frac{19}{2} = 9.5$$

Note that the slope of the line is the same as the rate she was running, 9.5 meters per second. The average speed of an object is related to slope.

Here is an example of slope taken from economics. According to the Department of Commerce, from 1987 to 1994, U.S. exports of goods to other countries had been increasing at a rate of approximately $2.5 billion per year. The graph at the right shows the value of exports for each year. From the graph, we learn that in 1988 exports were $27 billion and in 1992 exports were $37 billion. The slope of the line between these two points is

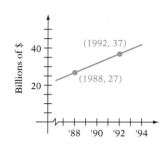

$$m = \frac{37 - 27}{1992 - 1988} = \frac{10}{4} = 2.5$$

Observe that the slope of the line is the same as the rate at which exports are increasing, $2.5 billion per year.

In general, any quantity that is expressed by using the word *per* is represented mathematically as slope. In the first example, slope was 9.5 meters *per* second; in the second example, slope was $2.5 billion *per* year.

Example 2
Find the slope of the line containing the points with coordinates $(-2, -3)$ and $(3, 4)$.

You Try It 2
Find the slope of the line containing the points with coordinates $(-3, 8)$ and $(1, 4)$.

Solution   Let $P_1 = (-2, -3)$ and $P_2 = (3, 4)$.

$$m = \frac{y_2 - y_1}{x_2 - x_1} = \frac{4 - (-3)}{3 - (-2)} = \frac{7}{5}$$

The slope is $\frac{7}{5}$.

Your solution

Example 3
Find the slope of the line containing the points with coordinates $(-1, 4)$ and $(-1, 0)$.

You Try It 3
Find the slope of the line containing the points with coordinates $(-1, 2)$ and $(4, 2)$.

Solution   Let $P_1 = (-1, 4)$ and $P_2 = (-1, 0)$.

$$m = \frac{y_2 - y_1}{x_2 - x_1} = \frac{0 - 4}{-1 - (-1)} = \frac{-4}{0}$$

The slope is undefined.

Your solution

Solutions on pp. A25–A26

Example 4

The graph below shows the price of a Macintosh IIci computer from January to June (shown as the numbers 1 through 6). Find the slope of the line. Write a sentence that states the meaning of the slope.

Solution

$$m = \frac{2125 - 2350}{5 - 2}$$

$$= \frac{-225}{3} = -75$$

A slope of -75 means that the price of a Macintosh IIci is *decreasing* at a rate of $75 per month.

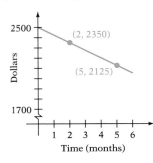

You Try It 4

The graph below shows the approximate decline in the value of a used car over a five-year period. Find the slope of the line. Write a sentence that states the meaning of the slope.

Your solution

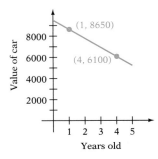

Solution on p. A26

Objective C   To graph a line using the slope and the y-intercept

The graph of the equation $y = \frac{2}{3}x + 1$ is shown at the right. The points $(-3, -1)$ and $(3, 3)$ are on the graph. The slope of the line between the two points is

$$m = \frac{3 - (-1)}{3 - (-3)} = \frac{4}{6} = \frac{2}{3}$$

Observe that the slope of the line is the coefficient of x in the equation $y = \frac{2}{3}x + 1$. Also recall that the y-intercept is $(0, 1)$, where 1 is the constant term of the equation.

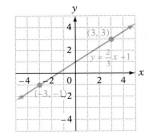

> **Slope–Intercept Equation of a Line**
>
> An equation of the form $y = mx + b$ is called the **slope–intercept** form of a straight line. The slope of the line is m, the coefficient of x. The y-intercept is $(0, b)$, where b is the constant term of the equation.

The following equations are written in slope–intercept form.

$$y = 2x - 3 \qquad \text{Slope} = 2,\ y\text{-intercept} = (0, -3)$$

$$y = -x + 2 \qquad \text{Slope} = -1(-x = -1x),\ y\text{-intercept} = (0, 2)$$

$$y = \frac{x}{2} \qquad \text{Slope} = \frac{1}{2}\left(\frac{x}{2} = \frac{1}{2}x\right),\ y\text{-intercept} = (0, 0)$$

When an equation of a straight line is written in slope–intercept form, the graph can be drawn using the slope and the y-intercept. First locate the y-intercept. Use the slope to find a second point on the line. Then draw a line through the two points.

➡ Graph $y = 2x - 3$.

y-intercept $= (0, b) = (0, -3)$

$m = 2 = \dfrac{2}{1} = \dfrac{\text{change in } y}{\text{change in } x}$

Beginning at the y-intercept, move right 1 unit (change in x) and then up 2 units (change in y).

$(1, -1)$ is a second point on the graph.

Draw a line through the two points $(0, -3)$ and $(1, -1)$.

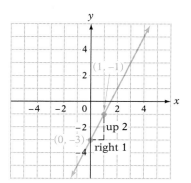

Example 5   Graph $y = -\dfrac{2}{3}x + 1$ by using the slope and y-intercept.

Solution y-intercept $= (0, b) = (0, 1)$

$m = -\dfrac{2}{3} = \dfrac{-2}{3}$

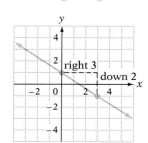

You Try It 5   Graph $y = -\dfrac{1}{4}x - 1$ by using the slope and y-intercept.

Your solution

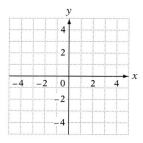

Example 6   Graph $2x - 3y = 6$ by using the slope and y-intercept.

Solution   Solve the equation for y.

$2x - 3y = 6$

$-3y = -2x + 6$

$y = \dfrac{2}{3}x - 2$

y-intercept $= (0, -2)$; $m = \dfrac{2}{3}$

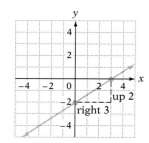

You Try It 6   Graph $x - 2y = 4$ by using the slope and y-intercept.

Your solution

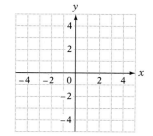

Solutions on p. A26

8.3 Exercises

· ·

Objective A

Find the *x*- and *y*-intercepts.

1. $x - y = 3$

2. $3x + 4y = 12$

3. $y = 3x - 6$

4. $y = 2x + 10$

5. $x - 5y = 10$

6. $3x + 2y = 12$

7. $y = 3x + 12$

8. $y = 5x + 10$

9. $2x - 3y = 0$

10. $3x + 4y = 0$

11. $y = -\dfrac{1}{2}x + 3$

12. $y = \dfrac{2}{3}x - 4$

Find the *x*- and *y*-intercepts and then graph.

13. $5x + 2y = 10$

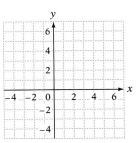

14. $x - 3y = 6$

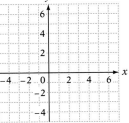

15. $y = \dfrac{3}{4}x - 3$

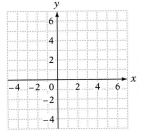

16. $y = \dfrac{2}{5}x - 2$

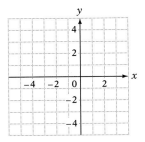

17. $5y - 3x = 15$

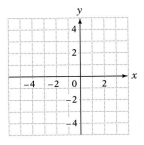

18. $9y - 4x = 18$

Objective B

Find the slope of the line containing the given points.

19. $P_1(4, 2)$, $P_2(3, 4)$

20. $P_1(2, 1)$, $P_2(3, 4)$

21. $P_1(-1, 3)$, $P_2(2, 4)$

22. $P_1(-2, 1)$, $P_2(2, 2)$

23. $P_1(2, 4)$, $P_2(4, -1)$

24. $P_1(1, 3)$, $P_2(5, -3)$

25. $P_1(-2, 3)$, $P_2(2, 1)$

26. $P_1(5, -2)$, $P_2(1, 0)$

27. $P_1(8, -3)$, $P_2(4, 1)$

28. $P_1(0, 3)$, $P_2(2, -1)$

29. $P_1(3, -4)$, $P_2(3, 5)$

30. $P_1(-1, 2)$, $P_2(-1, 3)$

31. $P_1(4, -2)$, $P_2(3, -2)$

32. $P_1(5, 1)$, $P_2(-2, 1)$

33. $P_1(0, -1)$, $P_2(3, -2)$

34. $P_1(3, 0)$, $P_2(2, -1)$

35. $P_1(-2, 3)$, $P_2(1, 3)$

36. $P_1(4, -1)$, $P_2(-3, -1)$

Objective C

Graph by using the slope and y-intercept.

37. $y = 3x + 1$

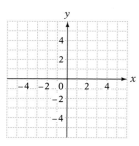

38. $y = -2x - 1$

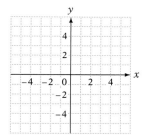

39. $y = \dfrac{2}{5}x - 2$

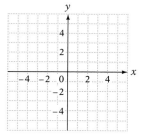

40. $y = \dfrac{3}{4}x + 1$

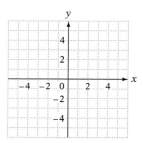

41. $2x + y = 3$

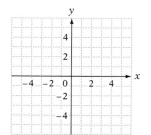

42. $3x - y = -1$

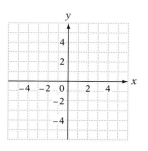

Graph by using the slope and *y*-intercept.

43. $x - 2y = 4$

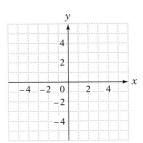

44. $x + 3y = 6$

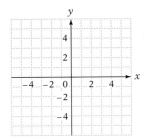

45. $y = \dfrac{2}{3}x$

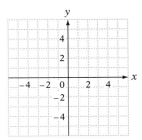

46. $y = \dfrac{1}{2}x$

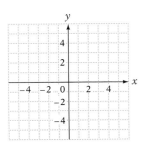

47. $y = -x + 1$

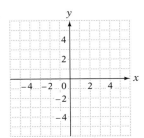

48. $y = -x - 3$

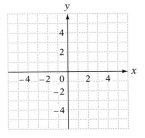

49. $3x - 4y = 12$

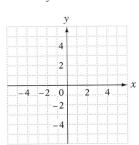

50. $5x - 2y = 10$

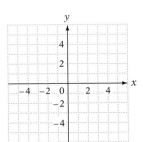

51. $y = -4x + 2$

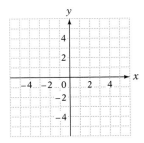

52. The graph at the right shows the cost, in dollars, to make a transatlantic telephone call. Find the slope of the line. Write a sentence that states the meaning of the slope.

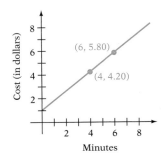

53. The graph in Fig. 1 shows the pressure, in pounds per square inch, on a diver. Find the slope of the line. Write a sentence that states the meaning of the slope.

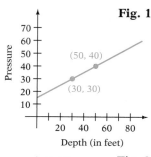

Fig. 1

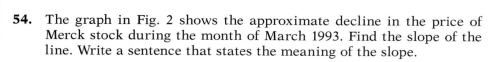

54. The graph in Fig. 2 shows the approximate decline in the price of Merck stock during the month of March 1993. Find the slope of the line. Write a sentence that states the meaning of the slope.

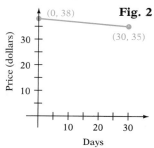

Fig. 2

55. The graph in Fig. 3 shows the decrease in the median price of a home in a city over a 6-month period. Find the slope of the line. Write a sentence that states the meaning of the slope.

ADDITIONAL EXERCISES

56. Do all straight lines have a y-intercept? If not, give an example of one that does not.

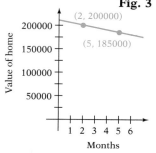

Fig. 3

57. If two lines have the same slope and same y-intercept, must the graphs of the lines be the same? If not, give an example.

58. Draw the graph of **a.** $\frac{x}{3} + \frac{y}{4} = 1$, of **b.** $\frac{x}{2} - \frac{y}{3} = 1$, and of **c.** $-\frac{x}{4} + \frac{y}{2} = 1$. What observations can you make about the x- and y-intercepts of these graphs and the coefficients of x and y? Use this observation to draw the graph of **d.** $\frac{x}{4} - \frac{y}{3} = 1$.

a. 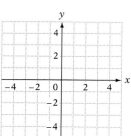 **b.** 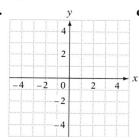 **c.** 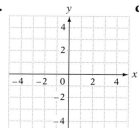 **d.**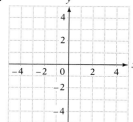

59. What does the highway sign at the right have to do with slope?

Equations of Straight Lines

Objective A   *To find the equation of a line given a point and the slope*

In earlier sections, the equation of a line was given and you were asked to determine some properties of the line, such as its intercepts and slope. Here, the process is reversed: Given properties of a line, determine its equation.

If the slope and y-intercept of the line are known, the equation of the line can be determined by using the slope–intercept form of a straight line.

➡ Find the equation of the line with slope $-\dfrac{1}{2}$ and y-intercept $(0, 3)$.

$$y = mx + b$$ • Use the slope–intercept formula.

$$y = -\frac{1}{2}x + 3$$ • $m = -\dfrac{1}{2}$; $(0, b) = (0, 3)$, so $b = 3$.

The equation of the line is $y = -\dfrac{1}{2}x + 3$.

When the coordinates of a point other than the y-intercept and the slope are known, the equation of the line can be found by using the formula for slope.

Suppose a line passes through the point $(3, 1)$ and has a slope of $\dfrac{2}{3}$. The equation of the line with these properties is determined by letting (x, y) be the coordinates of an unknown point on the line. Because the slope of the line is known, use the slope formula to write an equation. Then solve for y.

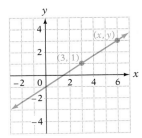

$$\frac{y - 1}{x - 3} = \frac{2}{3}$$ • $\dfrac{y_2 - y_1}{x_2 - x_1} = m$; $m = \dfrac{2}{3}$; $(x_2, y_2) = (x, y)$; $(x_1, y_1) = (3, 1)$

$$\frac{y - 1}{x - 3}(x - 3) = \frac{2}{3}(x - 3)$$ • Multiply each side by $(x - 3)$.

$$y - 1 = \frac{2}{3}x - 2$$ • Simplify.

$$y = \frac{2}{3}x - 1$$ • Solve for y.

The equation of the line is $y = \dfrac{2}{3}x - 1$.

The same procedure that was used above is used to derive the *point–slope formula*. We use this formula to determine the equation of a line when we are given the coordinates of a point on the line and the slope of the line.

Let (x_1, y_1) be the given coordinates of a point on a line, m the given slope of the line, and (x, y) the coordinates of an unknown point on the line. Then

$$\frac{y - y_1}{x - x_1} = m$$ • Formula for slope

$$\frac{y - y_1}{x - x_1}(x - x_1) = m(x - x_1)$$ • Multiply each side by $x - x_1$

$$y - y_1 = m(x - x_1)$$ • Simplify

> **Point–Slope Formula**
>
> If (x_1, y_1) is a point on a line with slope m, then $y - y_1 = m(x - x_1)$.

➡ Find the equation of the line that passes through point $(2, 3)$ and has slope -2.

$$y - y_1 = m(x - x_1)$$
$$y - 3 = -2(x - 2)$$
$$y - 3 = -2x + 4$$
$$y = -2x + 7$$

• Use the point–slope formula.
• $m = -2$; $(x_1, y_1) = (2, 3)$
• Solve for y.

The equation of the line is $y = -2x + 7$.

Example 1
Find the equation of the line whose slope is $-\dfrac{2}{3}$ and whose y-intercept is $(0, -1)$.

Solution
Because the slope and y-intercept are known, use the slope–intercept formula, $y = mx + b$.

$y = -\dfrac{2}{3}x - 1$ • $m = -\dfrac{2}{3}$; $b = -1$

You Try It 1
Find the equation of the line whose slope is $\dfrac{5}{3}$ and whose y-intercept is $(0, 2)$.

Your solution

Example 2
Use the point–slope formula to find the equation of the line that passes through the point $(-2, -1)$ and has slope $\dfrac{3}{2}$.

Solution
$$y - y_1 = m(x - x_1)$$
$$y - (-1) = \frac{3}{2}(x - (-2))$$
• $m = \dfrac{3}{2}$; $(x_1, y_1) = (-2, -1)$
$$y + 1 = \frac{3}{2}(x + 2)$$
$$y + 1 = \frac{3}{2}x + 3$$
$$y = \frac{3}{2}x + 2$$

The equation of the line is $y = \dfrac{3}{2}x + 2$.

You Try It 2
Use the point–slope formula to find the equation of the line that passes through the point $(4, -2)$ and has slope $\dfrac{3}{4}$.

Your solution

Solutions on p. A26

Objective B   **To find the equation of a line given two points**

The point–slope formula is used to find the equation of a line when a point on the line and the slope of the line are known. But this formula can also be used to find the equation of a line given two points on the line. In this case:

1. Use the slope formula to determine the slope of the line between the points.
2. Use the point–slope formula, the slope you just calculated, and one of the given points to find the equation of the line.

➡ Find the equation of the line that passes through the points whose coordinates are $(-3, -1)$ and $(3, 3)$.

Use the slope formula to determine the slope of the line between the points.

$$m = \frac{y_2 - y_1}{x_2 - x_1} = \frac{3 - (-1)}{3 - (-3)} = \frac{4}{6} = \frac{2}{3} \qquad (x_1, y_1) = (-3, -1); (x_2, y_2) = (3, 3)$$

Use the point–slope formula, the slope you just calculated, and one of the given points to find the equation of the line.

$$y - y_1 = m(x - x_1) \qquad \bullet \text{ Point–slope formula}$$

$$y - (-1) = \frac{2}{3}(x - (-3)) \qquad \bullet \ m = \frac{2}{3}; (x_1, y_1) = (-3, -1)$$

$$y + 1 = \frac{2}{3}(x + 3)$$

$$y + 1 = \frac{2}{3}x + 2$$

$$y = \frac{2}{3}x + 1 \qquad \bullet \text{ You can verify that the equation } y = \frac{2}{3}x + 1$$

$$\text{passes through the points } (-3, -1) \text{ and } (3, 3) \text{ by substituting the coordinates of these points into the equation.}$$

$$y = \frac{2}{3}x + 1 \qquad\qquad\qquad\qquad y = \frac{2}{3}x + 1$$

$$\begin{array}{c|c} -1 & \frac{2}{3}(-3) + 1 \\ \hline -1 & -2 + 1 \\ -1 = -1 \end{array} \quad \bullet \ (x, y) = (-3, -1) \qquad\qquad \begin{array}{c|c} 3 & \frac{2}{3}(3) + 1 \\ \hline 3 & 2 + 1 \\ 3 = 3 \end{array} \quad \bullet \ (x, y) = (3, 3)$$

The equation of the line that passes through the two points is $y = \frac{2}{3}x + 1$.

Example 3
Find the equation of the line that passes through the points $(-4, 0)$ and $(2, -3)$.

Solution
Find the slope of the line between the two points.

$$m = \frac{y_2 - y_1}{x_2 - x_1} = \frac{-3 - 0}{2 - (-4)} = \frac{-3}{6} = -\frac{1}{2}$$

Use the point–slope formula.

$$y - y_1 = m(x - x_1) \qquad \bullet \text{ Point–slope formula}$$

$$y - 0 = -\frac{1}{2}(x - (-4)) \qquad \bullet \ m = -\frac{1}{2}; (x_1, y_1) = (-4, 0)$$

$$y = -\frac{1}{2}(x + 4)$$

$$y = -\frac{1}{2}x - 2$$

The equation of the line is $y = -\frac{1}{2}x - 2$.

You Try It 3
Find the equation of the line that passes through the points $(-6, -1)$ and $(3, 1)$.

Your solution

Solution on p. A26

Objective C   *To solve application problems*

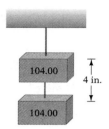

A **linear model** is a first-degree equation that is used to describe a relationship between quantities. In many cases, a linear model is used to approximate collected data. The data are graphed as points in a coordinate system, and then a line is drawn that approximates the data. The graph of the points is called a **scatter diagram;** the line is called a **line of best fit.**

Consider an experiment to determine the amount of weight that is necessary to stretch a spring a certain distance. Data from such an experiment are shown in the table below. Distance is in inches; weight is in pounds.

Distance	2.5	4	2	3.5	1	4.5
Weight	63	104	47	85	27	115

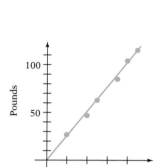

The accompanying graph shows the scatter diagram, which is the plotted points, and the line of best fit, which is the line that approximately goes through the plotted points. The equation of the line of best fit is $y = 25.7x - 1.3$, where x is the number of inches the spring is stretched and y is the weight in pounds.

The table below shows the actual recorded data for this experiment and the values that the model would predict. Good linear models should predict values that are close to the actual values. A more thorough analysis of lines of best fit is undertaken in statistics courses.

Distance, x	2.5	4	2	3.5	1	4.5
Actual weight	63	104	47	85	27	115
Weight predicted using $y = 25.7x - 1.3$	63	101.5	50.1	88.7	24.4	114.4

In each of the examples below, you will be given some data and the equation of the line of best fit. Base your answers on this information.

Example 4

The data in the table below show the size of a house in square feet and the cost to build the house. The line of best fit is $y = 70.3x + 41,100$, where x is the number of square feet and y is the cost of the house.

Square ft	1250	1400	1348	2675	2900
Cost	128,000	140,000	136,100	233,450	241,500

Graph the data and the line of best fit in the coordinate system below. Write a sentence that describes the meaning of the slope of the line of best fit.

Solution

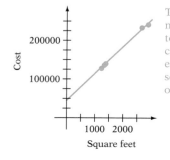

The slope of the line means that the cost to build the house increases $70.30 for each additional square foot in the size of the house.

You Try It 4

The data in the table below show a reading test grade and the final exam grade in a history class. The line of best fit is $y = 8.3x - 7.8$, where x is the reading test score and y is the history test score.

Reading	8.5	9.4	10.0	11.4	12.0
History	64	68	76	87	92

Graph the data and the line of best fit in the coordinate system below. Write a sentence that describes the meaning of the slope of the line of best fit.

Your solution

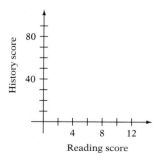

Solution on p. A26

8.4 Exercises

· ·

_____

Objective A

Solve.

1. Find the equation of the line that contains the point $(0, 2)$ and has slope 2.

2. Find the equation of the line that contains the point $(0, -1)$ and has slope -2.

3. Find the equation of the line that contains the point $(-1, 2)$ and has slope -3.

4. Find the equation of the line that contains the point $(2, -3)$ and has slope 3.

5. Find the equation of the line that contains the point $(3, 1)$ and has slope $\frac{1}{3}$.

6. Find the equation of the line that contains the point $(-2, 3)$ and has slope $\frac{1}{2}$.

7. Find the equation of the line that contains the point $(4, -2)$ and has slope $\frac{3}{4}$.

8. Find the equation of the line that contains the point $(2, 3)$ and has slope $-\frac{1}{2}$.

9. Find the equation of the line that contains the point $(5, -3)$ and has slope $-\frac{3}{5}$.

10. Find the equation of the line that contains the point $(5, -1)$ and has slope $\frac{1}{5}$.

11. Find the equation of the line that contains the point $(2, 3)$ and has slope $\frac{1}{4}$.

12. Find the equation of the line that contains the point $(-1, 2)$ and has slope $-\frac{1}{2}$.

Objective B

Solve.

13. Find the equation of the line that passes through the points $(1, -1)$ and $(-2, -7)$.

14. Find the equation of the line that passes through the points $(2, 3)$ and $(3, 2)$.

15. Find the equation of the line that passes through the points $(-2, 1)$ and $(1, -5)$.

16. Find the equation of the line that passes through the points $(-1, -3)$ and $(2, -12)$.

17. Find the equation of the line that passes through the points $(0, 0)$ and $(-3, -2)$.

18. Find the equation of the line that passes through the points $(0, 0)$ and $(-5, 1)$.

19. Find the equation of the line that passes through the points $(2, 3)$ and $(-4, 0)$.

20. Find the equation of the line that passes through the points $(3, -1)$ and $(0, -3)$.

21. Find the equation of the line that passes through the points $(-4, 1)$ and $(4, -5)$.

22. Find the equation of the line that passes through the points $(-5, 0)$ and $(10, -3)$.

23. Find the equation of the line that passes through the points $(-2, 1)$ and $(2, 4)$.

24. Find the equation of the line that passes through the points $(3, -2)$ and $(-3, -3)$.

Objective C   *Application Problems*

Solve.

25. The data in the table below show the tread depth of a tire and the number of miles that have been driven on that tire. The line of best fit is $y = -0.2x + 10$, where x is the number of miles driven in thousands and y is the depth of the tread in millimeters.

Miles driven, x	25	35	40	20	45
Tread depth, y	4.8	3.5	2.1	5.5	1.0

Graph the data and the line of best fit in the coordinate system at the right. Write a sentence that describes the meaning of the slope of the line of best fit.

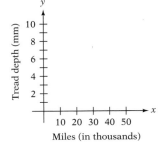

26. The data in the table below are estimates of the number of employed persons per retired person during the decade from 2020 to 2030. Information like this is important to the Social Security Administration as it plans for paying future retirement benefits. The line of best fit is $y = -0.13x + 3.35$, where x is the year (with 2020 as 0) and y is the number of employed persons per retired person.

Year, x	2 (2022)	4 (2024)	6 (2026)	8 (2028)	10 (2030)
Employed per retired, y	3.1	2.9	2.5	2.4	2.1

Graph the data and the line of best fit in the coordinate system at the right. Write a sentence that describes the meaning of the slope of the line of best fit.

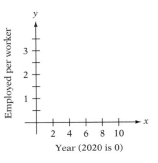

27. The data in the table below show the recorded and estimated sales of microprocessors by Texas Instruments for 1991 through 1995. The line of best fit is $y = 1.6x + 2.4$, where x is the year (with 1991 as 1) and y is the revenue in billions.

Year, x	1 (1991)	2 (1992)	3 (1993)	4 (1994)	5 (1995)
Revenue, y	3.9	5.5	7.7	8.7	10.3

Graph the data and the line of best fit in the coordinate system at the right. Write a sentence that describes the meaning of the slope of the line of best fit.

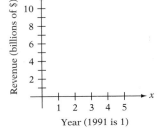

28. The data in the table below show the recorded and projected numbers of ATMs (automatic teller machines) in service from 1985 through 1995. The line of best fit is $y = 3.6x + 61.6$, where x is the year (with 1985 as 0) and y is the number of ATMs in thousands.

Year, x	0 (1985)	3 (1988)	5 (1990)	7 (1992)	10 (1995)
ATMs, y	61	73	81	86	98

Graph the data and the line of best fit in the coordinate system at the right. Write a sentence that describes the meaning of the slope of the line of best fit.

APPLYING THE CONCEPTS

In Exercises 29 to 32, the first two given points are on a line. Determine whether the third point is on the line.

29. $(-3, 2), (4, 1); (-1, 0)$

30. $(2, -2), (3, 4); (-1, 5)$

31. $(-3, -5), (1, 3); (4, 9)$

32. $(-3, 7), (0, -2); (1, -5)$

33. If $(-2, 4)$ are the coordinates of a point on the line whose equation is $y = mx + 1$, what is the slope of the line?

34. If $(3, 1)$ are the coordinates of a point on the line whose equation is $y = mx - 3$, what is the slope of the line?

35. If $(0, -3)$, $(6, -7)$ and $(3, n)$ are coordinates of points on the same line, determine n.

36. If $(-4, 11)$, $(2, -4)$ and $(6, n)$ are coordinates of points on the same line, determine n.

The formula $y - y_1 = \frac{y_2 - y_1}{x_2 - x_1}(x - x_1)$, where $x_1 \neq x_2$, is called the **two-point formula** for a straight line. This formula can be used to find the equation of a line given two points. Use this formula for Exercises 37 and 38.

37. Find the equation of the line passing through $(-2, 3)$ and $(4, -1)$.

38. Find the equation of the line passing through $(3, -1)$ and $(4, -3)$.

39. Explain why the condition $x_1 \neq x_2$ is placed on the two-point formula
[W]   given above.

40. Explain how the two-point formula given above can be derived from
[W]   the point–slope formula.

Project in Mathematics

. . . .

Graphing Linear Equations with a Graphing Utility

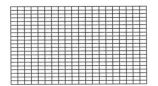

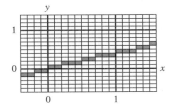

A computer or graphing calculator screen is divided into *pixels*. There are approximately 6000 to 790,000 pixels available on the screen (depending on the computer or calculator). The greater the number of pixels, the smoother a graph will appear. A portion of a screen is shown at the left. Each little rectangle represents one pixel.

The graphing utilities that are used by computers or calculators to graph an equation do basically what we have shown in the text: They choose values of x and, for each, calculate the corresponding value of y. The pixel corresponding to the ordered pair is then turned on. The graph is jagged because pixels are much larger than the dots we draw on paper.

The graph of $y = 0.45x$ is shown at the left as the calculator drew it (jagged) and as it would appear on paper. The x- and y-axes have been chosen so that each pixel represents $\frac{1}{10}$ of a unit. Consider the region of the graph where $x = 1$, 1.1, and 1.2.

The corresponding values of y are 0.45, 0.495, and 0.54. Because the y-axis is in tenths, the numbers 0.45, 0.495, and 0.54 are rounded to the nearest tenth before plotting. Rounding 0.45, 0.495 and 0.54 to the nearest tenth results in 0.5 for each number. Thus the ordered pairs $(1, 0.45)$, $(1.1, 0.495)$, and $(1.2, 0.54)$ are graphed as $(1, 0.5)$, $(1.1, 0.5)$, and $(1.2, 0.5)$. These points appear as three illuminated horizontal pixels. The graph of the line appears horizontal. However, if you use the TRACE feature of the calculator (see Appendix A), the actual y-coordinate for each value of x is displayed.

Here are the keystrokes to graph $y = \frac{2}{3}x + 1$. First the domain of the function (Xmin to Xmax) and the range (Ymin to Ymax) are entered. This is called the **viewing window** for the graph. Then the equation is entered. By changing the keystrokes that are shaded in color, you can graph different equations.

TI-*82*

| ZOOM | 6 | Y= | CLEAR | 2 |
| X,T,θ | ÷ | 3 | + | 1 | GRAPH |

SHARP *EL-9300*

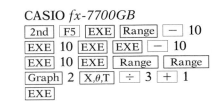

CASIO *fx-7700GB*

2nd	F5	EXE	Range	−	10
EXE	10	EXE	EXE	−	10
EXE	10	EXE	Range	Range
Graph	2	X,θ,T	÷	3	+	1
EXE

Exercises:

1. $y = 2x + 1$ For $2x$, you may enter $2 \times x$ or just $2x$. The times sign $\times$ is not necessary on many graphing calculators.

2. $y = -\dfrac{1}{2}x - 2$ Use the (−) key to enter a negative sign.

3. $3x + 2y = 6$ Solve for y. Then enter the equation.

4. $4x + 3y = 75$ You must adjust the viewing window.

Suggestion: Xmin = −25, Xmax = 25, Xsci = 5, Ymin = −35, Ymax = 35, Ysci = 5. See Appendix A for assistance.

Chapter Summary

Key Words   A *rectangular coordinate system* is formed by two number lines, one horizontal and one vertical, that intersect at the zero point of each line.

The number lines that make up a rectangular coordinate system are called the *coordinate axes* or simply *axes*.

The *origin* is the point of intersection of the two coordinate axes.

A rectangular coordinate system divides the plane into four regions called *quadrants*.

An *ordered pair* (a, b) is used to locate a point in a plane.

The first number in an ordered pair is called the *abscissa* or x-coordinate.

The second number in an ordered pair is called the *ordinate* or y-coordinate.

The *coordinates* of a point are the numbers in the ordered pair that is associated with the point.

A *relation* is any set of ordered pairs.

A *function* is a relation in which no two ordered pairs with the same first coordinate have different second coordinates.

A function designated by $f(x)$ is written in *functional notation*. The *value* of the function at x is $f(x)$.

An equation of the form $y = mx + b$, where m and b are constants, is a *linear equation in two variables* or a *linear function*.

The point at which a graph crosses the x-axis is called the *x-intercept*.

The point at which a graph crosses the y-axis is called the *y-intercept*.

The *slope* of a line is the measure of the slant of a line. The symbol for slope is m.

A line that slants upward to the right has a *positive slope*.

A line that slants downward to the right has a *negative slope*.

A horizontal line has *zero slope*.

The slope of a vertical line is *undefined*.

Essential Rules   *Slope of a straight line:*

$$Slope = m = \frac{y_2 - y_1}{x_2 - x_1}$$

Slope–intercept form of a straight line:

$$y = mx + b$$

Point–slope form of a straight line:

$$y - y_1 = m(x - x_1)$$

Chapter Review

. . . .

SECTION 8.1

1. **a.** Graph the ordered pairs $(-2, 4)$ and $(3, -2)$.
    b. Name the abscissa of point A.
    c. Name the ordinate of point B.

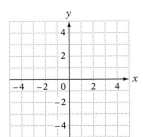

2. Graph the ordered-pair solutions of
$y = -\frac{1}{2}x - 2$ when $x \in \{-4, -2, 0, 2\}$.

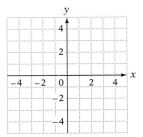

3. Does $y = -x + 3$, where $x \in \{-2, 0, 3, 5\}$, define y as a function of x?

4. Given $f(x) = x^2 - 2$, find $f(-1)$.

5. The height and weight of 8 seventh-grade students are shown in the following table. Write a relation where the first coordinate is the height, in inches, of the student and the second coordinate is the weight, in pounds, of the student. Is the relation a function?

Height	55	57	53	57	60	61	58	54
Weight	95	101	94	98	100	105	97	95

SECTION 8.2

6. Graph $y = \frac{1}{4}x + 3$.

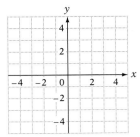

7. Graph $y = -2x - 1$.

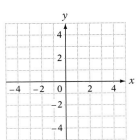

8. Graph $3x - 2y = -6$.

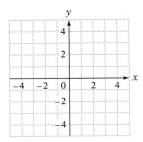

9. Graph $5x + 3y = 15$.

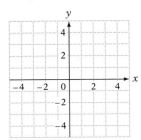

10. Graph $x = -3$.

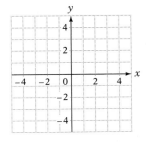

11. A computer bulletin board charges an access fee of $10 per month plus $.20 per minute to use the service. An equation that represents the monthly cost to use this bulletin board is $C = 0.20x + 10$, where C is the monthly cost and x is the number of minutes. Graph this equation for $0 \le x \le 200$. The point $(100, 30)$ is on the graph. Write a sentence that describes the meaning of this ordered pair.

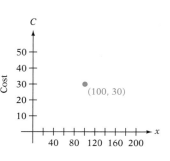

SECTION 8.3

12. Find the x- and y-intercepts of $3x - 2y = 24$.

13. Find the slope of the line containing the points $(9, 8)$ and $(-2, 1)$.

14. Find the slope of the line containing the points $(4, -3)$ and $(-2, -3)$.

15. Graph the line that has slope $\frac{1}{2}$ and y-intercept $(0, -1)$.

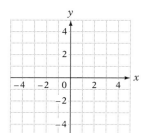

16. Graph the line that has slope 2 and y-intercept $(0, -4)$.

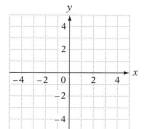

17. Graph the line that has slope $-\frac{2}{3}$ and y-intercept $(0, 2)$.

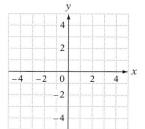

SECTION 8.4

18. Determine the equation of the line that passes through the point $(-3, 4)$ and has slope $\frac{2}{3}$.

19. Determine the equation of the line that passes through the point $(6, 1)$ and has slope $-\frac{5}{2}$.

20. Determine the equation of the line that passes through the points $(-2, 5)$ and $(4, 1)$.

21. Determine the equation of the line that passes through the points $(-1, 3)$ and $(2, -5)$.

22. The data in the table below show the age of a machine, in years, and its down time, in hours per month. (Down time is the number of hours a machine is not available because of a malfunction.) The line of best fit is $y = 10.4x - 1.1$, where x is the age of the machine in years and y is the down time each month.

Age, x	1.0	2.7	4.1	1.2	2.5	1.9
Down time, y	10	30	40	9	25	19

Graph the data and the line in the coordinate system at the right. Write a sentence that describes the meaning of the slope of the line of best fit.

Chapter Test

. . . .

1. Find the ordered-pair solution of $2x - 3y = 15$ corresponding to $x = 3$.

2. Does $y = \dfrac{1}{2}x - 3$ define y as a function of x for $x \in \{-2, 0, 4\}$?

4. Given $f(t) = t^2 + t$, find $t(2)$.

5. Given $f(x) = x^2 - 2x$, find $f(-1)$.

3. Graph the ordered-pair solutions of $y = -\dfrac{3}{2}x + 1$ for $x \in \{-2, 0, 4\}$.

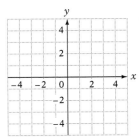

6. The distance a house is from a fire station and the amount of damage that the house sustained in a fire are given in the following table. Write a relation where the first coordinate of the ordered pair is the distance in miles from the fire station and the second coordinate is the amount of damage in thousands of dollars. Is the relation a function?

Distance	3.5	4.0	5.2	5.0	4.0	6.3	5.4
Damage	25	30	45	38	42	12	34

7. Graph $y = 3x + 1$.

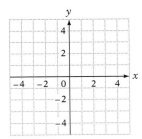

8. Graph $y = -\dfrac{3}{4}x + 3$.

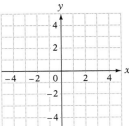

9. Graph $3x - 2y = 6$.

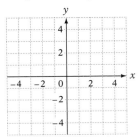

10. Graph $x + 3 = 0$.

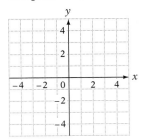

11. Graph the line that has slope $-\dfrac{2}{3}$ and y-intercept $(0, 4)$.

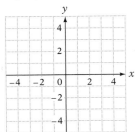

12. Graph the line that has slope 2 and y-intercept -2.

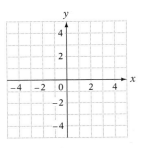

13. The equation for the speed of a ball that is thrown straight up with an initial speed of 128 feet per second is $v = 128 - 32t$, where v is the speed of the ball after t seconds. Graph this equation for $0 \le t \le 4$. The point whose coordinates are $(1, 96)$ is on the graph in Fig. 1. Write a sentence that describes this ordered pair.

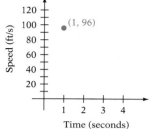

Fig. 1

14. The graph in Fig. 2 shows the increase in the cost of tuition for a college for the years 1989 through 1994 (with 1989 as 0). Find the slope of the line. Write a sentence that states the meaning of the slope.

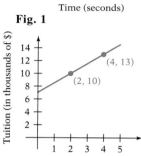

Fig. 2

15. The data in the following table show the number of mutual fund companies for selected years between 1983 and 1993. The line of best fit is $y = 330x + 1018$, where x is the year (with 1983 as 0) and y is the number of mutual fund companies.

Year, x	0 (1983)	3 (1986)	5 (1988)	8 (1991)	10 (1993)
No. of companies, y	1020	2000	2670	3660	4320

Graph the data and the line of best fit in the coordinate system in Fig. 3. Write a sentence that describes the meaning of the slope of the line of best fit.

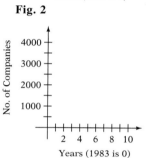

Fig. 3

16. Find the x- and y-intercepts for $6x - 4y = 12$.

17. Find the x- and y-intercepts for $y = \frac{1}{2}x + 1$.

18. Find the slope of the line containing the points $(2, -3)$ and $(4, 1)$.

19. Find the slope of the line containing the points $(3, -4)$ and $(1, -4)$.

20. Find the slope of the line containing the points $(-5, 2)$ and $(-5, 7)$.

21. Find the slope of the line whose equation is $2x + 3y = 6$.

22. Find the equation of the line that contains the point $(0, -1)$ and has slope 3.

23. Find the equation of the line that contains the point $(-3, 1)$ and has slope $\frac{2}{3}$.

24. Find the equation of the line that passes through the points $(5, -4)$ and $(-3, 1)$.

25. Find the equation of the line that passes through the points $(-2, 0)$ and $(5, -2)$.

Cumulative Review

1. Simplify $12 - 18 \div 3 \cdot (-2)^2$.

2. Evaluate $\dfrac{a - b}{a^2 - c}$ when $a = -2$, $b = 3$, and $c = -4$.

3. Given $f(x) = \dfrac{2}{x - 1}$, find $f(-2)$.

4. Solve: $2x - \dfrac{2}{3} = \dfrac{7}{3}$

5. Solve: $3x - 2[x - 3(2 - 3x)] = x - 7$

6. Write $6\dfrac{2}{3}\%$ as a fraction.

7. Simplify $(-2x^2y)^3(2xy^2)^2$.

8. Simplify $\dfrac{-15x^7}{5x^5}$.

9. Divide: $(x^2 - 4x - 21) \div (x - 7)$

10. Factor $5x^2 + 15x + 10$.

11. Factor $x(a + 2) + y(a + 2)$.

12. Solve: $x(x - 2) = 8$

13. Simplify: $\dfrac{x^5y^3}{x^2 - x - 6} \cdot \dfrac{x^2 - 9}{x^2y^4}$

14. Simplify: $\dfrac{3x}{x^2 + 5x - 24} - \dfrac{9}{x^2 + 5x - 24}$

15. Solve: $3 - \dfrac{1}{x} = \dfrac{5}{x}$

16. Solve $4x - 5y = 15$ for y.

17. Find the ordered-pair solution of $y = 2x - 1$ corresponding to $x = -2$.

18. Find the slope of the line that contains the points $(2, 3)$ and $(-2, 3)$.

19. Find the equation of the line that contains the point $(2, -1)$ and has slope $\frac{1}{2}$.

20. Find the equation of the line that contains the point $(0, 2)$ and has slope -3.

21. Find the equation of the line that contains the point $(-1, 0)$ and has slope 2.

22. Find the equation of the line that contains the point $(6, 1)$ and has slope $\frac{2}{3}$.

23. A suit that regularly sells for $89 is on sale for 30% off the regular price. Find the sale price.

24. The first angle of a triangle is 3° more than the measure of the second angle. The third angle is 5° more than twice the measure of the second angle. Find the measure of each angle

25. The real estate tax for a home that costs $50,000 is $625. At this rate, what is the value of a home for which the real estate tax is $1375?

26. An electrician requires 6 h to wire a garage. An apprentice can do the same job in 10 h. How long would it take to wire the garage if both the electrician and the apprentice were working?

27. Graph $y = \frac{1}{2}x - 1$.

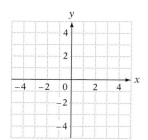

28. Graph the line that has slope $-\frac{2}{3}$ and y-intercept 2.

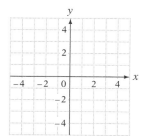

Chapter

Systems of Linear Equations

Objectives

Section 9.1
To solve a system of linear equations by graphing

Section 9.2
To solve a system of linear equations by the substitution method

Section 9.3
To solve a system of linear equations by the addition method

Section 9.4
To solve rate-of-wind or -current problems
To solve application problems using two variables

Input–Output Analysis

In 1973, the Nobel Prize in Economics was awarded for applications of mathematics to economics. The technique was to examine various sectors of an economy (the steel industry, oil, farms, autos, and many others) and determine how each sector interacted with the rest. Over 500 sectors of the economy were studied.

The interaction of each sector with the others was written as a series of equations. This series of equations is called a *system of equations*.

Using a computer, economists searched for a solution to the system of equations that would determine the output levels various sectors would have to meet to satisfy the requests from other sectors. The method is called Input–Output Analysis.

This chapter introduces the study of systems of equations.

Solving Systems of Linear Equations by Graphing

Objective A *To solve a system of linear equations by graphing* ....

Equations considered together are called **systems of equations.** A system of equations is shown at the right.

$$2x + 3y = 2$$
$$3x - 5y = 22$$

A **solution of a system of equations in two variables** is an ordered pair that is a solution of each equation of the system

➡ Is $(4, -2)$ a solution of the system

$$2x + 3y = 2$$
$$3x - 5y = 22?$$

$2x + 3y = 2$	
$2(4) + 3(-2)$	2
$8 + (-6)$	2
	$2 = 2$

$3x - 5y = 22$	
$3(4) - 5(-2)$	22
$12 - (-10)$	22
	$22 = 22$

Yes, because $(4, -2)$ is a solution of each equation, it is the solution of the system. However, $(7, -4)$ is not a solution, because

$2x + 3y = 2$	
$2(7) + 3(-4)$	2
$14 + (-12)$	2
	$2 = 2$

● $(7, -4)$ is a solution.

$3x - 5y = 22$	
$3(7) - 5(-4)$	22
$21 - (-20)$	22
	$41 \neq 22$

● $(7, -8)$ is not a solution.

➡ Is $(3, -3)$ a solution of the system

$$2x + y = 3$$
$$x + y = 1?$$

$2x + y = 3$	
$2(3) + (-3)$	3
$6 + (-3)$	3
	$3 = 3$

$x + y = 1$	
$3 + (-3)$	1
	$0 \neq 1$

Because $(3, -3)$ is not a solution of each equation, $(3, -3)$ is not a solution of the system of equations.

Graphing the equations in a system of linear equations is one method of finding a solution of the system of equations. The lines can intersect at one point, the lines can intersect at infinitely many points (the graphs are the same line), or the lines can be parallel and not intersect at all.

Such systems are called **independent, dependent, and inconsistent,** respectively.

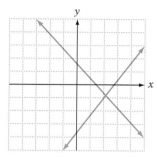

Independent: one solution

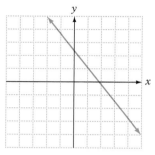

Dependent: infinitely many solutions

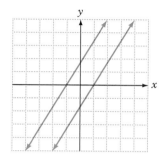

Inconsistent: no solutions

POINT OF INTEREST
The Project in Mathematics at the end of this chapter discusses using a calculator to approximate the solution of a system of equations.

➡ Solve by graphing: $2x + 3y = 6$
$ 2x + y = -2$

Graph each line.

Find the point of intersection.

$(-3, 4)$ is a solution of each equation. The system of equations is *independent*.

The solution is $(-3, 4)$.

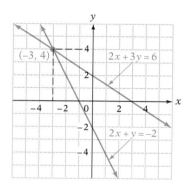

➡ Solve by graphing: $2x - y = 1$
$ 6x - 3y = 12$

Graph each line.

The lines are parallel and therefore do not intersect. The system of equations is *inconsistent* and has no solution.

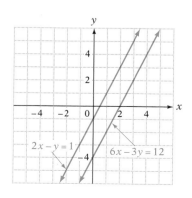

When a system of equations is *dependent,* the graphs of the two equations are the same line. Therefore, the lines intersect at infinitely many points. The solutions of the system of equations are the ordered pairs that satisfy either one (and hence both) of the two equations of the system of equations.

➡ Solve by graphing: $2x + 3y = 6$
$$6x + 9y = 18$$

Graph each line.

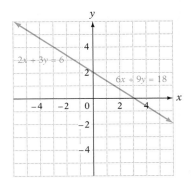

The two equations represent the same line. The system of equations is dependent, so there are an infinite number of solutions. The solutions are the ordered pairs that satisfy both equations. The solutions are stated by using the ordered pairs of one of the equations. Therefore, the solutions are the ordered pairs that satisfy the equation $2x + 3y = 6$.

By choosing values of x, we can find some specific ordered-pair solutions. For example, when $x = -3, 0,$ and 6, three of the infinite solutions of the system of equations are $(-3, 4), (0, 2)$ and $(6, -2)$.

Example 1

Is $(1, -3)$ a solution of the system
$$3x + 2y = -3$$
$$x - 3y = 6?$$

Solution

$$
\begin{array}{c|c}
3x + 2y = -3 & \\ \hline
3 \cdot 1 + 2(-3) & -3 \\
3 + (-6) & -3 \\
-3 = -3 &
\end{array}
\qquad
\begin{array}{c|c}
x - 3y = 6 & \\ \hline
1 - 3(-3) & 6 \\
1 - (-9) & 6 \\
10 \neq 6 &
\end{array}
$$

No, $(1, -3)$ is not a solution of the system of equations.

You Try It 1

Is $(-1, -2)$ a solution of the system
$$2x - 5y = 8$$
$$-x + 3y = -5?$$

Your solution

Solution on p. A27

Example 2

Solve by graphing:

$$x - 2y = 2$$
$$x + y = 5$$

Solution

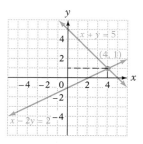

The solution is $(4, 1)$.

Example 3

Solve by graphing:

$$4x - 2y = 6$$
$$y = 2x - 3$$

Solution

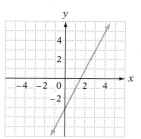

The solutions are the ordered pairs that satisfy the equation $y = 2x - 3$.

You Try It 2

Solve by graphing:

$$x + 3y = 3$$
$$-x + y = 5$$

Your solution

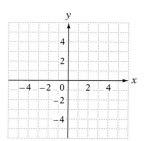

You Try It 3

Solve by graphing:

$$y = 3x - 1$$
$$6x - 2y = -6$$

Your solution

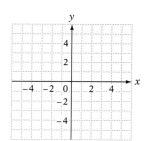

Solutions on p. A27

9.1 Exercises

· ·

Objective A

1. Is $(2, 3)$ a solution of the system
$$3x + 4y = 18$$
$$2x - y = 1?$$

2. Is $(2, -1)$ a solution of the system
$$x - 2y = 4$$
$$2x + y = 3?$$

3. Is $(1, -2)$ a solution of the system
$$3x - y = 5$$
$$2x + 5y = -8?$$

4. Is $(-1, -1)$ a solution of the system
$$x - 4y = 3$$
$$3x + y = 2?$$

5. Is $(4, 3)$ a solution of the system
$$5x - 2y = 14$$
$$x + y = 8?$$

6. Is $(2, 5)$ a solution of the system
$$3x + 2y = 16$$
$$2x - 3y = 4?$$

7. Is $(-1, 3)$ a solution of the system
$$4x - y = -5$$
$$2x + 5y = 13?$$

8. Is $(4, -1)$ a solution of the system
$$x - 4y = 9$$
$$2x - 3y = 11?$$

9. Is $(0, 0)$ a solution of the system
$$4x + 3y = 0$$
$$2x - y = 1?$$

10. Is $(2, 0)$ a solution of the system
$$3x - y = 6$$
$$x + 3y = 2?$$

11. Is $(2, -3)$ a solution of the system
$$y = 2x - 7$$
$$3x - y = 9?$$

12. Is $(-1, -2)$ a solution of the system
$$3x - 4y = 5$$
$$y = x - 1?$$

13. Is $(5, 2)$ a solution of the system
$$y = 2x - 8$$
$$y = 3x - 13?$$

14. Is $(-4, 3)$ a solution of the system
$$y = 2x + 11$$
$$y = 5x - 19?$$

15. Is $(-2, -3)$ a solution of the system
$$3x - 4y = 6$$
$$2x - 7y = 17?$$

16. Is $(0, 0)$ a solution of the system
$$y = 2x$$
$$3x + 5y = 0?$$

17. Is $(0, -3)$ a solution of the system
$$4x - 3y = 9$$
$$2x + 5y = 15?$$

18. Is $(4, 0)$ a solution of the system
$$2x + 3y = 8$$
$$x - 5y = 4?$$

Solve by graphing.

19. $x - y = 3$
$x + y = 5$

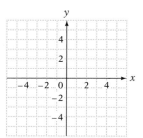

20. $2x - y = 4$
$x + y = 5$

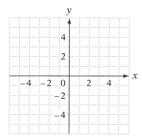

21. $x + 2y = 6$
$x - y = 3$

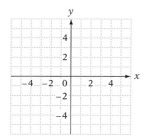

22. $3x - y = 3$
$2x + y = 2$

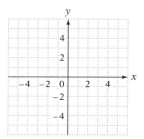

23. $3x - 2y = 6$
$y = 3$

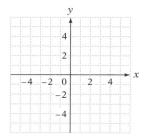

24. $x = 2$
$3x + 2y = 4$

25. $x = 3$
$y = -2$

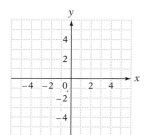

26. $x + 1 = 0$
$y - 3 = 0$

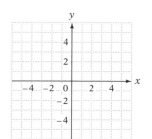

Solve by graphing.

27. $\quad y = 2x - 6$
$\quad x + y = 0$

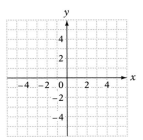

28. $5x - 2y = 11$
$\quad y = 2x - 5$

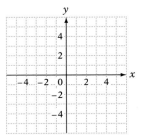

29. $\quad 2x + y = -2$
$\quad 6x + 3y = 6$

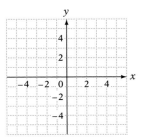

30. $\quad x + y = 5$
$\quad 3x + 3y = 6$

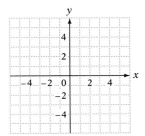

31. $\quad y = 2x - 2$
$\quad 4x - 2y = 4$

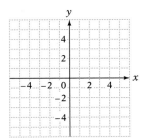

32. $\quad y = -\dfrac{1}{3}x + 1$
$\quad 2x + 6y = 6$

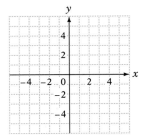

33. $\quad x - y = 5$
$\quad 2x - y = 6$

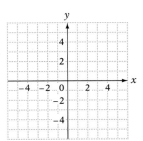

34. $5x - 2y = 10$
$\quad 3x + 2y = 6$

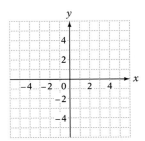

Solve by graphing.

35. $3x + 4y = 0$
         $2x - 5y = 0$

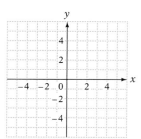

36. $2x - 3y = 0$
         $y = -\dfrac{1}{3}x$

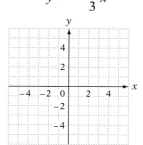

37. $x - 3y = 3$
         $2x - 6y = 12$

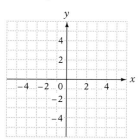

38. $4x + 6y = 12$
         $6x + 9y = 18$

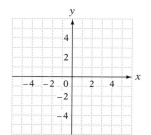

APPLYING THE CONCEPTS

39. Determine whether the statement is true, sometimes true, or never true.
  a. A solution of a system of two equations with two variables is a point in the plane.
  b. Two parallel lines have the same slope.
  c. Two different lines with the same y-intercept are parallel.
  d. Two different lines with the same slope are parallel.

40. [W] Explain how you can determine from the graph of a system of two equations in two variables whether it is an independent system of equations.

41. [W] Explain how you can determine from the graph of a system of two equations in two variables whether it is an inconsistent system of equations.

42. Write a system of equations that has $(-2, 4)$ as its only solution.

43. Write a system of equations for which there is no solution.

44. Write a system of equations that is a dependent system of equations.

45. [W] The graph at the right shows the Cumulative Total Return Performance for the Lipper Science and Technology Average versus the S&P 500 Stock Index from December 1990 to December 1991. Explain what the solution of this system of equations means.

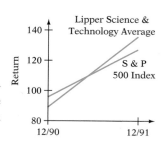

9.2 Solving Systems of Linear Equations by the Substitution Method

Objective A **To solve a system of linear equations by the substitution method** ....

A graphical solution of a system of equations is based on approximating the coordinates of a point of intersection. An algebraic method called the **substitution method** can be used to find an exact solution of a system of equations.

➡ Solve by the substitution method: (1) $2x + 5y = -11$
$\qquad\qquad\qquad\qquad\qquad\qquad\qquad$ (2) $y = 3x - 9$

Equation (2) states that $y = 3x - 9$. Substitute $3x - 9$ for y in Equation (1). Then solve for x.

$\qquad 2x + 5y = -11$ • This is Equation (1).

$\qquad 2x + 5(3x - 9) = -11$ • From Equation (2), substitute $3x - 9$ for y.

$\qquad 2x + 15x - 45 = -11$ • Solve for x.

$\qquad\qquad 17x - 45 = -11$

$\qquad\qquad\qquad 17x = 34$

$\qquad\qquad\qquad\quad x = 2$

Now substitute the value of x into Equation (2) and solve for y.

$y = 3x - 9$ • This is Equation (2).

$y = 3(2) - 9$ • Substitute 2 for x.

$y = 6 - 9 = -3$

The solution is the ordered pair $(2, -3)$.

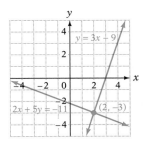

The graph of the equations in this system of equations is shown at the right. Note that the lines intersect at the point whose coordinates are $(2, -3)$, which is the algebraic solution we determined by the substitution method.

To solve a system of equations by the substitution method, we must solve one of the equations of the system of equations for one of its variables. For instance, the first step in solving the system of equations

$\qquad\qquad$ (1) $x + 2y = -3$
$\qquad\qquad$ (2) $2x - 3y = 5$

is to solve an equation of the system for one of its variables. Either equation can be used.

Solving Equation (1) for x:$\qquad\qquad\qquad$ Solving Equation (2) for x:

$\qquad x + 2y = -3 \qquad\qquad\qquad\qquad\quad 2x - 3y = 5$

$\qquad\quad x = -2y - 3 \qquad\qquad\qquad\qquad\quad 2x = 3y + 5$

$\qquad\qquad\qquad\qquad\qquad\qquad\qquad\qquad\qquad\quad x = \dfrac{3y + 5}{2}$

Because solving Equation (1) for x does not result in fractions, it is the easier of the two equations to use.

Here is the solution of the system of equations given on the previous page.

➡ Solve by the substitution method: (1) $x + 2y = -3$
  (2) $2x - 3y = 5$

To use the substitution method, we must solve an equation for one of its variables. Equation (1) is used here because solving it for x does not result in fractions.

$$x + 2y = -3$$
(3) $\qquad x = -2y - 3$ • Solve for x. This is Equation (3).

Now substitute $-2y - 3$ for x in Equation (2) and solve for y.

$$2x - 3y = 5$$ • This is Equation (2).
$$2(-2y - 3) - 3y = 5$$ • From Equation (3), substitute $-2y - 3$ for x.
$$-4y - 6 - 3y = 5$$ • Solve for y.
$$-7y - 6 = 5$$
$$-7y = 11$$
$$y = -\frac{11}{7}$$

Substitute the value of y into Equation (3) and solve for x.

$$x = -2y - 3$$ • This is Equation (3).
$$= -2\left(-\frac{11}{7}\right) - 3$$ • Substitute $-\frac{11}{7}$ for y.
$$= \frac{22}{7} - 3 = \frac{22}{7} - \frac{21}{7} = \frac{1}{7}$$

The solution is $\left(\frac{1}{7}, -\frac{11}{7}\right)$.

The graph of the system of equations given above is shown at the right. It would be difficult to determine the exact solution of this system of equations from the graphs of the equations.

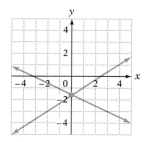

➡ Solve by the substitution method: (1) $y = 3x - 1$
  (2) $y = -2x - 6$

$$y = -2x - 6$$
$$3x - 1 = -2x - 6$$ • Substitute $3x - 1$ for y in Equation (2).
$$5x = -5$$ • Solve for x.
$$x = -1$$

Substitute this value of x into Equation (1) or Equation (2) and solve for y. Equation (1) is used here.

$$y = 3x - 1$$
$$y = 3(-1) - 1 = -4$$

The solution is $(-1, -4)$.

The substitution method can be used to analyze inconsistent and dependent systems of equations.

➡ Solve by the substitution method: (1) $2x + 3y = 3$

(2) $y = -\dfrac{2}{3}x + 3$

$$2x + 3y = 3$$
• This is Equation (1).

$$2x + 3\left(-\dfrac{2}{3}x + 3\right) = 3$$
• From Equation (2), replace y with $-\dfrac{2}{3}x + 3$.

$$2x + (-2x) + 9 = 3$$
• Solve for x.

$$9 = 3$$
• This is not a true equation.

Because $9 = 3$ is not a true equation, the system of equations has no solutions.

Solving Equation (1) above for y, we have $y = -\dfrac{2}{3}x + 1$. Comparing this with Equation (2) reveals that the slopes are equal and the y-intercepts are different. The graphs of the equations that make up this system of equations are parallel and thus never intersect. Because the graphs do not intersect, there are no solutions of the system of equations. The system of equations is inconsistent.

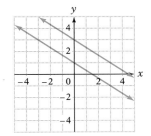

➡ Solve by the substitution method: (1) $x = 2y + 3$
(2) $4x - 8y = 12$

$$4x - 8y = 12$$
• This is Equation (2).

$$4(2y + 3) - 8y = 12$$
• From Equation (1), replace x by $2y + 3$.

$$8y + 12 - 8y = 12$$
• Solve for y.

$$12 = 12$$
• This is a true equation.

The true equation $12 = 12$ indicates that any ordered pair (x, y) that satisfies one equation of the system satisfies the other equation. Therefore, the system of equations has an infinite number of solutions. The solutions are the ordered pairs (x, y) that are solutions of $x = 2y + 3$.

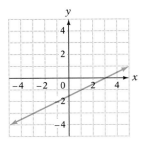

If we write Equation (1) and Equation (2) in slope–intercept form, we have

$$x = 2y + 3 \qquad\qquad 4x - 8y = 12$$
$$-2y = -x + 3 \qquad\qquad -8y = -4x + 12$$
$$y = \dfrac{1}{2}x - \dfrac{3}{2} \qquad\qquad y = \dfrac{1}{2}x - \dfrac{3}{2}$$

The slope–intercept forms of the equations are the same, and therefore the graphs are the same. If we graph these two equations, we essentially graph one over the other. Accordingly, the graphs intersect at an infinite number of points.

Example 1   Solve by substitution:
(1) $3x + 4y = -2$
(2) $-x + 2y = 4$

Solution   Solve Equation (2) for x.
$-x + 2y = 4$
$-x = -2y + 4$
$x = 2y - 4$

Substitute in Equation (1).
(1) $3x + 4y = -2$
$3(2y - 4) + 4y = -2$
$6y - 12 + 4y = -2$
$10y - 12 = -2$
$10y = 10$
$y = 1$

Substitute in Equation (2).
(2) $-x + 2y = 4$
$-x + 2(1) = 4$
$-x + 2 = 4$
$-x = 2$
$x = -2$

The solution is $(-2, 1)$.

You Try It 1   Solve by substitution:
$7x - y = 4$
$3x + 2y = 9$

Your solution

Example 2   Solve by substitution:
$4x + 2y = 5$
$y = -2x + 1$

Solution $4x + 2y = 5$
$4x + 2(-2x + 1) = 5$
$4x - 4x + 2 = 5$
$2 = 5$

This is not a true equation. The system of equations is inconsistent and therefore does not have a solution.

You Try It 2   Solve by substitution:
$3x - y = 4$
$y = 3x + 2$

Your solution

Example 3   Solve by substitution:
$y = 3x - 2$
$6x - 2y = 4$

Solution $6x - 2y = 4$
$6x - 2(3x - 2) = 4$
$6x - 6x + 4 = 4$
$4 = 4$

This is a true equation. The system of equations is dependent. The solutions are the ordered pairs that satisfy the equation $y = 3x - 2$.

You Try It 3   Solve by substitution:
$y = -2x + 1$
$6x + 3y = 3$

Your solution

Solutions on pp. A27–A28

9.2 Exercises

Objective A

Solve by substitution.

1. $2x + 3y = 7$
$x = 2$

2. $y = 3$
$3x - 2y = 6$

3. $y = x - 3$
$x + y = 5$

4. $y = x + 2$
$x + y = 6$

5. $x = y - 2$
$x + 3y = 2$

6. $x = y + 1$
$x + 2y = 7$

7. $y = 4 - 3x$
$3x + y = 5$

8. $y = 2 - 3x$
$6x + 2y = 7$

9. $x = 3y + 3$
$2x - 6y = 12$

10. $x = 2 - y$
$3x + 3y = 6$

11. $3x + 5y = -6$
$x = 5y + 3$

12. $y = 2x + 3$
$4x - 3y = 1$

13. $3x + y = 4$
$4x - 3y = 1$

14. $x - 4y = 9$
$2x - 3y = 11$

15. $3x - y = 6$
$x + 3y = 2$

16. $4x - y = -5$
$2x + 5y = 13$

17. $3x - y = 5$
$2x + 5y = -8$

18. $3x + 4y = 18$
$2x - y = 1$

19. $4x + 3y = 0$
$2x - y = 0$

20. $5x + 2y = 0$
$x - 3y = 0$

21. $2x - y = 2$
$6x - 3y = 6$

Solve by substitution.

22. $3x + y = 4$
$9x + 3y = 12$

23. $x = 3y + 2$
$y = 2x + 6$

24. $x = 4 - 2y$
$y = 2x - 13$

25. $y = 2x + 11$
$y = 5x - 19$

26. $y = 2x - 8$
$y = 3x - 13$

27. $y = -4x + 2$
$y = -3x - 1$

28. $x = 3y + 7$
$x = 2y - 1$

29. $x = 4y - 2$
$x = 6y + 8$

30. $x = 3 - 2y$
$x = 5y - 10$

APPLYING THE CONCEPTS

For what value of k does the system of equations have no solution?

31. $2x - 3y = 7$
$kx - 3y = 4$

32. $8x - 4y = 1$
$2x - ky = 3$

33. $x = 4y + 4$
$kx - 8y = 4$

34. Describe in your own words the process of solving a system of equa-
[W] tions by the substitution method.

35. When you solve a system of equations by the substitution method,
[W] how do you determine whether the system of equations is dependent?

36. When you solve a system of equations by the substitution method,
[W] how do you determine whether the system of equations is inconsistent?

37. The following was offered as a solution to the system of equations

(1) $y = \dfrac{1}{2}x + 2$

(2) $2x + 5y = 10$

$2x + 5y = 10$ • Equation (2)

$2x + 5\left(\dfrac{1}{2}x + 2\right) = 10$ • Substitute $\dfrac{1}{2}x + 2$ for y.

$2x + \dfrac{5}{2}x + 10 = 10$ • Solve for x.

$\dfrac{9}{2}x = 0$

$x = 0$

At this point the student stated that because $x = 0$, the system of
equations has no solution. If this assertion is correct, is the system of
equations independent, dependent, or inconsistent? If the assertion is
not correct, what is the correct solution?

Solving Systems of Equations by the Addition Method

Objective A *To solve a system of linear equations by the addition method* ....

Another method of solving a system of equations is called the **addition method.** This method is based on the Addition Property of Equations.

Note, for the system of equations at the right, the effect of adding Equation (2) to Equation (1). Because $2y$ and $-2y$ are opposites, adding the equations results in an equation with only one variable.

$$\begin{array}{ll}(1) & 5x + 2y = 11 \\ (2) & \underline{3x - 2y = 13} \\ & 8x + 0y = 24 \\ & 8x = 24\end{array}$$

Solving $8x = 24$ for x gives the first coordinate of the ordered-pair solution of the system of equations.

$$\frac{8x}{8} = \frac{24}{8}$$
$$x = 3$$

The second coordinate is found by substituting the value of x into Equation (1) or Equation (2) and then solving for y. Equation (1) is used here.

$$\begin{array}{ll}(1) & 5x + 2y = 11 \\ & 5(3) + 2y = 11 \\ & 15 + 2y = 11 \\ & 2y = -4 \\ & y = -2\end{array}$$

The solution is $(3, -2)$.

Sometimes adding the two equations does not eliminate one of the variables. In this case, use the Multiplication Property of Equations to rewrite one or both of the equations such that the coefficients of one of the variables are opposites.

➡ Solve by the addition method:
$$\begin{array}{ll}(1) & 4x + 3y = -1 \\ (2) & 2x - 5y = 19\end{array}$$

Multiply Equation (2) by -2. The coefficients of x will then be opposites.

$$\begin{array}{ll} & -2(2x - 5y) = -2 \cdot 19 & \bullet \text{ Multiply Equation (2) by } -2. \\ (3) & -4x + 10y = -38 & \bullet \text{ Simplify. This is Equation (3).}\end{array}$$

Add Equation (1) to Equation (3). Then solve for y.

$$\begin{array}{ll}(1) & 4x + 3y = -1 \\ (3) & \underline{-4x + 10y = -38} & \bullet \text{ Note that the coefficients of } x \text{ are opposites.} \\ & 13y = -39 & \bullet \text{ Add the two equations.} \\ & y = -3 & \bullet \text{ Solve for } y.\end{array}$$

Substitute the value of y into Equation (1) or Equation (2) and solve for x. Equation (1) will be used here.

$$\begin{array}{ll}(1) & 4x + 3y = -1 \\ & 4x + 3(-3) = -1 & \bullet \text{ Substitute } -3 \text{ for } y. \\ & 4x - 9 = -1 & \bullet \text{ Solve for } x. \\ & 4x = 8 \\ & x = 2\end{array}$$

The solution is $(2, -3)$.

Sometimes each equation of the system of equations must be multiplied by a constant so that the coefficients of one of the variable terms are opposites.

➡ Solve by the addition method:

(1) $\quad 3x + 7y = 2$
(2) $\quad 5x - 3y = -26$

To eliminate x, multiply Equation (1) by 5 and Equation (2) by -3. Note at the right how the constants are chosen.

$$5(3x + 7y) = 5 \cdot 2$$
$$-3(5x - 3y) = -3(-26)$$

• The negative is used so that the coefficients will be opposites.

$$\begin{array}{ll} 15x + 35y = 10 & \text{• 5 times Equation (1).} \\ \underline{-15x + 9y = 78} & \text{• } -3 \text{ times Equation (2).} \\ \qquad\quad 44y = 88 & \text{• Add the equations.} \\ \qquad\qquad y = 2 & \text{• Solve for } y. \end{array}$$

Substitute the value of y into Equation (1) or Equation (2) and solve for x. Equation (1) will be used here.

$$\begin{array}{ll} (1) \quad\quad 3x + 7y = 2 & \\ \quad\quad\quad 3x + 7(2) = 2 & \text{• Substitute 2 for } y. \\ \quad\quad\quad 3x + 14 = 2 & \text{• Solve for } x. \\ \quad\quad\quad\quad\quad 3x = -12 & \\ \quad\quad\quad\quad\quad\; x = -4 & \end{array}$$

The solution is $(-4, 2)$.

For the previous system of equations, the value of x was determined by substitution. This value can also be determined by eliminating y from the system.

$$\begin{array}{ll} 9x + 21y = 6 & \text{• 3 times Equation (1).} \\ \underline{35x - 21y = -182} & \text{• 7 times Equation (2).} \\ 44x \qquad\quad = -176 & \text{• Add the equations.} \\ \quad\;\; x = -4 & \text{• Solve for } x. \end{array}$$

Note that this is the same value of x as was determined by using substitution.

➡ Solve by the addition method:

(1) $\quad 5x - 2y = -4$
(2) $\quad\quad\quad 3y = 7x + 5$

Rewrite Equation (2) in the form $Ax + By = C$.

$$\begin{array}{ll} (2) \quad\quad\quad 3y = 7x + 5 & \text{• Subtract 7x from each side of the} \\ (3) \quad -7x + 3y = 5 & \text{equation. This is Equation (3).} \end{array}$$

Eliminate x or y. We will eliminate x by using Equations (1) and (3).

$$\begin{array}{ll} 35x - 14y = -28 & \text{• 7 times Equation (1).} \\ \underline{-35x + 15y = 25} & \text{• 5 times Equation (3).} \\ \qquad\quad y = -3 & \text{• Add the equations.} \end{array}$$

Substitute the value of y into Equation (1) or Equation (2) and solve for x. Equation (1) will be used here.

$$\begin{array}{ll} (1) \quad\quad 5x - 2y = -4 & \\ \quad\quad 5x - 2(-3) = -4 & \text{• Substitute } -3 \text{ for } y. \\ \quad\quad\quad 5x + 6 = -4 & \text{• Solve for } x. \\ \quad\quad\quad\quad\; 5x = -10 & \\ \quad\quad\quad\quad\;\; x = -2 & \end{array}$$

The solution is $(-2, -3)$.

➡️ Solve by the addition method: (1) $2x + y = 2$
                                        (2) $4x + 2y = -5$

Eliminate y. Multiply Equation (1) by -2.

(1) $-2(2x + y) = -2 \cdot 2$ • -2 times Equation (1).

(3) $-4x - 2y = -4$ • This is Equation (3).

Add Equation (1) to Equation (3) and solve for x.

(1) $-4x - 2y = -4$

(3) $\underline{4x + 2y = -5}$

       $0x + 0y = -9$ • Add Equation (1) to Equation (3).

           $0 = -9$ • This is not a true equation.

The system of equations is inconsistent and therefore does not have a solution.

The graphs of the two equations in the system of equations above are shown at the right. Note that the graphs are parallel and therefore do not intersect. Thus the system of equations has no solutions.

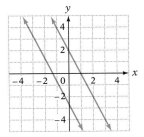

Example 1

Solve by the addition method:

(1) $2x + 4y = 7$

(2) $5x - 3y = -2$

Solution

Eliminate x.

$5(2x + 4y) = 5 \cdot 7$ • 5 times Equation (1).

$-2(5x - 3y) = -2(-2)$ • -2 times Equation (2).

$10x + 20y = 35$

$\underline{-10x + 6y = 4}$

        $26y = 39$ • Add the equations.

         $y = \dfrac{39}{26} = \dfrac{3}{2}$ • Solve for y.

Substitute $\dfrac{3}{2}$ for y in Equation (1).

(1) $2x + 4y = 7$

      $2x + 4\left(\dfrac{3}{2}\right) = 7$ • Replace y by $\dfrac{3}{2}$.

         $2x + 6 = 7$ • Solve for x.

             $2x = 1$

              $x = \dfrac{1}{2}$

The solution is $\left(\dfrac{1}{2}, \dfrac{3}{2}\right)$.

You Try It 1

Solve by the addition method:

(1) $x - 2y = 1$

(2) $2x + 4y = 0$

Your solution

Solution on p. A28

Example 2

Solve by the addition method:

(1) $6x + 9y = 15$
(2) $4x + 6y = 10$

Solution

Eliminate x.

$4(6x + 9y) = 4 \cdot 15$ • 4 times Equation (1).
$-6(4x + 6y) = -6 \cdot 10$ • −6 times Equation (2).

$\quad\ 24x + 36y = 60$
$\underline{-24x - 36y = -60}$
$\qquad\ 0x + 0y = 0$ • Add the equations.
$\qquad\qquad\ 0 = 0$

The system of equations is dependent. The solutions are the ordered pairs that satisfy the equation $6x + 9y = 15$.

You Try It 2

Solve by the addition method:

$\quad 2x - 3y = 4$
$-4x + 6y = -8$

Your solution

Example 3

Solve by the addition method:

(1) $2x = y + 8$
(2) $3x + 2y = 5$

Solution

Write Equation (1) in the form $Ax + By = C$.

$\qquad\quad 2x = y + 8$
(3) $2x - y = 8$ • This is Equation (3).

Eliminate y.

$2(2x - y) = 2 \cdot 8$ • 2 times Equation (3).
$\ 3x + 2y = 5$ • This is Equation (2).

$\quad 4x - 2y = 16$
$\underline{\ 3x + 2y = 5\ }$
$\qquad\ 7x = 21$ • Add the equations.
$\qquad\quad x = 3$

Replace x in Equation (1).
(1) $2x = y + 8$ • Replace x by 3.
$\quad 2 \cdot 3 = y + 8$
$\qquad 6 = y + 8$
$\quad\ -2 = y$

The solution is $(3, -2)$.

You Try It 3

Solve by the addition method:

$4x + 5y = 11$
$\quad 3y = x + 10$

Your solution

Solutions on pp. A28–A29

9.3 Exercises

Objective A

Solve by the addition method.

1. $x + y = 4$
$x - y = 6$

2. $2x + y = 3$
$x - y = 3$

3. $x + y = 4$
$2x + y = 5$

4. $x - 3y = 2$
$x + 2y = -3$

5. $2x - y = 1$
$x + 3y = 4$

6. $x - 2y = 4$
$3x + 4y = 2$

7. $4x - 5y = 22$
$x + 2y = -1$

8. $3x - y = 11$
$2x + 5y = 13$

9. $2x - y = 1$
$4x - 2y = 2$

10. $x + 3y = 2$
$3x + 9y = 6$

11. $4x + 3y = 15$
$2x - 5y = 1$

12. $3x - 7y = 13$
$6x + 5y = 7$

13. $2x - 3y = 1$
$4x - 6y = 2$

14. $2x + 4y = 6$
$3x + 6y = 9$

15. $5x - 2y = -1$
$x + 3y = -5$

16. $4x - 3y = 1$
$8x + 5y = 13$

17. $5x + 7y = 10$
$3x - 14y = 6$

18. $7x + 10y = 13$
$4x + 5y = 6$

19. $3x - 2y = 0$
$6x + 5y = 0$

20. $5x + 2y = 0$
$3x + 5y = 0$

21. $2x - 3y = 16$
$3x + 4y = 7$

Solve by the addition method.

22. $3x + 4y = 10$
$4x + 3y = 11$

23. $5x + 3y = 7$
$2x + 5y = 1$

24. $-2x + 7y = 9$
$3x + 2y = -1$

25. $7x - 2y = 13$
$5x + 3y = 27$

26. $12x + 5y = 23$
$2x - 7y = 39$

27. $8x - 3y = 11$
$6x - 5y = 11$

28. $4x - 8y = 36$
$3x - 6y = 27$

29. $5x + 15y = 20$
$2x + 6y = 8$

30. $y = 2x - 3$
$3x + 4y = -1$

31. $3x = 2y + 7$
$5x - 2y = 13$

32. $2y = 4 - 9x$
$9x - y = 25$

33. $2x + 9y = 16$
$5x = 1 - 3y$

34. $3x - 4 = y + 18$
$4x + 5y = -21$

35. $2x + 3y = 7 - 2x$
$7x + 2y = 9$

36. $5x - 3y = 3y + 4$
$4x + 3y = 11$

APPLYING THE CONCEPTS

37. Describe in your own words the process of solving a system of equa-
[W]   tions by the addition method.

38. The point of intersection of the graphs of the equations $Ax + 2y = 2$
and $2x + By = 10$ is $(2, -2)$. Find A and B.

39. The point of intersection of the graphs of the equations $Ax - 4y = 9$
and $4x + By = -1$ is $(-1, -3)$. Find A and B.

40. For what value of k is the system of equations dependent?

   a. $2x + 3y = 7$
   $4x + 6y = k$

   b. $y = \dfrac{2}{3}x - 3$
   $y = kx - 3$

   c. $x = ky - 1$
   $y = 2x + 2$

41. For what values of k is the system of equations independent?

   a. $x + y = 7$
   $kx + y = 3$

   b. $x + 2y = 4$
   $kx + 3y = 2$

   c. $2x + ky = 1$
   $x + 2y = 2$

Application Problems in Two Variables

Objective A   *To solve rate-of-wind or -current problems* ....

Motion problems that involve an object moving with or against a wind or current normally require two variables to solve.

➡ Flying with the wind, a small plane can fly 600 mi in 3 h. Against the wind, the plane can fly the same distance in 4 h. Find the rate of the plane in calm air and the rate of the wind.

> **Strategy for Solving Rate-of-Wind or -Current Problems**
>
> Choose one variable to represent the rate of the object in calm conditions and a second variable to represent the rate of the wind or current. Using these variables, express the rate of the object with and against the wind or current. Use the equation $d = rt$ to write expressions for the distance traveled by the object. The results can be recorded in a table.

Rate of plane in calm air: p
Rate of wind: w

	Rate	·	*Time*	=	*Distance*
With the wind	$p + w$	·	3	=	$3(p + w)$
Against the wind	$p - w$	·	4	=	$4(p - w)$

> Determine how the expressions for distance are related.

The distance traveled with the wind is 600 mi. $3(p + w) = 600$
The distance traveled against the wind is 600 mi. $4(p - w) = 600$

Solve the system of equations.

$3(p + w) = 600$ $\dfrac{1}{3} \cdot 3(p + w) = \dfrac{1}{3} \cdot 600$ $p + w = 200$

$\rightarrow$

$4(p - w) = 600$ $\dfrac{1}{4} \cdot 4(p - w) = \dfrac{1}{4} \cdot 600$ $p - w = 150$

$\rightarrow$

$$2p = 350$$
$$p = 175$$

$p + w = 200$
$175 + w = 200$
$w = 25$

The rate of the plane in calm air is 175 mph.
The rate of the wind is 25 mph.

Example 1

A 450-mile trip from one city to another takes 3 h when a plane is flying with the wind. The return trip, against the wind, takes 5 h. Find the rate of the plane in still air and the rate of the wind.

Strategy

● Rate of the plane in still air: p
 Rate of the wind: w

	Rate	Time	= Distance
With wind	$p + w$	3	$3(p + w)$
Against wind	$p - w$	5	$5(p - w)$

● The distance traveled with the wind is 450 mi. The distance traveled against the wind is 450 mi.

Solution

$$3(p + w) = 450 \quad \frac{1}{3} \cdot 3(p + w) = \frac{1}{3} \cdot 450$$

$$5(p - w) = 450 \quad \frac{1}{5} \cdot 5(p - w) = \frac{1}{5} \cdot 450$$

$$p + w = 150$$
$$p - w = 90$$

$$2p = 240$$
$$p = 120$$

$$p + w = 150$$
$$120 + w = 150$$
$$w = 30$$

The rate of the plane in still air is 120 mph.
The rate of the wind is 30 mph.

You Try It 1

A canoeist paddling with the current can travel 15 mi in 3 h. Against the current, it takes 5 h to travel the same distance. Find the rate of the current and the rate of the canoeist in calm water.

Your strategy

Your solution

Solution on p. A29

Objective B **To solve application problems using two variables** ....

The application problems in this section are varieties of those problems solved earlier in the text. Each of the strategies for the problems in this section will result in a system of equations.

➡ A jeweler purchased 5 oz of a gold alloy and 20 oz of a silver alloy for a total cost of $540. The next day, at the same prices per ounce, the jeweler purchased 4 oz of the gold alloy and 25 oz of the silver alloy for a total cost of $450. Find the cost per ounce of the gold and silver alloys.

> **Strategy for Solving an Application Problem in Two Variables**
>
> Choose one variable to represent one of the unknown quantities and a second variable to represent the other unknown quantity. Write numerical or variable expressions for all the remaining quantities. These results can be recorded in two tables, one for each of the conditions.

Cost per ounce of gold: g
Cost per ounce of silver: s

First Day	Amount	·	Unit Cost	=	Value
Gold	5	·	g	=	$5g$
Silver	20	·	s	=	$20s$

Second Day	Amount	·	Unit Cost	=	Value
Gold	4	·	g	=	$4g$
Silver	25	·	s	=	$25s$

> Determine a system of equations. Each table will give one equation of the system.

The total value of the purchase on the first day was $540. $5g + 20s = 540$
The total value of the purchase on the second day was $450. $4g + 25s = 450$

Solve the system of equations.

$$5g + 20s = 540 \qquad 4(5g + 20s) = 4 \cdot 540 \qquad 20g + 80s = 2160$$
$$4g + 25s = 450 \qquad -5(4g + 25s) = -5 \cdot 450 \qquad \underline{-20g - 125s = -2250}$$
$$-45s = -90$$
$$s = 2$$

$$5g + 20s = 540$$
$$5g + 20(2) = 540 \qquad \bullet \ s = 2$$
$$5g + 40 = 540$$
$$5g = 500$$
$$g = 100$$

The cost per ounce of the gold alloy was $100.
The cost per ounce of the silver alloy was $2.

Example 2

A store owner purchased 20 incandescent light bulbs and 30 fluorescent bulbs for a total cost of $40. A second purchase, at the same prices, included 30 incandescent bulbs and 10 fluorescent bulbs for a total cost of $25. Find the cost of an incandescent bulb and of a fluorescent bulb.

Strategy

Cost of an incandescent bulb: b
Cost of a fluorescent bulb: f

First purchase

	Amount	Unit Cost	Value
Incandescent	20	b	$20b$
Fluorescent	30	f	$30f$

Second purchase

	Amount	Unit Cost	Value
Incandescent	30	b	$30b$
Fluorescent	10	f	$10f$

The total of the first purchase was $40.
The total of the second purchase was $25.

Solution

$$20b + 30f = 40 \qquad 3(20b + 30f) = 3 \cdot 40$$
$$30b + 10f = 25 \qquad -2(30b + 10f) = -2 \cdot 25$$

$$60b + 90f = 120$$
$$\underline{-60b - 20f = -50}$$
$$70f = 70$$
$$f = 1$$

$$20b + 30f = 40$$
$$20b + 30(1) = 40$$
$$20b = 10$$
$$b = \frac{1}{2}$$

The cost of an incandescent bulb was $0.50.
The cost of a fluorescent bulb was $1.00.

You Try It 2

A citrus grower purchased 25 orange trees and 20 grapefruit trees for $290. The next week, at the same prices, the grower bought 20 orange trees and 30 grapefruit trees for $330. Find the cost of an orange tree and the cost of a grapefruit tree.

Your strategy

Your solution

Solution on p. A30

9.4 Exercises

. . . .

Objective A   *Application Problems*

Solve.

1. A plane flying with the jet stream flew from Los Angeles to Chicago, a distance of 2250 mi, in 5 h. Flying against the jet stream, the plane could fly only 1750 mi in the same amount of time. Find the rate of the plane in calm air and the rate of the wind.

2. A rowing team rowing with the current traveled 40 km in 2 h. Rowing against the current, the team could travel only 16 km in 2 h. Find the rate of rowing in calm water and the rate of the current.

3. A motorboat traveling with the current went 35 mi in 3.5 h. Traveling against the current, the boat went 12 mi in 3 h. Find the rate of the boat in calm water and the rate of the current.

4. A small plane flew 270 mi in 3 h into a headwind. Flying with the wind, the plane traveled 260 mi in 2 h. Find the rate of the plane in calm air and the rate of the wind.

5. A plane flying with a tailwind flew 300 mi in 2 h. Against the wind, it took 3 h to travel the same distance. Find the rate of the plane in calm air and the rate of the wind.

6. A rowing team rowing with the current traveled 17 mi in 2 h. Against the current, the team rowed 7 mi in the same amount of time. Find the rate of the rowing team in calm water and the rate of the current.

7. A seaplane pilot flying with the wind flew from an ocean port to a lake, a distance of 240 mi, in 2 h. Flying against the wind, the trip from the lake to the ocean port took 2 h and 40 min. Find the rate of the plane in calm air and the rate of the wind.

8. Rowing with the current, a canoeist paddled 14 mi in 2 h. Against the current, the canoeist could paddle only 10 mi in the same amount of time. Find the rate of the canoeist in calm water and the rate of the current.

Objective B   *Application Problems*

9. A computer software store received two shipments of software. The value of the first shipment, which contained 12 identical word processing programs and 10 identical spreadsheet programs, was $6190. The second shipment, at the same prices, contained 5 word processing programs and 8 spreadsheet programs. The value of the second shipment was $3825. Find the cost of a word processing program and of a spreadsheet program.

10. A baker purchased 12 lb of wheat flour and 15 lb of rye flour for a total cost of $18.30. A second purchase, at the same prices, included 15 lb of wheat flour and 10 lb of rye flour. The cost of the second purchase was $16.75. Find the cost per pound of the wheat flour and of the rye flour.

11. An investor owned 300 shares of an oil company and 200 shares of a movie company. The quarterly dividend from the two stocks was $165. After the investor sold 100 shares of the oil company and bought an additional 100 shares of the movie company, the quarterly dividend became $185. Find the dividend per share for each stock.

12. The charge for 25 min of prime time and 35 min of non-prime time to a customer for using a computerized financial news network was $10.75. A second customer used the system of 30 min of prime time and 45 min of non-prime time for a cost of $13.35. Find the cost per minute for using the financial news network during prime and during non-prime time.

13. A college football team scored 30 points in one game with only touchdowns and field goals. If the number of touchdowns had been field goals and the number of field goals had been touchdowns, the score would have been 33. Find the number of touchdowns and field goals that were actually scored. Use 6 points for a touchdown and 3 points for a field goal.

14. A professional basketball team scored 87 points in two-point baskets and three-point baskets. If the number of two-point baskets had been three-point baskets and the number of three-point baskets had been two-point baskets, the score would have been 93. Find the number of two-point and three-point baskets that were actually scored.

APPLYING THE CONCEPTS

15. Two angles are supplementary. The larger angle is 15° more than twice the measure of the smaller angle. Find the measure of the two angles. (Supplementary angles are two angles whose sum is 180°.)

16. The value of the nickels and dimes in a coin bank is $0.25. If the number of nickels and the number of dimes were doubled, the value of the coins would be $0.50. How many nickels and how many dimes are in the bank?

17. An investor has $5000 to invest in two accounts. The first account earns 8% annual simple interest and the second account earns 10% annual simple interest. How much money should be invested in each account so that the annual interest earned is $600?

18. Solve the following problem, which dates from a Chinese manuscript called the Jinzhang that is approximately 2100 years old. "The price of 1 acre of good land is 300 pieces of gold; the price of 7 acres of bad land is 500 pieces of gold. One has purchased altogether 100 acres. The price was 10,000 pieces of gold. How much good land and how much bad land was bought?" Adapted from *A History of Mathematics, An Introduction*, Victor J. Katz (New York: Harper Collins, 1993, page 15).

19. A coin bank contains only nickels or dimes, but there are no more than 27 coins. The value of the coins is $2.10. How many different combinations of nickels and dimes could be in the coin bank?

Content and Format © 1995 HMCo.

Project in Mathematics

. . . .

Calculator Solution of a System of Linear Equations

By using the addition method, it is possible to solve the system of equations

$$ax + by = c$$
$$dx + ey = f$$

The solution is

$$x = \frac{ce - bf}{ae - bd} \quad \text{and} \quad y = \frac{af - cd}{ae - bd}, \quad ae - bd \neq 0$$

Using this solution, a system of equations can be solved with a calculator. It is helpful to observe that the denominators for both expressions are identical. The calculation for the denominator is done first and then stored in the calculator's memory. If the value of the denominator is zero, then the system is dependent or inconsistent, and this calculator method cannot be used.

Solve: $2x - 5y = 9$
          $4x + 3y = 2$

Make a list of the values of a, b, c, d, e, and f.

$$a = 2 \quad\quad b = -5 \quad\quad c = 9$$
$$d = 4 \quad\quad e = 3 \quad\quad f = 2$$

Calculate the denominator $D = ae - bd$. $D = 2 \cdot 3 - (-5) \cdot 4 = 6 + 20 = 26$

Store the result in memory. Press $\boxed{M+}$.

Find x. Replace the letters by the given values. $x = \dfrac{ce - bf}{D} = \dfrac{9 \cdot 3 - (-5) \cdot 2}{26}$

Calculate x. $9 \boxed{\times} 3 \boxed{-} \boxed{(}\, 5 \boxed{+/-} \boxed{\times} 2 \boxed{)} \boxed{\div} \boxed{MR} \boxed{=}$

The result in the display should be 1.423077.

Find y. Replace the letters by the given values. $y = \dfrac{af - cd}{D} = \dfrac{2 \cdot 2 - 9 \cdot 4}{26}$

Calculate y. $2 \boxed{\times} 2 \boxed{-} \boxed{(}\, 9 \boxed{\times} 4 \boxed{)} \boxed{\div} \boxed{MR} \boxed{=}$

The result in the display should be -1.230769.

The solution of the system is $(1.423077, -1.230769)$.

The keys $\boxed{M+}$ (store in memory) and $\boxed{MR}$ (recall from memory) were used for this illustration. Some calculators have instead the keys $\boxed{STO}$ (store in memory) and $\boxed{RCL}$ (recall from memory). If your calculator has these keys, use them in place of the keys shown in the illustration.

A graphing calculator can also be used to approximate the solution of a system of equations in two variables. Graph each equation of the system of equations, and then approximate the coordinates of the point of intersection. The process by which you approximate the solution depends on the model of calculator you have.

Solve: $3x - 2y = 6$
$\qquad\quad 4x + 3y = 12$

Solve each equation for y. $y = \dfrac{3}{2}x - 3$ $y = -\dfrac{4}{3}x + 4$

Now graph each equation. The result is shown at the left.

TI-82

$\boxed{\text{ZOOM}}\,6\,\boxed{\text{Y=}}\,\boxed{\text{CLEAR}}\,3$
$\boxed{\text{X,T,}\theta}\,\boxed{\div}\,2\,\boxed{-}\,3\,\boxed{\text{ENTER}}$
$\boxed{\text{CLEAR}}\,\boxed{(-)}\,4\,\boxed{\text{X,T,}\theta}\,\boxed{\div}\,3$
$\boxed{+}\,4\,\boxed{\text{GRAPH}}$

SHARP EL-9300

$\boxed{\text{⋀}}\,\boxed{\text{CL}}\,\boxed{\text{RANGE}}\,\boxed{\text{MENU}}\,\boxed{(-)}\,10$
$\boxed{\text{ENTER}}\,10\,\boxed{\text{ENTER}}\,1\,\boxed{\text{ENTER}}$
$\boxed{(-)}\,10\,\boxed{\text{ENTER}}\,10\,\boxed{\text{ENTER}}$
$\boxed{\text{RANGE}}\,3\,\boxed{\text{X/}\theta\text{/T}}\,\boxed{\div}\,2\,\boxed{-}\,3$
$\boxed{\text{ENTER}}\,\boxed{(-)}\,4\,\boxed{\text{X/}\theta\text{/T}}\,\boxed{\div}\,3$
$\boxed{+}\,4\,\boxed{\text{ENTER}}\,\boxed{\text{⋀}}$

CASIO *fx-7700GB*

$\boxed{\text{2nd}}\,\boxed{\text{F5}}\,\boxed{\text{EXE}}\,\boxed{\text{Range}}\,\boxed{\text{SHIFT}}\,(-)\,10$
$\boxed{\text{EXE}}\,10\,\boxed{\text{EXE}}\,1\,\boxed{\text{EXE}}\,\boxed{\text{SHIFT}}\,(-)\,10$
$\boxed{\text{EXE}}\,10\,\boxed{\text{EXE}}\,1\,\boxed{\text{EXE}}\,\boxed{\text{Range}}$
$\boxed{\text{Graph}}\,3\,\boxed{\text{X,}\theta\text{,T}}$
$\boxed{\div}\,2\,\boxed{-}\,3\,\boxed{\text{EXE}}\,\boxed{\text{Graph}}\,\boxed{\text{SHIFT}}\,(-)\,4$
$\boxed{\text{X,}\theta\text{,T}}\,\boxed{\div}\,3\,\boxed{+}\,4\,\boxed{\text{EXE}}$

One method of determining the approximate coordinates of the point of intersection is to use the ZOOM and TRACE features of your calculator. ZOOM in on the point of intersection, and then use TRACE to determine the approximate coordinates.

Chapter Summary

Key Words   Equations considered together are called a *system of equations.*

A *solution of a system of equations* in two variables is an ordered pair that is a solution of each equation of the system.

An *independent system of equations* has one solution.

A *dependent system of equations* has an infinite number of solutions.

An *inconsistent system of equations* has no solution.

Essential Rules   A system of equations can be solved by *the graphing method, the substitution method,* or *the addition method.*

Chapter Review

. . . .

SECTION 9.1

1. Is $(-1, -3)$ a solution of the system
$$5x + 4y = -17$$
$$2x - y = 1?$$

2. Is $(-2, 0)$ a solution of the system
$$-x + 9y = 2$$
$$6x - 4y = 12?$$

3. Solve by graphing:
$$3x - y = 6$$
$$y = -3$$

4. Solve by graphing:
$$4x - 2y = 8$$
$$y = 2x - 4$$

5. Solve by graphing:
$$x + 2y = 3$$
$$y = -\frac{1}{2}x + 1$$

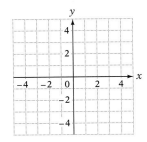

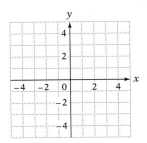

SECTION 9.2

6. Solve by substitution.
$$4x + 7y = 3$$
$$x = y - 2$$

7. Solve by substitution.
$$8x - y = 2$$
$$y = 5x + 1$$

8. Solve by substitution.
$$6x - y = 0$$
$$7x - y = 1$$

9. Solve by substitution.
$$7x + 3y = -16$$
$$x - 2y = 5$$

10. Solve by substitution.
$$12x - 9y = 18$$
$$y = \frac{4}{3}x - 3$$

11. Solve by substitution.
$$4x + 3y = 12$$
$$y = -\frac{4}{3}x + 4$$

SECTION 9.3

12. Solve by the addition method.
$$3x + 8y = -1$$
$$x - 2y = -5$$

13. Solve by the addition method.
$$4x - y = 9$$
$$2x + 3y = -13$$

14. Solve by the addition method.
$$6x + 4y = -3$$
$$12x - 10y = -15$$

15. Solve by the addition method.
$$5x + 7y = 21$$
$$20x + 28y = 63$$

16. Solve by the addition method.
$$3x + y = -2$$
$$-9x - 3y = 6$$

17. Solve by the addition method
$$5x + 2y = -9$$
$$12x - 7y = 2$$

18. Solve by the addition method.
$$6x - 18y = 7$$
$$9x + 24y = 2$$

SECTION 9.4

19. A canoeist traveling with the current traveled the 30 mi between two riverside camp sites in 3 h. The return trip took 5 h. Find the rate of the canoeist in still water and the rate of the current.

20. A flight crew flew 420 km in 3 h with a tailwind. Flying against the wind, the flight crew flew 440 km in 4 h. Find the rate of the flight crew in calm air and the rate of the wind.

21. A small plane flying with the wind flew 360 mi in 3 h. Against a head-wind, the plane took 4 h to fly the same distance. Find the rate of the plane in calm air and the rate of the wind.

22. A sculling team rowing with the current went 24 mi in 2 h. Rowing against the current, the sculling team went 18 mi in 3 h. Find the rate of the sculling team in calm water and the rate of the current.

23. A small wood carving company mailed 190 advertisements, some requiring 25 cents postage and others 45 cents. The total cost for mailing was $59.50. Find the number of advertisements mailed at each rate.

24. A silo contains a mixture of lentils and corn. If 50 bushels of lentils were added, there would be twice as many bushels of lentils as of corn; if 150 bushels of corn were added instead, there would be the same amount of corn as of lentils. How many bushels of each were originally in the silo?

25. An investor bought 1500 shares of stock, some at $6 per share and the rest at $25 per share. If $12,800 worth of stock was purchased, how many shares of each kind did the investor buy?

Content and Format © 1995 HMCo.

Chapter Test

. . . .

1. Is $(-2, 3)$ a solution of the system
$$2x + 5y = 11$$
$$x + 3y = 7?$$

2. Is $(1, -3)$ a solution of the system
$$3x - 2y = 9$$
$$4x + y = 1?$$

3. Solve by graphing: $3x + 2y = 6$
$$5x + 2y = 2$$

4. Solve by substitution.
$$4x - y = 11$$
$$y = 2x - 5$$

5. Solve by substitution.
$$x = 2y + 3$$
$$3x - 2y = 5$$

6. Solve by substitution.
$$3x + 5y = 1$$
$$2x - y = 5$$

7. Solve by substitution.
$$3x - 5y = 13$$
$$x + 3y = 1$$

8. Solve by substitution.
$$2x - 4y = 1$$
$$y = \frac{1}{2}x + 3$$

9. Solve by the addition method.
$$4x + 3y = 11$$
$$5x - 3y = 7$$

10. Solve by the addition method.
$$2x - 5y = 6$$
$$4x + 3y = -1$$

11. Solve by the addition method.

$$x + 2y = 8$$
$$3x + 6y = 24$$

12. Solve by the addition method.

$$7x + 3y = 11$$
$$2x - 5y = 9$$

13. Solve by the addition method.

$$5x + 6y = -7$$
$$3x + 4y = -5$$

14. With the wind, a plane flies 240 mi in 2 h. Against the wind, the plane requires 3 h to fly the same distance. Find the rate of the plane in calm air and the rate of the wind.

15. For the first performance of a play in a community theater, 50 reserved-seat tickets and 80 general-admission tickets were sold. The total receipts were $980. For the second performance, 60 reserved-seat tickets and 90 general-admission tickets were sold. The total receipts were $1140. Find the price of a reserved-seat ticket and the price of a general-admission ticket.

Cumulative Review

. . . .

1. Evaluate $\dfrac{a^2 - b^2}{2a}$ when $a = 4$ and $b = -2$.

2. Solve: $-\dfrac{3}{4}x = \dfrac{9}{8}$

3. Given $f(x) = x^2 + 2x - 1$, find $f(2)$.

4. Simplify: $(2a^2 - 3a + 1)(2 - 3a)$

5. Simplify: $\dfrac{(-2x^2y)^4}{-8x^3y^2}$

6. Simplify: $(4b^2 - 8b + 4) \div (2b - 3)$

7. Simplify: $\dfrac{8x^{-2}y^5}{-2xy^4}$

8. Factor $4x^2y^4 - 64y^2$.

9. Solve: $(x - 5)(x + 2) = -6$

10. Simplify: $\dfrac{x^2 - 6x + 8}{2x^3 + 6x^2} \div \dfrac{2x - 8}{4x^3 + 12x^2}$

11. Simplify: $\dfrac{x - 1}{x + 2} + \dfrac{2x + 1}{x^2 + x - 2}$

12. Simplify: $\dfrac{x + 4 - \dfrac{7}{x - 2}}{x + 8 + \dfrac{21}{x - 2}}$

13. Solve: $\dfrac{x}{2x - 3} + 2 = \dfrac{-7}{2x - 3}$

14. Solve $A = P + Prt$ for r.

15. Find the x- and y-intercepts for $2x - 3y = 12$.

16. Find the slope of the line that passes through the points $(2, -3)$ and $(-3, 4)$.

17. Find the equation of the line that passes through the point $(-2, 3)$ and has slope $-\dfrac{3}{2}$.

18. Is $(2, 0)$ a solution of the system
$$5x - 3y = 10$$
$$4x + 7y = 8?$$

19.   Solve by substitution.
$$3x - 5y = -23$$
$$x + 2y = -4$$

20.   Solve by the addition method.
$$5x - 3y = 29$$
$$4x + 7y = -5$$

21.   A total of $8750 is invested into two simple interest accounts. On one account, the annual simple interest rate is 9.6%; on the second account, the annual simple interest rate is 7.2%. How much should be invested in each account so that both accounts earn the same interest?

22.   A passenger train leaves a train depot $\frac{1}{2}$ h after a freight train leaves the same depot. The freight train is traveling 8 mph slower than the passenger train. Find the rate of each train if the passenger train overtakes the freight train in 3 h.

23.   The length of each side of a square is extended 4 in. The area of the resulting square is 144 in². Find the length of a side of the original square.

24.   A plane can travel 160 mph in calm air. Flying with the wind, the plane can fly 570 mi in the same amount of time as it takes to fly 390 mi against the wind. Find the rate of the wind.

25.   Graph $2x - 3y = 6$.

26.   Solve by graphing: $3x + 2y = 6$
$$3x - 2y = 6$$

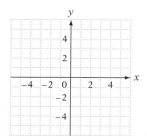

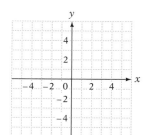

27.   With the current, a motorboat can travel 48 mi in 3 h. Against the current, the boat requires 4 h to travel the same distance. Find the rate of the boat in calm water.

28.   A child adds 8 g of sugar to a 50-g serving of a breakfast cereal that is 25% sugar. What is now the percent concentration of sugar in the mixture? Round to the nearest tenth of a percent.

10

Inequalities

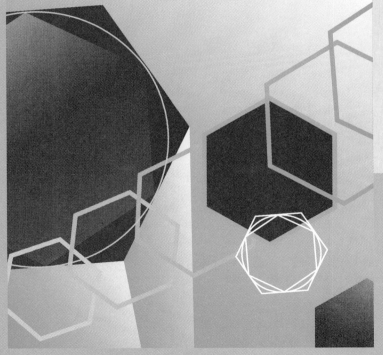

Objectives

Section 10.1
To write a set using the roster method
To write a set using set-builder notation
To graph an inequality on the number line

Section 10.2
To solve an inequality using the Addition
    Property of Inequalities
To solve an inequality using the Multiplication
    Property of Inequalities
To solve application problems

Section 10.3
To solve general inequalities
To solve application problems

Section 10.4
To graph an inequality in two variables

Calculations of Pi

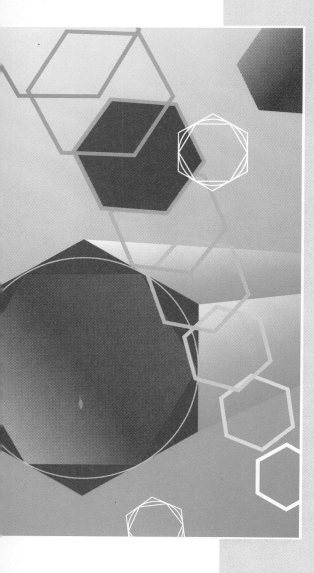

There are many early references to estimated values for pi. One of the earliest is from the Rhind Papyrus, which was found in Egypt in the 1800s. Scientists have estimated that these tablets were written around 1600 B.C. The Rhind Papyrus contains the estimate 3.1604 for pi.

One of the most famous calculations of pi occurred around 240 B.C. and was performed by Archimedes. The calculation was based on finding the perimeter of inscribed and circumscribed six-sided polygons (hexagons). Once the perimeter for a hexagon figure was calculated, known formulas could be used to calculate the perimeter of polygons with twice that number of sides. Continuing in this way, Archimedes calculated the perimeters for the polygons with 12, 24, 48, and 96 sides.

His calculations resulted in a value of pi between $3\frac{10}{71}$ and $3\frac{1}{7}$. You might recognize $3\frac{1}{7}$, or $\frac{22}{7}$, as an approximation for pi still used today.

Calculations to improve the accuracy of pi continued. One French mathematician, relying on Archimedes's method, estimated pi by using a polygon of 393,216 sides. A mathematician from the Netherlands estimated pi by using a polygon with over one million sides.

Around the 1650s, new mathematical methods were developed to estimate the value of pi. These methods started yielding estimates of pi that were accurate to over 70 places. By the 1850s, an estimate for pi was accurate to 200 places.

Today, thanks to more refined mathematical methods and computers, estimates of the value of pi now exceed one million places.

In 1914 an issue of *Scientific American* contained the following short note:

> "See, I have a rhyme assisting my feeble brain,
> its tasks oftimes resisting."

Can you see what this note has to do with the estimates for the value of pi?

(Each word length represents a digit in the approximation 3.141592653579.)

10.1 Sets

Objective A   To write a set using the roster method

Recall that a *set* is a collection of objects, which are called the *elements* of the set.

The **roster method** of writing a set encloses a list of the elements in braces.

The set of the last three letters of the alphabet is written $\{x, y, z\}$.

The set of the positive integers less than 5 is written $\{1, 2, 3, 4\}$.

➡ Use the roster method to write the set of integers between 0 and 10.

$$A = \{1, 2, 3, 4, 5, 6, 7, 8, 9\}$$

A set can be designated by a capital letter. Note that 0 and 10 are not elements of the set.

➡ Use the roster method to write the set of natural numbers.

$A = \{1, 2, 3, 4, \ldots\}$ • The three dots mean that the pattern of numbers continues without end.

The **empty set,** or **null set,** is the set that contains no elements. The symbol $\varnothing$ or $\{\ \}$ is used to represent the empty set.

The set of people who have run a two-minute mile is the empty set.

The **union** of two sets, written $A \cup B$, is the set that contains the elements of A and the elements of B.

➡ Find $A \cup B$, given $A = \{1, 2, 3, 4\}$ and $B = \{3, 4, 5, 6\}$.

$A \cup B = \{1, 2, 3, 4, 5, 6\}.$ • The union of *A* and *B* contains all the elements of *A* and all the elements of *B*. Any elements that are in both A and B are listed only once.

The **intersection** of two sets, written $A \cap B$, is the set that contains the elements that are common to both A and B.

➡ Find $A \cap B$, given $A = \{1, 2, 3, 4\}$ and $B = \{3, 4, 5, 6\}$.

$A \cap B = \{3, 4\}$ • The intersection of *A* and *B* contains the elements common to *A* and *B*.

Example 1
Use the roster method to write the set of the odd positive integers less than 12.

Solution
$A = \{1, 3, 5, 7, 9, 11\}$

You Try It 1
Use the roster method to write the set of the odd negative integers greater than -10.

Your solution

Solution on p. A30

Example 2
Use the roster method to write the set of the even positive integers.

Solution
$A = \{2, 4, 6, \ldots\}$

You Try It 2
Use the roster method to write the set of the odd positive integers.

Your solution

Example 3
Find $D \cup E$, given
$D = \{6, 8, 10, 12\}$ and
$E = \{-8, -6, 10, 12\}$.

Solution
$D \cup E = \{-8, -6, 6, 8, 10, 12\}$

You Try It 3
Find $A \cup B$, given
$A = \{-2, -1, 0, 1, 2\}$ and
$B = \{0, 1, 2, 3, 4\}$.

Your solution

Example 4
Find $A \cap B$, given
$A = \{5, 6, 9, 11\}$ and
$B = \{5, 9, 13, 15\}$.

Solution
$A \cap B = \{5, 9\}$

You Try It 4
Find $C \cap D$, given
$C = \{10, 12, 14, 16\}$ and
$D = \{10, 16, 20, 26\}$.

Your solution

Example 5
Find $A \cap B$, given
$A = \{1, 2, 3, 4\}$ and
$B = \{8, 9, 10, 11\}$.

Solution
$A \cap B = \varnothing$

You Try It 5
Find $A \cap B$, given
$A = \{-5, -4, -3, -2\}$ and
$B = \{2, 3, 4, 5\}$.

Your solution

Solutions on p. A30

Objective B   ***To write a set using set-builder notation*** ....

Another method of representing sets is called **set-builder notation.** Using set-builder notation, the set of all positive integers less than 10 is as follows:

$\{x \mid x < 10, x \text{ is a positive integer}\}$, which is read "the set of all x such that x is less than 10 and x is a positive integer."

➡ Use set-builder notation to write the set of real numbers greater than 4.

$\{x \mid x > 4, x \in \text{ real numbers}\}$ • "$x \in$ real numbers" is read "x is an element of the real numbers."

Example 6
Use set-builder notation to write the set of negative integers greater than −100.

Solution
$\{x \mid x > -100, x \text{ is a negative integer}\}$

You Try It 6
Use set-builder notation to write the set of positive even integers less than 59.

Your solution

Example 7
Use set-builder notation to write the set of real numbers less than 60.

Solution
$\{x \mid x < 60, x \in \text{real numbers}\}$

You Try It 7
Use set-builder notation to write the set of real numbers greater than −3.

Your solution

Solutions on p. A30

Objective C   *To graph an inequality on the number line* ....

An expression that contains the symbol >, <, ≥ (is greater than or equal to), or ≤ (is less than or equal to) is called an **inequality**. An inequality expresses the relative order of two mathematical expressions. The expressions can be either numerical or variable.

$$4 > 2$$
$$3x \le 7$$
$$x^2 - 2x > y + 4$$

$\left.\right\}$ Inequalities

An **inequality** can be graphed on the number line.

➡ Graph: $x > 1$

The graph is the real numbers greater than 1. The circle at 1 indicates that 1 is not included in the graph.

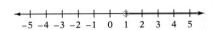

➡ Graph: $x \ge 1$

The dot at 1 indicates that 1 is included in the solution set.

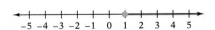

➡ Graph: $-1 > x$

$-1 > x$ is equivalent to $x < -1$. The numbers less than −1 are to the left of −1 on the number line.

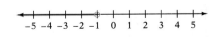

The union of two sets is the set that contains all the elements of each set.

➡ Graph: $(x > 4) \cup (x < 1)$

The graph is the numbers greater than 4 and the numbers less than 1.

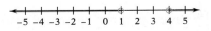

The intersection of two sets is the set that contains the elements common to both sets.

➡ Graph the solution set of $(x > -1) \cap (x < 2)$.

The graph is the numbers between -1 and 2.

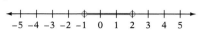

Example 8
Graph: $x < 5$

Solution
The graph is the numbers less than 5.

You Try It 8
Graph: $-2 < x$

Your solution

Example 9
Graph: $(x > -2) \cap (x < 1)$

Solution
The graph is the numbers between -2 and 1.

You Try It 9
Graph: $(x > -1) \cup (x < -3)$

Your solution

Example 10
Graph: $(x \le 5) \cup (x \ge -3)$

Solution
The graph is the real numbers.

You Try It 10
Graph: $(x < 5) \cup (x \ge -2)$

Your solution

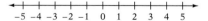

Example 11
Graph: $(x > 3) \cup (x < 1)$

Solution
The graph is the numbers greater than 3 or the numbers less than 1.

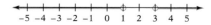

You Try It 11
Graph: $(x \le 4) \cap (x \ge -4)$

Your solution

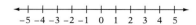

Solutions on p. A30

10.1 Exercises

· ·

Objective A

Use the roster method to write the set.

1. the integers between 15 and 22

2. the integers between -10 and -4

3. the odd integers between 8 and 18

4. the even integers between -11 and -1

5. the letters of the alphabet between a and d

6. the letters of the alphabet between p and v

Find $A \cup B$.

7. $A = \{3, 4, 5\}$ $B = \{4, 5, 6\}$

8. $A = \{-3, -2, -1\}$ $B = \{-2, -1, 0\}$

9. $A = \{-10, -9, -8\}$ $B = \{8, 9, 10\}$

10. $A = \{a, b, c\}$ $B = \{x, y, z\}$

11. $A = \{a, b, d, e\}$ $B = \{c, d, e, f\}$

12. $A = \{m, n, p, q\}$ $B = \{m, n, o\}$

13. $A = \{1, 3, 7, 9\}$ $B = \{7, 9, 11, 13\}$

14. $A = \{-3, -2, -1\}$ $B = \{-1, 1, 2\}$

Find $A \cap B$.

15. $A = \{3, 4, 5\}$ $B = \{4, 5, 6\}$

16. $A = \{-4, -3, -2\}$ $B = \{-6, -5, -4\}$

17. $A = \{-4, -3, -2\}$ $B = \{2, 3, 4\}$

18. $A = \{1, 2, 3, 4\}$ $B = \{1, 2, 3, 4\}$

19. $A = \{a, b, c, d, e\}$ $B = \{c, d, e, f, g\}$

20. $A = \{m, n, o, p\}$ $B = \{k, l, m, n\}$

Objective B

Use set-builder notation to write the set.

21. the negative integers greater than -5

22. the positive integers less than 5

23. the integers greater than 30

24. the integers less than -70

25. the even integers greater than 5

26. the odd integers less than -2

27. the real numbers greater than 8

28. the real numbers less than 57

Objective C

Graph.

29. $x > 2$

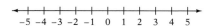

30. $x \geq -1$

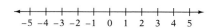

31. $0 \geq x$

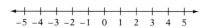

32. $4 > x$

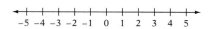

33. $(x > -2) \cup (x < -4)$

34. $(x > 4) \cup (x < -2)$

35. $(x > -2) \cap (x < 4)$

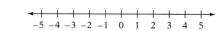

36. $(x > -3) \cap (x < 3)$

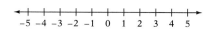

37. $(x \geq -2) \cup (x < 4)$

38. $(x > 0) \cup (x \leq 4)$

APPLYING THE CONCEPTS

39. Explain how to find the union of two sets.
[W]

40. Explain how to find the intersection of two sets.
[W]

41. Determine whether the statement is always true, sometimes true, or never true.
   a. Given that $a > 0$ and $b < 0$, then $ab > 0$.
   b. Given that $a < 0$, then $a^2 > 0$.
   c. Given that $a > 0$ and $b < 0$, then $a^2 > b$.

42. By trying various sets, make a conjecture as to whether the union of two sets is
   a. a commutative operation
   b. an associative operation

43. By trying various sets, make a conjecture as to whether the intersection of two sets is
   a. a commutative operation
   b. an associative operation

10.2 The Addition and Multiplication Properties of Inequalities

Objective A   *To solve an inequality using the Addition Property of Inequalities*

The **solution set of an inequality** is a set of numbers each element of which, when substituted for the variable, results in a true inequality.

The inequality at the right is true if the variable is replaced by 7, 9.3, or $\frac{15}{2}$.

$$\left. \begin{array}{c} x + 3 > 8 \\ 7 + 3 > 8 \\ 9.3 + 3 > 8 \\ \frac{15}{2} + 3 > 8 \end{array} \right\} \text{True inequalities}$$

The inequality $x + 3 > 8$ is false if the variable is replaced by 4, 1.5, or $-\frac{1}{2}$.

$$\left. \begin{array}{c} 4 + 3 > 8 \\ 1.5 + 3 > 8 \\ -\frac{1}{2} + 3 > 8 \end{array} \right\} \text{False inequalities}$$

There are many values of the variable x that will make the inequality $x + 5 > 8$ true. The solution set of $x + 5 > 8$ is any number greater than 3.

At the right is the graph of the solution set of $x + 5 > 8$.

$$-6\,-5\,-4\,-3\,-2\,-1\ \ 0\ \ 1\ \ 2\ \ 3\ \ 4\ \ 5\ \ 6$$

In solving an inequality, the goal is to rewrite the given inequality in the form *variable > constant* or *variable < constant*. The Addition Property of Inequalities is used to rewrite an inequality in this form.

Addition Property of Inequalities

The same term can be added to each side of an inequality without changing the solution set of the inequality.

If $a > b$, then $a + c > b + c$.

If $a < b$, then $a + c < b + c$.

The Addition Property of Inequalities also holds true for an inequality containing the symbol $\geq$ or $\leq$.

The Addition Property of Inequalities is used when, in order to rewrite an inequality in the form *variable > constant* or *variable < constant*, we must remove a term from one side of the inequality. Add the same term to each side of the inequality.

➡ Solve: $x - 4 < -3$

$x - 4 < -3$

$x - 4 + 4 < -3 + 4$ • Add 4 to each side of the inequality.

$x < 1$ • Simplify.

At the right is the graph of the solution set of $x - 4 < -3$.

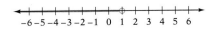

$$-6\,-5\,-4\,-3\,-2\,-1\ \ 0\ \ 1\ \ 2\ \ 3\ \ 4\ \ 5\ \ 6$$

Because subtraction is defined in terms of addition, the Addition Property of Inequalities allows the same term to be subtracted from each side of an inequality.

➡ Solve: $5x - 6 \le 4x - 4$

$$5x - 6 \le 4x - 4$$

$$5x - 4x - 6 \le 4x - 4x - 4$$ • Subtract $4x$ from each side of the equation.

$$x - 6 \le -4$$ • Simplify.

$$x - 6 + 6 \le -4 + 6$$ • Add 6 to each side of the equation.

$$x \le 2$$ • Simplify.

Example 1
Solve and graph the solution set of $3 < x + 5$.

Solution
$$3 < x + 5$$
$$3 - 5 < x + 5 - 5$$
$$-2 < x$$

$$\text{graph}: -5\ -4\ -3\ -2\ -1\ \ 0\ \ 1\ \ 2\ \ 3\ \ 4\ \ 5$$

You Try It 1
Solve and graph the solution set of $x + 2 < -2$.

Your solution

$$-5\ -4\ -3\ -2\ -1\ \ 0\ \ 1\ \ 2\ \ 3\ \ 4\ \ 5$$

Example 2
Solve: $7x - 14 \le 6x - 16$

Solution
$$7x - 14 \le 6x - 16$$
$$7x - 6x - 14 \le 6x - 6x - 16$$
$$x - 14 \le -16$$
$$x - 14 + 14 \le -16 + 14$$
$$x \le -2$$

You Try It 2
Solve: $5x + 3 > 4x + 5$

Your solution

Solutions on p. A31

Objective B   *To solve an inequality using the Multiplication Property of Inequalities*

In solving an inequality, the goal is to rewrite the given inequality in the form *variable > constant* or *variable < constant*. The Multiplication Property of Inequalities is used when, in order to rewrite an inequality in this form, we must remove a coefficient from one side of the inequality.

> **Multiplication Property of Inequalities**
>
> Each side of an inequality can be multiplied by the same positive number without changing the solution set of the inequality.
>
> If $a > b$ and $c > 0$, then $ac > bc$.
>
> If $a < b$ and $c > 0$, then $ac < bc$.

$$5 > 4$$
$$5(2) > 4(2)$$
$$10 > 8$$

- Multiply by *positive* 2.
- Still a true inequality

LOOK CLOSELY

Any time an inequality is multiplied or divided by a negative number, the inequality symbol must be reversed. Compare the next two examples.

$2x < -4$ Divide each side
$\dfrac{2x}{2} < \dfrac{-4}{2}$ by *positive* 2. Inequality *is*
$x < -2$ *not* reversed.

$-2x < 4$ Divide each
$\dfrac{-2x}{2} > \dfrac{4}{-2}$ side by *negative* 2.
$x > -2$ Inequality *is* reversed.

> If each side of an inequality is multiplied by the same negative number and the inequality symbol is reversed, then the solution set of the inequality is not changed.
>
> If $a > b$ and $c < 0$, then $ac < bc$.
>
> If $a < b$ and $c < 0$, then $ac > bc$.

$$6 < 9$$
$$6(-3) > 9(-3)$$
$$-18 > -27$$

- Multiply by *negative* 3 and *reverse* the inequality.
- Still a true inequality

The Multiplication Property of Inequalities also holds true for an inequality containing the symbol $\geq$ or $\leq$.

➡ Solve $-\dfrac{3}{2}x \leq 6$ and graph the solution set.

$$-\dfrac{3}{2}x \leq 6$$

$$-\dfrac{2}{3}\left(-\dfrac{3}{2}x\right) \geq -\dfrac{2}{3}(6)$$

$$x \geq -4$$

- Multiply each side of the inequality by $-\dfrac{2}{3}$.

  Because $-\dfrac{2}{3}$ is a negative number, the inequality symbol must be reversed.

The graph of the solution set is shown at the right.

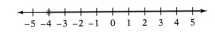

Because division is defined in terms of multiplication, the Multiplication Property of Inequalities allows each side of an inequality to be divided by a non-zero constant.

➡ Solve: $-4 < 6x$

$$-4 < 6x$$

$$\dfrac{-4}{6} < \dfrac{6x}{6}$$

- Divide each side of the inequality by 6.

$$-\dfrac{2}{3} < x$$

- Simplify: $\dfrac{-4}{6} = -\dfrac{2}{3}$

Example 3 Solve and graph the solution set of $-7x > 14$.

Solution $-7x > 14$

$$\frac{-7x}{-7} < \frac{14}{-7}$$

$x < -2$

You Try It 3 Solve and graph the solution set of $-3x > -9$.

Your solution

Example 4 Solve: $-\frac{5}{8}x \le \frac{5}{12}$

Solution $-\frac{5}{8}x \le \frac{5}{12}$

$$-\frac{8}{5}\left(-\frac{5}{8}x\right) \ge -\frac{8}{5}\left(\frac{5}{12}\right)$$

$$x \ge -\frac{2}{3}$$

You Try It 4 Solve: $-\frac{3}{4}x \ge 18$

Your solution

Solutions on p. A31

Objective C **To solve application problems**

Example 5

A student must have at least 450 points out of 500 points on five tests to receive an A in a course. One student's results on the first four tests were 94, 87, 77, and 95. What scores on the last test will enable this student to receive an A in the course?

Strategy

To find the scores, write and solve an inequality using N to represent the possible scores on the last test.

Solution

Total number of points on the 5 tests	is greater than or equal to	450

$94 + 87 + 77 + 95 + N \ge 450$

$353 + N \ge 450$

$353 - 353 + N \ge 450 - 353$

$N \ge 97$

The student's score on the last test must be equal to or greater than 97.

You Try It 5

An appliance dealer will make a profit on the sale of a television set if the cost of the new set is less than 70% of the selling price. What selling prices will enable the dealer to make a profit on a television set that costs the dealer $314?

Your strategy

Your solution

Solution on p. A31

10.2 Exercises

Objective A

Solve and graph the solution set.

1. $x + 1 < 3$

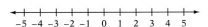

2. $y + 2 < 2$

3. $x - 5 > -2$

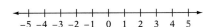

4. $x - 3 > -2$

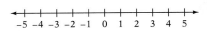

5. $7 \leq n + 4$

6. $3 \leq 5 + x$

7. $x - 6 \leq -10$

8. $y - 8 \leq -11$

9. $4 \leq x + 5$

10. $0 \leq n - 2$

Solve.

11. $y - 3 \geq -12$

12. $x + 8 \geq -14$

13. $3x - 5 < 2x + 7$

14. $5x + 4 < 4x - 10$

15. $8x - 7 \geq 7x - 2$

16. $3n - 9 \geq 2n - 8$

17. $2x + 4 < x - 7$

18. $9x + 7 < 8x - 7$

19. $4x - 8 \leq 2 + 3x$

20. $5b - 9 < 3 + 4b$

21. $6x + 4 \geq 5x - 2$

22. $7x - 3 \geq 6x - 2$

23. $2x - 12 > x - 10$

24. $3x + 9 > 2x + 7$

25. $d + \dfrac{1}{2} < \dfrac{1}{3}$

26. $x - \dfrac{3}{8} < \dfrac{5}{6}$

27. $x + \dfrac{5}{8} \geq -\dfrac{2}{3}$

28. $y + \dfrac{5}{12} \geq -\dfrac{3}{4}$

Solve.

29. $x - \dfrac{3}{8} < \dfrac{1}{4}$

30. $y + \dfrac{5}{9} \leq \dfrac{5}{6}$

31. $2x - \dfrac{1}{2} < x + \dfrac{3}{4}$

32. $6x - \dfrac{1}{3} \leq 5x - \dfrac{1}{2}$

33. $3x + \dfrac{5}{8} > 2x + \dfrac{5}{6}$

34. $4b - \dfrac{7}{12} \geq 3b - \dfrac{9}{16}$

35. $3.8x < 2.8x - 3.8$

36. $1.2x < 0.2x - 7.3$

37. $x + 5.8 \leq 4.6$

38. $n - 3.82 \leq 3.95$

39. $x - 3.5 < 2.1$

40. $x - 0.23 \leq 0.47$

41. $1.33x - 1.62 > 0.33x - 3.1$

42. $2.49x + 1.35 \geq 1.49x - 3.45$

Objective B

Solve and graph the solution set.

43. $3x < 12$

44. $8x \leq -24$

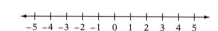

45. $15 \leq 5y$

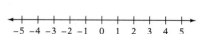

46. $-48 < 24x$

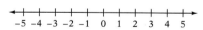

47. $16x \leq 16$

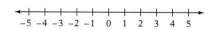

48. $3x > 0$

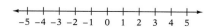

49. $-8x > 8$

50. $-2n \leq -8$

51. $-6b > 24$

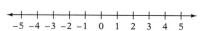

52. $-4x < 8$

Solve.

53. $-5y \geq 0$

54. $-3z < 0$

55. $7x > 2$

56. $6x \leq -1$

57. $2x \leq -5$

58. $\dfrac{5}{6}n < 15$

59. $\dfrac{3}{4}x < 12$

60. $\dfrac{2}{3}y \geq 4$

61. $10 \leq \dfrac{5}{8}x$

62. $4 \geq \dfrac{2}{3}x$

63. $-\dfrac{3}{7}x \leq 6$

64. $-\dfrac{2}{11}b \geq -6$

65. $-\dfrac{4}{7}x \geq -12$

66. $\dfrac{2}{3}n < \dfrac{1}{2}$

67. $-\dfrac{3}{5}x < 0$

68. $-\dfrac{2}{3}x \geq 0$

69. $-\dfrac{3}{8}x \geq \dfrac{9}{14}$

70. $-\dfrac{3}{5}x < -\dfrac{6}{7}$

71. $-\dfrac{4}{5}x < -\dfrac{8}{15}$

72. $-\dfrac{3}{4}y \geq -\dfrac{5}{8}$

73. $-\dfrac{8}{9}x \geq -\dfrac{16}{27}$

74. $1.5x \leq 6.30$

75. $2.3x \leq 5.29$

76. $-3.5d > 7.35$

77. $-0.24x > 0.768$

78. $4.25m > -34$

79. $-3.9x \geq -19.5$

80. $0.035x < -0.0735$

81. $0.07x < -0.378$

82. $-11.7x \leq 4.68$

Objective C   *Application Problems*

Solve.

83. The circumference of an official major league baseball is between 9.00 in. and 9.25 in. Find the possible diameters of a major league baseball to the nearest hundredth of an inch. Recall that $C = \pi d$ and use $\pi \approx 3.14$.

84. To be eligible for a basketball tournament, a basketball team must win at least 60% of its remaining games. If the team has 17 games remaining, how many games must the team win to qualify for the tournament?

85. To avoid a tax penalty, at least 90% of a self-employed person's total annual income tax liability must be paid by April 15. What amount of income tax must a person with an annual income tax liability of $3500 pay?

Solve.

86. Computer software engineers are fond of saying that software takes at least twice as long to develop as they think it will. Applying that saying, how many hours will it take to develop a software product that an engineer thinks can be finished in 50 hours?

87. A health official recommends a maximum cholesterol level of 220 units. How many units must a patient with a cholesterol level of 275 units reduce her cholesterol level to satisfy the recommended maximum level?

88. A government agency recommends a minimum daily allowance of vitamin C of 60 mg. How many additional milligrams of vitamin C does a person who drank a glass of orange juice with 10 mg of vitamin C need in order to satisfy the recommended daily allowance?

89. To pass a course with a B grade, a student must have an average of 80 points on five tests. The student's grades on the first four tests were 75, 83, 86, and 78. What scores can the student receive on the fifth test to earn a B grade?

90. A professor scores all tests with a maximum of 100 points. To earn an A grade in this course, a student must have an average of 92 on four tests. The student's grades on the first three tests were 89, 86, and 90. Can this student earn an A grade?

91. A car sales representative receives a commission that is the greater of $250 or 8% of the selling price of a car. What dollar amounts for the sale price of a car will make the commission offer more attractive?

92. A sales representative for a stereo store has the option of a monthly salary of $2000 or a 35% commission on the selling price of each item sold by the representative. What dollar amounts in sales will make the commission more attractive than the monthly salary?

APPLYING THE CONCEPTS

Given that $a > b$ and that a and b are real numbers, determine for which real numbers c the statement is true. Use set-builder notation to write the answer.

93. $ac > bc$ 94. $ac < bc$

95. $a + c > b + c$ 96. $a + c < b + c$

97. $\dfrac{a}{c} > \dfrac{b}{c}$ 98. $\dfrac{a}{c} < \dfrac{b}{c}$

99. In your own words, state the Addition Property of Inequalities.
[W]

100. In your own words, state the Multiplication Property of Inequalities.
[W]

10.3 General Inequalities

Objective A   To solve general inequalities

Solving an inequality frequently requires application of both the Addition and the Multiplication Properties of Inequalities.

➡ Solve: $4y - 3 \geq 6y + 5$

$$4y - 3 \geq 6y + 5$$

$4y - 6y - 3 \geq 6y - 6y + 5$ • Subtract 6y from each side of the inequality.

$-2y - 3 \geq 5$ • Simplify.

$-2y - 3 + 3 \geq 5 + 3$ • Add 3 to each side of the inequality.

$-2y \geq 8$ • Simplify.

$\dfrac{-2y}{-2} \leq \dfrac{8}{-2}$ • Divide each side of the inequality by −2. Because −2 is a negative number, the inequality symbol must be reversed.

$y \leq -4$

When an inequality contains parentheses, one of the steps in solving the inequality requires the use of the Distributive Property.

➡ Solve: $-2(x - 7) > 3 - 4(2x - 3)$

$$-2(x - 7) > 3 - 4(2x - 3)$$

$-2x + 14 > 3 - 8x + 12$ • Use the Distributive Property to remove parentheses.

$-2x + 14 > -8x + 15$ • Simplify.

$-2x + 8x + 14 > -8x + 8x + 15$ • Add 8x to each side of the inequality.

$6x + 14 > 15$ • Simplify.

$6x + 14 - 14 > 15 - 14$ • Subtract 14 from each side of the inequality.

$6x > 1$ • Simplify.

$\dfrac{6x}{6} > \dfrac{1}{6}$ • Divide each side of the inequality by 6.

$x > \dfrac{1}{6}$

Example 1   Solve: $7x - 3 \leq 3x + 17$

Solution
$$7x - 3 \leq 3x + 17$$
$$7x - 3x - 3 \leq 3x - 3x + 17$$
$$4x - 3 \leq 17$$
$$4x - 3 + 3 \leq 17 + 3$$
$$4x \leq 20$$
$$\frac{4x}{4} \leq \frac{20}{4}$$
$$x \leq 5$$

You Try It 1   Solve: $5 - 4x > 9 - 8x$

Your solution

Solution on p. A31

Example 2
Solve:
$3(3 - 2x) \geq -5x - 2(3 - x)$

You Try It 2
Solve:
$8 - 4(3x + 5) \leq 6(x - 8)$

Solution

$$3(3 - 2x) \geq -5x - 2(3 - x)$$
$$9 - 6x \geq -5x - 6 + 2x$$
$$9 - 6x \geq -3x - 6$$
$$9 - 6x + 3x \geq -3x + 3x - 6$$
$$9 - 3x \geq -6$$
$$9 - 9 - 3x \geq -6 - 9$$
$$-3x \geq -15$$
$$\frac{-3x}{-3} \leq \frac{-15}{-3}$$
$$x \leq 5$$

Your solution

Solution on p. A31

Objective B   To solve application problems ....

Example 3
A rectangle is 10 ft wide and $(2x + 4)$ ft long. Express as an integer the maximum length of the rectangle when the area is less than 200 ft². (The area of a rectangle is equal to its length times its width.)

You Try It 3
Company A rents cars for $8 a day and 10¢ for every mile driven. Company B rents cars for $10 a day and 8¢ per mile driven. You want to rent a car for one week. What is the maximum number of miles you can drive a Company A car if it is to cost you less than a Company B car?

Strategy
To find the maximum length:

- Replace the variables in the area formula by the given values and solve for x.

- Replace the variable in the expression $2x + 4$ with the value found for x.

Your strategy

Solution

| Length times width | is less than | 200 ft² |

$$10(2x + 4) < 200$$
$$20x + 40 < 200$$
$$20x + 40 - 40 < 200 - 40$$
$$20x < 160$$
$$\frac{20x}{20} < \frac{160}{20}$$
$$x < 8$$

The length is $(2x + 4)$ ft. Because $x < 8$, $2x + 4 < 2(8) + 4 = 20$. Therefore, the length is less than 20 ft. The maximum length is 19 ft.

Your solution

Solution on p. A31

10.3 Exercises

· ·

Objective A

Solve.

1. $4x - 8 < 2x$

2. $7x - 4 < 3x$

3. $2x - 8 > 4x$

4. $3y + 2 > 7y$

5. $8 - 3x \leq 5x$

6. $10 - 3x \leq 7x$

7. $3x + 2 > 5x - 8$

8. $2n - 9 \geq 5n + 4$

9. $5x - 2 < 3x - 2$

10. $8x - 9 > 3x - 9$

11. $0.1(180 + x) > x$

12. $x > 0.2(50 + x)$

13. $2(2y - 5) \leq 3(5 - 2y)$

14. $2(5x - 8) \leq 7(x - 3)$

15. $5(2 - x) > 3(2x - 5)$

16. $4(3d - 1) > 3(2 - 5d)$

17. $5(x - 2) > 9x - 3(2x - 4)$

18. $3x - 2(3x - 5) > 4(2x - 1)$

19. $4 - 3(3 - n) \leq 3(2 - 5n)$

20. $15 - 5(3 - 2x) \leq 4(x - 3)$

21. $2x - 3(x - 4) \geq 4 - 2(x - 7)$

22. $4 + 2(3 - 2y) \leq 4(3y - 5) - 6y$

Objective B   *Application Problems*

Solve.

23. The sales agent for a jewelry company is offered a flat monthly salary of $3200 or a salary of $1000 plus an 11% commission on the selling price of each item sold by the agent. If the agent chooses the $3200, what dollar amount does the agent expect to sell in one month?

24. A baseball player is offered an annual salary of $200,000 or a base salary of $100,000 plus a bonus of $1000 for each hit over 100 hits. How many hits must the baseball player make to earn more than $200,000?

Solve.

25. A computer bulletin board service offers service for a flat fee of $10 per month or $4 per month plus $.10 for each minute the service is used. How many minutes must a person use this service to exceed $10?

26. A site licensing fee for a computer program is $1500. Paying this fee allows the company to use the program at any computer terminal within the company. Alternatively, the company can choose to pay $200 for each individual computer it has. How many individual computers must a company have for the site license to be more economical for the company?

27. For a product to be labeled orange juice, a state agency requires that at least 80% of the drink be real orange juice. How many ounces of artificial flavors can be added to 32 ounces of real orange juice and it still be legal to label the drink orange juice?

28. Grade A hamburger cannot contain more than 20% fat. How much fat can a butcher mix with 300 lb of lean meat to meet the 20% requirement?

29. A shuttle service taking skiers to a ski area charges $8 per person each way. Four skiers are debating whether to take the shuttle bus or rent a car for $45 plus $.25 per mile. Assuming that the skiers will share the cost of the car and that they want the least expensive method of transportation, find how far away the ski area is if they should choose the shuttle service.

30. Company A rents a car for $25 per day and $.08 per mile. Company B rents a car for $15 per day and $.14 per mile. Find the maximum number of miles you can drive company B's car before the cost exceeds the cost of company A's car.

APPLYING THE CONCEPTS

31. Determine whether the statement is always true, sometimes true, or never true, given that a, b, and c are real numbers.
  a. If $a > b$, then $-a > -b$.
  b. If $a < b$, then $ac < bc$.
  c. If $a > b$, then $a + c > b + c$.
  d. If $a \neq 0$, $b \neq 0$, and $a > b$, then $\frac{1}{a} > \frac{1}{b}$.

Use the roster method to list the set of positive integers that are solutions of the inequality.

32. $7 - 2b \leq 15 - 5b$ **33.** $-6(2 - d) \geq (4d - 9)$

Use the roster method to list the set of integers that are common to the solution sets of the two inequalities.

34. $5x - 12 \leq x + 8$ **35.** $3(x + 2) > 9x - 2$
  $3x - 4 \geq 2 + x$ $4(x + 5) > 3(x + 6)$

36. Determine the solution set of $2 - 3(x + 4) < 5 - 3x$.

37. Determine the solution set of $3x + 2(x - 1) > 5(x + 1)$.

10.4 Graphing Linear Inequalities

Objective A   *To graph an inequality in two variables* ....

The graph of the linear equation $y = x - 2$ separates a plane into three sets:

the set of points on the line
the set of points above the line
the set of points below the line

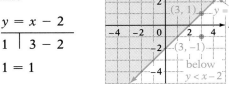

The point $(3, 1)$ is a solution of $y = x - 2$.

$$y = x - 2$$
$$\frac{1 \mid 3 - 2}{1 = 1}$$

The point $(3, 3)$ is a solution of $y > x - 2$.

$$y > x - 2$$
$$\frac{3 \mid 3 - 2}{3 > 1}$$

Any point above the line is a solution of $y > x - 2$.

The point $(3, -1)$ is a solution of $y < x - 2$.

$$y < x - 2$$
$$\frac{-1 \mid 3 - 2}{-1 < 1}$$

Any point below the line is a solution of $y < x - 2$.

The solution set of $y = x - 2$ is all points on the line. The solution set of $y > x - 2$ is all points above the line. The solution set of $y < x - 2$ is all points below the line. The solution set of an inequality in two variables is a **half-plane.**

The following illustrates the procedure for graphing a linear inequality.

➡ Graph the solution set of $2x + 3y \leq 6$.

Solve the inequality for y.

$$2x + 3y \leq 6$$
$$2x - 2x + 3y \leq -2x + 6 \qquad \bullet \text{ Substract } x \text{ from each side.}$$
$$3y \leq -2x + 6 \qquad \bullet \text{ Simplify}$$
$$\frac{3y}{3} \leq \frac{-2x + 6}{3} \qquad \bullet \text{ Divide each side by 3.}$$
$$y \leq -\frac{2}{3}x + 2 \qquad \bullet \text{ Simplify}$$

➡ Change the inequality to an equality and graph $y = -\frac{2}{3}x + 2$. If the inequality is $\geq$ **or** $\leq$, the line is in the solution set and is shown by a **solid line.** If the inequality is $>$ **or** $<$, the line is not a part of the solution set and is shown by a **dotted line.**

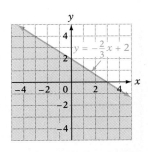

If the inequality is $>$ **or** $\geq$, shade the **upper half-plane.** If the inequality is $<$ **or** $\leq$, shade the **lower half-plane.**

Example 1
Graph the solution set of $3x + y > -2$.

Solution
$$3x + y > -2$$
$$3x - 3x + y > -3x - 2$$
$$y > -3x - 2$$

Graph $y = -3x - 2$ as a dotted line.
Shade the upper half-plane.

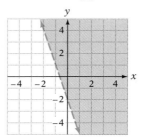

Example 2
Graph the solution set of $2x - y \geq 2$.

Solution
$$2x - y \geq 2$$
$$2x - 2x - y \geq -2x + 2$$
$$-y \geq -2x + 2$$
$$-1(-y) \leq -1(-2x + 2)$$
$$y \leq 2x - 2$$

Graph $y = 2x - 2$ as a solid line. Shade
the lower half-plane.

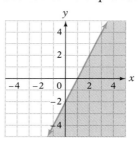

Example 3
Graph the solution set of $y > 3$.

Solution
Graph $y = 3$ as a dotted line.
Shade the upper half-plane.

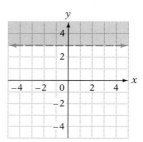

You Try It 1
Graph the solution set of $x - 3y < 2$.

Your solution

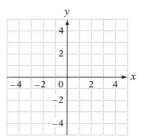

You Try It 2
Graph the solution set of $2x - 4y \leq 8$.

Your solution

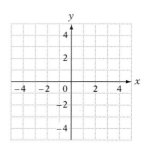

You Try It 3
Graph the solution set of $x < 3$.

Your solution

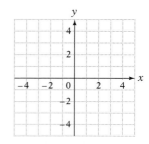

Solutions on p. A32

10.4 Exercises

Objective A

Graph the solution set.

1. $x + y > 4$

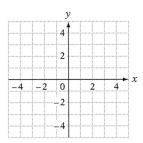

2. $x - y > -3$

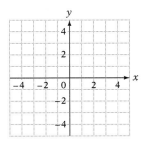

3. $2x - y < -3$

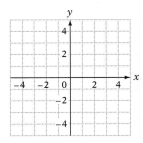

4. $3x - y < 9$

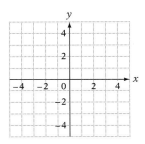

5. $2x + y \geq 4$

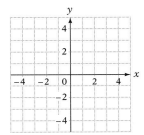

6. $3x + y \geq 6$

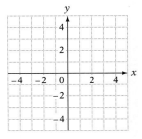

7. $y \leq -2$

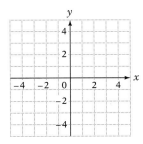

8. $y > 3$

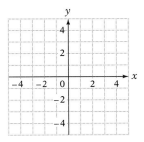

9. $3x - 2y < 8$

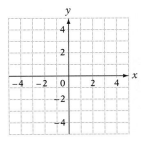

10. $5x + 4y > 4$

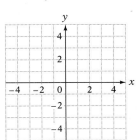

11. $-3x - 4y \geq 4$

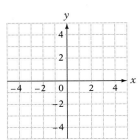

12. $-5x - 2y \geq 8$

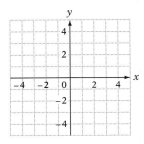

Graph the solution set.

13. $6x + 5y \leq -10$

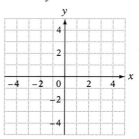

14. $2x + 2y \leq -4$

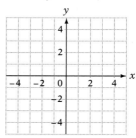

15. $-4x + 3y < -12$

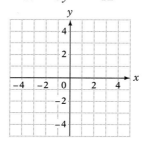

16. $-4x + 5y < 15$

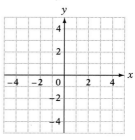

17. $-2x + 3y \leq 6$

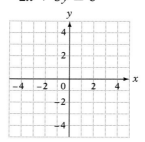

18. $3x - 4y > 12$

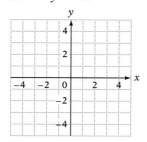

APPLYING THE CONCEPTS

Graph the solution set.

19. $\dfrac{x}{4} + \dfrac{y}{2} > 1$

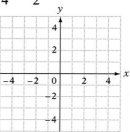

20. $2x - 3(y + 1) > y - (4 - x)$

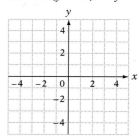

Write the inequality given its graph.

21.

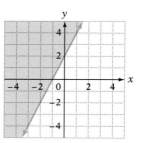

22.

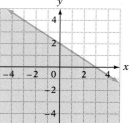

23.

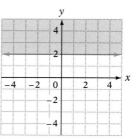

24.

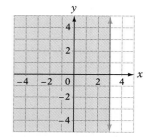

Project in Mathematics

. . . .

**Measurements as
Approximations**

From arithmetic, you know the rules for rounding decimals.

If the digit to the right of the given place value is less than 5, drop that digit and all digits to the right.

6.31 rounded to the nearest tenth is 6.3

If the digit to the right of the given place value is greater than or equal to 5, increase the given place value by one and drop all digits to its right.

6.28 rounded to the nearest tenth is 6.3

Given the rules for rounding numbers, what range of values can the number 6.3 represent? The smallest possible value of 6.3 is 6.25; any number smaller than that would not have been rounded up to 6.3. What about the largest possible value of 6.3? The number 6.34 would be rounded down to 6.3. So would the numbers 6.349, 6.3499, 6.349999, and so on. Therefore, we cannot name the largest possible value of 6.3. We can say that the number must be less than 6.35. Any number less than 6.35 would be rounded down to 6.3. The exact value of 6.3 is greater than or equal to 6.25 and less than 6.35.

The dimensions of a rectangle are given as 4.3 cm by 3.2 cm. Using the smallest and the largest possible values of the length and of the width, we can represent the possible values of the area, A, of the rectangle as follows:

$$4.25(3.15) \leq A < 4.35(3.25)$$
$$13.3875 \quad \leq A < 14.1375$$

The area is greater than or equal to 13.3875 cm^2 and less than 14.1375 cm^2.

Exercises:

Solve.

1. The measurements of the three sides of a triangle are given as 8.37 m, 5.42 m, and 9.61 m. Find the possible lengths of the perimeter of the triangle.

2. The length of a side of a square is given as 4.7 cm. Find the possible values for the area of the square.

3. What are the possible values for the area of a rectangle whose dimensions are 6.5 cm by 7.8 cm?

4. The length of a box is 40 cm, the width is 25 cm, and the height is 8 cm. What are the possible values of the volume of the box?

Chapter Summary

Key Words   A *set* is a collection of objects. The objects of a set are called the *elements* of the set.

The *roster method* of writing a set encloses a list of the elements in braces.

The *empty set* or *null set*, written $\varnothing$ or { }, is the set that contains no elements.

The *union* of two sets, written $A \cup B$, is the set that contains all the elements of A and all the elements of B (any elements that are in both set A and set B are listed only once).

The *intersection* of two sets, written $A \cap B$, is the set that contains the elements that are common to both A and B.

An *inequality* is an expression that contains the symbol $<$, $>$, $\leq$, or $\geq$.

The *solution set of an inequality* is a set of numbers each element of which, when substituted for the variable, results in a true inequality. The solution set of an inequality can be graphed on the number line.

The solution set of an inequality in two variables is a *half-plane*.

Essential Rules   *Addition Property of Inequalities*

The same term can be added to each side of an inequality without changing the solution set of the inequality.

$$\text{If } a > b, \text{ then } a + c > b + c.$$
$$\text{If } a < b, \text{ then } a + c < b + c.$$

The Addition Property of Inequalities also holds true for an inequality containing the symbol $\geq$ or $\leq$.

Multiplication Property of Inequalities

Each side of an inequality can be multiplied by the same **positive number** without changing the solution set of the inequality.

$$\text{If } a > b \text{ and } c > 0, \text{ then } ac > bc.$$
$$\text{If } a < b \text{ and } c > 0, \text{ then } ac < bc.$$

If each side of an inequality is multiplied by the same **negative number** and the inequality symbol is reversed, then the solution set of the inequality is not changed.

$$\text{If } a > b \text{ and } c < 0, \text{ then } ac < bc.$$
$$\text{If } a < b \text{ and } c < 0, \text{ then } ac > bc.$$

The Multiplication Property of Inequalities also holds true for an inequality containing the symbol $\geq$ or $\leq$.

Chapter Review

. . . .

SECTION 10.1

1. Use the roster method to write the set of odd positive integers less than 8.

2. Find $A \cup B$, given $A = \{6, 8, 10\}$ and $B = \{2, 4, 6\}$.

3. Find $A \cap B$, given $A = \{0, 2, 4, 6, 8\}$ and $B = \{-2, -4\}$.

4. Find $A \cap B$, given $A = \{1, 5, 9, 13\}$ and $B = \{1, 3, 5, 7, 9\}$.

5. Use set-builder notation to write the set of odd integers greater than -8.

6. Use set-builder notation to write the set of real numbers greater than 3.

7. Graph: $x > 3$

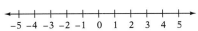

8. Graph: $(x > -1) \cap (x \le 2)$

9. Graph: $(x < 2) \cup (x > 5)$

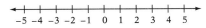

SECTION 10.2

10. Solve and graph the solution set of $x - 3 > -1$.

11. Solve: $2x - 3 > x + 15$

12. Solve: $-15x \le 45$

13. Solve: $-\dfrac{3}{4}x > \dfrac{2}{3}$

14. Six less than a number is greater than twenty-five. Find the smallest integer that will satisfy the inequality.

15. A student's grades on five sociology tests were 68, 82, 90, 73, 95. What is the lowest score the student can receive on the next test and still be able to attain a minimum of 480 points?

SECTION 10.3

16. Solve: $3x + 4 \geq -8$

17. Solve: $12 - 4(x - 1) \leq 5(x - 4)$

18. Solve: $7x - 2(x + 3) \geq x + 10$

19. Solve: $6x - 9 < 4x + 3(x + 3)$

20. Solve: $5 - 4(x + 9) > 11(12x - 9)$

21. The width of a rectangular garden is 12 ft. The length of the garden is $(3x + 5)$ ft. Express as an integer the minimum length of the garden when the area is greater than 276 ft². (The area of a rectangle is equal to its length times its width.)

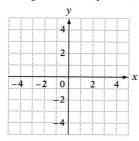

12 ft

$(3x + 5)$ ft

22. Florist A charges a $3 delivery fee plus $21 per bouquet delivered. Florist B charges a $15 delivery fee plus $18 per bouquet delivered. A church wants to supply each resident of a small nursing home with a bouquet for Grandparent's Day. Find the number of residents of the nursing home if Florist B is more economical than Florist A.

SECTION 10.4

23. Graph: $3x + 2y \leq 12$

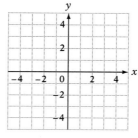

24. Graph: $2x - 3y < 9$

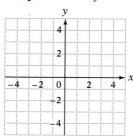

25. Graph: $5x + 2y < 6$

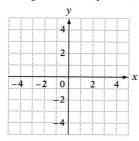

Chapter Test

1. Use the roster method to write the set of the even positive integers between 3 and 9.

2. Find $A \cap B$, given $A = \{6, 8, 10, 12\}$ and $B = \{12, 14, 16\}$.

3. Use set-builder notation to write the set of the positive integers less than 50.

4. Use set-builder notation to write the set of the real numbers greater than -23.

5. Graph: $x > -2$

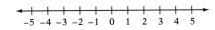
$$-5 \ -4 \ -3 \ -2 \ -1 \ 0 \ 1 \ 2 \ 3 \ 4 \ 5$$

6. Graph: $(x < 5) \cap (x > 0)$

$$-5 \ -4 \ -3 \ -2 \ -1 \ 0 \ 1 \ 2 \ 3 \ 4 \ 5$$

7. Solve and graph the solution set of $4 + x < 1$.

$$-5 \ -4 \ -3 \ -2 \ -1 \ 0 \ 1 \ 2 \ 3 \ 4 \ 5$$

8. Solve: $x + \dfrac{1}{2} > \dfrac{5}{8}$

9. Solve and graph the solution set of $\dfrac{2}{3}x \geq 2$.

$$-5 \ -4 \ -3 \ -2 \ -1 \ 0 \ 1 \ 2 \ 3 \ 4 \ 5$$

10. Solve: $-\dfrac{3}{8}x \leq 5$

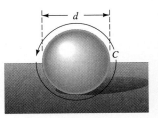

48 in.

43 in.

11. To ride a certain roller coaster at an amusement park, a person must be at least 48 inches tall. How many inches must a child who is 43 inches tall grow to be eligible for the roller coaster?

12. A ball bearing for a rotary engine must have a circumference between 0.1220 inches and 0.1240 inches. What are the allowable diameters for the bearings to the nearest ten-thousandths of an inch? Recall that $c = \pi d$ and use $\pi \approx 3.14$.

13. Solve: $5 - 3x > 8$

14. Solve: $2x - 7 \leq 6x + 9$

15. Solve: $3(2x - 5) \geq 8x - 9$

16. Solve: $6x - 3(2 - 3x) < 4(2x - 7)$

17. A rectangle is 15 ft long and $(2x - 4)$ ft wide. Express, using an integer, how wide the rectangle can be for the area to be less than 180 ft². (The area of a rectangle is equal to its length times its width.)

$(2x - 4)$ ft

15 ft

18. A stock broker receives a monthly salary that is the greater of $2500 or $1000 plus 2% of the total value of all stock transactions the broker processes during the month. What dollar amounts of transactions did the broker process in a month for which the broker's salary was $2500?

19. Graph the solution set of $3x + y > 4$.

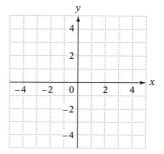

20. Graph the solution set of $4x - 5y \geq 15$.

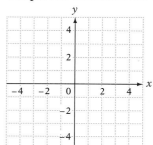

Cumulative Review

1. Simplify: $2[5a - 3(2 - 5a) - 8]$

2. Solve: $\dfrac{5}{8} - 4x = \dfrac{1}{8}$

3. Solve: $2x - 3[x - 2(x - 3)] = 2$

4. Simplify: $(-3a)(-2a^3b^2)^2$

5. Simplify: $\dfrac{27a^3b^2}{(-3ab^2)^3}$

6. Simplify: $(16x^2 - 12x - 2) \div (4x - 1)$

7. Given $f(x) = x^2 - 4x - 5$, find $f(-1)$.

8. Factor $27a^2x^2 - 3a^2$.

9. Simplify: $\dfrac{x^2 - 2x}{x^2 - 2x - 8} \div \dfrac{x^3 - 5x^2 + 6x}{x^2 - 7x + 12}$

10. Simplify: $\dfrac{4a}{2a - 3} - \dfrac{2a}{a + 3}$

11. Solve: $\dfrac{5y}{6} - \dfrac{5}{9} = \dfrac{y}{3} - \dfrac{5}{6}$

12. Solve $R = \dfrac{C - S}{t}$ for C.

13. Find the slope of the line that passes through the points $(2, -3)$ and $(-1, 4)$.

14. Find the equation of the line that passes through the points $(1, -3)$ and has slope $-\dfrac{3}{2}$.

15. Solve by substitution.
$$x = 3y + 1$$
$$2x + 5y = 13$$

16. Solve by the addition method.
$$9x - 2y = 17$$
$$5x + 3y = -7$$

17. Find $A \cup B$, given $A = \{0, 1, 2\}$ and $B = \{-2, -10\}$.

18. Use set-builder notation to write the set of the real numbers less than 48.

19. Graph: $(x > 1) \cup (x < -1)$

20. Graph the solution set of $\frac{3}{8}x > -\frac{3}{4}$.

−5 −4 −3 −2 −1 0 1 2 3 4 5

21. Solve: $-\frac{4}{5}x > 12$

22. Solve: $15 - 3(5x - 7) < 2(7 - 2x)$

23. Three-fifths of a number is less than negative fifteen. What integers N satisfy this inequality?

24. Company A rents cars for $6 a day and 25¢ for every mile driven. Company B rents cars for $15 a day and 10¢ per mile. You want to rent a car for 6 days. What is the maximum number of miles you can drive a Company A car if it is to cost you less than a Company B car?

25. In a lake, 100 fish are caught, tagged, and then released. Later 150 fish are caught. Three of the 150 fish are found to have tags. Estimate the number of fish in the lake.

26. The first angle of a triangle is 30 degrees more than a second angle. The third angle is 10 degrees more than twice the second angle. Find the measure of each angle.

27. Graph: $y = 2x - 1$

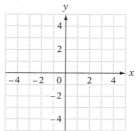

28. Graph the solution set of $6x - 3y \geq 6$.

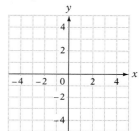

11

Radical Expressions

Objectives

Section 11.1
To simplify numerical radical expressions
To simplify variable radical expressions

Section 11.2
To add and subtract radical expressions

Section 11.3
To multiply radical expressions
To divide radical expressions

Section 11.4
To solve an equation containing a radical
   expression
To solve application problems

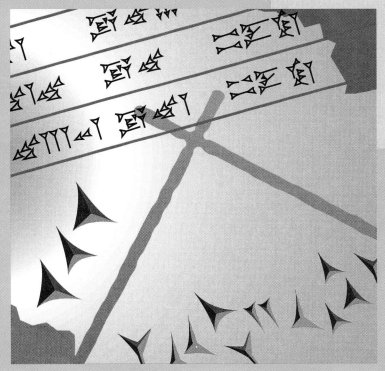

A Table of Square Roots

The practice of finding the square root of a number has existed for at least two thousand years. Because the process of finding a square root is tedious and time-consuming, it is convenient to have tables of square roots. There is one such table in the back of this book.

But this table is not the first square root table ever written (nor is it likely to be the last). The table shown below is part of an old Babylonian clay tablet that was written around 350 B.C. It is an incomplete table of square roots written in a style called *cuneiform*.

The number base of the Babylonians was 60 instead of 10, as we use today. The symbol $\triangledown$ was used for 1, and 10 was written as $\triangleleft$. Some examples of numbers written in this system are given below.

$$= 9 \qquad\qquad = 40$$

Translations of the first couple of lines of the table are given below. The number given in parentheses is the equivalent base-10 number that would be used today. You might try to translate the third line. The answer is given at the bottom of this page.

$40 \times 60 + 1$ (= 2401) which is the square of 49

$41 \times 60 + 40$ (= 2500) which is the square of 50

Answer: $45 \times 60 + 21$ (= 2601), which is the square of 51

Introduction to Radical Expressions

Objective A   *To simplify numerical radical expressions* ....

A **square root** of a positive number x is a number whose square is x.

A square root of 16 is 4 because $4^2 = 16$.
A square root of 16 is -4 because $(-4)^2 = 16$.

Every positive number has two square roots, one a positive and one a negative number. The symbol "$\sqrt{}$," called a **radical**, is used to indicate the positive or **principal square root** of a number. For example, $\sqrt{16} = 4$ and $\sqrt{25} = 5$. The number under the radical sign is called the **radicand**.

When the negative square root of a number is to be found, a negative sign is placed in front of the radical. For example, $-\sqrt{16} = -4$ and $-\sqrt{25} = -5$.

The square of an integer is a **perfect square.** 49, 81, and 144 are examples of perfect squares.

$7^2 = 49$
$9^2 = 81$
$12^2 = 144$

An integer that is a perfect square can be written as the product of prime factors, each of which has an even exponent when expressed in exponential form.

$49 = 7 \cdot 7 = 7^2$
$81 = 3 \cdot 3 \cdot 3 \cdot 3 = 3^4$
$144 = 2 \cdot 2 \cdot 2 \cdot 2 \cdot 3 \cdot 3 = 2^4 3^2$

➡ Simplify $\sqrt{625}$.

$\sqrt{625} = \sqrt{5^4}$ • Write the prime factorization of the radicand in exponential form.

$= 5^2$ • Remove the radical sign and multiply the exponent by $\dfrac{1}{2}$.

$= 25$ • Simplify.

If a number is not a perfect square, its square root can only be approximated. For example, 2 and 7 are not perfect squares. The square roots of these numbers are **irrational numbers.** Their decimal representations never terminate or repeat.

$$\sqrt{2} \approx 1.4142135\ldots \qquad \sqrt{7} \approx 2.6457513\ldots$$

The approximate square roots of the positive integers up to 200 can be found in the Appendix. The square roots have been rounded to the nearest thousandth.

A radical expression is in simplest form when the radicand contains no factor greater than 1 that is a perfect square. The Product Property of Square Roots is used to simplify radical expressions.

> **The Product Property of Square Roots**
>
> If a and b are positive real numbers, then $\sqrt{ab} = \sqrt{a} \cdot \sqrt{b}$

➡ Simplify: $\sqrt{96}$

$\sqrt{96} = \sqrt{2^5 \cdot 3}$ • Write the prime factorization of the radicand in simplest form.

$= \sqrt{2^4(2 \cdot 3)}$ • Write the radicand as a product of a perfect square and factors that do not contain perfect-square factors.

$= \sqrt{2^4} \sqrt{2 \cdot 3}$ • Use the Product Property of Square Roots.

$= 2^2\sqrt{6}$ • Simplify.

$= 4\sqrt{6}$

➡ Simplify: $\sqrt{360}$

$\sqrt{360} = \sqrt{2^3 \cdot 3^2 \cdot 5}$ • Write the prime factorization of the radicand in simplest form.

$= \sqrt{(2^2 \cdot 3^2)(2 \cdot 5)}$ • Write the radicand as a product of a perfect square and factors that do not contain perfect-square factors.

$= \sqrt{2^2 \cdot 3^2} \sqrt{2 \cdot 5}$ • Use the Product Property of Square Roots.

$= (2 \cdot 3)\sqrt{10}$ • Simplify.

$= 6\sqrt{10}$

From the last example, note that $\sqrt{360} = 6\sqrt{10}$. The two expressions are different representations of the same number. Using a calculator, we find that $\sqrt{360} \approx 18.973666$ and $6\sqrt{10} \approx 6(3.1622777) = 18.973666$.

➡ Simplify: $\sqrt{-16}$

Because the square of any real number is positive, there is no real number whose square is -16. $\sqrt{-16}$ is not a real number.

Example 1   Simplify: $3\sqrt{90}$

Solution $3\sqrt{90} = 3\sqrt{2 \cdot 3^2 \cdot 5}$
$= 3\sqrt{3^2(2 \cdot 5)}$
$= 3\sqrt{3^2} \sqrt{2 \cdot 5}$
$= 3 \cdot 3\sqrt{10} = 9\sqrt{10}$

You Try It 1   Simplify: $-5\sqrt{32}$

Your solution

Example 2   Simplify: $\sqrt{252}$

Solution $\sqrt{252} = \sqrt{2^2 \cdot 3^2 \cdot 7}$
$= \sqrt{2^2 \cdot 3^2} \sqrt{7}$
$= 2 \cdot 3\sqrt{7}$
$= 6\sqrt{7}$

You Try It 2   Simplify: $\sqrt{216}$

Your solution

Solutions on p. A33

Objective B   _To simplify variable radical expressions_ ....

Variable expressions that contain radicals do not always represent real numbers. For example, if $a = -4$, then

$$\sqrt{a^3} = \sqrt{(-4)^3} = \sqrt{-64}$$

and $\sqrt{-64}$ is not a real number.

Now consider the expression $\sqrt{x^2}$. Evaluate this expression for $x = -2$ and $x = 2$.

$$\sqrt{x^2}$$ $$\sqrt{x^2}$$
$$\sqrt{(-2)^2} = \sqrt{4} = 2 = |-2|$$ $$\sqrt{2^2} = \sqrt{4} = 2 = |2|$$

This suggests the following:

For any real number a, $\sqrt{a^2} = |a|$. If $a \geq 0$, then $\sqrt{a^2} = a$.

In order to avoid variable expressions that do not represent real numbers, and so that absolute-value signs are not needed for certain expressions, the variables in this chapter will represent _positive_ numbers unless otherwise stated.

A variable or a product of variables written in exponential form is a perfect square when each exponent is an even number.

To find the square root of a perfect square, remove the radical sign and multiply each exponent by $\frac{1}{2}$.

➡ Simplify: $\sqrt{a^6}$

$$\sqrt{a^6} = a^3$$ • Remove the radical sign and multiply the exponent by $\frac{1}{2}$.

A variable radical expression is in simplest form when the radicand contains no factor greater than 1 that is a perfect square.

➡ Simplify: $\sqrt{x^7}$

$$\sqrt{x^7} = \sqrt{x^6 \cdot x}$$ • Write x^7 as the product of a perfect square and x.
$$= \sqrt{x^6}\sqrt{x}$$ • Use the Product Property of Square Roots.
$$= x^3\sqrt{x}$$ • Simplify the perfect square.

➡ Simplify: $3x\sqrt{8x^3y^{13}}$

$$3x\sqrt{8x^3y^{13}} = 3x\sqrt{2^3x^3y^{13}}$$ • Write the prime factorization of the coefficient of the radicand in exponential form.

$$= 3x\sqrt{2^2x^2y^{12}(2xy)}$$ • Write the radicand as a product of a perfect square and factors that do not contain a perfect square.

$$= 3x\sqrt{2^2x^2y^{12}}\ \sqrt{2xy}$$ • Use the Product Property of Square Roots.
$$= 3x \cdot 2xy^6\sqrt{2xy}$$ • Simplify.
$$= 6x^2y^6\sqrt{2xy}$$

➡ Simplify: $\sqrt{25(x+2)^2}$

$$\sqrt{25(x+2)^2} = \sqrt{5^2(x+2)^2}$$
$$= 5(x+2)$$
$$= 5x + 10$$

● **Write the prime factorization of the coefficient in exponential form.**

Example 3
Simplify: $\sqrt{b^{15}}$

Solution
$\sqrt{b^{15}} = \sqrt{b^{14} \cdot b} = \sqrt{b^{14}} \cdot \sqrt{b} = b^7\sqrt{b}$

You Try It 3
Simplify: $\sqrt{y^{19}}$

Your solution

Example 4
Simplify: $\sqrt{24x^5}$

Solution
$$\sqrt{24x^5} = \sqrt{2^3 \cdot 3 \cdot x^5} = \sqrt{2^2 x^4(2 \cdot 3x)}$$
$$= \sqrt{2^2 \cdot x^4}\sqrt{2 \cdot 3x}$$
$$= 2x^2\sqrt{6x}$$

You Try It 4
Simplify: $\sqrt{45b^7}$

Your solution

Example 5
Simplify: $2a\sqrt{18a^3b^{10}}$

Solution
$$2a\sqrt{18a^3b^{10}} = 2a\sqrt{2 \cdot 3^2 \cdot a^3b^{10}}$$
$$= 2a\sqrt{3^2 a^2 b^{10}(2a)}$$
$$= 2a\sqrt{3^2 a^2 b^{10}}\sqrt{2a}$$
$$= 2a \cdot 3ab^5\sqrt{2a}$$
$$= 6a^2 b^5\sqrt{2a}$$

You Try It 5
Simplify: $3a\sqrt{28a^9b^{18}}$

Your solution

Example 6
Simplify: $\sqrt{16(x+5)^2}$

Solution
$$\sqrt{16(x+5)^2} = \sqrt{2^4(x+5)^2} = 2^2(x+5)$$
$$= 4(x+5) = 4x + 20$$

You Try It 6
Simplify: $\sqrt{25(a+3)^2}$

Your solution

Example 7
Simplify: $\sqrt{x^2 + 10x + 25}$

Solution
$\sqrt{x^2 + 10x + 25} = \sqrt{(x+5)^2} = x + 5$

You Try It 7
Simplify: $\sqrt{x^2 + 14x + 49}$

Your solution

Solutions on p. A33

11.1 Exercises

Objective A

Simplify.

1. $\sqrt{16}$

2. $\sqrt{64}$

3. $\sqrt{49}$

4. $\sqrt{144}$

5. $\sqrt{32}$

6. $\sqrt{50}$

7. $\sqrt{8}$

8. $\sqrt{12}$

9. $6\sqrt{18}$

10. $-3\sqrt{48}$

11. $5\sqrt{40}$

12. $2\sqrt{28}$

13. $\sqrt{15}$

14. $\sqrt{21}$

15. $\sqrt{29}$

16. $\sqrt{13}$

17. $-9\sqrt{72}$

18. $11\sqrt{80}$

19. $\sqrt{45}$

20. $\sqrt{225}$

21. $\sqrt{0}$

22. $\sqrt{210}$

23. $6\sqrt{128}$

24. $9\sqrt{288}$

Find the decimal approximation rounded to the nearest thousandth.

25. $\sqrt{240}$

26. $\sqrt{300}$

27. $\sqrt{288}$

28. $\sqrt{600}$

29. $\sqrt{256}$

30. $\sqrt{324}$

31. $\sqrt{275}$

32. $\sqrt{450}$

33. $\sqrt{245}$

34. $\sqrt{525}$

35. $\sqrt{352}$

36. $\sqrt{363}$

Objective B

Simplify.

37. $\sqrt{x^6}$

38. $\sqrt{x^{12}}$

39. $\sqrt{y^{15}}$

40. $\sqrt{y^{11}}$

41. $\sqrt{a^{20}}$

42. $\sqrt{a^{16}}$

43. $\sqrt{x^4 y^4}$

44. $\sqrt{x^{12} y^8}$

45. $\sqrt{4x^4}$

46. $\sqrt{25y^8}$

47. $\sqrt{24x^2}$

48. $\sqrt{x^3 y^{15}}$

49. $\sqrt{60x^5}$

50. $\sqrt{72y^7}$

51. $\sqrt{49a^4 b^8}$

52. $\sqrt{144x^2 y^8}$

53. $\sqrt{18x^5 y^7}$

54. $\sqrt{32a^5 b^{15}}$

55. $\sqrt{40x^{11} y^7}$

56. $\sqrt{72x^9 y^3}$

57. $\sqrt{80a^9 b^{10}}$

58. $\sqrt{96a^5 b^7}$

59. $2\sqrt{16a^2 b^3}$

60. $5\sqrt{25a^4 b^7}$

Simplify.

61. $x\sqrt{x^4y^2}$

62. $y\sqrt{x^3y^6}$

63. $4\sqrt{20a^4b^7}$

64. $5\sqrt{12a^3b^4}$

65. $3x\sqrt{12x^2y^7}$

66. $4y\sqrt{18x^5y^4}$

67. $2x^2\sqrt{8x^2y^3}$

68. $3y^2\sqrt{27x^4y^3}$

69. $\sqrt{25(a+4)^2}$

70. $\sqrt{81(x+y)^4}$

71. $\sqrt{4(x+2)^4}$

72. $\sqrt{9(x+2)^8}$

73. $\sqrt{x^2+4x+4}$

74. $\sqrt{b^2+8b+16}$

75. $\sqrt{y^2+2y+1}$

76. $\sqrt{a^2+6a+9}$

APPLYING THE CONCEPTS

77. If a and b are positive real numbers, does $\sqrt{a+b} = \sqrt{a} + \sqrt{b}$? If not, give an example when the expressions are not equal.

78. **a.** Find the two-digit perfect square that has exactly nine factors.
 b. Find two whole numbers such that their difference is 10, the smaller number is a perfect square, and the larger number is two less than a perfect square.

79. Describe in your own words how to simplify a radical expression.
[W]

80. Explain why $2\sqrt{2}$ is in simplest form and $\sqrt{8}$ is not in simplest
[W] form.

81. Use the roster method to list the whole numbers between $\sqrt{8}$ and $\sqrt{90}$.

82. Simplify. Assume that no radicand is negative.
 a. $\sqrt{x^2y^3 + x^3y^2}$ **b.** $\sqrt{4a^5b^4 - 4a^4b^5}$ **c.** $\sqrt{(x^2-y^2)(x-y)}$

83. You are to grade the following solution to the problem "Write $\sqrt{72}$ in simplest form."

$$\sqrt{72} = \sqrt{4}\,\sqrt{18}$$
$$= 2\sqrt{18}$$

Is the solution correct? If not, what error was made? What is the correct solution?

84. Simplify.
 a. $\sqrt{\sqrt{16}}$ **b.** $\sqrt{\sqrt{81}}$

85. Use a calculator to approximate
 a. $\sqrt{\sqrt{77}}$ **b.** $\sqrt{\sqrt{17}}$

86. Given $f(x) = \sqrt{2x-1}$, find each of the following. Write your answer in simplest form.
 a. $f(1)$ **b.** $f(5)$ **c.** $f(14)$

11.2 Addition and Subtraction of Radical Expressions

· ·

Objective A   *To add and subtract radical expressions* ·

The Distributive Property is used to simplify the sum or difference of radical expressions with like radicands.

$5\sqrt{2} + 3\sqrt{2} = (5 + 3)\sqrt{2} = 8\sqrt{2}$
$6\sqrt{2x} - 4\sqrt{2x} = (6 - 4)\sqrt{2x} = 2\sqrt{2x}$

Radical expressions that are in simplest form and have unlike radicands cannot be simplified by the Distributive Property.

$2\sqrt{3} + 4\sqrt{2}$ cannot be simplified by the Distributive Property.

➡ Simplify: $4\sqrt{8} - 10\sqrt{2}$

$$\begin{aligned}
4\sqrt{8} - 10\sqrt{2} &= 4\sqrt{2^3} - 10\sqrt{2} \\
&= 4\sqrt{2^2 \cdot 2} - 10\sqrt{2} \\
&= 4\sqrt{2^2}\,\sqrt{2} - 10\sqrt{2} \\
&= 4 \cdot 2\sqrt{2} - 10\sqrt{2} \\
&= 8\sqrt{2} - 10\sqrt{2} \\
&= (8 - 10)\sqrt{2} \\
&= -2\sqrt{2}
\end{aligned}$$

- Simplify each term.
- Write prime factorization of coefficients of radicands.
- Use the Product Property of Radicals.
- Simplify the expression by using the Distributive Property. This step is usually done mentally.

➡ Simplify: $8\sqrt{18x} - 2\sqrt{32x}$

$$\begin{aligned}
8\sqrt{18x} - 2\sqrt{32x} &= 8\sqrt{2 \cdot 3^2 x} - 2\sqrt{2^5 x} \\
&= 8\sqrt{3^2 \cdot 2x} - 2\sqrt{2^4 \cdot 2x} \\
&= 8\sqrt{3^2}\,\sqrt{2x} - 2\sqrt{2^4}\,\sqrt{2x} \\
&= 8 \cdot 3\sqrt{2x} - 2 \cdot 2^2\sqrt{2x} \\
&= 24\sqrt{2x} - 8\sqrt{2x} \\
&= (24 - 8)\sqrt{2x} \\
&= 16\sqrt{2x}
\end{aligned}$$

- Simplify each term.
- Write prime factorization of coefficients of radicands.
- Use the Product Property of Radicals.
- Simplify the expression by using the Distributive Property. Do this step mentally.

Example 1
Simplify: $5\sqrt{2} - 3\sqrt{2} + 12\sqrt{2}$

Solution
$5\sqrt{2} - 3\sqrt{2} + 12\sqrt{2} = 14\sqrt{2}$

You Try It 1
Simplify: $9\sqrt{3} + 3\sqrt{3} - 18\sqrt{3}$

Your solution

Example 2
Simplify: $3\sqrt{12} - 5\sqrt{27}$

Solution
$$3\sqrt{12} - 5\sqrt{27} = 3\sqrt{2^2 \cdot 3} - 5\sqrt{3^3}$$
$$= 3\sqrt{2^2}\sqrt{3} - 5\sqrt{3^2}\sqrt{3}$$
$$= 3 \cdot 2\sqrt{3} - 5 \cdot 3\sqrt{3}$$
$$= 6\sqrt{3} - 15\sqrt{3}$$
$$= -9\sqrt{3}$$

You Try It 2
Simplify: $2\sqrt{50} - 5\sqrt{32}$

Your solution

Example 3
Simplify: $3\sqrt{12x^3} - 2x\sqrt{3x}$

Solution
$$3\sqrt{12x^3} - 2x\sqrt{3x}$$
$$= 3\sqrt{2^2 \cdot 3 \cdot x^3} - 2x\sqrt{3x}$$
$$= 3\sqrt{2^2 \cdot x^2}\sqrt{3x} - 2x\sqrt{3x}$$
$$= 3 \cdot 2 \cdot x\sqrt{3x} - 2x\sqrt{3x}$$
$$= 6x\sqrt{3x} - 2x\sqrt{3x}$$
$$= 4x\sqrt{3x}$$

You Try It 3
Simplify: $y\sqrt{28y} + 7\sqrt{63y^3}$

Your solution

Example 4
Simplify: $2x\sqrt{8y} - 3\sqrt{2x^2y} + 2\sqrt{32x^2y}$

Solution
$$2x\sqrt{8y} - 3\sqrt{2x^2y} + 2\sqrt{32x^2y}$$
$$= 2x\sqrt{2^3y} - 3\sqrt{2x^2y} + 2\sqrt{2^5x^2y}$$
$$= 2x\sqrt{2^2}\sqrt{2y} - 3\sqrt{x^2}\sqrt{2y} + 2\sqrt{2^4x^2}\sqrt{2y}$$
$$= 2x \cdot 2\sqrt{2y} - 3 \cdot x\sqrt{2y} + 2 \cdot 2^2 \cdot x\sqrt{2y}$$
$$= 4x\sqrt{2y} - 3x\sqrt{2y} + 8x\sqrt{2y}$$
$$= 9x\sqrt{2y}$$

You Try It 4
Simplify: $2\sqrt{27a^5} - 4a\sqrt{12a^3} + a^2\sqrt{75a}$

Your solution

Solutions on p. A33

Content and Format © 1995 HMCo.

11.2 Exercises

Objective A

Simplify.

1. $2\sqrt{2} + \sqrt{2}$ **2.** $3\sqrt{5} + 8\sqrt{5}$ **3.** $-3\sqrt{7} + 2\sqrt{7}$ **4.** $4\sqrt{5} - 10\sqrt{5}$

5. $-3\sqrt{11} - 8\sqrt{11}$ **6.** $-3\sqrt{3} - 5\sqrt{3}$ **7.** $2\sqrt{x} + 8\sqrt{x}$ **8.** $3\sqrt{y} + 2\sqrt{y}$

9. $8\sqrt{y} - 10\sqrt{y}$ **10.** $-5\sqrt{2a} + 2\sqrt{2a}$ **11.** $-2\sqrt{3b} - 9\sqrt{3b}$ **12.** $-7\sqrt{5a} - 5\sqrt{5a}$

13. $3x\sqrt{2} - x\sqrt{2}$ **14.** $2y\sqrt{3} - 9y\sqrt{3}$ **15.** $2a\sqrt{3a} - 5a\sqrt{3a}$

16. $-5b\sqrt{3x} - 2b\sqrt{3x}$ **17.** $3\sqrt{xy} - 8\sqrt{xy}$ **18.** $-4\sqrt{xy} + 6\sqrt{xy}$

19. $\sqrt{45} + \sqrt{125}$ **20.** $\sqrt{32} - \sqrt{98}$ **21.** $2\sqrt{2} + 3\sqrt{8}$

22. $4\sqrt{128} - 3\sqrt{32}$ **23.** $5\sqrt{18} - 2\sqrt{75}$ **24.** $5\sqrt{75} - 2\sqrt{18}$

25. $5\sqrt{4x} - 3\sqrt{9x}$ **26.** $-3\sqrt{25y} + 8\sqrt{49y}$ **27.** $3\sqrt{3x^2} - 5\sqrt{27x^2}$

28. $-2\sqrt{8y^2} + 5\sqrt{32y^2}$ **29.** $2x\sqrt{xy^2} - 3y\sqrt{x^2y}$ **30.** $4a\sqrt{b^2a} - 3b\sqrt{a^2b}$

31. $3x\sqrt{12x} - 5\sqrt{27x^3}$ **32.** $2a\sqrt{50a} + 7\sqrt{32a^3}$ **33.** $4y\sqrt{8y^3} - 7\sqrt{18y^5}$

34. $2a\sqrt{8ab^2} - 2b\sqrt{2a^3}$ **35.** $b^2\sqrt{a^5b} + 3a^2\sqrt{ab^5}$ **36.** $y^2\sqrt{x^5y} + x\sqrt{x^3y^5}$

Simplify.

37. $4\sqrt{2} - 5\sqrt{2} + 8\sqrt{2}$

38. $3\sqrt{3} + 8\sqrt{3} - 16\sqrt{3}$

39. $5\sqrt{x} - 8\sqrt{x} + 9\sqrt{x}$

40. $\sqrt{x} - 7\sqrt{x} + 6\sqrt{x}$

41. $8\sqrt{2} - 3\sqrt{y} - 8\sqrt{2}$

42. $8\sqrt{3} - 5\sqrt{2} - 5\sqrt{3}$

43. $8\sqrt{8} - 4\sqrt{32} - 9\sqrt{50}$

44. $2\sqrt{12} - 4\sqrt{27} + \sqrt{75}$

45. $-2\sqrt{3} + 5\sqrt{27} - 4\sqrt{45}$

46. $-2\sqrt{8} - 3\sqrt{27} + 3\sqrt{50}$

47. $4\sqrt{75} + 3\sqrt{48} - \sqrt{99}$

48. $2\sqrt{75} - 5\sqrt{20} + 2\sqrt{45}$

49. $\sqrt{25x} - \sqrt{9x} + \sqrt{16x}$

50. $\sqrt{4x} - \sqrt{100x} - \sqrt{49x}$

51. $3\sqrt{3x} + \sqrt{27x} - 8\sqrt{75x}$

52. $5\sqrt{5x} + 2\sqrt{45x} - 3\sqrt{80x}$

53. $2a\sqrt{75b} - a\sqrt{20b} + 4a\sqrt{45b}$

54. $2b\sqrt{75a} - 5b\sqrt{27a} + 2b\sqrt{20a}$

55. $x\sqrt{3y^2} - 2y\sqrt{12x^2} + xy\sqrt{3}$

56. $a\sqrt{27b^2} + 3b\sqrt{147a^2} - ab\sqrt{3}$

57. $3a\sqrt{ab^3} + 4b\sqrt{a^3b} - 5\sqrt{a^3b^3}$

58. $5\sqrt{a^3b} + a\sqrt{4ab} - 3\sqrt{49a^3b}$

APPLYING THE CONCEPTS

59. Given $G(x) = \sqrt{x + 5} + \sqrt{5x + 3}$, write $G(3)$ in simplest form.

60. Is the equation $\sqrt{a^2 + b^2} = \sqrt{a} + \sqrt{b}$ true for all real numbers a and b?

61. Use complete sentences to explain the steps in simplifying $4\sqrt{2a^3b} + 5\sqrt{5a^3b}$.
[W]

62. For each problem, write "ok" if the answer is correct. If the answer is incorrect, write the correct answer.
  a. $3\sqrt{ab} + 5\sqrt{ab} = 8\sqrt{2ab}$
  b. $7\sqrt{x^3} - 3x\sqrt{x} - x\sqrt{16x} = 0$
  c. $5 - 2\sqrt{y} = 3\sqrt{y}$

Simplify.

63. $2\sqrt{8x + 4y} - 5\sqrt{18x + 9y}$

64. $6\sqrt{16x - 16} + \sqrt{25x - 25}$

65. $3\sqrt{a^3 + a^2} + 5\sqrt{4a^3 + 4a^2}$

66. $3\sqrt{x^3y^2 + x^2y^3} + xy\sqrt{4x + 4y}$

11.3 Multiplication and Division of Radical Expressions

Objective A *To multiply radical expressions* . . . .

The Product Property of Square Roots is used to multiply variable radical expressions.

$$\sqrt{2x}\,\sqrt{3y} = \sqrt{2x \cdot 3y} = \sqrt{6xy}$$

➡ Simplify: $\sqrt{2x^2}\,\sqrt{32x^5}$

$$\sqrt{2x^2}\,\sqrt{32x^5} \;\boxed{= \sqrt{2x^2 \cdot 32x^5}}$$

- Use the Product Property of Square Roots. Do this step mentally.

$$= \sqrt{64x^7}$$

- Multiply the radicands.

$$= \sqrt{2^6 x^7}$$

- Simplify.

$$= \sqrt{2^6 x^6}\,\sqrt{x}$$
$$= 2^3 x^3 \sqrt{x}$$
$$= 8x^3 \sqrt{x}$$

➡ Simplify: $\sqrt{2x}(x + \sqrt{2x})$

$$\sqrt{2x}\,(x + \sqrt{2x}) \;\boxed{= \sqrt{2x}(x) + \sqrt{2x}\,\sqrt{2x}}$$

- Use the Distributive Property to remove parentheses. Do this step mentally.

$$= x\sqrt{2x} + \sqrt{4x^2}$$
$$= x\sqrt{2x} + \sqrt{2^2 x^2}$$

- Simplify.

$$= x\sqrt{2x} + 2x$$

➡ Simplify: $(\sqrt{2} - 3x)(\sqrt{2} + x)$

$$(\sqrt{2} - 3x)(\sqrt{2} + x) = \sqrt{2 \cdot 2} + x\sqrt{2} - 3x\sqrt{2} - 3x^2$$

- Use the FOIL method to remove parentheses.

$$= \sqrt{2^2} + (x - 3x)\sqrt{2} - 3x^2$$
$$= 2 - 2x\sqrt{2} - 3x^2$$

The expressions $a + b$ and $a - b$, which are the sum and difference of two terms, are called **conjugates** of each other. You will recall that $(a + b)(a - b) = a^2 - b^2$.

➡ Simplify: $(2 + \sqrt{7})(2 - \sqrt{7})$

$$(2 + \sqrt{7})(2 - \sqrt{7}) = 2^2 - \sqrt{7^2}$$

- $(2 + \sqrt{7})(2 - \sqrt{7})$ is the product of conjugates.

$$= 4 - 7$$
$$= -3$$

➡ Simplify: $(3 + \sqrt{y})(3 - \sqrt{y})$

$$(3 + \sqrt{y})(3 - \sqrt{y}) = 3^2 - \sqrt{y^2}$$

- $(3 + \sqrt{y})(3 - \sqrt{y})$ is the product of conjugates.

$$= 9 - y$$

Example 1

Simplify: $\sqrt{3x^4}\,\sqrt{2x^2y}\,\sqrt{6xy^2}$

Solution

$$\sqrt{3x^4}\,\sqrt{2x^2y}\,\sqrt{6xy^2} = \sqrt{36x^7y^3}$$
$$= \sqrt{2^2 3^2 x^7 y^3}$$
$$= \sqrt{2^2 3^2 x^6 y^2}\,\sqrt{xy}$$
$$= 2 \cdot 3x^3 y\sqrt{xy}$$
$$= 6x^3 y\sqrt{xy}$$

You Try It 1

Simplify: $\sqrt{5a}\,\sqrt{15a^3b^4}\,\sqrt{3b^5}$

Your solution

Example 2

Simplify: $\sqrt{3ab}\,(\sqrt{3a} + \sqrt{9b})$

Solution

$$\sqrt{3ab}\,(\sqrt{3a} + \sqrt{9b})$$
$$= \sqrt{3^2 a^2 b} + \sqrt{3^3 ab^2}$$
$$= \sqrt{3^2 a^2}\,\sqrt{b} + \sqrt{3^2 b^2}\,\sqrt{3a}$$
$$= 3a\sqrt{b} + 3b\sqrt{3a}$$

You Try It 2

Simplify: $\sqrt{5x}\,(\sqrt{5x} - \sqrt{25y})$

Your solution

Example 3

Simplify: $(\sqrt{a} - \sqrt{b})(\sqrt{a} + \sqrt{b})$

Solution

$$(\sqrt{a} - \sqrt{b})(\sqrt{a} + \sqrt{b}) = \sqrt{a^2} - \sqrt{b^2}$$
$$= a - b$$

You Try It 3

Simplify: $(2\sqrt{x} + 7)(2\sqrt{x} - 7)$

Your solution

Example 4

Simplify: $(2\sqrt{x} - \sqrt{y})(5\sqrt{x} - 2\sqrt{y})$

Solution

$$(2\sqrt{x} - \sqrt{y})(5\sqrt{x} - 2\sqrt{y})$$
$$= 10\sqrt{x^2} - 4\sqrt{xy} - 5\sqrt{xy} + 2\sqrt{y^2}$$
$$= 10x - 9\sqrt{xy} + 2y$$

You Try It 4

Simplify: $(3\sqrt{x} - \sqrt{y})(5\sqrt{x} - 2\sqrt{y})$

Your solution

Solutions on p. A34

Objective B **To divide radical expressions** ....

The Quotient Property of Square Roots

If *a* and *b* are positive real numbers, then

$$\sqrt{\dfrac{a}{b}} = \dfrac{\sqrt{a}}{\sqrt{b}} \quad \text{and} \quad \dfrac{\sqrt{a}}{\sqrt{b}} = \sqrt{\dfrac{a}{b}}.$$

The square root of a quotient is equal to the quotient of the square roots.

POINT OF INTEREST

A radical expression that occurs in Einstein's Theory of Relativity is

$$\frac{1}{\sqrt{1 - \dfrac{v^2}{c^2}}}$$

where v is the velocity of an object and c is the speed of light.

➡ Simplify: $\sqrt{\dfrac{4x^2}{z^6}}$

$$\sqrt{\frac{4x^2}{z^6}} = \frac{\sqrt{4x^2}}{\sqrt{z^6}}$$

• Rewrite the radical expression as the quotient of the square roots.

$$= \frac{\sqrt{2^2x^2}}{\sqrt{z^6}} = \frac{2x}{z^3}$$

• Simplify.

➡ Simplify: $\sqrt{\dfrac{24x^3y^7}{3x^7y^2}}$

$$\sqrt{\frac{24x^3y^7}{3x^7y^2}} = \sqrt{\frac{8y^5}{x^4}}$$

• Simplify the radicand.

$$= \frac{\sqrt{8y^5}}{\sqrt{x^4}}$$

• Rewrite the radical expression as the quotient of the square roots.

$$= \frac{\sqrt{2^3y^5}}{\sqrt{x^4}}$$

• Simplify.

$$= \frac{\sqrt{2^2y^4}\,\sqrt{2y}}{\sqrt{x^4}}$$

$$= \frac{2y^2\sqrt{2y}}{x^2}$$

➡ Simplify: $\dfrac{\sqrt{4x^2y}}{\sqrt{xy}}$

$$\frac{\sqrt{4x^2y}}{\sqrt{xy}} = \sqrt{\frac{4x^2y}{xy}}$$

• Use the Quotient Property of Square Roots.

$$= \sqrt{4x}$$

• Simplify the radicand.

$$= \sqrt{2^2}\,\sqrt{x}$$

• Simplify the radical expression.

$$= 2\sqrt{x}$$

A radical expression is not considered to be in simplest form if a radical remains in the denominator. The procedure used to remove a radical from the denominator is called **rationalizing the denominator.**

➡ Simplify: $\dfrac{2}{\sqrt{3}}$

$$\frac{2}{\sqrt{3}} = \frac{2}{\sqrt{3}} \cdot \boxed{\frac{\sqrt{3}}{\sqrt{3}}}$$

• Multiply the expression by $\dfrac{\sqrt{3}}{\sqrt{3}}$, which equals 1.

$$= \frac{2\sqrt{3}}{\sqrt{3^2}}$$

• The radicand in the denominator is a perfect square. Do this step mentally.

$$= \frac{2\sqrt{3}}{3}$$

• Simplify.

The radical expression is in simplest form, because no radical remains in the denominator and the numerator radical contains no perfect-square factors other than 1.

When the denominator contains a binomial radical expression, simplify the radical expression by multiplying the numerator and denominator by the conjugate of the denominator.

Simplify: $\dfrac{\sqrt{2y}}{\sqrt{y}+3}$

$$\dfrac{\sqrt{2y}}{\sqrt{y}+3} = \dfrac{\sqrt{2y}}{\sqrt{y}+3} \cdot \dfrac{\sqrt{y}-3}{\sqrt{y}-3}$$

$$= \dfrac{\sqrt{2y^2}-3\sqrt{2y}}{\sqrt{y^2}-3^2}$$

$$= \dfrac{y\sqrt{2}-3\sqrt{2y}}{y-9}$$

- Multiply the numerator and denominator by $\sqrt{y}-3$, the conjugate of $\sqrt{y}+3$.
- Simplify.

Example 5

Simplify: $\dfrac{\sqrt{4x^2y^5}}{\sqrt{3x^4y}}$

Solution

$$\dfrac{\sqrt{4x^2y^5}}{\sqrt{3x^4y}} = \sqrt{\dfrac{2^2x^2y^5}{3x^4y}} = \sqrt{\dfrac{2^2y^4}{3x^2}} = \dfrac{2y^2}{x\sqrt{3}}$$

$$= \dfrac{2y^2}{x\sqrt{3}} \cdot \dfrac{\sqrt{3}}{\sqrt{3}} = \dfrac{2y^2\sqrt{3}}{3x}$$

You Try It 5

Simplify: $\dfrac{\sqrt{15x^6y^7}}{\sqrt{3x^7y^9}}$

Your solution

Example 6

Simplify: $\dfrac{\sqrt{2}}{\sqrt{2}+\sqrt{6}}$

Solution

The denominator contains a binomial expression. Multiply the numerator and denominator by the conjugate of the denominator.

$$\dfrac{\sqrt{2}}{\sqrt{2}+\sqrt{6}} = \dfrac{\sqrt{2}}{\sqrt{2}+\sqrt{6}} \cdot \dfrac{\sqrt{2}-\sqrt{6}}{\sqrt{2}-\sqrt{6}}$$

$$= \dfrac{\sqrt{4}-\sqrt{12}}{2-6} = \dfrac{2-2\sqrt{3}}{-4} = \dfrac{2(1-\sqrt{3})}{-4}$$

$$= \dfrac{1-\sqrt{3}}{-2} = -\dfrac{1-\sqrt{3}}{2}$$

You Try It 6

Simplify: $\dfrac{\sqrt{3}}{\sqrt{3}-\sqrt{6}}$

Your solution

Example 7

Simplify: $\dfrac{3-\sqrt{5}}{2+3\sqrt{5}}$

Solution

$$\dfrac{3-\sqrt{5}}{2+3\sqrt{5}} = \dfrac{3-\sqrt{5}}{2+3\sqrt{5}} \cdot \dfrac{2-3\sqrt{5}}{2-3\sqrt{5}}$$

$$= \dfrac{6-9\sqrt{5}-2\sqrt{5}+3 \cdot \sqrt{5^2}}{4-9 \cdot 5}$$

$$= \dfrac{6-11\sqrt{5}+15}{4-45}$$

$$= \dfrac{21-11\sqrt{5}}{-41} = -\dfrac{21-11\sqrt{5}}{41}$$

You Try It 7

Simplify: $\dfrac{5+\sqrt{y}}{1-2\sqrt{y}}$

Your solution

Solutions on p. A34

11.3 Exercises

. . . .

Objective A

Simplify.

1. $\sqrt{5} \cdot \sqrt{5}$ **2.** $\sqrt{11} \cdot \sqrt{11}$ **3.** $\sqrt{3} \cdot \sqrt{12}$ **4.** $\sqrt{2} \cdot \sqrt{8}$

5. $\sqrt{x} \cdot \sqrt{x}$ **6.** $\sqrt{y} \cdot \sqrt{y}$ **7.** $\sqrt{xy^3} \cdot \sqrt{x^5y}$ **8.** $\sqrt{a^3b^5} \cdot \sqrt{ab^5}$

9. $\sqrt{3a^2b^5} \cdot \sqrt{6ab^7}$ **10.** $\sqrt{5x^3y} \cdot \sqrt{10x^2y}$ **11.** $\sqrt{6a^3b^2} \cdot \sqrt{24a^5b}$

12. $\sqrt{8ab^5} \cdot \sqrt{12a^7b}$ **13.** $\sqrt{2}\,(\sqrt{2} - \sqrt{3})$ **14.** $3(\sqrt{12} - \sqrt{3})$

15. $\sqrt{x}\,(\sqrt{x} - \sqrt{y})$ **16.** $\sqrt{b}\,(\sqrt{a} - \sqrt{b})$ **17.** $\sqrt{5}\,(\sqrt{10} - \sqrt{x})$

18. $\sqrt{6}\,(\sqrt{y} - \sqrt{18})$ **19.** $\sqrt{8}\,(\sqrt{2} - \sqrt{5})$ **20.** $\sqrt{10}\,(\sqrt{20} - \sqrt{a})$

21. $(\sqrt{x} - 3)^2$ **22.** $(2\sqrt{a} - y)^2$ **23.** $\sqrt{3a}\,(\sqrt{3a} - \sqrt{3b})$

24. $\sqrt{5x}\,(\sqrt{10x} - \sqrt{x})$ **25.** $\sqrt{2ac} \cdot \sqrt{5ab} \cdot \sqrt{10cb}$ **26.** $\sqrt{3xy} \cdot \sqrt{6x^3y} \cdot \sqrt{2y^2}$

27. $(\sqrt{5} + 3)(2\sqrt{5} - 4)$ **28.** $(2 - 3\sqrt{7})(5 + 2\sqrt{7})$ **29.** $(4 + \sqrt{8})(3 + \sqrt{2})$

30. $(6 - \sqrt{27})(2 + \sqrt{3})$ **31.** $(2\sqrt{x} + 4)(3\sqrt{x} - 1)$ **32.** $(5 + \sqrt{y})(6 - 3\sqrt{y})$

33. $(3\sqrt{x} - 2y)(5\sqrt{x} - 4y)$ **34.** $(5\sqrt{x} + 2\sqrt{y})(3\sqrt{x} - \sqrt{y})$ **35.** $(\sqrt{x} - \sqrt{y})(\sqrt{x} + \sqrt{y})$

36. $(\sqrt{2a} - \sqrt{b})(\sqrt{2a} + \sqrt{b})$

Objective B

Simplify.

37. $\dfrac{\sqrt{32}}{\sqrt{2}}$　　**38.** $\dfrac{\sqrt{45}}{\sqrt{5}}$　　**39.** $\dfrac{\sqrt{98}}{\sqrt{2}}$　　**40.** $\dfrac{\sqrt{48}}{\sqrt{3}}$　　**41.** $\dfrac{\sqrt{27a}}{\sqrt{3a}}$

42. $\dfrac{\sqrt{72x^5}}{\sqrt{2x}}$　　**43.** $\dfrac{\sqrt{15x^3y}}{\sqrt{3xy}}$　　**44.** $\dfrac{\sqrt{40x^5y^2}}{\sqrt{5xy}}$　　**45.** $\dfrac{\sqrt{2a^5b^4}}{\sqrt{98ab^4}}$　　**46.** $\dfrac{\sqrt{48x^5y^2}}{\sqrt{3x^3y}}$

47. $\dfrac{\sqrt{9xy^2}}{\sqrt{27x}}$　　**48.** $\dfrac{\sqrt{4x^2y}}{\sqrt{3xy^3}}$　　**49.** $\dfrac{\sqrt{16x^3y^2}}{\sqrt{8x^3y}}$　　**50.** $\dfrac{\sqrt{2}}{\sqrt{8}+4}$

51. $\dfrac{1}{\sqrt{2}-3}$　　**52.** $\dfrac{5}{\sqrt{7}-3}$　　**53.** $\dfrac{3}{5+\sqrt{5}}$　　**54.** $\dfrac{\sqrt{3}}{5-\sqrt{27}}$

55. $\dfrac{7}{\sqrt{2}-7}$　　**56.** $\dfrac{3-\sqrt{6}}{5-2\sqrt{6}}$　　**57.** $\dfrac{6-2\sqrt{3}}{4+3\sqrt{3}}$　　**58.** $\dfrac{-6}{4+\sqrt{2}}$

59. $\dfrac{\sqrt{2}+2\sqrt{6}}{2\sqrt{2}-3\sqrt{6}}$　　**60.** $\dfrac{2\sqrt{3}-\sqrt{6}}{5\sqrt{3}+2\sqrt{6}}$　　**61.** $\dfrac{3+\sqrt{x}}{2-\sqrt{x}}$　　**62.** $\dfrac{-\sqrt{15}}{3-\sqrt{12}}$

63. $\dfrac{\sqrt{a}-4}{2\sqrt{a}+2}$　　**64.** $\dfrac{\sqrt{xy}}{\sqrt{x}-\sqrt{y}}$　　**65.** $\dfrac{\sqrt{x}}{\sqrt{x}-\sqrt{y}}$　　**66.** $\dfrac{-12}{\sqrt{6}-3}$

APPLYING THE CONCEPTS

Simplify.

67. In your own words, describe the process of rationalizing the
[W]　denominator.

68. Show that $(1+\sqrt{6})$ and $(1-\sqrt{6})$ are solutions of the equation $x^2-2x-5=0$.

69. Answer true or false. If the answer is false, write the correct answer.
   a. $(\sqrt{y})^4 = y^2$　　　　　　　　**b.** $(2\sqrt{x})^3 = 8x\sqrt{x}$

   c. $(\sqrt{x}+1)^2 = x+1$　　　　　**d.** $\dfrac{1}{2-\sqrt{3}} = 2+\sqrt{3}$

70. The number $\dfrac{\sqrt{5}+1}{2}$ is called the golden ratio. Research the golden ratio
[W]　and write a few paragraphs about this number and its applications.

11.4

Solving Equations Containing Radical Expressions

Objective A *To solve an equation containing a radical expression* ....

An equation that contains a variable expression in a radicand is a **radical equation.**

$$\left.\begin{array}{l} \sqrt{x} = 4 \\ \sqrt{x} = 2 = \sqrt{x - 7} \end{array}\right\} \begin{array}{l} \text{Radical} \\ \text{equations} \end{array}$$

The following property of equality states that if two numbers are equal, the squares of the numbers are equal. This property is used to solve radical equations.

Property of Squaring Both Sides of an Equation

If a and b are real numbers and $a = b$, then $a^2 = b^2$

The first step when solving a radical equation is to isolate a radical in the equation.

➡ Solve $\sqrt{x - 2} - 7 = 0$.

$$\sqrt{x - 2} - 7 = 0$$
$$\sqrt{x - 2} = 7$$

• Rewrite the equation with the radical on one side of the equation and the constant on the other side.

$$(\sqrt{x - 2})^2 = 7^2$$

• Square both sides of the equation.

$$x - 2 = 49$$

• Solve the resulting equation.

$$x = 51$$

Check: $\sqrt{x - 2} - 7 = 0$

$$\begin{array}{r|l} \sqrt{51 - 2} - 7 & 0 \\ \sqrt{49} - 7 & 0 \\ \sqrt{7^2} - 7 & 0 \\ 7 - 7 & 0 \\ 0 = 0 & \text{A true equation} \end{array}$$

The solution is 51.

When both sides of an equation are squared, the resulting equation may have a solution that is not a solution of the original equation. Checking a proposed solution of a radical equation, as we did in the previous example, is a necessary step.

➡ Solve $\sqrt{2x - 5} + 3 = 0$.

$$\sqrt{2x - 5} + 3 = 0$$
$$\sqrt{2x - 5} = -3$$

• Rewrite the equation with the radical on one side of the equation and the constant on the other side.

$$(\sqrt{2x - 5})^2 = (-3)^2$$

• Square each side of the equation.

$$2x - 5 = 9$$

• Solve for x.

$$2x = 14$$
$$x = 7$$

LOOK CLOSELY
Any time each side of an equation is squared, you must check the proposed solution of the equation.

Here is the check for the equation on the previous page.

Check: $\sqrt{2x - 5} + 3 = 0$

$$\begin{array}{r|l} \sqrt{2 \cdot 7 - 5} + 3 & 0 \\ \sqrt{14 - 5} + 3 & 0 \\ \sqrt{9} + 3 & 0 \\ 3 + 3 \neq & 0 \end{array}$$

7 does not check as a solution. The equation has no solution.

Example 1
Solve $\sqrt{3x} + 2 = 5$.

Solution

$$\begin{aligned} \sqrt{3x} + 2 &= 5 \\ \sqrt{3x} &= 3 \\ (\sqrt{3x})^2 &= 3^2 \\ 3x &= 9 \\ x &= 3 \end{aligned}$$

Check: $\sqrt{3x} + 2 = 5$

$$\begin{array}{r|l} \sqrt{3 \cdot 3} + 2 & 5 \\ \sqrt{3^2} + 2 & 5 \\ 3 + 2 & 5 \\ 5 = & 5 \end{array}$$

The solution is 3.

You Try It 1
Solve $\sqrt{4x} + 3 = 7$.

Your solution

Example 2
Solve $\sqrt{x} - \sqrt{x - 5} = 1$.

Solution

$$\begin{aligned} \sqrt{x} - \sqrt{x - 5} &= 1 \\ \sqrt{x} &= 1 + \sqrt{x - 5} \\ (\sqrt{x})^2 &= (1 + \sqrt{x - 5})^2 \\ x &= 1 + 2\sqrt{x - 5} + (x - 5) \\ 4 &= 2\sqrt{x - 5} \\ 2 &= \sqrt{x - 5} \\ 2^2 &= (\sqrt{x - 5})^2 \\ 4 &= x - 5 \\ 9 &= x \end{aligned}$$

Check:

$\sqrt{x} - \sqrt{x - 5} = 1$

$$\begin{array}{r|l} \sqrt{9} - \sqrt{9 - 5} & 1 \\ \sqrt{9} - \sqrt{4} & 1 \\ 3 - 2 & 1 \\ 1 = & 1 \end{array}$$

9 checks as the solution.

You Try It 2
Solve $\sqrt{x} + \sqrt{x + 9} = 9$.

Your solution

Solutions on p. A34

Objective B   To solve application problems ....

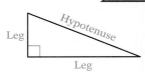

A right triangle contains one 90° angle. The side opposite the 90° angle is called the **hypotenuse**. The other two sides are called legs.

Pythagoras, a Greek mathematician who lived around 550 B.C., is given credit for this theorem. It states that the square of the hypotenuse of a right triangle is equal to the sum of the squares of the two legs. Actually, this theorem was known to the Babylonians around 1200 B.C.

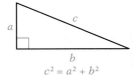

POINT OF INTEREST
The first known proof of this theorem occurs in a Chinese text, *Arithmetic Classic* which was first written around 600 B.C. (but there are no existing copies) and revised over a period of 500 years. The earliest known copy of this text dates from approximately 100 B.C.

Pythagorean Theorem

If a and b are the lengths of the legs of a right triangle and c is the length of the hypotenuse, then $c^2 = a^2 + b^2$.

Using this theorem, we can find the hypotenuse of a right triangle when we know the two legs. Use the formula

$$\text{Hypotenuse} = \sqrt{(\text{leg})^2 + (\text{leg})^2}$$
$$c = \sqrt{a^2 + b^2}$$
$$= \sqrt{(5)^2 + (12)^2}$$
$$= \sqrt{25 + 144}$$
$$= \sqrt{169}$$
$$= 13$$

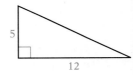

The leg of a right triangle can be found when one leg and the hypotenuse are known. Use the formula

$$\text{Leg} = \sqrt{(\text{hypotenuse})^2 - (\text{leg})^2}$$
$$a = \sqrt{c^2 - b^2}$$
$$= \sqrt{(25)^2 - (20)^2}$$
$$= \sqrt{625 - 400}$$
$$= \sqrt{225}$$
$$= 15$$

Example 3 and You Try It 3 illustrate the use of the Pythagorean Theorem. Example 4 and You Try It 4 illustrate other applications of radical equations.

Example 3

A guy wire is attached to a point 20 m above the ground on a telephone pole. The wire is anchored to the ground at a point 8 m from the base of the pole. Find the length of the guy wire, to the nearest tenth of a meter.

Strategy

To find the length of the guy wire, use the Pythagorean Theorem. One leg is 20 m. The other leg is 8 m. The guy wire is the hypotenuse. Solve the Pythagorean Theorem for the hypotenuse.

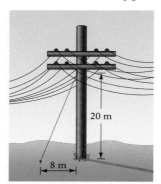

Solution

$c = \sqrt{a^2 + b^2}$

$\quad = \sqrt{(20)^2 + (8)^2}$

$\quad = \sqrt{400 + 64} = \sqrt{464} \approx 21.5$

The guy wire has a length of 21.5 m.

Example 4

How far would a submarine periscope have to be above the water to locate a ship 4 mi away? The equation for the distance in miles that the lookout can see is $d = \sqrt{1.5h}$, where h is the height in feet above the surface of the water. Round to the nearest hundredth.

Strategy

To find the height above the water, replace d in the equation with the given value and solve for h.

Solution

$\sqrt{1.5h} = d$

$\sqrt{1.5h} = 4$

$(\sqrt{1.5h})^2 = 4^2$

$\quad\quad 1.5h = 16$

$\quad\quad\quad h = \dfrac{16}{1.5} \approx 10.67$

The periscope must be 10.67 ft above the water.

You Try It 3

A ladder 8 ft long is resting against a building. How high on the building will the ladder reach when the bottom of the ladder is 3 ft from the building? Round to the nearest hundredth of a foot.

Your strategy

Your solution

You Try It 4

Find the length of a pendulum that makes one swing in 2.5 s. The equation for the time for one swing is $T = 2\pi\sqrt{\dfrac{L}{32}}$, where t is the time in seconds and L is the length in feet. Use 3.14 for π. Round to the nearest hundredth.

Your strategy

Your solution

Solutions on p. A35

11.4 Exercises

Objective A

Solve and check.

1. $\sqrt{x} = 5$ **2.** $\sqrt{y} = 7$ **3.** $\sqrt{a} = 12$ **4.** $\sqrt{a} = 9$ **5.** $\sqrt{5x} = 5$

6. $\sqrt{4x} + 5 = 2$ **7.** $\sqrt{3x} + 9 = 4$ **8.** $\sqrt{3x - 2} = 4$ **9.** $\sqrt{5x + 6} = 1$

10. $\sqrt{2x + 1} = 7$ **11.** $\sqrt{5x + 4} = 3$ **12.** $0 = 2 - \sqrt{3 - x}$ **13.** $0 = 5 - \sqrt{10 + x}$

14. $\sqrt{5x + 2} = 0$ **15.** $\sqrt{3x - 7} = 0$ **16.** $\sqrt{3x} - 6 = -4$ **17.** $\sqrt{5x} + 8 = 23$

18. $\sqrt{4x - 2} + 8 = 5$ **19.** $\sqrt{5x} + 8 = 2$ **20.** $\sqrt{2x - 4} = 12$

21. $0 = 3 - \sqrt{3x - 9}$ **22.** $\sqrt{x + 2} = \sqrt{x + 1}$ **23.** $\sqrt{3x - 2} = \sqrt{5 - 2x}$

Objective B   *Application Problems*

Solve.

24. The infield of a baseball diamond is a square. The distance between successive bases is 90 ft. The pitcher's mound is on the diagonal between home plate and second base at a distance of 60.5 ft from home plate. Is the pitcher's mound more or less than halfway between home plate and second base?

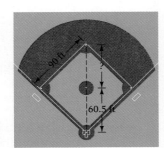

25. The infield of a softball diamond is a square. The distance between successive bases is 60 ft. The pitcher's mound is on the diagonal between home plate and second base at a distance of 46 ft from home plate. Is the pitcher's mound more or less than halfway between home plate and second base?

26. How far would a submarine periscope have to be above the water to locate a ship 5 mi away? The equation for the distance in miles that the lookout can see is $d = \sqrt{1.5h}$, where h is the height in feet above the surface of the water. Round to the nearest hundredth.

27. How far would a submarine periscope have to be above the water to locate a ship 6 mi away? The equation for the distance in miles that the lookout can see is $d = \sqrt{1.5h}$, where h is the height in feet above the surface of the water. Round to the nearest hundredth.

Solve.

28. The measure of a big-screen television is given by the length of a diagonal across the screen. A 36-in. television has a width of 28.8 in. Find the height of the screen to the nearest tenth of an inch.

29. The measure of a television screen is given by the length of a diagonal across the screen. A 33-in. big-screen television has a width of 26.4 in. Find the height of the screen to the nearest tenth of an inch.

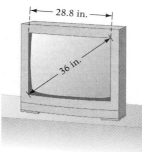

30. Find the length of a pendulum that makes one swing in 2 s. The equation for the time of one swing of a pendulum is $T = 2\pi\sqrt{\dfrac{L}{32}}$, where T is the time in seconds and L is the length in feet. Use 3.14 for π. Round to the nearest hundredth.

31. Find the length of a pendulum that makes one swing in 1.5 s. The equation for the time of one swing of a pendulum is $T = 2\pi\sqrt{\dfrac{L}{32}}$, where T is the time in seconds and L is the length in feet. Use 3.14 for π. Round to the nearest hundredth.

APPLYING THE CONCEPTS

Solve.

32. $\sqrt{\dfrac{3x-2}{4}} = 2$ **33.** $\sqrt{\dfrac{4x}{3}} - 1 = 1$ **34.** $\sqrt{\dfrac{3y}{5}} - 1 = 2$ **35.** $\sqrt{9x^2 + 49} + 1 = 3x + 2$

36. In the coordinate plane, a triangle is formed by drawing lines between the points $(0, 0)$ and $(5, 0)$, $(5, 0)$ and $(5, 12)$, and $(5, 12)$ and $(0, 0)$. Find the perimeter of the triangle.

37. The hypotenuse of a right triangle is $5\sqrt{2}$ cm, and one leg is $4\sqrt{2}$ cm.
   a. Find the perimeter of the triangle.
   b. Find the area of the triangle.

38. If a and b are real numbers and $a^2 = b^2$, does $a = b$? Explain your
[W]   answer.

39. Can the Pythagorean Theorem be used to find the length of side c of the triangle at the right? If so, determine c. If not, explain why the theorem cannot be used.

40. A circular fountain is being designed for a triangular plaza in a cultural center. The fountain is placed so that each side of the triangle touches the fountain as shown in the diagram at the right. Find the area of the fountain. The formula for the radius of the circle is given by

$$r = \sqrt{\dfrac{(s - a)(s - b)(s - c)}{s}}$$

where $s = \dfrac{1}{2}(a + b + c)$ and a, b, and c are the lengths of the sides of the triangle.

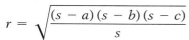

41. A farmer owns a triangular piece of land that measures 100 ft, 120 ft, and 150 ft, as shown at the right. Find the maximum area that can be irrigated with a circular irrigation system in which the irrigation does not go outside the triangular shape. Use the formulas in Exercise 40.

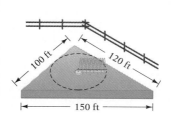

Project in Mathematics

· ·

Distance to the Horizon

In Section 11.4, we used the formula $d = \sqrt{1.5h}$ to calculate the approximate distance d (in miles) that a person could see who used a periscope h feet above the water. That formula is derived by using the Pythagorean Theorem.

Consider the diagram (not to scale) at the right, which shows the earth as a sphere and the periscope extending h feet above the surface. From geometry, because AB is tangent to the circle and OA is a radius, triangle AOB is a right triangle. Therefore,

$$(OA)^2 + (AB)^2 = (OB)^2$$

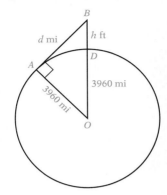

Substituting into this formula, we have

$$3960^2 + d^2 = \left(3960 + \frac{h}{5280}\right)^2$$

● **Because h is in feet, $h/5280$ is in miles.**

$$3960^2 + d^2 = 3960^2 + \frac{2 \cdot 3960}{5280}h + \left(\frac{h}{5280}\right)^2$$

$$d^2 = \frac{3}{2}h + \left(\frac{h}{5280}\right)^2$$

$$d = \sqrt{\frac{3}{2}h + \left(\frac{h}{5280}\right)^2}$$

At this point, an assumption is made that $\sqrt{\dfrac{3}{2}h + \left(\dfrac{h}{5280}\right)^2} \approx \sqrt{1.5h}$, where we have written $\dfrac{3}{2}$ as 1.5. Thus $d \approx \sqrt{1.5h}$ is used to approximate the distance that can be seen using a periscope h feet above the water.

Exercises:

1. Write a paragraph that justifies the assumption that

$$\sqrt{\frac{3}{2}h + \left(\frac{h}{5280}\right)^2} \approx \sqrt{1.5h}.$$

   (*Suggestion:* Evaluate each expression for various values of h. Because h is the height of a periscope above water, it is unlikely that $h > 25$ ft.)

2. The distance d is the distance from the top of the periscope to A. The distance along the surface of the water is given by arc AD. This distance can be approximated by the equation

$$D \approx \sqrt{1.5h} + 0.306186(\sqrt{h})^3$$

   Using this formula, calculate D when $h = 10$.

Chapter Summary

. . . .

Key Words   A *square root* of a positive number x is a number whose square is x.

The *principal square root* of a number is the positive square root.

The symbol $\sqrt{}$ is called a *radical* and is used to indicate the principal square root of a number.

The *radicand* is the number under the radical sign.

The square of an integer is a *perfect square*.

If a number is not a perfect square, its square root can only be approximated. Such numbers are *irrational numbers*. Their decimal representations never terminate or repeat.

Conjugates are binomial expressions that differ only in the sign of a term. (The expressions $a + b$ and $a - b$ are conjugates.)

Rationalizing the denominator is the procedure used to remove a radical from the denominator of a fraction.

A *radical equation* is an equation that contains a variable expression in a radicand.

Essential Rules   *The Product Property of Square Roots*

If a and b are positive real numbers, then $\sqrt{ab} = \sqrt{a}\,\sqrt{b}$.

The Quotient Property of Square Roots

If a and b are positive real numbers, then $\sqrt{\dfrac{a}{b}} = \dfrac{\sqrt{a}}{\sqrt{b}}$ and $\dfrac{\sqrt{a}}{\sqrt{b}} = \sqrt{\dfrac{a}{b}}$.

Property of Squaring Both Sides of An Equation

If a and b are real numbers and $a = b$, then $a^2 = b^2$.

Pythagorean Theorem

$$c^2 = a^2 + b^2$$

Chapter Review

· ·

SECTION 11.1

1. Simplify: $2\sqrt{36}$

2. Simplify: $5\sqrt{48}$

3. Simplify: $-3\sqrt{120}$

4. Simplify: $4\sqrt{250}$

5. Simplify: $3\sqrt{18a^5b}$

6. Simplify: $y\sqrt{24y^6}$

7. Simplify: $4y\sqrt{243x^{17}y^9}$

SECTION 11.2

8. Simplify: $3\sqrt{12x} + 5\sqrt{48x}$

9. Simplify: $2x\sqrt{60x^3y^3} + 3x^2y\sqrt{15xy}$

10. Simplify: $6a\sqrt{80b} - \sqrt{180a^2b} + 5a\sqrt{b}$

11. Simplify:
$2x^2\sqrt{18x^2y^5} + 6y\sqrt{2x^6y^3} - 9xy^2\sqrt{8x^4y}$

SECTION 11.3

12. Simplify: $\sqrt{3}\,(\sqrt{12} - \sqrt{3})$

13. Simplify: $\sqrt{6a}\,(\sqrt{3a} + \sqrt{2a})$

14. Simplify: $(4\sqrt{y} - \sqrt{5})(2\sqrt{y} + 3\sqrt{5})$

15. Simplify: $\dfrac{\sqrt{98x^7y^9}}{\sqrt{2x^3y}}$

16. Simplify: $\dfrac{16}{\sqrt{a}}$

17. Simplify: $\dfrac{8}{\sqrt{x}-3}$

18. Simplify: $\dfrac{2x}{\sqrt{3}-\sqrt{5}}$

SECTION 11.4

19. Solve: $\sqrt{5x}=10$

20. Solve: $3-\sqrt{7x}=5$

21. Solve: $\sqrt{2x-3}+4=0$

22. Solve: $\sqrt{5x+1}=\sqrt{20x-8}$

23. The weight of an object is related to the distance the object is above the surface of the earth. An equation for this relationship is $d=4000\sqrt{\dfrac{W_o}{W_a}}$, where W_o is an object's weight on the surface of the earth and W_a is the object's weight at a distance of d miles above the earth's surface. If a space explorer weighs 36 lb 8000 mi above the surface of the earth, how much does the explorer weigh on the surface of the earth?

24. A tsunami is a great sea wave produced by underwater earthquakes or volcanic eruption. The velocity of a tsunami as it approaches land depends on the depth of the water and can be approximated by the equation $v=3\sqrt{d}$, where d is the depth of the water in feet and v is the velocity of the tsunami in feet per second. Find the depth of the water if the velocity is 30 ft per sec.

25. A bicycle will overturn if it rounds a corner too sharply or too fast. An equation for the maximum velocity at which a cyclist can turn a corner without tipping over is $v=4\sqrt{r}$, where v is the velocity of the bicycle in miles per hour and r is the radius of the corner in feet. What is the radius of the sharpest corner that a cyclist can safely turn if riding at 20 mph?

Chapter Test

. . . .

1. Simplify: $\sqrt{45}$

2. Simplify: $\sqrt{75}$

3. Simplify: $\sqrt{121x^8y^2}$

4. Simplify: $\sqrt{72x^7y^2}$

5. Simplify: $\sqrt{32a^5b^{11}}$

6. Simplify: $5\sqrt{8} - 3\sqrt{50}$

7. Simplify: $3\sqrt{8y} - 2\sqrt{72x} + 5\sqrt{18y}$

8. Simplify: $2x\sqrt{3xy^3} - 2y\sqrt{12x^3y} - 3xy\sqrt{xy}$

9. Simplify: $\sqrt{8x^3y}\ \sqrt{10xy^4}$

10. Simplify: $\sqrt{3x^2y}\ \sqrt{6xy^2}\ \sqrt{2x}$

11. Simplify: $\sqrt{a}\,(\sqrt{a} - \sqrt{b})$

12. Simplify: $(\sqrt{y} - 3)(\sqrt{y} + 5)$

13. Simplify: $\dfrac{\sqrt{162}}{\sqrt{2}}$

14. Simplify: $\dfrac{\sqrt{98a^6b^4}}{\sqrt{2a^3b^2}}$

15. Simplify: $\dfrac{2}{\sqrt{3} - 1}$

16. Simplify: $\dfrac{2 - \sqrt{5}}{6 + \sqrt{5}}$

17. Solve: $\sqrt{9x} + 3 = 18$

18. Solve: $\sqrt{2x - 4} = \sqrt{3x - 5}$

19. The square root of the sum of two consecutive odd integers is equal to 10. Find the larger integer.

20. Find the length of a pendulum that makes one swing in 3 s. The equation for the time of one swing of a pendulum is $T = 2\pi \sqrt{\dfrac{L}{32}}$, where T is the time in seconds and L is the length in feet. Use 3.14 for π. Round to the nearest hundredth.

Content and Format © 1995 HMCo.

Cumulative Review

. . . .

1. Simplify:
$$\left(\frac{2}{3}\right)^2 \cdot \left(\frac{3}{4} - \frac{3}{2}\right) + \left(\frac{1}{2}\right)^2$$

2. Simplify:
$$-3[x - 2(3 - 2x) - 5x] + 2x$$

3. Solve:
$$2x - 4[3x - 2(1 - 3x)] = 2(3 - 4x)$$

4. Simplify: $(-3x^2y)(-2x^3y^4)$

5. Simplify: $\dfrac{12b^4 - 6b^2 + 2}{-6b^2}$

6. Given $f(x) = \dfrac{2x}{x - 3}$, find $f(-3)$.

7. Factor: $2a^3 - 16a^2 + 30a$

8. Simplify: $\dfrac{3x^3 - 6x^2}{4x^2 + 4x} \cdot \dfrac{3x - 9}{9x^3 - 45x^2 + 54x}$

9. Simplify: $\dfrac{x + 2}{x - 4} - \dfrac{6}{(x - 4)(x - 3)}$

10. Solve: $\dfrac{x}{2x - 5} - 2 = \dfrac{3x}{2x - 5}$

11. Find the equation of the line that contains the point $(-2, -3)$, and has slope $\dfrac{1}{2}$.

12. Solve by substitution:
$$4x - 3y = 1$$
$$2x + y = 3$$

13. Solve by the addition method:
$$5x + 4y = 7$$
$$3x - 2y = 13$$

14. Solve: $3(x - 7) \geq 5x - 12$

15. Simplify: $\sqrt{108}$

16. Simplify: $3\sqrt{32} - 2\sqrt{128}$

17. Simplify: $2a\sqrt{2ab^3} + b\sqrt{8a^3b} - 5ab\sqrt{ab}$

18. Simplify: $\sqrt{2a^9b} \, \sqrt{98ab^3} \, \sqrt{2a}$

19.   Simplify: $\sqrt{3}(\sqrt{6} - \sqrt{x^2})$

20.   Simplify: $\dfrac{\sqrt{320}}{\sqrt{5}}$

21.   Simplify: $\dfrac{3}{2 - \sqrt{5}}$

22.   Solve: $\sqrt{3x - 2} - 4 = 0$

23.   The selling price for a book is $29.40. The markup rate used by the bookstore is 20%. Find the cost of the book.

24.   How many ounces of pure water must be added to 40 oz of a 12% salt solution to make a salt solution that is 5% salt?

25.   The sum of two numbers is twenty-one. The product of the two numbers is one hundred four. Find the two numbers.

26.   A small water pipe takes twice as long to fill a tank as does a larger water pipe. With both pipes open it takes 16 h to fill the tank. Find the time it would take the small pipe working alone to fill the tank.

27.   Solve by graphing: $3x - 2y = 8$
$\qquad\qquad\qquad\qquad 4x + 5y = 3$

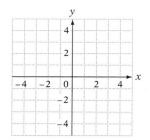

28.   Graph the solution set of $3x + y \le 2$.

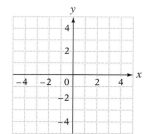

29.   The square root of the sum of two consecutive integers is equal to 9. Find the smaller integer.

30.   A stone is dropped from a building and hits the ground 5 s later. How high is the building? The equation for the distance an object falls in T seconds is $T = \sqrt{\dfrac{d}{16}}$, where d is the distance in feet.

Quadratic Equations

Objectives

Section 12.1
To solve a quadratic equation by factoring
To solve a quadratic equation by taking square roots

Section 12.2
To solve a quadratic equation by completing the square

Section 12.3
To solve a quadratic equation by using the quadratic formula

Section 12.4
To graph a quadratic equation of the form $y = ax^2 + bx + c$

Section 12.5
To solve application problems

Algebraic Symbolism

The way in which an algebraic expression or equation is written has gone through several stages of development. First there was the *rhetoric*, which was in vogue until the late 13th century. In this method, expressions were written out in sentences. The word *res* was used to represent an unknown.

Rhetoric: From the additive *res* in the additive *res* results in a square *res*. From the three in an additive *x* come three additive *res* and from the subtractive four in the additive *res* come subtractive four *res*. From three in subtractive four comes subtractive twelve.

Modern: $(x + 3)(x - 4) = x^2 - x - 12$

The second stage was *syncoptic*, which was a shorthand in which abbreviations were used for words.

Syncoptic: *a* 6 in *b* quad − *c* plano 4 in *b* + *b* cub

Modern: $6ab^2 - 4cb + b^3$

The current modern stage, called the *symbolic* stage, began with the use of exponents rather than words to symbolize exponential expressions. This occurred near the beginning of the 17th century with the publication of the book *La Geometrie* by René Descartes. Modern notation is still evolving as mathematicians continue to search for convenient methods to symbolize concepts.

Solving Quadratic Equations by Factoring or by Taking Square Roots

Objective A   *To solve a quadratic equation by factoring* ....

An equation of the form $ax^2 + bx + c = 0$, where a, b, and c are constants and $a \neq 0$, is a **quadratic equation.**

$4x^2 - 3x + 1 = 0, a = 4, b = -3, c = 1$

$3x^2 - 4 = 0, a = 3, b = 0, c = -4$

$\dfrac{x^2}{2} - 2x + 4 = 0, a = \dfrac{1}{2}, b = -2, c = 4$

A quadratic equation is also called a **second-degree equation.**

A quadratic equation is in **standard form** when the polynomial is in descending order and equal to zero.

Recall that the Principle of Zero Products states that if the product of two factors is zero, then at least one of the factors must be zero.

If $a \cdot b = 0$, then $a = 0$ or $b = 0$.

The Principle of Zero Products can be used in solving quadratic equations.

➡ Solve by factoring: $2x^2 - x = 1$

$$2x^2 - x = 1$$
$$2x^2 - x - 1 = 0$$
• Write the equation in standard form.

$$(2x + 1)(x - 1) = 0$$
• Factor.

$$2x + 1 = 0 \qquad x - 1 = 0$$
• Use the Principle of Zero Products to set each factor equal to zero.

$$2x = -1 \qquad x = 1$$
• Rewrite each equation in the form *variable = constant.*

$$x = -\dfrac{1}{2}$$
• Write the solutions.

Check:

$$\begin{array}{c|c} 2x^2 - x = 1 \\ \hline 2\left(-\dfrac{1}{2}\right)^2 - \left(-\dfrac{1}{2}\right) & 1 \\ 2 \cdot \dfrac{1}{4} + \dfrac{1}{2} & 1 \\ \dfrac{1}{2} + \dfrac{1}{2} & 1 \\ 1 = 1 \end{array}$$

$$\begin{array}{c|c} 2x^2 - x = 1 \\ \hline 2(1)^2 - 1 & 1 \\ 2 \cdot 1 - 1 & 1 \\ 2 - 1 & 1 \\ 1 = 1 \end{array}$$

The solutions are $-\dfrac{1}{2}$ and 1.

➡ Solve by factoring: $3x^2 - 4x + 8 = (4x + 1)(x - 2)$

$3x^2 - 4x + 8 = (4x + 1)(x - 2)$

$3x^2 - 4x + 8 = 4x^2 - 7x - 2$	• Simplify the right side of the equation.
$0 = x^2 - 3x - 10$	• Write the equation in standard form.
$0 = (x - 5)(x + 2)$	• Factor.
$x - 5 = 0 \qquad x + 2 = 0$	• Use the Principle of Zero Products to set each factor equal to zero.
$x = 5 \qquad\qquad x = -2$	• Rewrite each equation in the form *variable = constant*.

Check:

$3x^2 - 4x + 8 = (4x + 1)(x - 2)$	$3x^2 - 4x + 8 = (4x + 1)(x - 2)$
$3(5)^2 - 4(5) + 8 \mid (4[5] + 1)(5 - 2)$	$3(-2)^2 - 4(-2) + 8 \mid (4[-2] + 1)(-2 - 2)$
$3(25) - 20 + 8 \mid (20 + 1)(3)$	$3(4) + 8 + 8 \mid (-8 + 1)(-4)$
$75 - 12 \mid (21)(3)$	$12 + 16 \mid (-7)(-4)$
$63 = 63$	$28 = 28$

The solutions are 5 and −2.

➡ Solve by factoring: $x^2 - 10x + 25 = 0$

$x^2 - 10x + 25 = 0$	
$(x - 5)(x - 5) = 0$	• Factor.
$x - 5 = 0 \qquad x - 5 = 0$	• Use the Principle of Zero Products.
$x = 5 \qquad\qquad x = 5$	• Solve each equation for *x*.

The solution is 5.

In this last example, 5 is called a **double root** of the quadratic equation.

Example 1

Solve by factoring: $\dfrac{z^2}{2} - \dfrac{z}{4} - \dfrac{1}{4} = 0$

Solution

$\dfrac{z^2}{2} - \dfrac{z}{4} - \dfrac{1}{4} = 0$

$4\left(\dfrac{z^2}{2} - \dfrac{z}{4} - \dfrac{1}{4}\right) = 4(0)$ • Multiply each side by 4.

$2z^2 - z - 1 = 0$
$(2z + 1)(z - 1) = 0$
$2z + 1 = 0 \qquad z - 1 = 0$
$2z = -1 \qquad\quad z = 1$
$z = -\dfrac{1}{2}$

The solutions are $-\dfrac{1}{2}$ and 1.

You Try It 1

Solve by factoring: $\dfrac{3y^2}{2} + y - \dfrac{1}{2} = 0$

Your solution

Solution on p. A35

Objective B   ***To solve a quadratic equation by taking square roots*** ···········

Consider a quadratic equation of the form $x^2 = a$. This equation can be solved by factoring.

$$x^2 = 25$$
$$x^2 - 25 = 0$$
$$(x - 5)(x + 5) = 0$$
$$x = 5; \quad x = -5$$

The solutions are 5 and -5. The fact that the solutions are plus or minus the same number is frequently written by using $\pm$; for example, "the solutions are ± 5." Because ± 5 can be written as $\pm\sqrt{25}$, an alternative method of solving this equation is suggested.

> **The Square Root Property of an Equality**
>
> If $x^2 = a$, then $x = \pm\sqrt{a}$.

➡ Solve by taking square roots: $x^2 = 25$

$$x^2 = 25$$
$$\sqrt{x^2} = \sqrt{25}$$
$$x = \pm\sqrt{25} = \pm 5$$

- Take the square root of each side of the equation. Then simplify.

The solutions are 5 and -5.

➡ Solve by taking square roots: $3x^2 = 36$

$$3x^2 = 36$$
$$x^2 = 12$$
$$\sqrt{x^2} = \sqrt{12}$$
$$x = \pm\sqrt{12} = \pm 2\sqrt{3}$$

- Solve for x^2.
- Take the square root of each side.
- Simplify.

The solutions are $2\sqrt{3}$ and $-2\sqrt{3}$.

➡ Solve by taking square roots: $49y^2 - 25 = 0$

$$49y^2 - 25 = 0$$
$$49y^2 = 25$$
$$y^2 = \frac{25}{49}$$
$$\sqrt{y^2} = \sqrt{\frac{25}{49}}$$
$$y = \pm\frac{5}{7}$$

- Solve for y^2.
- Take the square root of each side.
- Simplify.

The solutions are $\frac{5}{7}$ and $-\frac{5}{7}$.

An equation that contains the square of a binomial can be solved by taking square roots.

➡ Solve by taking square roots: $2(x - 1)^2 - 36 = 0$

$$2(x - 1)^2 - 36 = 0$$
$$2(x - 1)^2 = 36$$ • Solve for $(x - 1)^2$.
$$(x - 1)^2 = 18$$
$$\sqrt{(x - 1)^2} = \sqrt{18}$$ • Take the square root of each side of the equation.

$$x - 1 = \pm\sqrt{18} = \pm 3\sqrt{2}$$ • Simplify.

$$x - 1 = 3\sqrt{2} \qquad x - 1 = -3\sqrt{2}$$ • Solve for x.
$$x = 1 + 3\sqrt{2} \qquad x = 1 - 3\sqrt{2}$$

The solutions are $1 + 3\sqrt{2}$ and $1 - 3\sqrt{2}$.

Example 2
Solve by taking square roots:
$x^2 + 16 = 0$

Solution
$$x^2 + 16 = 0$$
$$x^2 = -16$$
$$\sqrt{x^2} = \sqrt{-16}$$

$\sqrt{-16}$ is not a real number.

The equation has no real number solution.

Example 3
Solve by taking square roots:
$5(y - 4)^2 = 25$

Solution
$$5(y - 4)^2 = 25$$
$$(y - 4)^2 = 5$$
$$\sqrt{(y - 4)^2} = \sqrt{5}$$
$$y - 4 = \pm\sqrt{5}$$
$$y = 4 \pm\sqrt{5}$$

The solutions are $4 + \sqrt{5}$ and $4 - \sqrt{5}$.

You Try It 2
Solve by taking square roots:
$x^2 + 81 = 0$

Your solution

You Try It 3
Solve by taking square roots:
$7(z + 2)^2 = 21$

Your solution

Solutions on pp. A35–A36

12.1 Exercises

. . . .

Objective A

Solve by factoring.

1. $x^2 + 2x - 15 = 0$ **2.** $t^2 + 3t - 10 = 0$ **3.** $z^2 - 4z + 3 = 0$ **4.** $s^2 - 5s + 4 = 0$

5. $p^2 + 3p + 2 = 0$ **6.** $v^2 + 6v + 5 = 0$ **7.** $x^2 - 6x + 9 = 0$ **8.** $y^2 - 8y + 16 = 0$

9. $12y^2 + 8y = 0$ **10.** $6x^2 - 9x = 0$ **11.** $r^2 - 10 = 3r$ **12.** $t^2 - 12 = 4t$

13. $3v^2 - 5v + 2 = 0$ **14.** $2p^2 - 3p - 2 = 0$ **15.** $3s^2 + 8s = 3$

16. $3x^2 + 5x = 12$ **17.** $\dfrac{3}{4}z^2 - z = -\dfrac{1}{3}$ **18.** $\dfrac{r^2}{2} = 1 - \dfrac{r}{12}$

19. $4t^2 = 4t + 3$ **20.** $5y^2 + 11y = 12$ **21.** $4v^2 - 4v + 1 = 0$

22. $9s^2 - 6s + 1 = 0$ **23.** $x^2 - 9 = 0$ **24.** $t^2 - 16 = 0$

25. $4y^2 - 1 = 0$ **26.** $9z^2 - 4 = 0$ **27.** $x + 15 = x(x - 1)$

28. $p + 18 = p(p - 2)$ **29.** $r^2 - r - 2 = (2r - 1)(r - 3)$ **30.** $s^2 + 5s - 4 = (2s + 1)(s - 4)$

Objective B

Solve by taking square roots.

31. $x^2 = 36$ **32.** $y^2 = 49$ **33.** $v^2 - 1 = 0$

34. $z^2 - 64 = 0$ **35.** $4x^2 - 49 = 0$ **36.** $9w^2 - 64 = 0$

Solve by taking square roots.

37. $9y^2 = 4$

38. $4z^2 = 25$

39. $16v^2 - 9 = 0$

40. $25x^2 - 64 = 0$

41. $y^2 + 81 = 0$

42. $z^2 + 49 = 0$

43. $w^2 - 24 = 0$

44. $v^2 - 48 = 0$

45. $(x - 1)^2 = 36$

46. $(y + 2)^2 = 49$

47. $2(x + 5)^2 = 8$

48. $4(z - 3)^2 = 100$

49. $9(x - 1)^2 - 16 = 0$

50. $4(y + 3)^2 - 81 = 0$

51. $49(v + 1)^2 - 25 = 0$

52. $81(y - 2)^2 - 64 = 0$

53. $(x - 4)^2 - 20 = 0$

54. $(y + 5)^2 - 50 = 0$

55. $(x + 1)^2 + 36 = 0$

56. $2\left(z - \dfrac{1}{2}\right)^2 = 12$

57. $3\left(v + \dfrac{3}{4}\right)^2 = 36$

APPLYING THE CONCEPTS

Solve for x.

58. $(x^2 - 1)^2 = 9$

59. $(x^2 + 3)^2 = 25$

60. $(6x^2 - 5)^2 = 1$

61. $ax^2 - bx = 0$, $a > 0$ and $b > 0$

62. $ax^2 - b = 0$, $a > 0$ and $b > 0$

63. $x^2 = x$

64. The value P of an initial investment of A dollars after 2 years is given by $P = A(1 + r)^2$, where r is the annual percentage rate earned by the investment. If an initial investment of \$1500 grew to a value of \$1782.15 in 2 years, what was the annual percentage rate?

65. An initial investment of \$5000 grew to a value of \$5832 in 2 years. Use the formula in Exercise 64 to find the annual percentage rate.

66. The kinetic energy of a moving body is given by $E = \dfrac{1}{2}mv^2$, where E is the kinetic energy, m is the mass, and v is the velocity. What is the velocity of a moving body whose mass is 5 kg and whose kinetic energy is 250 newton-meters?

67. On a certain type of street surface, the equation $d = 0.0074v^2$ can be used to approximate the distance d a car traveling v miles per hour will slide when its brakes are applied. After applying the brakes, the owner of a car involved in an accident skidded 40 feet. Did the traffic officer investigating the accident issue the car owner a ticket for speeding if the speed limit is 65 mph?

12.2 Solving Quadratic Equations by Completing the Square

Objective A   *To solve a quadratic equation by completing the square* ....

Recall that a perfect-square trinomial is the square of a binomial.

Perfect-Square Trinomial		Square of a Binomial
$x^2 + 6x + 9$	$=$	$(x + 3)^2$
$x^2 - 10x + 25$	$=$	$(x - 5)^2$
$x^2 + 8x + 16$	$=$	$(x + 4)^2$

For each perfect-square trinomial, the square of $\frac{1}{2}$ of the coefficient of x equals the constant term.

$$x^2 + 6x + 9, \qquad \left(\frac{1}{2} \cdot 6\right)^2 = 9$$

$$x^2 - 10x + 25, \qquad \left[\frac{1}{2}(-10)\right]^2 = 25$$

$$x^2 + 8x + 16, \qquad \left(\frac{1}{2} \cdot 8\right)^2 = 16$$

Adding to a binomial the constant term that makes it a perfect-square trinomial is called **completing the square.**

➡ Complete the square of $x^2 - 8x$. Write the resulting perfect-square trinomial as the square of a binomial.

$$\left[\frac{1}{2}(-8)\right]^2 = 16$$ • Find the constant term.

$$x^2 - 8x + 16$$ • Complete the square of $x^2 - 8x$ by adding the constant term.

$$x^2 - 8x + 16 = (x - 4)^2$$ • Write the resulting perfect-square trinomial as the square of a binomial.

➡ Complete the square of $y^2 + 5y$. Write the resulting perfect-square trinomial as the square of a binomial.

$$\left(\frac{1}{2} \cdot 5\right)^2 = \left(\frac{5}{2}\right)^2 = \frac{25}{4}$$ • Find the constant term.

$$y^2 + 5y + \frac{25}{4}$$ • Complete the square of $y^2 + 5y$ by adding the constant term.

$$y^2 + 5y + \frac{25}{4} = \left(y + \frac{5}{2}\right)^2$$ • Write the resulting perfect-square trinomial as the square of a binomial.

POINT OF INTEREST

Early mathematicians solved quadratic equations by literally *completing the square*. For these mathematicians, all equations had geometric interpretations. They found that a quadratic equation could be solved by making certain figures into squares. See the first Project in Mathematics at the end of this chapter for an idea of how this was done.

A quadratic equation that cannot be solved by factoring can be solved by completing the square. Add to each side of the equation the term that completes the square. Rewrite the quadratic equation in the form $(x + a)^2 = b$. Take the square root of each side of the equation and then solve for x.

➡ Solve by completing the square: $x^2 + 8x - 2 = 0$

$$x^2 + 8x - 2 = 0$$

$$x^2 + 8x = 2$$ ● Add 2 to each side of the equation.

$$x^2 + 8x + \left(\frac{1}{2} \cdot 8\right)^2 = 2 + \left(\frac{1}{2} \cdot 8\right)^2$$ ● Complete the square of $x^2 + 8x$. Add $\left(\frac{1}{2} \cdot 8\right)^2$ to each side of the equation.

$$x^2 + 8x + 16 = 2 + 16$$ ● Simplify.

$$(x + 4)^2 = 18$$ ● Factor the perfect-square trinomial.

$$\sqrt{(x + 4)^2} = \sqrt{18}$$ ● Take the square root of each side of the equation.

$$x + 4 = \pm\sqrt{18} = \pm 3\sqrt{2}$$ ● Solve for x.

$$x + 4 = -3\sqrt{2} \qquad x + 4 = 3\sqrt{2}$$
$$x = -4 - 3\sqrt{2} \qquad x = -4 + 3\sqrt{2}$$

Check:

$$x^2 + 8x - 2 = 0 \qquad\qquad\qquad x^2 + 8x - 2 = 0$$

$(-4 - 3\sqrt{2})^2 - 8(-4 - 3\sqrt{2}) - 2$	0	$(-4 + 3\sqrt{2})^2 - 8(-4 + 3\sqrt{2}) - 2$	0
$16 + 24\sqrt{2} + 18 - 32 - 24\sqrt{2} - 2$	0	$16 - 24\sqrt{2} + 18 - 32 + 24\sqrt{2} - 2$	0
	$0 = 0$		$0 = 0$

The solutions are $-4 - 3\sqrt{2}$ and $-4 + 3\sqrt{2}$.

If the coefficient of the second-degree term is not 1, a step in completing the square is to multiply each side of the equation by the reciprocal of that coefficient.

➡ Solve by completing the square: $2x^2 - 3x + 1 = 0$

$$2x^2 - 3x + 1 = 0$$

$$2x^2 - 3x = -1$$ ● Subtract 1 from each side of the equation.

$$\frac{1}{2}(2x^2 - 3x) = \frac{1}{2} \cdot (-1)$$ ● To complete the square, the coefficient of x^2 must be 1. Multiply each side of the equation by $\frac{1}{2}$.

$$x^2 - \frac{3}{2}x = -\frac{1}{2}$$

$$x^2 - \frac{3}{2}x + \left[\frac{1}{2}\left(-\frac{3}{2}\right)\right]^2 = -\frac{1}{2} + \left[\frac{1}{2}\left(-\frac{3}{2}\right)\right]^2$$ ● Complete the square. Add $\left[\frac{1}{2}\left(-\frac{3}{2}\right)\right]^2$ to each side of the equation.

$$x^2 - \frac{3}{2}x + \frac{9}{16} = -\frac{1}{2} + \frac{9}{16}$$ ● Simplify.

$$\left(x - \frac{3}{4}\right)^2 = \frac{1}{16}$$ ● Factor the perfect-square trinomial.

$$\sqrt{\left(x - \frac{3}{4}\right)^2} = \sqrt{\frac{1}{16}}$$ ● Take the square root of each side of the equation.

$$x - \frac{3}{4} = \pm\frac{1}{4}$$ ● Solve for x.

$$x - \frac{3}{4} = -\frac{1}{4} \qquad x - \frac{3}{4} = \frac{1}{4}$$

$$x = \frac{1}{2} \qquad\qquad x = 1$$

The solutions are $\frac{1}{2}$ and 1.

Example 1
Solve by completing the square:
$2x^2 - 4x - 1 = 0$

Solution

$$2x^2 - 4x - 1 = 0$$
$$2x^2 - 4x = 1$$
$$\frac{1}{2}(2x^2 - 4x) = \frac{1}{2} \cdot 1$$
$$x^2 - 2x = \frac{1}{2}$$

Complete the square.

$$x^2 - 2x + 1 = \frac{1}{2} + 1$$
$$(x - 1)^2 = \frac{3}{2}$$
$$\sqrt{(x - 1)^2} = \sqrt{\frac{3}{2}}$$
$$x - 1 = \pm\sqrt{\frac{3}{2}} = \pm\frac{\sqrt{6}}{2}$$
$$x - 1 = \frac{\sqrt{6}}{2} \qquad x - 1 = -\frac{\sqrt{6}}{2}$$
$$x = 1 + \frac{\sqrt{6}}{2} = \frac{2 + \sqrt{6}}{2}$$
$$x = 1 - \frac{\sqrt{6}}{2} = \frac{2 - \sqrt{6}}{2}$$

The solutions are $\dfrac{2 + \sqrt{6}}{2}$ and $\dfrac{2 - \sqrt{6}}{2}$.

Check.

$$2x^2 - 4x - 1 = 0$$

$$
\begin{array}{c|c}
2\left(\dfrac{2 + \sqrt{6}}{2}\right)^2 - 4\left(\dfrac{2 + \sqrt{6}}{2}\right) - 1 & 0 \\
2\left(\dfrac{4 + 4\sqrt{6} + 6}{4}\right) - 2(2 + \sqrt{6}) - 1 & 0 \\
2 + 2\sqrt{6} + 3 - 4 - 2\sqrt{6} - 1 & 0 \\
0 = 0
\end{array}
$$

$$2x^2 - 4x - 1 = 0$$

$$
\begin{array}{c|c}
2\left(\dfrac{2 - \sqrt{6}}{2}\right)^2 - 4\left(\dfrac{2 - \sqrt{6}}{2}\right) - 1 & 0 \\
2\left(\dfrac{4 - 4\sqrt{6} + 6}{4}\right) - 2(2 - \sqrt{6}) - 1 & 0 \\
2 - 2\sqrt{6} + 3 - 4 + 2\sqrt{6} - 1 & 0 \\
0 = 0
\end{array}
$$

You Try It 1
Solve by completing the square:
$3x^2 - 6x - 2 = 0$

Your solution

Solutions on p. A36

Example 2

Solve by completing the square:
$x^2 + 4x + 5 = 0$

Solution

$x^2 + 4x + 5 = 0$
$\qquad x^2 + 4x = -5$

Complete the square.

$x^2 + 4x + 4 = -5 + 4$
$\qquad (x + 2)^2 = -1$
$\qquad \sqrt{(x + 2)^2} = \sqrt{-1}$

$\sqrt{-1}$ is not a real number.

The quadratic equation has no real number solution.

You Try It 2

Solve by completing the square:
$x^2 + 6x + 12 = 0$

Your solution

Example 3

Solve $\dfrac{x^2}{4} + \dfrac{3x}{2} + 1 = 0$ by completing the square. Approximate the solutions to the nearest thousandth.

Solution

$$\frac{x^2}{4} + \frac{3x}{2} + 1 = 0$$
$$4\left(\frac{x^2}{4} + \frac{3x}{2} + 1\right) = 4(0)$$
$$x^2 + 6x + 4 = 0$$
$$x^2 + 6x = -4$$

Complete the square.

$x^2 + 6x + 9 = -4 + 9$
$\qquad (x + 3)^2 = 5$
$\qquad \sqrt{(x + 3)^2} = \sqrt{5}$
$\qquad x + 3 = \pm\sqrt{5}$

$$x + 3 = \sqrt{5} \qquad\qquad x + 3 = -\sqrt{5}$$
$$x = -3 + \sqrt{5} \qquad\qquad x = -3 - \sqrt{5}$$
$$\approx -3 + 2.236 \qquad\qquad \approx -3 - 2.236$$
$$\approx -0.764 \qquad\qquad \approx -5.236$$

The solutions are approximately -0.764 and -5.236.

You Try It 3

Solve $\dfrac{x^2}{8} + x + 1 = 0$ by completing the square. Approximate the solutions to the nearest thousandth.

Your solution

Solutions on pp. A36–A37

12.2 Exercises

. . . .

Objective A

Solve by completing the square.

1. $x^2 + 2x - 3 = 0$ **2.** $y^2 + 4y - 5 = 0$ **3.** $z^2 - 6z - 16 = 0$ **4.** $w^2 + 8w - 9 = 0$

5. $x^2 = 4x - 4$ **6.** $z^2 = 8z - 16$ **7.** $v^2 - 6v + 13 = 0$ **8.** $x^2 + 4x + 13 = 0$

9. $y^2 + 5y + 4 = 0$ **10.** $v^2 - 5v - 6 = 0$ **11.** $w^2 + 7w = 8$ **12.** $y^2 + 5y = -4$

13. $v^2 + 4v + 1 = 0$ **14.** $y^2 - 2y - 5 = 0$ **15.** $x^2 + 6x = 5$

16. $w^2 - 8w = 3$ **17.** $\dfrac{z^2}{2} = z + \dfrac{1}{2}$ **18.** $\dfrac{y^2}{10} = y - 2$

19. $p^2 + 3p = 1$ **20.** $r^2 + 5r = 2$ **21.** $t^2 - 3t = -2$

22. $z^2 - 5z = -3$ **23.** $v^2 + v - 3 = 0$ **24.** $x^2 - x = 1$

25. $y^2 = 7 - 10y$ **26.** $v^2 = 14 + 16v$ **27.** $r^2 - 3r = 5$

28. $s^2 + 3s = -1$ **29.** $t^2 - t = 4$ **30.** $y^2 + y - 4 = 0$

31. $x^2 - 3x + 5 = 0$ **32.** $z^2 + 5z + 7 = 0$ **33.** $2t^2 - 3t + 1 = 0$

Solve by completing the square.

34. $2x^2 - 7x + 3 = 0$ **35.** $2r^2 + 5r = 3$ **36.** $2y^2 - 3y = 9$ **37.** $2s^2 = 7s - 6$

38. $2x^2 = 3x + 20$ **39.** $2v^2 = v + 1$ **40.** $2z^2 = z + 3$ **41.** $3r^2 + 5r = 2$

42. $3t^2 - 8t = 3$ **43.** $3y^2 + 8y + 4 = 0$ **44.** $3z^2 - 10z - 8 = 0$ **45.** $4x^2 + 4x - 3 = 0$

46. $4v^2 + 4v - 15 = 0$ **47.** $6s^2 + 7s = 3$ **48.** $6z^2 = z + 2$ **49.** $6p^2 = 5p + 4$

50. $6t^2 = t - 2$ **51.** $4v^2 - 4v - 1 = 0$ **52.** $2s^2 - 4s - 1 = 0$

Solve by completing the square. Approximate the solutions to the nearest thousandth.

53. $y^2 + 3y = 5$ **54.** $w^2 + 5w = 2$ **55.** $2z^2 - 3z = 7$

56. $2x^2 + 3x = 11$ **57.** $4x^2 + 6x - 1 = 0$ **58.** $4x^2 + 2x - 3 = 0$

APPLYING THE CONCEPTS

59. Explain why the equation $(x - 2)^2 = -4$ does not have a real-number
[W]   solution.

Solve.

60. $\dfrac{x^2}{6} - \dfrac{x}{3} = 1$ **61.** $\sqrt{x + 2} = x - 4$ **62.** $\sqrt{3x + 4} - x = 2$

63. $\dfrac{x}{3} + \dfrac{3}{x} = \dfrac{8}{3}$ **64.** $\dfrac{x + 1}{2} + \dfrac{3}{x - 1} = 4$ **65.** $\dfrac{x - 2}{3} + \dfrac{2}{x + 2} = 4$

66. $4\sqrt{x + 1} - x = 4$ **67.** $\sqrt{2x^2 + 7} = x + 2$ **68.** $3\sqrt{x - 1} + 3 = x$

69. A basketball player shoots at a basket 25 ft away. The height of the ball above the ground at time t is given by $h = 16t^2 - 32t + 6.5$. How many seconds after the ball is released does it hit the basket? *Hint:* When it hits the basket, $h = 11$ ft.

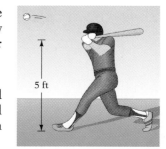
5 ft

70. A ball player hits a ball. The height of the ball above the ground can be approximated by the equation $h = 16t^2 - 76t + 5$. When will the ball hit the ground? *Hint:* The ball strikes the ground when $h = 0$ ft.

12.3

Solving Quadratic Equations by Using the Quadratic Formula

Objective A   *To solve a quadratic equation by using the quadratic formula* ....

Any quadratic equation can be solved by completing the square. Applying this method to the standard form of a quadratic equation produces a formula that can be used to solve any quadratic equation.

Solve $ax^2 + bx + c = 0$ by completing the square.

$$ax^2 + bx + c = 0$$

Add the opposite of the constant term to each side of the equation.

$$ax^2 + bx + c + (-c) = 0 + (-c)$$
$$ax^2 + bx = -c$$

Multiply each side of the equation by the reciprocal of a, the coefficient of x^2.

$$\frac{1}{a}(ax^2 + bx) = \frac{1}{a}(-c)$$
$$x^2 + \frac{b}{a}x = -\frac{c}{a}$$

Complete the square by adding $\left(\frac{1}{2} \cdot \frac{b}{a}\right)^2$ to each side of the equation.

$$x^2 + \frac{b}{a}x + \left(\frac{1}{2} \cdot \frac{b}{a}\right)^2 = \left(\frac{1}{2} \cdot \frac{b}{a}\right)^2 - \frac{c}{a}$$
$$x^2 + \frac{b}{a}x + \frac{b^2}{4a^2} = \frac{b^2}{4a^2} - \frac{c}{a}$$

Simplify the right side of the equation.

$$x^2 + \frac{b}{a}x + \frac{b^2}{4a^2} = \frac{b^2}{4a^2} - \left(\frac{c}{a} \cdot \frac{4a}{4a}\right)$$
$$x^2 + \frac{b}{a}x + \frac{b^2}{4a^2} = \frac{b^2}{4a^2} - \frac{4ac}{4a^2}$$
$$x^2 + \frac{b}{a}x + \frac{b^2}{4a^2} = \frac{b^2 - 4ac}{4a^2}$$

Factor the perfect-square trinomial on the left side of the equation.

$$\left(x + \frac{b}{2a}\right)^2 = \frac{b^2 - 4ac}{4a^2}$$

Take the square root of each side of the equation.

$$\sqrt{\left(x + \frac{b}{2a}\right)^2} = \sqrt{\frac{b^2 - 4ac}{4a^2}}$$
$$\left(x + \frac{b}{2a}\right) = \pm\frac{\sqrt{b^2 - 4ac}}{2a}$$

Solve for x.

$$x + \frac{b}{2a} = \frac{\sqrt{b^2 - 4ac}}{2a} \qquad x + \frac{b}{2a} = -\frac{\sqrt{b^2 - 4ac}}{2a}$$
$$x = -\frac{b}{2a} + \frac{\sqrt{b^2 - 4ac}}{2a} \qquad x = -\frac{b}{2a} - \frac{\sqrt{b^2 - 4ac}}{2a}$$
$$= \frac{-b + \sqrt{b^2 - 4ac}}{2a} \qquad = \frac{-b - \sqrt{b^2 - 4ac}}{2a}$$

The Quadratic Formula

The solutions of $ax^2 + bx + c = 0$, $a \neq 0$, are

$$\frac{-b \pm \sqrt{b^2 - 4ac}}{2a}$$

➡ Solve by using the quadratic formula: $2x^2 = 4x - 1$.

First write the equation in standard form. Then use the quadratic formula.

$$2x^2 = 4x - 1$$
$$2x^2 - 4x + 1 = 0$$

- Subtract 4x and add 1 to each side of the equation.

$$x = \frac{-b \pm \sqrt{b^2 - 4ac}}{2a}$$

- The quadratic formula.

$$= \frac{-(-4) \pm \sqrt{(-4)^2 - (4 \cdot 2 \cdot 1)}}{2 \cdot 2}$$

- $a = 2$, $b = -4$, $c = 1$. Replace a, b, and c by their values.

$$= \frac{4 \pm \sqrt{16 - 8}}{4} = \frac{4 \pm \sqrt{8}}{4}$$

- Simplify.

$$= \frac{4 \pm 2\sqrt{2}}{4} = \frac{2 \pm \sqrt{2}}{2}$$

The solutions are $\frac{2 - \sqrt{2}}{2}$ and $\frac{2 + \sqrt{2}}{2}$.

Example 1
Solve by using the quadratic formula:
$2x^2 - 3x + 1 = 0$

Solution
$2x^2 - 3x + 1 = 0$ • $a = 2$, $b = -3$, $c = 1$
$$x = \frac{-(-3) \pm \sqrt{(-3)^2 - 4(2)(1)}}{2 \cdot 2}$$
$$= \frac{3 \pm \sqrt{9 - 8}}{4} = \frac{3 \pm \sqrt{1}}{4} = \frac{3 \pm 1}{4}$$
$$x = \frac{3 + 1}{4} \qquad x = \frac{3 - 1}{4}$$
$$= \frac{4}{4} = 1 \qquad = \frac{2}{4} = \frac{1}{2}$$

The solutions are 1 and $\frac{1}{2}$.

You Try It 1
Solve by using the quadratic formula:
$3x^2 + 4x - 4 = 0$

Your solution

Example 2
Solve by using the quadratic formula:
$\frac{x^2}{2} = 2x - \frac{5}{4}$

Solution $\frac{x^2}{2} = 2x - \frac{5}{4}$
$$4\left(\frac{x^2}{2}\right) = 4\left(2x - \frac{5}{4}\right)$$
$$2x^2 = 8x - 5$$
$2x^2 - 8x + 5 = 0$ • $a = 2$, $b = -8$, $c = 5$
$$x = \frac{-(-8) \pm \sqrt{(-8)^2 - 4(2)(5)}}{2 \cdot 2}$$
$$= \frac{8 \pm \sqrt{64 - 40}}{4} = \frac{8 \pm \sqrt{24}}{4}$$
$$= \frac{8 \pm 2\sqrt{6}}{4} = \frac{4 \pm \sqrt{6}}{2}$$

The solutions are $\frac{4 + \sqrt{6}}{2}$ and $\frac{4 - \sqrt{6}}{2}$.

You Try It 2
Solve by using the quadratic formula:
$\frac{x^2}{4} + \frac{x}{2} = \frac{1}{4}$

Your solution

Solutions on p. A37

12.3 Exercises

· ·

Objective A

Solve by using the quadratic formula.

1. $x^2 - 4x - 5 = 0$ **2.** $y^2 + 3y + 2 = 0$ **3.** $z^2 - 2z - 15 = 0$ **4.** $v^2 + 5v + 4 = 0$

5. $y^2 = 2y + 3$ **6.** $w^2 = 3w + 18$ **7.** $r^2 = 5 - 4r$ **8.** $z^2 = 3 - 2z$

9. $2y^2 - y - 1 = 0$ **10.** $2t^2 - 5t + 3 = 0$ **11.** $w^2 + 3w + 5 = 0$

12. $x^2 - 2x + 6 = 0$ **13.** $p^2 - p = 0$ **14.** $2v^2 + v = 0$

15. $4t^2 - 9 = 0$ **16.** $4s^2 - 25 = 0$ **17.** $4y^2 + 4y = 15$

18. $6y^2 + 5y - 4 = 0$ **19.** $2x^2 + x + 1 = 0$ **20.** $3r^2 - r + 2 = 0$

21. $\dfrac{1}{2}t^2 - t = \dfrac{5}{2}$ **22.** $y^2 - 4y = 6$ **23.** $\dfrac{1}{3}t^2 + 2t - \dfrac{1}{3} = 0$

24. $z^2 + 4z + 1 = 0$ **25.** $w^2 = 4w + 9$ **26.** $y^2 = 8y + 3$

27. $9y^2 + 6y - 1 = 0$ **28.** $9s^2 - 6s - 2 = 0$ **29.** $4p^2 + 4p + 1 = 0$

30. $9z^2 + 12z + 4 = 0$ **31.** $\dfrac{x^2}{2} = x - \dfrac{5}{4}$ **32.** $r^2 = \dfrac{5}{3}r - 2$

33. $4p^2 + 16p = -11$ **34.** $4y^2 - 12y = -1$ **35.** $4x^2 = 4x + 11$

Solve by using the quadratic formula.

36. $4s^2 + 12s = 3$ **37.** $9v^2 = -30v - 23$ **38.** $9t^2 = 30t + 17$

Solve by using the quadratic formula. Approximate the solutions to the nearest thousandth.

39. $x^2 - 2x - 21 = 0$ **40.** $y^2 + 4y - 11 = 0$ **41.** $s^2 - 6s - 13 = 0$

42. $w^2 + 8w - 15 = 0$ **43.** $2p^2 - 7p - 10 = 0$ **44.** $3t^2 - 8t - 1 = 0$

45. $4z^2 + 8z - 1 = 0$ **46.** $4x^2 + 7x + 1 = 0$ **47.** $5v^2 - v - 5 = 0$

APPLYING THE CONCEPTS

48. Factoring, completing the square, and using the quadratic formula
[W] are three methods of solving quadratic equations. Describe each method, and cite the advantages and disadvantages of using each.

49. Explain why the equation $0x^2 + 3x + 4 = 0$ cannot be solved by the
[W] quadratic formula.

50. Solve $x^2 + ax + b = 0$ for x.

51. True or false?
   a. The equations $x = \sqrt{12 - x}$ and $x^2 = 12 - x$ have the same solution.
   b. If $\sqrt{a} + \sqrt{b} = c$, then $a + b = c^2$.
   c. $\sqrt{9} = \pm 3$
   d. $\sqrt{x^2} = |x|$

Solve.

52. $\sqrt{x + 3} = x - 3$ **53.** $\sqrt{x + 4} = x + 4$ **54.** $\sqrt{x + 1} = x - 1$

55. $\sqrt{x^2 + 2x + 1} = x - 1$ **56.** $\dfrac{x}{4} + \dfrac{3}{x} = \dfrac{5}{2}$ **57.** $\dfrac{x + 1}{5} - \dfrac{4}{x - 1} = 2$

58. An L-shaped sidewalk from the parking lot to a memorial is shown in the figure at the right. The distance directly across the grass to the memorial is 650 ft. The distance to the corner is 600 ft. Find the distance from the corner to the memorial.

59. A commuter plane leaves an airport traveling due south at 400 mph. Another plane leaving at the same time travels due east at 300 mph. Find the distance between the two planes after two hours.

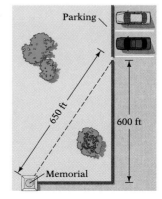

Parking

650 ft

600 ft

Memorial

12.4 Graphing Quadratic Equations in Two Variables

Objective A   *To graph a quadratic equation of the form $y = ax^2 + bx + c$*

An equation of the form $y = ax^2 + bx + c$, $a \neq 0$, is a **quadratic equation in two variables.** Examples of quadratic equations in two variables are shown at the right.

$$y = 3x^2 - x + 1$$
$$y = -x^2 - 3$$
$$y = 2x^2 - 5x$$

For these equations, y is a function of x, and we can write $f(x) = ax^2 + bx + c$. This represents a **quadratic function.**

The graph of a quadratic equation in two variables is a **parabola**. The graph is "cup shaped" and opens either up or down. The graphs of two parabolas are shown below.

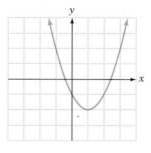

Parabola that opens up

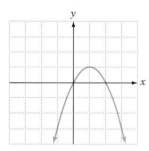

Parabola that opens down

➡ Graph $y = x^2 - 2x - 3$.

x	y
0	−3
1	−4
−1	0
2	−3
3	0

● Find several solutions of the equation. Because the graph is not a straight line, several solutions must be found in order to determine the cup shape. Display the ordered-pair solutions in a table.

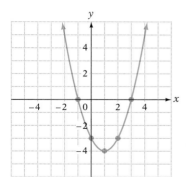

● Graph the ordered-pair solutions on a rectangular coordinate system. Draw a parabola through the points.

➡ Graph $y = -2x^2 + 1$.

x	y
0	1
1	−1
−1	−1
2	−7
−2	−7

● Find enough solutions of the equation to determine the cup shape. Display the ordered-pair solutions in a table.

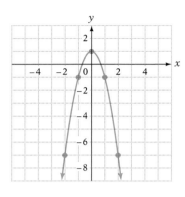

● Graph the ordered-pair solutions on a rectangular coordinate system. Draw a parabola through the points.

Note in the example on page 467 that the coefficient of x^2 is **positive** and the graph **opens up.** In the example above, the coefficient of x^2 is **negative** and the graph **opens down.**

Example 1   Graph $y = x^2 - 2x$.

Solution

x	y
0	0
1	−1
−1	3
2	0
3	3

You Try It 1   Graph $y = x^2 + 2$.

Your solution

Example 2   Graph $y = -x^2 + 4x - 4$.

Solution

x	y
0	−4
1	−1
2	0
3	−1
4	−4

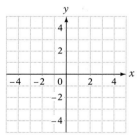

You Try It 2   Graph $y = -x^2 - 2x - 1$.

Your solution

Solutions on p. A37

12.4 Exercises

Objective A

Graph.

1. $y = x^2$

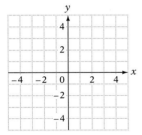

2. $y = -x^2$

3. $y = -x^2 + 1$

4. $y = x^2 - 1$

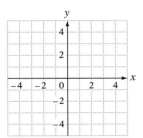

5. $y = 2x^2$

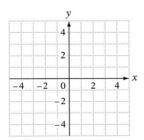

6. $y = \frac{1}{2}x^2$

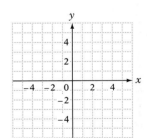

7. $y = -\frac{1}{2}x^2 + 1$

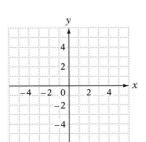

8. $y = 2x^2 - 1$

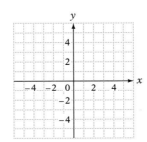

9. $y = x^2 - 4x$

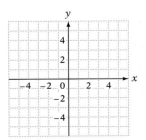

10. $y = x^2 + 4x$

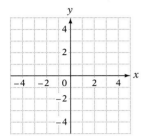

11. $y = x^2 - 2x + 3$

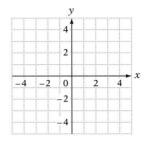

12. $y = x^2 - 4x + 2$

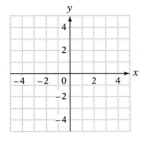

Graph.

13. $y = -x^2 + 2x + 3$

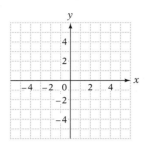

14. $y = -x^2 - 2x + 3$

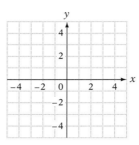

15. $y = -x^2 + 4x - 4$

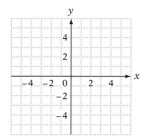

16. $y = -x^2 + 6x - 9$

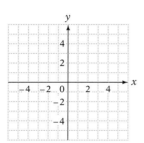

17. $y = (x - 2)^2$

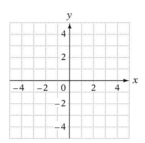

18. $y = -(x + 1)^2$

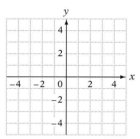

APPLYING THE CONCEPTS

Determine whether the graph of the equation opens up or down.

19. $y = -\dfrac{1}{3}x^2 + 5$

20. $y = x^2 - 2x + 3$

21. $y = -x^2 + 4x - 1$

Show that the equation is a quadratic equation in two variables by writing it in the form $y = ax^2 + bx + c$.

22. $y + 1 = (x - 4)^2$

23. $y - 2 = 3(x + 1)^2$

24. $y - 4 = 2(x - 3)^2$

The x-intercepts of the graph of $y = ax^2 + bx + c$ occur when $y = 0$. Therefore, the x-coordinate of an x-intercept is a solution of $ax^2 + bx + c = 0$. Determine the x-intercepts of the graphs of the following equations.

25. $y = x^2 - 9$

26. $y = x^2 - 4x$

27. $y = 2x^2 - x - 1$

Evaluate the function.

28. Find $f(3)$ when $f(x) = x^2 + 3$.

29. Find $g(2)$ when $g(x) = x^2 - 2x - 3$.

30. Find $S(-2)$ when $S(t) = 2t^2 - 3t - 1$.

31. Find $P(-3)$ when $P(x) = 3x^2 - 6x - 7$.

12.5

Application Problems

. . . .

Objective A *To solve application problems* ....

The application problems in this section are varieties of those problems solved earlier in the text. Each of the strategies for the problems in this section will result in a quadratic equation.

➡ In 5 h, two campers rowed 12 mi down a stream and then rowed back to their campsite. The rate of the stream's current was 1 mph. Find the rate at which the campers rowed.

> **Strategy for Solving an Application Problem**
>
> 1. Determine the type of problem. For example, is it a distance-rate problem, a geometry problem, a work problem, or an age problem?

The problem is a distance-rate problem.

> 2. Choose a variable to represent the unknown quantity. Write numerical or variable expressions for all the remaining quantities. These results can be recorded in a table.

The unknown rate of the campers: r

	Distance	÷	Rate	=	Time
Downstream	12	÷	$r + 1$	=	$\dfrac{12}{r + 1}$
Upstream	12	÷	$r - 1$	=	$\dfrac{12}{r - 1}$

> 3. Determine how the quantities are related.

The total time of the trip was 5 h.

$$\frac{12}{r + 1} + \frac{12}{r - 1} = 5$$

$$(r + 1)(r - 1)\left(\frac{12}{r + 1} + \frac{12}{r - 1}\right) = (r + 1)(r - 1)5$$

$$(r - 1)12 + (r + 1)12 = (r^2 - 1)5$$

$$12r - 12 + 12r + 12 = 5r^2 - 5$$

$$24r = 5r^2 - 5$$

$$0 = 5r^2 - 24r - 5$$

$$0 = (5r + 1)(r - 5)$$

$$5r + 1 = 0 \qquad\qquad r - 5 = 0$$

$$5r = -1 \qquad\qquad\quad r = 5$$

$$r = -\frac{1}{5}$$

The rowing rate was 5 mph. The solution $r = -\dfrac{1}{5}$ is not possible, because the rate cannot be a negative number.

Example 1

A painter and the painter's apprentice working together can paint a room in 2 h. The apprentice working alone requires 3 more hours to paint the room than the painter requires working alone. How long does it take the painter working alone to paint the room?

You Try It 1

The length of a rectangle is 2 m more than the width. The area is 15 m². Find the width.

Strategy

- This is a work problem.
- Time for the painter to paint the room: t
  Time for the apprentice to paint the room: $t + 3$

Your strategy

	Rate	Time	Part
Painter	$\dfrac{1}{t}$	2	$\dfrac{2}{t}$
Apprentice	$\dfrac{1}{t+3}$	2	$\dfrac{2}{t+3}$

- The sum of the parts of the task completed must equal 1.

Solution

$$\frac{2}{t} + \frac{2}{t+3} = 1$$

$$t(t+3)\left(\frac{2}{t} + \frac{2}{t+3}\right) = t(t+3) \cdot 1$$

$$(t+3)2 + t(2) = t(t+3)$$
$$2t + 6 + 2t = t^2 + 3t$$
$$0 = t^2 - t - 6$$
$$0 = (t-3)(t+2)$$

$$t - 3 = 0 \qquad t + 2 = 0$$
$$t = 3 \qquad\quad t = -2$$

The time is 3 h.

The solution $t = -2$ is not possible.

Your solution

Solution on p. A38

12.5 Exercises

· ·

Objective A   *Application Problems*

Solve.

1. The area of the batter's box on a major league baseball field is 24 ft^2. The length of the batter's box is 2 ft more than the width. Find the length and width of the batter's box. (Area = lw)

2. The length of the batter's box on a softball field is 1 ft less than twice the width. The area of the batter's box is 15 ft^2. Find the length and width of the batter's box.

3. The length of a swimming pool is twice the width. The area of the pool is 5000 ft^2. Find the length and width of the pool. (Area = lw)

4. The length of the singles tennis court is 24 ft more than twice the width. The area of the tennis court is 2106 ft^2. Find the length and width of the court. (Area = lw)

5. The sum of the squares of two positive odd integers is 130. Find the two integers.

6. The sum of the squares of two consecutive positive even integers is 164. Find the two integers.

7. The sum of two integers is 12. The product of the two integers is 35. Find the two integers.

8. The difference between two integers is 4. The product of the two integers is 60. Find the integers.

9. Twice an integer equals the square of the integer. Find the integer.

10. The square of an integer equals the integer. Find the integer.

11. The sum of the squares of three consecutive odd integers is 371. Find the integers.

12. The sum of the squares of three consecutive even integers is 56. Find the integers.

13. The height of a triangle is 2 m more than twice the length of the base. The area of the triangle is 20 m^2. Find the height of the triangle and the length of the base.

Solve.

14. The length of a rectangle is 4 ft more than twice the width. The area of the rectangle is 160 ft². Find the length and width of the rectangle.

15. One computer takes 21 min longer to calculate the value of a complex equation than a second computer. Working together, these computers complete the calculation in 10 min. How long would it take each computer, working separately, to calculate the value?

16. A tank has two drains. One drain takes 16 minutes longer to empty the tank than does a second drain. With both drains open, the tank is emptied in 6 minutes. How long would it take each drain, working alone, to empty the tank?

17. Using one engine of a ferryboat, it takes 6 h longer to cross a channel than it does using a second engine alone. Using both engines, the ferryboat can make the crossing in 4 h. How long would it take each engine, working alone, to power the ferryboat across the channel?

18. An apprentice mason takes 8 h longer to build a small fireplace than an experienced mason. Working together, they can build the fireplace in 3 h. How long would it take each mason, working alone, to complete the fireplace?

19. It took a small plane 2 h more to fly 375 mi against the wind than it took to fly the same distance with the wind. The rate of the wind was 25 mph. Find the rate of the plane in calm air.

20. It took a motorboat 1 h more to travel 36 mi against the current than it took to go 36 mi with the current. The rate of the current was 3 mph. Find the rate of the boat in calm water.

APPLYING THE CONCEPTS

21. The sum of the squares of four consecutive integers is 86. Find the four integers.

22. The hypotenuse of a right triangle is $\sqrt{13}$ cm. One leg is 1 cm shorter than twice the length of the other leg. Find the lengths of the legs of the right triangle.

23. The radius of a large pizza is 1 in. less than twice the radius of a small pizza. The difference between the areas of the two pizzas is 33π in². Find the radius of the large pizza.

24. The perimeter of a rectangular garden is 54 ft. The area of the garden is 180 ft². Find the length and width of the garden.

Projects in Mathematics

Geometric Construction of Completing the Square

Completing the square as a method of solving a quadratic equation has been known for centuries. The Persian mathematician Al-Khwarismi used this method in a textbook written around A.D. 825. The method was very geometric. That is, Al-Khwarismi literally completed a square. To understand how this method works, consider the following geometric shapes: a square whose area is x^2, a rectangle whose area is x, and another square whose area is 1.

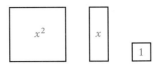

Now consider the expression $x^2 + 6x$. From our discussion in this chapter, to complete the square, we added $\left(\dfrac{1}{2} \cdot 6\right)^2 = 3^2 = 9$ to the expression. Here is the geometric construction that Al-Khwarismi used.

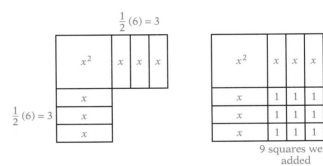

9 squares were added

Note that it is necessary to add 9 squares to the figure to "complete the square." One of the difficulties of using a geometric method such as this is that it cannot easily be extended to $x^2 - 6x$. There is no way to draw an area of $-6x$! That really did not bother Al-Khwarismi much. Negative numbers were not a significant part of mathematics until well into the 13th century.

Exercises:

1. Show how Al-Khwarismi would have completed the square for $x^2 + 4x$.
2. Show how Al-Khwarismi would have completed the square for $x^2 + 10x$.
3. Do the geometric constructions for Exercises 1 and 2 correspond to the algebraic method shown in this chapter?

Graphing a Quadratic Function by Using a Graphing Calculator

A parabola can be graphed by using a graphing calculator. Here are the keystrokes to graph $y = \frac{2}{5}x^2 + 3x - 2$. By changing the keystrokes that are highlighted in blue below, you can enter your own equations to graph. The graph is shown below.

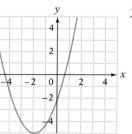

$$y = \frac{2}{5}x^2 + 3x - 2$$

TI-82

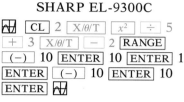

SHARP EL-9300C

CASIO fx-7700GB

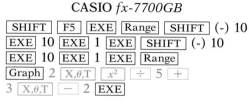

The keystrokes given above used the $[x^2]$ key. A variable to a power also can be entered by using the exponent key. This would be necessary if you were trying to graph a higher-degree polynomial. Here are the keystrokes for using the exponent key.

TI-82

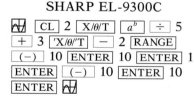

SHARP EL-9300C

CASIO fx-7700GB

Use a graphing calculator to verify some of the graphs you made for the exercises in Section 12.4.

Chapter Summary

Key Words

A *quadratic equation* is an equation of the form $ax^2 + bx + c = 0$, where $a \neq 0$. A quadratic equation is also called a *second-degree equation*.

A quadratic equation is in *standard form* when the polynomial is in descending order and equal to zero.

A *quadratic equation in two variables* is given by $y = ax^2 + bx + c$, where $a \neq 0$.

A *quadratic function* is given by $f(x) = ax^2 + bx + c$, where $a \neq 0$. The graph of a quadratic function is a *parabola*.

Essential Rules

The Quadratic Formula

$$x = \frac{-b \pm \sqrt{b^2 - 4ac}}{2a}$$

The Square Root Property

If $x^2 = a$, then $x = \pm\sqrt{a}$.

Chapter Review

. . . .

SECTION 12.1

1. Solve by factoring: $6x^2 + 13x - 28 = 0$

2. Solve by factoring: $12x^2 + 10 = 29x$

3. Solve by factoring: $(x + 9)^2 = x + 11$

4. Solve by factoring: $6x(x + 1) = x - 1$

5. Solve by taking square roots: $49x^2 = 25$

6. Solve by taking square roots: $4y^2 + 9 = 0$

7. Solve by taking square roots:
$(y + 4)^2 - 25 = 0$

8. Solve by taking square roots:
$(x + 2)^2 - 24 = 0$

9. Solve by taking square roots:
$\left(x - \dfrac{1}{2}\right)^2 = \dfrac{9}{4}$

SECTION 12.2

10. Solve by completing the square:
$x^2 + 2x - 24 = 0$

11. Solve by completing the square:
$2x^2 + 5x = 12$

12. Solve by completing the square:
$x^2 - 4x + 1 = 0$

13. Solve by completing the square:
$x^2 + 6x + 12 = 0$

14. Solve by completing the square:
$4x^2 + 16x = 7$

SECTION 12.3

15. Solve by using the quadratic formula:
$x^2 + 5x - 6 = 0$

16. Solve by using the quadratic formula:
$2x^2 + 3 = 5x$

17. Solve by using the quadratic formula: $x^2 - 4x + 8 = 0$

18. Solve by using the quadratic formula: $x^2 - 3x - 5 = 0$

19. Solve by using the quadratic formula: $2x^2 + 5x + 2 = 0$

SECTION 12.4

20. Graph $y = -3x^2$.

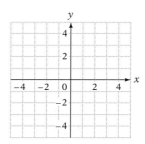

21. Graph $y = -\frac{1}{4}x^2$.

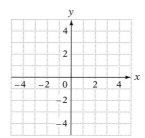

22. Graph $y = 2x^2 + 1$.

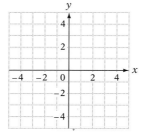

23. Graph $y = x^2 - 4x + 3$.

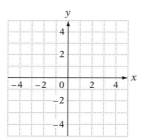

24. Graph $y = -x^2 + 4x - 5$.

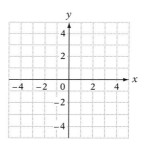

SECTION 12.5

25. It took a hawk half an hour more to fly 70 mi against the wind than to go 40 mi with the wind. The rate of the wind was 5 mph. Find the rate of the hawk in calm air.

Chapter Test

1. Solve by factoring: $x^2 - 5x - 6 = 0$

2. Solve by factoring: $3x^2 + 7x = 20$

3. Solve by taking square roots:
   $2(x - 5)^2 - 50 = 0$

4. Solve by taking square roots:
   $3(x + 4)^2 - 60 = 0$

5. Solve by completing the square:
   $x^2 + 4x - 16 = 0$

6. Solve by completing the square:
   $x^2 + 3x = 8$

7. Solve by completing the square:
   $2x^2 - 6x + 1 = 0$

8. Solve by completing the square:
   $2x^2 + 8x = 3$

9. Solve by using the quadratic formula:
   $x^2 + 4x + 2 = 0$

10. Solve by using the quadratic formula:
    $x^2 - 3x = 6$

11. Solve by using the quadratic formula:
$2x^2 - 5x - 3 = 0$

12. Solve by using the quadratic formula:
$3x^2 - x = 1$

13. Graph $y = x^2 + 2x - 4$.

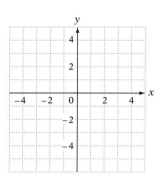

14. The length of a rectangle is 2 ft less than twice the width. The area of the rectangle is 40 ft². Find the length and width of the rectangle.

15. It took a motorboat 1 h more to travel 60 mi against a current than it took the boat to travel 60 mi with the current. The rate of the current was 1 mph. Find the rate of the boat in calm water.

Cumulative Review

. . . .

1. Simplify: $2x - 3[2x - 4(3 - 2x) + 2] - 3$

2. Solve: $-\dfrac{3}{5}x = -\dfrac{9}{10}$

3. Solve: $2x - 3(4x - 5) = -3x - 6$

4. Simplify: $(2a^2b)^2(-3a^4b^2)$

5. Simplify: $(x^2 - 8) \div (x - 2)$

6. Factor $3x^3 + 2x^2 - 8x$.

7. Simplify: $\dfrac{3x^2 - 6x}{4x - 6} \div \dfrac{2x^2 + x - 6}{6x^3 - 24x}$

8. Simplify: $\dfrac{x}{2(x - 1)} - \dfrac{1}{(x - 1)(x + 1)}$

9. Simplify: $\dfrac{1 - \dfrac{7}{x} + \dfrac{12}{x^2}}{2 - \dfrac{1}{x} - \dfrac{15}{x^2}}$

10. Find the x- and y-intercepts for the graph of the line $4x - 3y = 12$.

11. Find the equation of the line that contains the point $(-3, 2)$ and has slope $-\dfrac{4}{3}$.

12. Solve by substitution:
$$3x - y = 5$$
$$y = 2x - 3$$

13. Solve by the addition method:
$$3x + 2y = 2$$
$$5x - 2y = 14$$

14. Solve: $2x - 3(2 - 3x) > 2x - 5$

15. Simplify: $(\sqrt{a} - \sqrt{2})(\sqrt{a} + \sqrt{2})$

16. Simplify: $\dfrac{\sqrt{108a^7b^3}}{\sqrt{3a^4b}}$

17. Simplify: $\dfrac{\sqrt{3}}{5 + 2\sqrt{3}}$

18. Solve: $3 = 8 - \sqrt{5x}$

19. Solve by factoring: $6x^2 - 17x = -5$

20. Solve by taking square roots: $2(x - 5)^2 = 36$

21. Solve by completing the square: $3x^2 + 7x = -3$

22. Solve by using the quadratic formula: $2x^2 - 3x - 2 = 0$

23. Find the selling price per pound of a mixture made from 20 lb of cashews that cost $3.50 per pound and 50 lb of peanuts that cost $1.75 per pound.

24. A stock investment of 100 shares paid a dividend of $215. At this rate, how many additional shares are required for the investor to earn a dividend of $752.50?

25. A 720-mi trip from one city to another takes 3 h when a plane is flying with the wind. The return trip, against the wind, takes 4.5 h. Find the rate of the plane in still air and the rate of the wind.

26. A student received a 70, a 91, an 85, and a 77 on four tests in a math class. What scores on the last test will enable the student to receive a minimum of 400 points?

27. The sum of the squares of three consecutive odd integers is 83. Find the middle odd integer.

28. A jogger ran 7 mi at a constant rate and then reduced the rate by 3 mph. An additional 8 mi was run at the reduced rate. The total time spent jogging the 15 mi was 3 h. Find the rate for the last 8 mi.

29. Graph $2x - 3y > 6$.

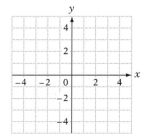

30. Graph $y = x^2 - 2x - 3$.

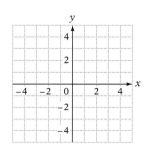

Final Exam

· ·

1. Evaluate $-|-3|$.

2. Subtract: $-15 - (-12) - 3$

3. Simplify: $-2^4 \cdot (-2)^4$

4. Simplify: $-7 - \dfrac{12 - 15}{2 - (-1)} \cdot (-4)$

5. Evaluate $\dfrac{a^2 - 3b}{2a - 2b^2}$ when $a = 3$ and $b = -2$.

6. Simplify: $6x - (-4y) - (-3x) + 2y$

7. Simplify: $(-15z)\left(-\dfrac{2}{5}\right)$

8. Simplify: $-2[5 - 3(2x - 7) - 2x]$

9. Solve: $20 = -\dfrac{2}{5}x$

10. Solve: $4 - 2(3x + 1) = 3(2 - x) + 5$

11. Write $\dfrac{1}{8}$ as a percent.

12. Find 19% of 80.

13. Simplify: $(2x^2 - 5x + 1) - (5x^2 - 2x - 7)$

14. Simplify: $(-3xy^3)^4$

15. Simplify: $(3x^2 - x - 2)(2x + 3)$

16. Simplify: $\dfrac{(-2x^2y^3)^3}{(-4xy^4)^2}$

17. Simplify: $\dfrac{12x^2y - 16x^3y^2 - 20y^2}{4xy^2}$

18. Simplify: $(5x^2 - 2x - 1) \div (x + 2)$

19. Simplify: $(4x^{-2}y)^2(2xy^{-2})^{-2}$

20. Given $f(t) = \dfrac{t}{t + 1}$, find $f(3)$.

21. Factor: $x^2 - 5x - 6$

22. Factor: $6x^2 - 5x - 6$

23. Factor: $8x^3 - 28x^2 + 12x$

24. Factor: $25x^2 - 16$

25. Factor: $2a(4 - x) - 6(x - 4)$

26. Factor: $75y - 12x^2y$

27. Solve: $2x^2 = 7x - 3$

28. Simplify: $\dfrac{2x^2 - 3x + 1}{4x^2 - 2x} \cdot \dfrac{4x^2 + 4x}{x^2 - 2x + 1}$

29. Simplify: $\dfrac{5}{x + 3} - \dfrac{3x}{2x - 5}$

30. Simplify: $x - \dfrac{1}{1 - \dfrac{1}{x}}$

31. Solve: $\dfrac{5x}{3x - 5} - 3 = \dfrac{7}{3x - 5}$

32. Solve $a = 3a - 2b$ for a.

33. Find the slope of the line that contains the points $(-1, -3)$ and $(2, -1)$.

34. Find the equation of the line that contains the point $(3, -4)$ and has slope $-\dfrac{2}{3}$.

35. Solve by substitution.
$$y = 4x - 7$$
$$y = 2x + 5$$

36. Solve by the addition method.
$$4x - 3y = 11$$
$$2x + 5y = -1$$

37. Solve: $4 - x \geq 7$

38. Solve: $2 - 2(y - 1) \leq 2y - 6$

39. Simplify: $\sqrt{49x^6}$

40. Simplify: $2\sqrt{27a} + 8\sqrt{48a}$

41. Simplify: $\dfrac{\sqrt{3}}{\sqrt{5} - 2}$

42. Solve: $\sqrt{2x - 3} + 4 = 5$

43. Solve by factoring:
$$3x^2 - x = 4$$

44. Solve by using the quadratic formula:
$$4x^2 - 2x - 1 = 0$$

45. Translate and simplify "the sum of twice a number and three times the difference of the number and two."

46. Because of depreciation, the value of an office machine is now $2400. This is 80% of its original value. Find the original value.

47. The manufacturer's cost for a laser printer is $900. The manufacturer then sells the printer for $1485. What is the markup rate?

48. An investment of $3000 is made at an annual simple interest rate of 8%. How much additional money must be invested at 11% so that the total interest earned is 10% of the total investment?

49. A grocer mixes 4 lb of peanuts that cost $2 per pound with 2 lb of walnuts that cost $5 per pound. What is the cost per pound of the resulting mixture?

50. A pharmacist mixes together 20 L of a solution that is 60% acid and 30 L of a solution that is 20% acid. What is the percent concentration of the acid in the mixture?

51. At 2:00 P.M. a small plane had been flying 1 h when a change of wind direction doubled its average ground speed. The pilot completed the 860-km trip in 2.5 h. How far did the plane travel in the first hour?

52. The angles of a triangle are such that the second angle is 10° more than the first angle and the third angle is 10° more than the second angle. Find the measure of each of the three angles.

53. The sum of the squares of three consecutive integers is 50. Find the middle integer.

54. The length of a rectangle is 5 m more than the width. The area of the rectangle is 50 m². Find the dimensions of the rectangle.

55. A paint formula requires 2 oz of dye for every 15 oz of base paint. How many ounces of dye are required for 120 oz of base paint?

56. It takes a chef 1 h to prepare a dinner. The chef's apprentice can prepare the dinner in 1.5 h. How long would it take the chef and the apprentice, working together, to prepare the dinner?

57. With the current, a motorboat travels 50 mi in 2.5 h. Against the current, it takes twice as long to travel 50 mi. Find the rate of the boat in calm water and the rate of the current.

58. Flying against the wind, it took a pilot $\frac{1}{2}$ h longer to travel 500 miles than it took flying with the wind. The rate of the plane in calm air is 225 mph. Find the rate of the wind.

59. Graph the line that has slope $-\frac{1}{2}$ and y-intercept $(0, -3)$.

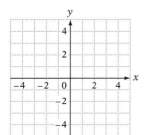

60. Graph $y = x^2 - 4x + 3$.

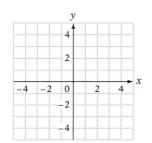

You Try It 9   The product is negative.

$$-\frac{7}{12} \times \frac{9}{14} = -\frac{7 \cdot 9}{12 \cdot 14}$$

$$= -\frac{7 \cdot \overset{1}{\cancel{3}} \cdot 3}{2 \cdot 2 \cdot \underset{1}{\cancel{3}} \cdot 2 \cdot \underset{1}{\cancel{7}}}$$

$$= -\frac{3}{8}$$

You Try It 10   The quotient is positive.

$$-\frac{3}{8} \div \left(-\frac{5}{12}\right) = \frac{3}{8} \times \frac{12}{5}$$

$$= \frac{3 \cdot 12}{8 \cdot 5} = \frac{3 \cdot \overset{1}{\cancel{2}} \cdot \overset{1}{\cancel{2}} \cdot 3}{\underset{1}{\cancel{2}} \cdot \underset{1}{\cancel{2}} \cdot 2 \cdot 5} = \frac{9}{10}$$

You Try It 11

$$\begin{array}{r} 5.44 \\ \times \ \ 3.8 \\ \hline 4352 \\ 1632 \ \ \\ \hline 20.672 \end{array}$$

$$-5.44 \times 3.8 = -20.672$$

You Try It 12

$$\begin{array}{r} 0.231 \\ 1.7\overline{)0.3\,940} \\ \underline{-3\ 4} \\ 54 \\ \underline{-51} \\ 30 \\ \underline{-17} \\ 13 \end{array}$$

$$-0.394 \div 1.7 \approx -0.23$$

You Try It 13

Strategy   **a.** To find the total spent, add the 5 amounts shown in the circle graph.
b. To find the percent spent on hypertension drugs:
   • Find the total spent on Procardia, Cardizen and Vasotec.
   • Divide the amount spent on hypertension drugs by the total amount spent on drugs and convert the decimal to a percent.

Solution   **a.** $0.8 + 1.7 + 1.1 + 1.0 + 0.9 = 5.5$
$5.5 billion was spent on the drugs.
b. $1.1 + 0.9 + 0.8 = 2.8$

$$\frac{2.8}{5.5} \approx 0.50909$$

$$0.50909(100\%) \approx 50.9$$

Approximately 50.9% was spent on hypertension drugs.

You Try It 1 $-6^3 = -(6 \cdot 6 \cdot 6) = -216$

You Try It 2
$(-3)^4 = (-3)(-3)(-3)(-3) = 81$

You Try It 3
$(3^3)(-2)^3 = (3)(3)(3) \cdot (-2)(-2)(-2)$
$= 27(-8) = -216$

You Try It 4 $\left(-\dfrac{2}{5}\right)^2 = \left(-\dfrac{2}{5}\right)\left(-\dfrac{2}{5}\right) = \dfrac{4}{25}$

You Try It 5 $-3(0.3)^3 = -3(0.3)(0.3)(0.3)$
$= -0.9(0.3)(0.3)$
$= -0.27(0.3) = -0.081$

You Try It 6 $18 - 5[8 - 2(2 - 5)] \div 10$
$18 - 5[8 - 2(-3)] \div 10$
$18 - 5[8 + 6] \div 10$
$18 - 5[14] \div 10$
$18 - 70 \div 10$
$18 - 7$
11

You Try It 7 $36 \div (8 - 5)^2 - (-3)^2 \cdot 2$
$36 \div (3)^2 - (-3)^2 \cdot 2$
$36 \div 9 - 9 \cdot 2$
$4 - 9 \cdot 2$
$4 - 18$
-14

You Try It 8 $(6.97 - 4.72)^2 \times 4.5 \div 0.05$
$(2.25)^2 \times 4.5 \div 0.05$
$5.0625 \times 4.5 \div 0.05$
$22.78125 \div 0.05$
455.625

You Try It 9 $\dfrac{5}{8} \div \left(\dfrac{1}{3} - \dfrac{3}{4}\right) + \dfrac{7}{12}$

$\dfrac{5}{8} \div \left(\dfrac{-5}{12}\right) + \dfrac{7}{12}$

$\dfrac{5}{8} \cdot \left(-\dfrac{12}{5}\right) + \dfrac{7}{12}$

$-\dfrac{3}{2} + \dfrac{7}{12}$

$-\dfrac{18}{12} + \dfrac{7}{12}$

$-\dfrac{11}{12}$

SOLUTIONS to Chapter 2 "You Try It"

SECTION 2.1   *pages 49–50*

You Try It 1 -4 is the constant term.

You Try It 2 $2xy + y^2$

$2(-4)(2) + (2)^2$
$2(-4)(2) + 4$
$(-8)(2) + 4$
$(-16) + 4$
-12

You Try It 3 $\dfrac{a^2 + b^2}{a + b}$

$\dfrac{5^2 + (-3)^2}{5 + (-3)}$

$\dfrac{25 + 9}{5 + (-3)}$

$\dfrac{34}{2}$

17

You Try It 4 $x^3 - 2(x + y) + z^2$

$(2)^3 - 2[2 + (-4)] + (-3)^2$
$8 - 2(-2) + 9$
$8 + 4 + 9$
$12 + 9$
21

SECTION 2.2   *pages 53–58*

You Try It 1
$3a - 2b - 5a + 6b = -2a + 4b$

You Try It 2
$-3y^2 + 7 + 8y^2 - 14 = 5y^2 - 7$

You Try It 3 $-5(4y^2) = -20y^2$

You Try It 4 $-7(-2a) = 14a$

You Try It 5 $(-5x)(-2) = 10x$

You Try It 6 $5(3 + 7b) = 15 + 35b$

You Try It 7 $(3a - 1)5 = 15a - 5$

You Try It 8 $-8(-2a + 7b) = 16a - 56b$

You Try It 9 $12 - (x - 8) = 12 - x + 8$
$\qquad\qquad\qquad\quad = -x + 20$

You Try It 10
$3(-a^2 - 6a + 7) = -3a^2 - 18a + 21$

You Try It 11 $3y - 2(y - 7x) = 3y - 2y + 14x$
$\qquad\qquad\qquad\qquad\quad = 14x + y$

You Try It 12
$-2(x - 2y) - (-x + 3y) = -2x + 4y + x - 3y$
$\qquad\qquad\qquad\qquad\qquad\quad = -x + y$

You Try It 13
$-5(-2y - 3x) + 4y = 10y + 15x + 4y$
$\qquad\qquad\qquad\qquad\quad = 15x + 14y$

You Try It 14
$3y - 2[x - 4(2 - 3y)] = 3y - 2[x - 8 + 12y]$
$\qquad\qquad\qquad\qquad\quad\; = 3y - 2x + 16 - 24y$
$\qquad\qquad\qquad\qquad\quad\; = -2x - 21y + 16$

You Try It 1 the <u>difference between</u> twice n and <u>one third of</u> n

$$2n - \frac{1}{3}n$$

You Try It 2 the <u>quotient of</u> <u>7 less than</u> b and 15

$$\frac{b - 7}{15}$$

You Try It 3 an unknown number: n
the cube of the number: n^3
the total of ten and the cube of
the number: $10 + n^3$

$$-4(10 + n^3)$$

You Try It 4 the first integer: x
the second integer: $x + 1$
the third integer: $x + 2$

$$x + (x + 1) + (x + 2); \; 3x + 3$$

You Try It 5 the unknown number: x
the difference between the number
and sixty: $x - 60$

$$5(x - 60); \; 5x - 300$$

You Try It 6 the speed of the older model: s
the new jet operates at twice the
speed of the older model: $2s$

You Try It 7 the length of the longer piece: y
the length of the shorter piece: $6 - y$

SOLUTIONS to Chapter 3 "You Try It"

You Try It 1

$$5 - 4x = 8x + 2$$

$5 - 4\left(\dfrac{1}{4}\right)$	$8\left(\dfrac{1}{4}\right) + 2$
$5 - 1$	$2 + 2$

$$4 = 4$$

Yes, $\dfrac{1}{4}$ is a solution.

You Try It 2

$$10x - x^2 = 3x - 10$$

$10(5) - (5)^2$	$3(5) - 10$
$50 - 25$	$15 - 10$

$$25 \neq 5$$

No, 5 is not a solution.

You Try It 3

$$\frac{5}{6} = y - \frac{3}{8}$$
$$\frac{5}{6} + \frac{3}{8} = y - \frac{3}{8} + \frac{3}{8}$$
$$\frac{29}{24} = y$$

The solution is $\dfrac{29}{24}$.

You Try It 4

$$-\frac{2}{5}x = 6$$
$$\left(-\frac{5}{2}\right)\left(-\frac{2}{5}x\right) = \left(-\frac{5}{2}\right)(6)$$
$$x = -15$$

The solution is -15.

You Try It 5

$$4x - 8x = 16$$
$$-4x = 16$$
$$\frac{-4x}{-4} = \frac{16}{-4}$$
$$x = -4$$

The solution is -4.

You Try It 6

$$P \cdot B = A$$
$$0.19(125) = A$$
$$23.75 = A$$

23.75 is 19% of 125.

You Try It 7

Strategy
To find the percent decrease, solve the basic percent equation using $B = 2500$ and $A = 500$. The percent is unknown.

Solution
$$P \cdot B = A$$
$$P(2500) = 500$$
$$\frac{P(2500)}{2500} = \frac{500}{2500}$$
$$P = 0.2$$

The percent decrease is 20%.

You Try It 8

Strategy
To find the percent, use the basic equation, where $B = 171$, $A = 223$, and P is unknown.

$$P \cdot B = A$$
$$P(171) = 223$$
$$P = \frac{223}{171}$$
$$P \approx 1.304 \approx 130\%$$

The 1994 federal deficit is approximately 130% of the 1995 deficit.

SECTION 3.2   *pages 93–94*

You Try It 1
$$5x + 7 = 10$$
$$5x + 7 - 7 = 10 - 7$$
$$5x = 3$$
$$\frac{5x}{5} = \frac{3}{5}$$
$$x = \frac{3}{5}$$
The solution is $\frac{3}{5}$.

You Try It 3
$$x - 5 + 4x = 25$$
$$5x - 5 = 25$$
$$5x - 5 + 5 = 25 + 5$$
$$5x = 30$$
$$\frac{5x}{5} = \frac{30}{5}$$
$$x = 6$$
The solution is 6.

You Try It 2
$$2 = 11 + 3x$$
$$2 - 11 = 11 - 11 + 3x$$
$$-9 = 3x$$
$$\frac{-9}{3} = \frac{3x}{3}$$
$$-3 = x$$
The solution is -3.

You Try It 4

Strategy   To find the depth, replace P with the given value and solve for D.

Solution
$$P = 15 + \frac{1}{2}D$$
$$45 = 15 + \frac{1}{2}D$$
$$45 - 15 = 15 - 15 + \frac{1}{2}D$$
$$30 = \frac{1}{2}D$$
$$2(30) = 2 \cdot \frac{1}{2}D$$
$$60 = D$$
The depth is 60 ft.

You Try It 1

$$5x + 4 = 6 + 10x$$
$$5x - 10x + 4 = 6 + 10x - 10x$$
$$-5x + 4 = 6$$
$$-5x + 4 - 4 = 6 - 4$$
$$-5x = 2$$
$$\frac{-5x}{-5} = \frac{2}{-5}$$
$$x = -\frac{2}{5}$$

The solution is $-\frac{2}{5}$.

You Try It 2

$$5x - 10 - 3x = 6 - 4x$$
$$2x - 10 = 6 - 4x$$
$$2x + 4x - 10 = 6 - 4x + 4x$$
$$6x - 10 = 6$$
$$6x - 10 + 10 = 6 + 10$$
$$6x = 16$$
$$\frac{6x}{6} = \frac{16}{6}$$
$$x = \frac{8}{3}$$

The solution is $\frac{8}{3}$.

You Try It 3

$$5x - 4(3 - 2x) = 2(3x - 2) + 6$$
$$5x - 12 + 8x = 6x - 4 + 6$$
$$13x - 12 = 6x + 2$$
$$13x - 6x - 12 = 6x - 6x + 2$$
$$7x - 12 = 2$$
$$7x - 12 + 12 = 2 + 12$$
$$7x = 14$$
$$\frac{7x}{7} = \frac{14}{7}$$
$$x = 2$$

The solution is 2.

You Try It 4

$$-2[3x - 5(2x - 3)] = 3x - 8$$
$$-2[3x - 10x + 15] = 3x - 8$$
$$-2[-7x + 15] = 3x - 8$$
$$14x - 30 = 3x - 8$$
$$14x - 3x - 30 = 3x - 3x - 8$$
$$11x - 30 = -8$$
$$11x - 30 + 30 = -8 + 30$$
$$11x = 22$$
$$\frac{11x}{11} = \frac{22}{11}$$
$$x = 2$$

The solution is 2.

You Try It 5

Solve $2x = 5x + 6$ for x.

$$2x = 5x + 6$$
$$-3x = 6$$
$$x = -2$$

Evaluate $-2x + 7$ for $x = -2$

$$-2x + 7$$
$$-2(-2) + 7 = 4 + 7$$
$$= 11$$

You Try It 6

Strategy
To find the location of the fulcrum when the system balances, replace the variables F_1, F_2, and d in the lever system equation by the given values and solve for x.

Solution

$$F_1 \cdot x = F_2 \cdot (d - x)$$
$$45x = 80(25 - x)$$
$$45x = 2000 - 80x$$
$$45x + 80x = 2000 - 80x + 80x$$
$$125x = 2000$$
$$\frac{125x}{125} = \frac{2000}{125}$$
$$x = 16$$

The fulcrum is 16 feet from the 45-pound force.

You Try It 1
The unknown number: n

four less than one third of a number	equals	five minus two thirds of the number

$$\frac{1}{3}n - 4 = 5 - \frac{2}{3}n$$
$$\frac{1}{3}n + \frac{2}{3}n - 4 = 5 - \frac{2}{3}n + \frac{2}{3}n$$
$$n - 4 = 5$$
$$n - 4 + 4 = 5 + 4$$
$$n = 9$$

The number is 9.

You Try It 2
The smaller number: n
The larger number: $12 - n$

the total of three times the smaller and six	amounts to	seven less than the product of four and the larger

$$3n + 6 = 4(12 - n) - 7$$
$$3n + 6 = 48 - 4n - 7$$
$$3n + 6 = 41 - 4n$$
$$3n + 4n + 6 = 41 - 4n + 4n$$
$$7n + 6 = 41$$
$$7n + 6 - 6 = 41 - 6$$
$$7n = 35$$
$$\frac{7n}{7} = \frac{35}{7}$$
$$n = 5$$

$$12 - n = 12 - 5 = 7$$

The smaller number is 5.
The larger number is 7.

You Try It 3

Strategy
- First consecutive integer: n
  Second consecutive integer: $n + 1$
  Third consecutive integer: $n + 2$
- The sum of the three integers is -6.

Solution
$$n + (n + 1) + (n + 2) = -6$$
$$3n + 3 = -6$$
$$3n = -9$$
$$n = -3$$
$$n + 1 = -3 + 1 = -2$$
$$n + 2 = -3 + 2 = -1$$

The three consecutive integers are -3, -2, and -1.

You Try It 4

Strategy
To find the Fahrenheit temperature, write and solve an equation using F to represent the Fahrenheit temperature.

Solution

20	is	$\frac{5}{9}$ of the difference between the Fahrenheit temperature and 32

$$20 = \frac{5}{9}(F - 32)$$
$$20 = \frac{5}{9}F - \frac{160}{9}$$
$$20 + \frac{160}{9} = \frac{5}{9}F - \frac{160}{9} + \frac{160}{9}$$
$$\frac{340}{9} = \frac{5}{9}F$$
$$\frac{9}{5} \cdot \frac{340}{9} = \frac{9}{5} \cdot \frac{5}{9}F$$
$$68 = F$$

The Fahrenheit temperature is 68°.

You Try It 5

Strategy

To find the number of color TV's made each day, write and solve an equation, using x to represent the number of color TV's and $140 - x$ to represent the number of black and white TV's.

Solution

three times the number of black and white TV's	equals	20 less than the number of color TV's

$$3(140 - x) = x - 20$$
$$420 - 3x = x - 20$$
$$420 - 3x - x = x - x - 20$$
$$420 - 4x = -20$$
$$420 - 420 - 4x = -20 - 420$$
$$-4x = -440$$
$$\frac{-4x}{-4} = \frac{-440}{-4}$$
$$x = 110$$

There are 110 color TV's made each day.

SOLUTIONS to Chapter 4 "You Try It"

SECTION 4.1 *pages 125–126*

You Try It 1

Strategy
Given: $C = \$40$
$\quad\quad S = \$60$
Unknown: r
Use the equation $S = C + rC$.

Solution
$S = C + rC$
$60 = 40 + 40r$
$20 = 40r$
$0.5 = r$

The markup rate is 50%.

You Try It 2

Strategy
Given: $R = \$27.60$
$\quad\quad S = \$20.70$
Unknown: r
Use the equation $S = R - rR$.

Solution
$S = R - rR$
$20.70 = 27.60 - 27.60r$
$-6.90 = -27.60r$
$0.25 = r$

The discount rate is 25%.

SECTION 4.2   *pages 129–130*

You Try It 1

Strategy
- Additional amount: x

	Principal	Rate	Interest
6%	5000	0.06	0.06(5000)
9%	x	0.09	0.09x
8%	5000 + x	0.08	0.08(5000 + x)

- The sum of the interest earned by the two investments equals the interest earned on the total investment.

Solution

$$0.06(5000) + 0.09x = 0.08(5000 + x)$$
$$300 + 0.09x = 400 + 0.08x$$
$$300 + 0.01x = 400$$
$$0.01x = 100$$
$$x = 10,000$$

$10,000 more must be invested at 9%

SECTION 4.3   *pages 133–136*

You Try It 1

Strategy
- Pounds of $.55 fertilizer: x

	Amount	Cost	Value
$.80 fertilizer	20	$.80	0.80(20)
$.55 fertilizer	x	$.55	0.55x
$.75 fertilizer	20 + x	$.75	0.75(20 + x)

- The sum of the values before mixing equals the value after mixing.

Solution

$$0.80(20) + 0.55x = 0.75(20 + x)$$
$$16 + 0.55x = 15 + 0.75x$$
$$16 - 0.20x = 15$$
$$-0.20x = -1$$
$$x = 5$$

5 lb of the $.55 fertilizer must be added.

You Try It 2

Strategy
- Liters of water: x

	Amount	Percent	Quantity
6%	x	0.06	0.06x
12%	5	0.12	5(0.12)
8%	$x + 5$	0.08	0.08(x + 5)

- The sum of the quantities before mixing equals the quantity after mixing.

Solution

$$0.06x + 5(0.12) = 0.08(x + 5)$$
$$0.06x + 0.60 = 0.08x + 0.40$$
$$0.06x + 0.20 = 0.08x$$
$$0.20 = 0.02x$$
$$10 = x$$

The pharmacist adds 10 L of the 6% solution to the 12% solution to get an 8% solution.

You Try It 1

Strategy
- Rate of the first train: r
  Rate of the second train: $2r$

	Rate	Time	Distance
1st train	r	3	$3r$
2nd train	$2r$	3	$3(2r)$

- The sum of the distances traveled by each train equals 288 mi.

Solution
$$3r + 3(2r) = 288$$
$$3r + 6r = 288$$
$$9r = 288$$
$$r = 32$$

$$2r = 2(32) = 64$$

The first train is traveling at 32 mph.
The second train is traveling at 64 mph.

You Try It 2

Strategy
- Time spent flying out: t
  Time spent flying back: $5 - t$

	Rate	Time	Distance
Out	150	t	$150t$
Back	100	$5 - t$	$100(5 - t)$

- The distance out equals the distance back.

Solution
$$150t = 100(5 - t)$$
$$150t = 500 - 100t$$
$$250t = 500$$
$$t = 2 \text{ (The time out was 2 h.)}$$

The distance $= 150t = 150(2) = 300$ mi.

The parcel of land was 300 mi away.

You Try It 1

Strategy
- Width of the rectangle: w
  Length of the rectangle: $w + 3$
- Use the equation for the perimeter of a rectangle.

Solution
$$2l + 2w = P$$
$$2(w + 3) + 2w = 34 \quad \bullet \; l + w + 3$$
$$2w + 6 + 2w = 34$$
$$4w + 6 = 34$$
$$4w = 28$$
$$w = 7$$

The width of the rectangle is 7 m.

You Try It 2

Strategy
- Measure of the first angle: $2x$
  Measure of the second angle: x
  Measure of the third angle: $x - 4$
- Use the equation $A + B + C = 180°$.

Solution
$$A + B + C = 180$$
$$2x + x + (x - 4) = 180$$
$$4x - 4 = 180$$
$$4x = 184$$
$$x = 46$$

$$2x = 2(46) = 92$$

$$x - 4 = 46 - 4 = 42$$

The measure of the first angle is 92°.
The measure of the second angle is 46°.
The measure of the third angle is 42°.

SOLUTIONS to Chapter 5 "You Try It"

SECTION 5.1   *pages 159–160*

You Try It 1
$(-4x^3 + 2x^2 - 8) + (4x^3 + 6x^2 - 7x + 5)$
$= (-4x^3 + 4x^3) + (2x^2 + 6x^2)$
$\quad + (-7x) + (-8 + 5)$
$= 8x^2 - 7x - 3$

You Try It 2
$\begin{array}{r} 6x^3 \qquad\quad + 2x + 8 \\ -9x^3 + 2x^2 - 12x - 8 \\ \hline -3x^3 + 2x^2 - 10x \end{array}$

You Try It 3
$(-4w^3 + 8w - 8) - (3w^3 - 4w^2 - 2w - 1)$
$= (-4w^3 + 8w - 8)$
$\quad + (-3w^3 + 4w^2 - 2w + 1)$
$= -7w^3 + 4w^2 + 10w - 7$

You Try It 4
$\begin{array}{r} 13y^3 \qquad\quad -6y - 7 \\ -4y^2 + 6y + 9 \\ \hline 13y^3 - 4y^2 \qquad + 2 \end{array}$

SECTION 5.2   *pages 163–164*

You Try It 1
$(8m^3n)(-3n^5) = [8(-3)](m^3)(n \cdot n^5)$
$\qquad\qquad = -24m^3n^6$

You Try It 2
$(12p^4q^3)(-3p^5q^2) = [12(-3)](p^4 \cdot p^5)(q^3 \cdot q^2)$
$\qquad\qquad = -36p^9q^5$

You Try It 3
$(-3a^4bc^2)^3 = (-3)^{1\cdot3}a^{4\cdot3}b^{1\cdot3}c^{2\cdot3}$
$\qquad = (-3)^3a^{12}b^3c^6$
$\qquad = -27a^{12}b^3c^6$

You Try It 4
$(-xy^4)(-2x^3y^2)^2 = (-xy^4)[(-2)^{1\cdot2}x^{3\cdot2}y^{2\cdot2}]$
$\qquad = (-xy^4)[(-2)^2x^6y^4]$
$\qquad = (-xy^4)(4x^6y^4)$
$\qquad = -4x^7y^8$

SECTION 5.3   *pages 167–170*

You Try It 1
$(-2y + 3)(-4y) = 8y^2 - 12y$

You Try It 2
$-a^2(3a^2 + 2a - 7) = -3a^4 - 2a^3 + 7a^2$

You Try It 3
$\begin{array}{r} 2y^3 + 2y^2 \qquad - 3 \\ 3y - 1 \\ \hline -2y^3 - 2y^2 \qquad + 3 \\ 6y^4 + 6y^3 \qquad - 9y \\ \hline 6y^4 + 4y^3 - 2y^2 - 9y + 3 \end{array}$

You Try It 4
$(4y - 5)(2y - 3) = 8y^2 - 12y - 10y + 15$
$\qquad = 8y^2 - 22y + 15$

You Try It 5
$(3b + 2)(3b - 5) = 9b^2 - 15b + 6b - 10$
$\qquad = 9b^2 - 9b - 10$

You Try It 6
$(2a + 5c)(2a - 5c) = 4a^2 - 25c^2$

You Try It 7 $(3x + 2y)^2 = 9x^2 + 12xy + 4y^2$

You Try It 8

Strategy To find the area, replace the variable r in the equation $A = \pi r^2$ by the given value and solve for A.

Solution $A = \pi r^2$
$A = 3.14(x - 4)^2$
$A = 3.14(x^2 - 8x + 16)$
$A = 3.14x^2 - 25.12x + 50.24$

The area is
$(3.14x^2 - 25.12x + 50.24) \text{ ft}^2$.

SECTION 5.4   *pages 175–180*

You Try It 1 $\dfrac{18y^3}{-27y^7} = \dfrac{9 \cdot 2y^{3-7}}{-9 \cdot 3} = -\dfrac{2y^{-4}}{3} = -\dfrac{2}{3y^4}$

You Try It 2 $(-2x^2)(x^{-3}y^{-4})^{-2} = (-2x^2)(x^6y^8)$
$= -2x^8y^8$

You Try It 3 $\dfrac{(6a^{-2}b^3)^{-1}}{(4a^3b^{-2})^{-2}} = \dfrac{6^{-1}a^2b^{-3}}{4^{-2}a^{-6}b^4}$
$= 4^2 \cdot 6^{-1}a^8b^{-7}$
$= \dfrac{16a^8}{6b^7} = \dfrac{8a^8}{3b^7}$

You Try It 4 $\left[\dfrac{6r^3s^{-3}}{9r^3s^{-1}}\right]^{-2} = \left[\dfrac{2r^0s^{-2}}{3}\right]^{-2}$
$= \dfrac{2^{-2}s^4}{3^{-2}} = \dfrac{9s^4}{4}$

You Try It 5 $16{,}000{,}000{,}000{,}000 = 1.6 \times 10^{13}$

SECTION 5.5   *pages 185–186*

You Try It 1
$\dfrac{24x^2y^2 - 18xy + 6y}{6xy} = \dfrac{24x^2y^2}{6xy} - \dfrac{18xy}{6xy} + \dfrac{6y}{6xy}$
$= 4xy - 3 + \dfrac{1}{x}$

You Try It 3

$$\begin{array}{r} x^2 + x - 1 \\ x - 1\overline{)x^3 + 0x^2 - 2x + 1} \\ \underline{x^3 - x^2} \\ x^2 - 2x \\ \underline{x^2 - x} \\ -x + 1 \\ \underline{-x + 1} \\ 0 \end{array}$$

$(x^3 - 2x + 1) \div (x - 1) = x^2 + x - 1$

You Try It 2

$$\begin{array}{r} x^2 + 2x - 1 \\ 2x - 3\overline{)2x^3 + x^2 - 8x - 3} \\ \underline{2x^3 - 3x^2} \\ 4x^2 - 8x \\ \underline{4x^2 - 6x} \\ -2x - 3 \\ \underline{-2x + 3} \\ -6 \end{array}$$

$(2x^3 + x^2 - 8x - 3) \div (2x - 3)$
$= x^2 + 2x - 1 - \dfrac{6}{2x - 3}$

SOLUTIONS to Chapter 6 "You Try It"

You Try It 1
The GCF is $7a^2$.

$$14a^2 - 21a^4b = 7a^2(2) + 7a^2(-3a^2b)$$
$$= 7a^2(2 - 3a^2b)$$

You Try It 2
The GCF is 9.

$$27b^2 + 18b + 9$$
$$= 9(3b^2) + 9(2b) + 9(1)$$
$$= 9(3b^2 + 2b + 1)$$

You Try It 3
The GCF is $3x^2y^2$.

$$6x^4y^2 - 9x^3y^2 + 12x^2y^4$$
$$= 3x^2y^2(2x^2) + 3x^2y^2(-3x) + 3x^2y^2(4y^2)$$
$$= 3x^2y^2(2x^2 - 3x + 4y^2)$$

You Try It 4
$$2y(5x - 2) - 3(2 - 5x)$$
$$= 2y(5x - 2) + 3(5x - 2)$$
$$= (5x - 2)(2y + 3)$$

You Try It 5
$$a^2 - 3a + 2ab - 6b = (a^2 - 3a) + (2ab - 6b)$$
$$= a(a - 3) + 2b(a - 3)$$
$$= (a - 3)(a + 2b)$$

You Try It 6
$$2mn^2 - n + 8mn - 4$$
$$= (2mn^2 - n) + (8mn - 4)$$
$$= n(2mn - 1) + 4(2mn - 1)$$
$$= (2mn - 1)(n + 4)$$

You Try It 1
Find the positive factors of 20 whose sum is 9.

Factors	Sums
1, 20	21
2, 10	12
4, 5	9

$$x^2 + 9x + 20 = (x + 4)(x + 5)$$

You Try It 2
Find the factors of -18 whose sum is 7.

Factors	Sums
+1, −18	−17
−1, +18	17
+2, −9	−7
−2, +9	7
+3, −6	−3
−3, +6	+3

$$x^2 + 7x - 18 = (x + 9)(x - 2)$$

You Try It 3
The GCF is $-2x$.

$$-2x^3 + 14x^2 - 12x = -2x(x^2 - 7x + 6)$$

Factor the trinomial $x^2 - 7x + 6$. Find two negative factors of 6 whose sum is -7.

Factors	Sum
−3, −2	−5
−6, −1	−7

$$-2x^3 + 14x^2 - 12x = -2x(x - 6)(x - 1)$$

You Try It 4
The GCF is 3.

$$3x^2 - 9xy - 12y^2 = 3(x^2 - 3xy - 4y^2)$$

Factor the trinomial.

Find the factors of -4 whose sum is -3.

Factors	Sums
+1, −4	−3
−1, +4	3
+2, −2	0

$$3x^2 - 9xy - 12y^2 = 3(x + y)(x - 4y)$$

You Try It 1

Factor the trinomial $2x^2 - x - 3$.

Positive	Factors of -3: $+1, -3$
factors of 2: 1, 2	$-1, +3$

Trial Factors	*Middle Term*
$(1x + 1)(2x - 3)$	$-3x + 2x = -x$
$(1x - 3)(2x + 1)$	$x - 6x = -5x$
$(1x - 1)(2x + 3)$	$3x - 2x = x$
$(1x + 3)(2x - 1)$	$-x + 6x = 5x$

$2x^2 - x - 3 = (x + 1)(2x - 3)$

You Try It 2

The GCF is $-3y$.

$-45y^3 + 12y^2 + 12y = -3y(15y^2 - 4y - 4)$

Factor the trinomial $15y^2 - 4y - 4$.

Positive	Factors of -4: $-1,\ \ 4$
factors of 15: 1, 15	$1, -4$
3, 5	$-2,\ \ 2$

Trial Factors	*Middle Term*
$(1y - 1)(15y + 4)$	$4y - 15y = -11y$
$(1y + 4)(15y - 1)$	$-y + 60y = 59y$
$(1y + 1)(15y - 4)$	$-4y + 15y = 11y$
$(1y - 4)(15y + 1)$	$y - 60y = -59y$
$(1y - 2)(15y + 2)$	$2y - 30y = -28y$
$(1y + 2)(15y - 2)$	$-2y + 30y = 28y$
$(3y - 1)(5y + 4)$	$12y - 5y = 7y$
$(3y + 4)(5y - 1)$	$-3y + 20y = 17y$
$(3y + 1)(5y - 4)$	$-12y + 5y = -7y$
$(3y - 4)(5y + 1)$	$3y - 20y = -17y$
$(3y - 2)(5y + 2)$	$6y - 10y = -4y$
$(3y + 2)(5y - 2)$	$-6y + 10y = 4y$

$-45y^3 + 12y^2 + 12y = -3y(3y - 2)(5y + 2)$

You Try It 3

Factors of -14 [2(−7)]	*Sum*
$-1, +14$	13
$1, -14$	-13
$2, -7$	-5
$-2, 7$	5

$$\begin{aligned}
2a^2 + 13a - 7 &= 2a^2 - a + 14a - 7 \\
&= (2a^2 - a) + (14a - 7) \\
&= a(2a - 1) + 7(2a - 1) \\
&= (2a - 1)(a + 7)
\end{aligned}$$

$2a^3 + 13a - 7 = (2a - 1)(a + 7)$

You Try It 4

The GCF is $5x$.

$15x^3 + 40x^2 - 80x = 5x(3x^2 + 8x - 16)$

Factors of -48 [3(−16)]	*Sum*
$-1, +48$	47
$+1, -48$	-47
$-2, +24$	22
$+2, -24$	-22
$-3, +16$	13
$+3, -16$	-13
$-4, +12$	8

$$\begin{aligned}
3x^2 + 8x - 16 &= 3x^2 - 4x + 12x - 16 \\
&= (3x^2 - 4x) + (12x - 16) \\
&= x(3x - 4) + 4(3x - 4) \\
&= (3x - 4)(x + 4)
\end{aligned}$$

$$\begin{aligned}
15x^3 + 40x^2 - 80x &= 5x(3x^2 + 8x - 16) \\
&= 5x(3x - 4)(x + 4)
\end{aligned}$$

SECTION 6.4   *pages 221–224*

You Try It 1
$25a^2 - b^2 = (5a)^2 - b^2 = (5a + b)(5a - b)$

You Try It 2
$n^4 - 81 = (n^2)^2 - 9^2 = (n^2 + 9)(n^2 - 9)$
$= (n^2 + 9)(n + 3)(n - 3)$

You Try It 3
Because $16y^2 = (4y)^2$, $1 = 1^2$, and $8y = 2(4y)(1)$ the trinomial is a perfect square trinomial.

$16y^2 + 8y + 1 = (4y + 1)^2$

You Try It 4
Because $x^2 = (x)^2$, $36 = 6^2$, and $15x \neq 2(x)(6)$, the trinomial is not a perfect square trinomial. Try to factor the trinomial by another method.

$x^2 + 15x + 36 = (x + 3)(x + 12)$

You Try It 5
$(x^2 - 6x + 9) - y^2 = (x - 3)^2 - y^2$
$= (x - 3 - y)(x - 3 + y)$

You Try It 6
The GCF is $3x$.

$12x^3 - 75x = 3x(4x^2 - 25)$
$= 3x(2x + 5)(2x - 5)$

You Try It 7
Factor by grouping.

$a^3b - 7a^2 - b + 7 = a^2(b - 7) - (b - 7)$
$= (b - 7)(a^2 - 1)$
$= (b - 7)(a + 1)(a - 1)$

You Try It 8
The GCF is $4x$.

$4x^3 + 28x^2 - 120x = 4x(x^2 + 7x - 30)$
$= 4x(x + 10)(x - 3)$

SECTION 6.5   *pages 229–232*

You Try It 1
$2x(x + 7) = 0$

$2x = 0 \qquad x + 7 = 0$
$x = 0 \qquad\quad x = -7$

The solutions are 0 and -7.

You Try It 2
$4x^2 - 9 = 0$
$(2x - 3)(2x + 3) = 0$

$2x - 3 = 0 \qquad 2x + 3 = 0$
$\quad 2x = 3 \qquad\quad 2x = -3$
$\qquad x = \dfrac{3}{2} \qquad\qquad x = -\dfrac{3}{2}$

The solutions are $\dfrac{3}{2}$ and $-\dfrac{3}{2}$.

You Try It 3
$(x + 2)(x - 7) = 52$
$x^2 - 5x - 14 = 52$
$x^2 - 5x - 66 = 0$
$(x + 6)(x - 11) = 0$

$x + 6 = 0 \qquad x - 11 = 0$
$\quad x = -6 \qquad\quad x = 11$

The solutions are -6 and 11.

You Try It 4

Strategy
First positive consecutive integer: n
Second positive consecutive integer: $n + 1$

The sum of the squares of two positive consecutive integers is 61.

Solution
$n^2 + (n + 1)^2 = 61$
$n^2 + n^2 + 2n + 1 = 61$
$2n^2 + 2n + 1 = 61$
$2n^2 + 2n - 60 = 0$
$2(n^2 + n - 30) = 0$
$2(n - 5)(n + 6) = 0$

$n - 5 = 0 \qquad n + 6 = 0$
$\qquad n = 5 \qquad\quad n = -6$

Since -6 is not a positive integer, it is not a solution.

$n = 5$
$n + 1 = 5 + 1 = 6$

The two integers are 5 and 6.

You Try It 5

Strategy
Width $= x$
Length $= 2x + 4$

The area of a rectangle is 96 in^2.
Use the equation $A = l \cdot w$.

Solution
$A = l \cdot w$
$96 = (2x + 4)x$
$96 = 2x^2 + 4x$
$0 = 2x^2 + 4x - 96$
$0 = 2(x^2 + 2x - 48)$
$0 = 2(x + 8)(x - 6)$

$x + 8 = 0 \qquad x - 6 = 0$
$\quad x = -8 \qquad\quad x = 6$

Since the width cannot be a negative number, -8 is not a solution.

$x = 6$
$2x + 4 = 2(6) + 4 = 12 + 4 = 16$

The width is 6 in.
The length is 16 in.

SOLUTIONS to Chapter 7 "You Try It"

SECTION 7.1 *pages 247–250*

You Try It 1

$$\frac{6x^5y}{12x^2y^3} = \frac{\overset{1}{\cancel{2}} \cdot \overset{1}{\cancel{3}} \cdot x^5y}{\underset{1}{\cancel{2}} \cdot 2 \cdot \underset{1}{\cancel{3}} \cdot x^2y^3} = \frac{x^3}{2y^2}$$

You Try It 2

$$\frac{x^2 + 2x - 24}{16 - x^2} = \frac{\overset{-1}{(\cancel{x - 4})}(x + 6)}{\underset{1}{(\cancel{4 - x})}(4 + x)} = -\frac{x + 6}{x + 4}$$

You Try It 3

$$\frac{x^2 + 4x - 12}{x^2 - 3x + 2} = \frac{\overset{1}{(\cancel{x - 2})}(x + 6)}{(x - 1)\underset{1}{(\cancel{x - 2})}} = \frac{x + 6}{x - 1}$$

You Try It 4

$$\frac{12x^2 + 3x}{10x - 15} \cdot \frac{8x - 12}{9x + 18} = \frac{3x(4x + 1)}{5(2x - 3)} \cdot \frac{4(2x - 3)}{9(x + 2)}$$

$$= \frac{\overset{1}{\cancel{3}}x(4x + 1) \cdot 2 \cdot 2\overset{1}{(\cancel{2x - 3})}}{5\underset{1}{(\cancel{2x - 3})} \cdot \underset{1}{\cancel{3}} \cdot 3(x + 2)}$$

$$= \frac{4x(4x + 1)}{15(x + 2)}$$

You Try It 5

$$\frac{x^2 + 2x - 15}{9 - x^2} \cdot \frac{x^2 - 3x - 18}{x^2 - 7x + 6}$$

$$= \frac{(x - 3)(x + 5)}{(3 - x)(3 + x)} \cdot \frac{(x + 3)(x - 6)}{(x - 1)(x - 6)}$$

$$= \frac{\overset{-1}{(\cancel{x - 3})}(x + 5) \cdot \overset{1}{(\cancel{x + 3})}\overset{1}{(\cancel{x - 6})}}{\underset{1}{(\cancel{3 - x})}\underset{1}{(\cancel{3 + x})} \cdot (x - 1)\underset{1}{(\cancel{x - 6})}} = -\frac{x + 5}{x - 1}$$

You Try It 6

$$\frac{a^2}{4bc^2 - 2b^2c} \div \frac{a}{6bc - 3b^2} = \frac{a^2}{4bc^2 - 2b^2c} \cdot \frac{6bc - 3b^2}{a}$$

$$= \frac{a^2 \cdot 3\cancel{b}(2\cancel{c} - \cancel{b})}{2\cancel{b}c(2\cancel{c} - \cancel{b}) \cdot a} = \frac{3a}{2c}$$

You Try It 7

$$\frac{3x^2 + 26x + 16}{3x^2 - 7x - 6} \div \frac{2x^2 + 9x - 5}{x^2 + 2x - 15}$$

$$= \frac{3x^2 + 26x + 16}{3x^2 - 7x - 6} \cdot \frac{x^2 + 2x - 15}{2x^2 + 9x - 5}$$

$$= \frac{(3x + 2)(x + 8) \cdot (x + 5)(x - 3)}{(3x + 2)(x - 3) \cdot (2x - 1)(x + 5)} = \frac{x + 8}{2x - 1}$$

SECTION 7.2   *pages 255–256*

You Try It 1

$8uv^2 = 2 \cdot 2 \cdot 2 \cdot u \cdot v \cdot v$

$12uw = 2 \cdot 2 \cdot 3 \cdot u \cdot w$

$\text{LCM} = 2 \cdot 2 \cdot 2 \cdot 3 \cdot u \cdot v \cdot v \cdot w = 24uv^2w$

You Try It 2

$m^2 - 6m + 9 = (m - 3)(m - 3)$

$m^2 - 2m - 3 = (m + 1)(m - 3)$

$\text{LCM} = (m - 3)(m - 3)(m + 1)$

You Try It 3

The LCM is $36xy^2z$.

$$\frac{x - 3}{4xy^2} = \frac{x - 3}{4xy^2} \cdot \frac{9z}{9z} = \frac{9xz - 27z}{36xy^2z}$$

$$\frac{2x + 1}{9y^2z} = \frac{2x + 1}{9y^2z} \cdot \frac{4x}{4x} = \frac{8x^2 + 4x}{36xy^2z}$$

You Try It 4

The LCM is $(x + 2)(x - 5)(x + 5)$.

$$\frac{x + 4}{x^2 - 3x - 10} = \frac{x + 4}{(x + 2)(x - 5)} \cdot \frac{x + 5}{x + 5}$$

$$= \frac{x^2 + 9x + 20}{(x + 2)(x - 5)(x + 5)}$$

$$\frac{2x}{25 - x^2} = \frac{2x}{-(x^2 - 25)} = -\frac{2x}{(x - 5)(x + 5)} \cdot \frac{x + 2}{x + 2}$$

$$= -\frac{2x^2 + 4x}{(x + 2)(x - 5)(x + 5)}$$

SECTION 7.3   *pages 259–262*

You Try It 1

$$\frac{3}{xy} + \frac{12}{xy} = \frac{3 + 12}{xy} = \frac{15}{xy}$$

You Try It 2

$$\frac{2x^2}{x^2 - x - 12} - \frac{7x + 4}{x^2 - x - 12}$$

$$= \frac{2x^2 - (7x + 4)}{x^2 - x - 12} = \frac{2x^2 - 7x - 4}{x^2 - x - 12}$$

$$= \frac{(2x + 1)(x - 4)}{(x + 3)(x - 4)} = \frac{2x + 1}{x + 3}$$

You Try It 3

$$\frac{x^2 - 1}{x^2 - 8x + 12} - \frac{2x + 1}{x^2 - 8x + 12} + \frac{x}{x^2 - 8x + 12}$$

$$= \frac{(x^2 - 1) - (2x + 1) + x}{x^2 - 8x + 12} = \frac{x^2 - 1 - 2x - 1 + x}{x^2 - 8x + 12}$$

$$= \frac{x^2 - x - 2}{x^2 - 8x + 12} = \frac{(x + 1)(x - 2)}{(x - 2)(x - 6)} = \frac{x + 1}{x - 6}$$

You Try It 4

The LCM of the denominators is $24y$.

$$\frac{z}{8y} - \frac{4z}{3y} + \frac{5z}{4y} = \frac{z}{8y} \cdot \frac{3}{3} - \frac{4z}{3y} \cdot \frac{8}{8} + \frac{5z}{4y} \cdot \frac{6}{6}$$

$$= \frac{3z}{24y} - \frac{32z}{24y} + \frac{30z}{24y}$$

$$= \frac{3z - 32z + 30z}{24y} = \frac{z}{24y}$$

You Try It 5

$$2 - x = -(x - 2)$$

Therefore, $\dfrac{3}{2 - x} = \dfrac{-3}{x - 2}$.

The LCM is $x - 2$.

$$\dfrac{5x}{x - 2} - \dfrac{3}{2 - x} = \dfrac{5x}{x - 2} - \dfrac{-3}{x - 2}$$
$$= \dfrac{5x - (-3)}{x - 2} = \dfrac{5x + 3}{x - 2}$$

You Try It 6

The LCM is $(3x - 1)(x + 4)$.

$$\dfrac{4x}{3x - 1} - \dfrac{9}{x + 4} = \dfrac{4x}{3x - 1} \cdot \dfrac{x + 4}{x + 4} - \dfrac{9}{x + 4} \cdot \dfrac{3x - 1}{3x - 1}$$
$$= \dfrac{4x^2 + 16x}{(3x - 1)(x + 4)} - \dfrac{27x - 9}{(3x - 1)(x + 4)}$$
$$= \dfrac{(4x^2 + 16x) - (27x - 9)}{(3x - 1)(x + 4)}$$
$$= \dfrac{4x^2 + 16x - 27x + 9}{(3x - 1)(x + 4)} = \dfrac{4x^2 - 11x + 9}{(3x - 1)(x + 4)}$$

You Try It 7

The LCM is $x - 3$.

$$2 - \dfrac{1}{x - 3} = 2 \cdot \dfrac{x - 3}{x - 3} - \dfrac{1}{x - 3}$$
$$= \dfrac{2x - 6}{x - 3} - \dfrac{1}{x - 3}$$
$$= \dfrac{2x - 6 - 1}{x - 3}$$
$$= \dfrac{2x - 7}{x - 3}$$

You Try It 8

The LCM is $(x + 5)(x - 5)$.

$$\dfrac{2}{5 - x} = \dfrac{-2}{x - 5}$$

$$\dfrac{2x - 1}{x^2 - 25} + \dfrac{2}{5 - x} = \dfrac{2x - 1}{(x + 5)(x - 5)} + \dfrac{-2}{x - 5}$$
$$= \dfrac{2x - 1}{(x + 5)(x - 5)} + \dfrac{-2}{x - 5} \cdot \dfrac{x + 5}{x + 5}$$
$$= \dfrac{2x - 1}{(x + 5)(x - 5)} + \dfrac{-2(x + 5)}{(x + 5)(x - 5)}$$
$$= \dfrac{2x - 1 + (-2)(x + 5)}{(x + 5)(x - 5)}$$
$$= \dfrac{2x - 1 - 2x - 10}{(x + 5)(x - 5)}$$
$$= \dfrac{-11}{(x + 5)(x - 5)}$$
$$= -\dfrac{11}{(x + 5)(x - 5)}$$

You Try It 9

The LCM is $(3x + 2)(x - 1)$.

$$\dfrac{2x - 3}{3x^2 - x - 2} + \dfrac{5}{3x + 2} - \dfrac{1}{x - 1}$$
$$= \dfrac{2x - 3}{(3x + 2)(x - 1)} + \dfrac{5}{3x + 2} \cdot \dfrac{x - 1}{x - 1} - \dfrac{1}{x - 1} \cdot \dfrac{3x + 2}{3x + 2}$$
$$= \dfrac{2x - 3}{(3x + 2)(x - 1)} + \dfrac{5x - 5}{(3x + 2)(x - 1)} - \dfrac{3x + 2}{(3x + 2)(x - 1)}$$
$$= \dfrac{(2x - 3) + (5x - 5) - (3x + 2)}{(3x + 2)(x - 1)}$$
$$= \dfrac{2x - 3 + 5x - 5 - 3x - 2}{(3x + 2)(x - 1)}$$
$$= \dfrac{4x - 10}{(3x + 2)(x - 1)} = \dfrac{2(2x - 5)}{(3x + 2)(x - 1)}$$

SECTION 7.4 *pages 267–270*

You Try It 1

The LCM of 3, x, 9, and x^2 is $9x^2$.

$$\dfrac{\dfrac{1}{3} - \dfrac{1}{x}}{\dfrac{1}{9} - \dfrac{1}{x^2}} = \dfrac{\dfrac{1}{3} - \dfrac{1}{x}}{\dfrac{1}{9} - \dfrac{1}{x^2}} \cdot \dfrac{9x^2}{9x^2} = \dfrac{\dfrac{1}{3} \cdot 9x^2 - \dfrac{1}{x} \cdot 9x^2}{\dfrac{1}{9} \cdot 9x^2 - \dfrac{1}{x^2} \cdot 9x^2}$$

$$= \dfrac{3x^2 - 9x}{x^2 - 9} = \dfrac{3x(x - 3)}{(x - 3)(x + 3)} = \dfrac{3x}{x + 3}$$

Content and Format © 1995 HMCo.

You Try It 2
The LCM of x and x^2 is x^2.

$$\dfrac{1 + \dfrac{4}{x} + \dfrac{3}{x^2}}{1 + \dfrac{10}{x} + \dfrac{21}{x^2}} = \dfrac{1 + \dfrac{4}{x} + \dfrac{3}{x^2}}{1 + \dfrac{10}{x} + \dfrac{21}{x^2}} \cdot \dfrac{x^2}{x^2}$$

$$= \dfrac{1 \cdot x^2 + \dfrac{4}{x} \cdot x^2 + \dfrac{3}{x^2} \cdot x^2}{1 \cdot x^2 + \dfrac{10}{x} \cdot x^2 + \dfrac{21}{x^2} \cdot x^2}$$

$$= \dfrac{x^2 + 4x + 3}{x^2 + 10x + 21} = \dfrac{(x + 1)\overset{1}{\cancel{(x + 3)}}}{\underset{1}{\cancel{(x + 3)}}(x + 7)} = \dfrac{x + 1}{x + 7}$$

You Try It 3
The LCM is $x - 5$.

$$\dfrac{x + 3 - \dfrac{20}{x - 5}}{x + 8 + \dfrac{30}{x - 5}} = \dfrac{x + 3 - \dfrac{20}{x - 5}}{x + 8 + \dfrac{30}{x - 5}} \cdot \dfrac{x - 5}{x - 5}$$

$$= \dfrac{(x + 3)(x - 5) - \dfrac{20}{x - 5} \cdot (x - 5)}{(x + 8)(x - 5) + \dfrac{30}{x - 5} \cdot (x - 5)}$$

$$= \dfrac{x^2 - 2x - 15 - 20}{x^2 + 3x - 40 + 30} = \dfrac{x^2 - 2x - 35}{x^2 + 3x - 10}$$

$$= \dfrac{\overset{1}{\cancel{(x + 5)}}(x - 7)}{(x - 2)\underset{1}{\cancel{(x + 5)}}} = \dfrac{x - 7}{x - 2}$$

SECTION 7.5   *pages 271–272*

You Try It 1

$$\dfrac{x}{x + 6} = \dfrac{3}{x} \quad \text{The LCM is } x(x + 6).$$

$$\dfrac{\overset{1}{x\cancel{(x + 6)}}}{1} \cdot \dfrac{x}{\underset{1}{\cancel{x + 6}}} = \dfrac{x(x + 6)}{1} \cdot \dfrac{3}{x}$$

$$x^2 = (x + 6)3$$
$$x^2 = 3x + 18$$
$$x^2 - 3x - 18 = 0$$
$$(x + 3)(x - 6) = 0$$

$$x + 3 = 0 \qquad x - 6 = 0$$
$$x = -3 \qquad x = 6$$

Both -3 and 6 check as solutions.
The solutions are -3 and 6.

You Try It 2

$$\dfrac{5x}{x + 2} = 3 - \dfrac{10}{x + 2} \quad \text{The LCM is } x + 2.$$

$$\dfrac{(x + 2)}{1} \cdot \dfrac{5x}{x + 2} = \dfrac{(x + 2)}{1}\left(3 - \dfrac{10}{x + 2}\right)$$

$$\dfrac{\overset{1}{\cancel{x + 2}}}{1} \cdot \dfrac{5x}{\underset{1}{\cancel{x + 2}}} = \dfrac{x + 2}{1} \cdot 3 - \dfrac{\overset{1}{\cancel{x + 2}}}{1} \cdot \dfrac{10}{\underset{1}{\cancel{x + 2}}}$$

$$5x = (x + 2)3 - 10$$
$$5x = 3x + 6 - 10$$
$$5x = 3x - 4$$
$$2x = -4$$
$$x = -2$$

-2 does not check as a solution.
The equation has no solution.

You Try It 1

$$\frac{2}{x+3} = \frac{6}{5x+5}$$

$$\frac{(x+3)(5x+5)}{1} \cdot \frac{2}{x+3} = \frac{(x+3)(5x+5)}{1} \cdot \frac{6}{5x+5}$$

$$\frac{\overset{1}{\cancel{(x+3)}}(5x+5)}{1} \cdot \frac{2}{\underset{1}{\cancel{x+3}}} = \frac{(x+3)\overset{1}{\cancel{(5x+5)}}}{1} \cdot \frac{6}{\underset{1}{\cancel{5x+5}}}$$

$$(5x+5)2 = (x+3)6$$
$$10x + 10 = 6x + 18$$
$$4x + 10 = 18$$
$$4x = 8$$
$$x = 2$$

The solution is 2.

You Try It 3

Strategy
To find the additional amount of medication required for a 200-pound adult, write and solve a proportion using x to represent the additional medication. Then $3 + x$ is the total amount required for a 200-pound adult.

Solution

$$\frac{150}{3} = \frac{200}{3+x}$$

$$\frac{50}{1} = \frac{200}{3+x}$$

$$(3+x) \cdot 50 = (3+x) \cdot \frac{200}{3+x}$$

$$(3+x) \cdot 50 = 200$$
$$150 + 50x = 200$$
$$50x = 50$$
$$x = 1$$

One additional ounce is required for a 200-pound adult.

You Try It 2

Strategy
To find the total area that 256 ceramic tiles will cover, write and solve a proportion using x to represent the number of square feet that 256 tiles will cover.

Solution

$$\frac{9}{16} = \frac{x}{256}$$

$$256\left(\frac{9}{16}\right) = 256\left(\frac{x}{256}\right)$$

$$144 = x$$

A 144-square-foot area can be tiled using 256 ceramic tiles.

SECTION 7.7   *pages 279–280*

You Try It 1
$$5x - 2y = 10$$
$$5x - 5x - 2y = -5x + 10$$
$$-2y = -5x + 10$$
$$\frac{-2y}{-2} = \frac{-5x + 10}{-2}$$
$$y = \frac{5}{2}x - 5$$

You Try It 2
$$s = \frac{A + L}{2}$$
$$2 \cdot s = 2\left(\frac{A + L}{2}\right)$$
$$2s = A + L$$
$$2s - A = A - A + L$$
$$2s - A = L$$

You Try It 3
$$S = a + (n - 1)d$$
$$S = a + nd - d$$
$$S - a = a - a + nd - d$$
$$S - a = nd - d$$
$$S - a + d = nd - d + d$$
$$S - a + d = nd$$
$$\frac{S - a + d}{d} = \frac{nd}{d}$$
$$\frac{S - a + d}{d} = n$$

You Try It 4
$$S = C + rC$$
$$S = (1 + r)C$$
$$\frac{S}{1 + r} = \frac{(1 + r)C}{1 + r}$$
$$\frac{S}{1 + r} = C$$

SECTION 7.8   *pages 283–286*

You Try It 1

Strategy
• Time for one printer to complete the job: t

	Rate	*Time*	*Part*
1st printer	$\frac{1}{t}$	2	$\frac{2}{t}$
2nd printer	$\frac{1}{t}$	5	$\frac{5}{t}$

• The sum of the parts of the task completed must equal 1.

Solution
$$\frac{2}{t} + \frac{5}{t} = 1$$
$$t\left(\frac{2}{t} + \frac{5}{t}\right) = t \cdot 1$$
$$2 + 5 = t$$
$$7 = t$$

Working alone, one printer takes 7 h to print the payroll.

You Try It 2

Strategy
• Rate sailing across the lake: r
  Rate sailing back: $3r$

	Distance	*Rate*	*Time*
Across	6	r	$\frac{6}{r}$
Back	6	$3r$	$\frac{6}{3r}$

• The total time for the trip was 2 h.

Solution
$$\frac{6}{r} + \frac{6}{3r} = 2$$
$$3r\left(\frac{6}{r} + \frac{6}{3r}\right) = 3r(2)$$
$$3r \cdot \frac{6}{r} + 3r \cdot \frac{6}{3r} = 6r$$
$$18 + 6 = 6r$$
$$24 = 6r$$
$$4 = r$$

The rate across the lake was 4 km/h.

SOLUTIONS to Chapter 8 "You Try It"

You Try It 1

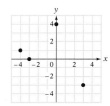

You Try It 2 $A(4, -2)$, $B(-2, 4)$.
The abscissa of D is 0.
The ordinate of C is 0.

You Try It 3

$$x - 3y = -14$$

$-2 - 3(4)$	-14
$-2 - 12$	-14
$-14 = -14$	

Yes, $(-2, 4)$ is a solution of
$x - 3y = -14$.

You Try It 4

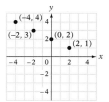

You Try It 5
$\{(82, 91), (78, 86), (81, 96), (87, 79), (81, 87)\}$

No the relation is not a function. The two
ordered pairs, $(81, 96)$ and $(81, 87)$ have
the same first coordinate but different
second coordinates.

You Try It 6
Determine the ordered pairs
defined by the equation. Replace
x in $y = \dfrac{1}{2}x + 1$ by the given values and
solve for y. $\{(-4, -1), (0, 1), (2, 2)\}$ Yes, y is
a function of x.

You Try It 7 $H(x) = \dfrac{x}{x - 4}$

$$H(8) = \frac{8}{8 - 4}$$

$$H(8) = \frac{8}{4} = 2$$

You Try It 1

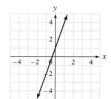

You Try It 2

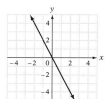

You Try It 3

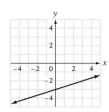

You Try It 4 $5x - 2y = 10$
$$-2y = -5x + 10$$
$$y = \frac{5}{2}x - 5$$

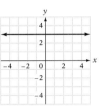

You Try It 5 $x - 3y = 9$
$$-3y = -x + 9$$
$$y = \frac{1}{3}x - 3$$

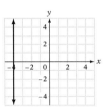

You Try It 6

You Try It 7

You Try It 8

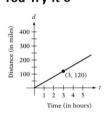

The ordered pair $(3, 120)$ means that in 3 h the car will travel 120 mi.

SECTION 8.3   *pages 323–328*

You Try It 1

x-intercept	*y*-intercept
$y = 2x - 4$	$(0, b)$
$0 = 2x - 4$	$b = -4$
$-2x = -4$	$(0, -4)$
$x = 2$	
$(2, 0)$	

You Try It 2   Let $P_2 = (-3, 8)$ and $P_1 = (1, 4)$.

$$m = \frac{y_2 - y_1}{x_2 - x_2} = \frac{8 - 4}{-3 - 1} = \frac{4}{-4} = -1$$

The slope is -1

You Try It 3   Let $P_1 = (-1, 2)$ and $P_2 = (4, 2)$.

$$m = \frac{y_2 - y_1}{x_2 - x_1} = \frac{2 - 2}{4 - (-1)} = \frac{0}{5} = 0$$

The slope is 0.

You Try It 4

$$m = \frac{8650 - 6100}{1 - 4} = \frac{2550}{-3}$$
$$= -850$$

A slope of -850 means that the value of the car is decreasing at a rate of $850 per year.

You Try It 5 y-intercept $= (0, b) = (0, -1)$

$$m = -\frac{1}{4}$$

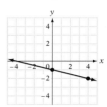

You Try It 6   Solve the equation for y.

$$x - 2y = 4$$
$$-2y = -x + 4$$
$$y = \frac{1}{2}x - 2$$

y-intercept $= (0, b) = (0, -2)$

$$m = \frac{1}{2}$$

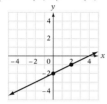

SECTION 8.4   *pages 333–336*

You Try It 1
Because the slope and y-intercept are known, use the slope–intercept formula, $y = mx + b$.

$$y = mx + b$$
$$y = \frac{5}{3}x + 2$$

You Try It 2

$$m = \frac{3}{4} \qquad (x_1, y_1) = (4, -2)$$
$$y - y_1 = m(x - x_1)$$
$$y - (-2) = \frac{3}{4}(x - 4)$$
$$y + 2 = \frac{3}{4}x - 3$$
$$y = \frac{3}{4}x - 5$$

The equation of the line is $y = \frac{3}{4}x - 5$.

You Try It 3
Find the slope of the line between the two points.

$$m = \frac{y_2 - y_1}{x_2 - x_1} = \frac{1 - (-1)}{3 - (-6)} = \frac{2}{9}$$

Use the point slope formula.

$$y - y_1 = m(x - x_1)$$
$$y - (-1) = \frac{2}{9}[x - (-6)]$$
$$y + 1 = \frac{2}{9}x + \frac{4}{3}$$
$$y = \frac{2}{9}x + \frac{1}{3}$$

You Try It 4

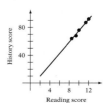

The slope of the line means that the grade on the history test increases 8.3 points for each 1 point increase in the grade on the reading test.

SOLUTIONS to Chapter 9 "You Try It"

You Try It 1

$$2x - 5y = 8 \qquad\qquad -x + 3y = -5$$

$2(-1) - 5(-2)$	8		$-(-1) + 3(-2)$	-5
$-2 + 10$	8		$1 + (-6)$	-5
	$8 = 8$			$-5 = -5$

Yes, $(-1, -2)$ is a solution of the system of equations.

You Try It 2

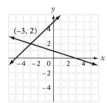

The solution is $(-3, 2)$.

You Try It 3

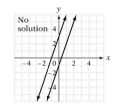

The lines are parallel. The system of equations is inconsistent and therefore does not have a solution.

You Try It 1

$$(1) \quad 7x - y = 4$$
$$(2) \quad 3x + 2y = 9$$

Solve equation (1) for y.

$$7x - y = 4$$
$$-y = -7x + 4$$
$$y = 7x - 4$$

Substitute in equation (2).

$$3x + 2y = 9$$
$$3x + 2(7x - 4) = 9$$
$$3x + 14x - 8 = 9$$
$$17x - 8 = 9$$
$$17x = 17$$
$$x = 1$$

Substitute in equation (1).

$$7x - y = 4$$
$$7(1) - y = 4$$
$$7 - y = 4$$
$$-y = -3$$
$$y = 3$$

The solution is $(1, 3)$.

You Try It 2

$$(1) \quad 3x - y = 4$$
$$(2) \qquad\quad y = 3x + 2$$

$$3x - y = 4$$
$$3x - (3x + 2) = 4$$
$$3x - 3x - 2 = 4$$
$$-2 = 4$$

This is not a true equation. The system of equations is inconsistent and therefore does not have a solution.

You Try It 3
$$(1) \qquad\qquad y = -2x + 1$$
$$(2) \quad 6x + 3y = 3$$

$$6x + 3y = 3$$
$$6x + 3(-2x + 1) = 3$$
$$6x - 6x + 3 = 3$$
$$3 = 3$$

The system of equations is dependent. The solutions are the ordered pairs that satisfy the equation $y = -2x + 1$.

SECTION 9.3   *pages 365–368*

You Try It 1
$$(1) \quad x - 2y = 1$$
$$(2) \quad 2x + 4y = 0$$

Eliminate y.
$$2(x - 2y) = 2 \cdot 1$$
$$2x + 4y = 0$$

$$2x - 4y = 2$$
$$2x + 4y = 0$$

Add the equations.
$$4x = 2$$
$$x = \frac{2}{4} = \frac{1}{2}$$

Replace x in equation (2).
$$2\left(\frac{1}{2}\right) + 4y = 0$$
$$1 + 4y = 0$$
$$4y = -1$$
$$y = -\frac{1}{4}$$

The solution is $\left(\frac{1}{2}, -\frac{1}{4}\right)$.

You Try It 2
$$(1) \quad 2x - 3y = 4$$
$$(2) \quad -4x + 6y = -8$$

Eliminate y.
$$2(2x - 3y) = 2 \cdot 4$$
$$-4x + 6y = -8$$

$$4x - 6y = 8$$
$$-4x + 6y = -8$$

Add the equations.
$$0 + 0 = 0$$
$$0 = 0$$

The system of equations is dependent. The solutions are the ordered pairs that satisfy the equation $2x - 3y = 4$.

You Try It 3 (1) $4x + 5y = 11$
(2) $\quad 3y = x + 10$

Write equation (2) in the form
$Ax + By = C$.
$$3y = x + 10$$
$$-x + 3y = 10$$

Eliminate x.
$$4x + 5y = 11$$
$$4(-x + 3y) = 4 \cdot 10$$

$$4x + 5y = 11$$
$$-4x + 12y = 40$$

Add the equations.
$$17y = 51$$
$$y = 3$$

Replace y in equation (1).
$$4x + 5y = 11$$
$$4x + 5 \cdot 3 = 11$$
$$4x + 15 = 11$$
$$4x = -4$$
$$x = -1$$

The solution is $(-1, 3)$.

SECTION 9.4 *pages 371–374*

You Try It 1

Strategy

• Rate of the current: c
  Rate of the canoeist in calm water: r

	Rate	Time	Distance
With current	$r + c$	3	$3(r + c)$
Against current	$r - c$	5	$5(r - c)$

• The distance traveled with the current is 15 mi.
  The distance traveled against the current is 15 mi.

Solution

$$3(r + c) = 15 \qquad \frac{1}{3} \cdot 3(r + c) = \frac{1}{3} \cdot 15$$

$$5(r - c) = 15 \qquad \frac{1}{5} \cdot 5(r - c) = \frac{1}{5} \cdot 15$$

$$r + c = 5$$
$$r - c = 3$$

$$2r = 8$$
$$r = 4$$

$$r + c = 5$$
$$4 + c = 5$$
$$c = 1$$

The rate of the current is 1 mph.
The rate of the canoeist in calm water is 4 mph.

You Try It 2

Strategy

- Cost of an orange tree: x
  Cost of a grapefruit tree: y

First purchase:

	Amount	Unit Cost	Value
Orange trees	25	x	$25x$
Grapefruit trees	20	y	$20y$

Second purchase:

	Amount	Unit Cost	Value
Orange trees	20	x	$20x$
Grapefruit trees	30	y	$30y$

- The total of the first purchase was $290.
  The total of the second purchase was $330.

Solution

$$25x + 20y = 290 \qquad 4(25x + 20y) = 4 \cdot 290$$
$$20x + 30y = 330 \qquad -5(20x + 30y) = -5 \cdot 330$$

$$100x + 80y = 1160$$
$$-100x - 150y = -1650$$
$$-70y = -490$$
$$y = 7$$

$$25x + 20y = 290$$
$$25x + 20(7) = 290$$
$$25x + 140 = 290$$
$$25x = 150$$
$$x = 6$$

The cost of an orange tree is $6.
The cost of a grapefruit tree is $7.

SOLUTIONS to Chapter 10 "You Try It"

You Try It 1 $A = \{-9, -7, -5, -3, -1\}$

You Try It 2 $A = \{1, 3, 5, \ldots\}$

You Try It 3 $A \cup B = \{-2, -1, 0, 1, 2, 3, 4\}$

You Try It 4 $C \cap D = \{10, 16\}$

You Try It 5 $A \cap B = \varnothing$

You Try It 6 $\{x \mid x < 59, x \text{ is a positive even integer}\}$

You Try It 7 $\{x \mid x > -3, x \in \text{real numbers}\}$

You Try It 8   The solution set is the numbers greater than -2.

You Try It 9   The solution set is the numbers greater than -1 and the numbers less than -3.

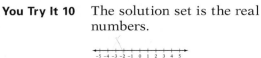

You Try It 10   The solution set is the real numbers.

You Try It 11   The solution set is the numbers which are less than or equal to 4 and greater than or equal to -4.

SECTION 10.2   *pages 393–396*

You Try It 1
$$x + 2 < -2$$
$$x + 2 - 2 < -2 - 2$$
$$x < -4$$

You Try It 2
$$5x + 3 > 4x + 5$$
$$5x - 4x + 3 > 4x - 4x + 5$$
$$x + 3 > 5$$
$$x + 3 - 3 > 5 - 3$$
$$x > 2$$

You Try It 3
$$-3x > -9$$
$$\frac{-3x}{-3} < \frac{-9}{-3}$$
$$x < 3$$

You Try It 4
$$-\frac{3}{4}x \geq 18$$
$$-\frac{4}{3}\left(-\frac{3}{4}x\right) \leq -\frac{4}{3}(18)$$
$$x \leq -24$$

You Try It 5

Strategy   To find the selling prices, write and solve an inequality using p to represent the possible selling prices.

Solution
$$0.70p \geq 314$$
$$p \geq 448.571$$

The dealer will make a profit if the selling price is greater than or equal to $448.58

SECTION 10.3   *pages 401–402*

You Try It 1
$$5 - 4x > 9 - 8x$$
$$5 - 4x + 8x > 9 - 8x + 8x$$
$$5 + 4x > 9$$
$$5 - 5 + 4x > 9 - 5$$
$$4x > 4$$
$$\frac{4x}{4} > \frac{4}{4}$$
$$x > 1$$

You Try It 2
$$8 - 4(3x + 5) \leq 6(x - 8)$$
$$8 - 12x - 20 \leq 6x - 48$$
$$-12 - 12x \leq 6x - 48$$
$$-12 - 12x - 6x \leq 6x - 6x - 48$$
$$-12 - 18x \leq -48$$
$$-12 + 12 - 18x \leq -48 + 12$$
$$-18x \leq -36$$
$$\frac{-18x}{-18} \geq \frac{-36}{-18}$$
$$x \geq 2$$

You Try It 3

Strategy   To find the maximum number
of miles:
- Write an expression for the
  cost of each car, using x to
  represent the number of
  miles driven during the week.
- Write and solve an inequality.

Solution

Cost of a Company A car	is less than	Cost of a Company B car

$$8(7) + 0.10x < 10(7) + 0.08x$$
$$56 + 0.10x < 70 + 0.08x$$
$$56 + 0.10x - 0.08x < 70 + 0.08x - 0.08x$$
$$56 + 0.02x < 70$$
$$56 - 56 + 0.02x < 70 - 56$$
$$0.02x < 14$$
$$\frac{0.02x}{0.02} < \frac{14}{0.02}$$
$$x < 700$$

The maximum number of miles
is 699.

SECTION 10.4   *pages 405–406*

You Try It 1

$$x - 3y < 2$$
$$x - x - 3y < -x + 2$$
$$-3y < -x + 2$$
$$\frac{-3y}{-3} > \frac{-x + 2}{-3}$$
$$y > \frac{1}{3}x - \frac{2}{3}$$

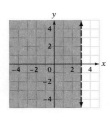

You Try It 2

$$2x - 4y \le 8$$
$$2x - 2x - 4y \le -2x + 8$$
$$-4y \le -2x + 8$$
$$\frac{-4y}{-4} \ge \frac{-2x + 8}{-4}$$
$$y \ge \frac{1}{2}x - 2$$

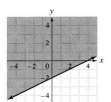

You Try It 3 $x < 3$

SOLUTIONS to Chapter 11 "You Try It"

SECTION 11.1   *pages 419–422*

You Try It 1
$$-5\sqrt{32} = -5\sqrt{2^5} = -5\sqrt{2^4 \cdot 2} = -5\sqrt{2^4}\sqrt{2}$$
$$= -5 \cdot 2^2\sqrt{2} = -20\sqrt{2}$$

You Try It 2
$$\sqrt{216} = \sqrt{2^3 \cdot 3^3} = \sqrt{2^2 \cdot 3^2(2 \cdot 3)}$$
$$= \sqrt{2^2 \cdot 3^2}\sqrt{2 \cdot 3} = 2 \cdot 3\sqrt{2 \cdot 3} = 6\sqrt{6}$$

You Try It 3
$$\sqrt{y^{19}} = \sqrt{y^{18} \cdot y} = \sqrt{y^{18}}\sqrt{y} = y^9\sqrt{y}$$

You Try It 4
$$\sqrt{45b^7} = \sqrt{3^2 \cdot 5 \cdot b^7} = \sqrt{3^2b^6(5 \cdot b)}$$
$$= \sqrt{3^2b^6}\sqrt{5b} = 3b^3\sqrt{5b}$$

You Try It 5
$$3a\sqrt{28a^9b^{18}} = 3a\sqrt{2^2 \cdot 7 \cdot a^9b^{18}}$$
$$= 3a\sqrt{2^2a^8b^{18}(7a)}$$
$$= 3a\sqrt{2^2a^8b^{18}}\sqrt{7a}$$
$$= 3a \cdot 2 \cdot a^4b^9\sqrt{7a} = 6a^5b^9\sqrt{7a}$$

You Try It 6
$$\sqrt{25(a+3)^2} = \sqrt{5^2(a+3)^2} = 5(a+3)$$
$$= 5a + 15$$

You Try It 7
$$\sqrt{x^2 + 14x + 49} = \sqrt{(x+7)^2} = x + 7$$

SECTION 11.2   *pages 425–426*

You Try It 1
$$9\sqrt{3} + 3\sqrt{3} - 18\sqrt{3} = -6\sqrt{3}$$

You Try It 2
$$2\sqrt{50} - 5\sqrt{32} = 2\sqrt{2 \cdot 5^2} - 5\sqrt{2^5}$$
$$= 2\sqrt{5^2}\sqrt{2} - 5\sqrt{2^4}\sqrt{2}$$
$$= 2 \cdot 5\sqrt{2} - 5 \cdot 2^2\sqrt{2}$$
$$= 10\sqrt{2} - 20\sqrt{2}$$
$$= -10\sqrt{2}$$

You Try It 3
$$y\sqrt{28y} + 7\sqrt{63y^3}$$
$$= y\sqrt{2^2 \cdot 7y} + 7\sqrt{3^2 \cdot 7 \cdot y^3}$$
$$= y\sqrt{2^2}\sqrt{7y} + 7\sqrt{3^2 \cdot y^2}\sqrt{7y}$$
$$= y \cdot 2\sqrt{7y} + 7 \cdot 3 \cdot y\sqrt{7y}$$
$$= 2y\sqrt{7y} + 21y\sqrt{7y}$$
$$= 23y\sqrt{7y}$$

You Try It 4
$$2\sqrt{27a^5} - 4a\sqrt{12a^3} + a^2\sqrt{75a}$$
$$= 2\sqrt{3^3 \cdot a^5} - 4a\sqrt{2^2 \cdot 3 \cdot a^3} + a^2\sqrt{3 \cdot 5^2 \cdot a}$$
$$= 2\sqrt{3^2 \cdot a^4}\sqrt{3a} - 4a\sqrt{2^2 \cdot a^2}\sqrt{3a}$$
$$\quad + a^2\sqrt{5^2}\sqrt{3a}$$
$$= 2 \cdot 3 \cdot a^2\sqrt{3a} - 4a \cdot 2 \cdot a\sqrt{3a} + a^2 \cdot 5\sqrt{3a}$$
$$= 6a^2\sqrt{3a} - 8a^2\sqrt{3a} + 5a^2\sqrt{3a} = 3a^2\sqrt{3a}$$

SECTION 11.3 *pages 429–432*

You Try It 1

$\sqrt{5a}\,\sqrt{15a^3b^4}\,\sqrt{3b^5}$
$= \sqrt{225a^4b^9} = \sqrt{3^25^2a^4b^9}$
$= \sqrt{3^25^2a^4b^8}\,\sqrt{b} = 3 \cdot 5a^2b^4\sqrt{b}$
$= 15a^2b^4\sqrt{b}$

You Try It 2

$\sqrt{5x}\,(\sqrt{5x} - \sqrt{25y})$
$= \sqrt{5^2x^2} - \sqrt{5^3xy}$
$= \sqrt{5^2x^2} - \sqrt{5^2}\,\sqrt{5xy} = 5x - 5\sqrt{5xy}$

You Try It 3

$(2\sqrt{x} + 7)(2\sqrt{x} - 7) = 4\sqrt{x^2} - 7^2$
$\qquad\qquad\qquad\qquad = 4x - 49$

You Try It 4

$(3\sqrt{x} - \sqrt{y})(5\sqrt{x} - 2\sqrt{y})$
$= 15\sqrt{x^2} - 6\sqrt{xy} - 5\sqrt{xy} + 2\sqrt{y^2}$
$= 15\sqrt{x^2} - 11\sqrt{xy} + 2\sqrt{y^2}$
$= 15x - 11\sqrt{xy} + 2y$

You Try It 5

$\dfrac{\sqrt{15x^6y^7}}{\sqrt{3x^7y^9}} = \sqrt{\dfrac{15x^6y^7}{3x^7y^9}} = \sqrt{\dfrac{5}{xy^2}}$
$= \dfrac{\sqrt{5}}{y\sqrt{x}} = \dfrac{\sqrt{5}}{y\sqrt{x}} \cdot \dfrac{\sqrt{x}}{\sqrt{x}} = \dfrac{\sqrt{5x}}{xy}$

You Try It 6

$\dfrac{\sqrt{3}}{\sqrt{3} - \sqrt{6}} = \dfrac{\sqrt{3}}{\sqrt{3} - \sqrt{6}} \cdot \dfrac{\sqrt{3} + \sqrt{6}}{\sqrt{3} + \sqrt{6}}$
$= \dfrac{3 + \sqrt{18}}{3 - 6} = \dfrac{3 + 3\sqrt{2}}{-3}$
$= \dfrac{3(1 + \sqrt{2})}{-3} = -1(1 + \sqrt{2})$
$= -1 - \sqrt{2}$

You Try It 7

$\dfrac{5 + \sqrt{y}}{1 - 2\sqrt{y}} = \dfrac{5 + \sqrt{y}}{1 - 2\sqrt{y}} \cdot \dfrac{1 + 2\sqrt{y}}{1 + 2\sqrt{y}} = \dfrac{5 + 10\sqrt{y} + \sqrt{y} + 2\sqrt{y^2}}{1 - 4y}$
$= \dfrac{5 + 11\sqrt{y} + 2y}{1 - 4y}$

SECTION 11.4 *pages 435–438*

You Try It 1

$\sqrt{4x} + 3 = 7$
$\sqrt{4x} = 4$
$\qquad (\sqrt{4x})^2 = 4^2$
$4x = 16$
$x = 4$

Check: $\quad\dfrac{\sqrt{4x} + 3 = 7}{}$

$\begin{array}{c|c}\sqrt{4 \cdot 4} + 3 & 7 \\ \sqrt{4^2} + 3 & 7 \\ 4 + 3 & 7 \\ & 7 = 7\end{array}$

The solution is 4.

You Try It 2

$\sqrt{x} + \sqrt{x + 9} = 9$
$\sqrt{x} = 9 - \sqrt{x + 9}$
$(\sqrt{x})^2 = (9 - \sqrt{x + 9})^2$
$x = 81 - 18\sqrt{x + 9} + (x + 9)$
$-90 = -18\sqrt{x + 9}$
$5 = \sqrt{x + 9}$
$5^2 = (\sqrt{x + 9})^2$
$25 = x + 9$
$16 = x$

Check:
$\sqrt{16} + \sqrt{16 + 9} = 9 \qquad \sqrt{x} + \sqrt{x + 9} = 9$
$4 + 5 = 9 \qquad\qquad \dfrac{\sqrt{16} + \sqrt{16 + 9}}{}\Big| 9$
$9 = 9 \qquad\qquad\qquad\qquad 4 + 5 \,\Big|\, 9$
$\qquad\qquad\qquad\qquad\qquad\qquad\quad 9 = 9$

The solution is 16.

Content and Format © 1995 HMCo.

You Try It 3

Strategy
To find the distance, use the Pythagorean Theorem. The hypotenuse is the length of the ladder. One leg is the distance from the bottom of the ladder to the base of the building. The distance along the building from the ground to the top of the ladder is the unknown leg.

Solution
$$a^2 = \sqrt{c^2 - b^2}$$
$$= \sqrt{(8)^2 - (3)^2}$$
$$= \sqrt{64 - 9}$$
$$= \sqrt{55}$$
$$\approx 7.416$$

The distance is 7.416 ft.

You Try It 4

Strategy
To find the length of the pendulum, replace T in the equation with the given value and solve for L.

Solution
$$T = 2\pi\sqrt{\frac{L}{32}}$$
$$2.5 = 2(3.14)\sqrt{\frac{L}{32}}$$
$$2.5 = 6.28\sqrt{\frac{L}{32}}$$
$$\frac{2.5}{6.28} = \sqrt{\frac{L}{32}}$$
$$\left(\frac{2.5}{6.28}\right)^2 = \left(\sqrt{\frac{L}{32}}\right)^2$$
$$\frac{6.25}{39.4384} = \frac{L}{32}$$
$$(32)\left(\frac{6.25}{39.4384}\right) = (32)\left(\frac{L}{32}\right)$$
$$\frac{200}{39.4384} = L$$
$$5.07 \approx L$$

The length of the pendulum is 5.07 ft.

SOLUTIONS to Chapter 12 "You Try It"

SECTION 12.1 *pages 451–454*

You Try It 1
$$\frac{3y^2}{2} + y - \frac{1}{2} = 0$$
$$2\left(\frac{3y^2}{2} + y - \frac{1}{2}\right) = 2(0)$$
$$3y^2 + 2y - 1 = 0$$
$$(3y - 1)(y + 1) = 0$$
$$3y - 1 = 0 \qquad y + 1 = 0$$
$$3y = 1 \qquad\quad y = -1$$
$$y = \frac{1}{3}$$

The solutions are $\frac{1}{3}$ and -1.

You Try It 2
$$x^2 + 81 = 0$$
$$x^2 = -81$$
$$\sqrt{x^2} = \sqrt{-81}$$

$\sqrt{-81}$ is not a real number.

The equation has no real number solution.

You Try It 3

$$7(z + 2)^2 = 21$$
$$(z + 2)^2 = 3$$
$$\sqrt{(z + 2)^2} = \sqrt{3}$$
$$z + 2 = \pm\sqrt{3}$$
$$z = -2 \pm \sqrt{3}$$

The solutions are $-2 + \sqrt{3}$ and $-2 - \sqrt{3}$.

SECTION 12.2 *pages 457–460*

You Try It 1

$$3x^2 - 6x - 2 = 0$$
$$3x^2 - 6x = 2$$
$$\frac{1}{3}(3x^2 - 6x) = \frac{1}{3} \cdot 2$$
$$x^2 - 2x = \frac{2}{3}$$

Complete the square.

$$x^2 - 2x + 1 = \frac{2}{3} + 1$$
$$(x - 1)^2 = \frac{5}{3}$$
$$\sqrt{(x - 1)^2} = \sqrt{\frac{5}{3}}$$
$$x - 1 = \pm\sqrt{\frac{5}{3}} = \pm\frac{\sqrt{15}}{3}$$
$$x - 1 = \frac{\sqrt{15}}{3} \qquad x - 1 = -\frac{\sqrt{15}}{3}$$
$$x = 1 + \frac{\sqrt{15}}{3} \qquad x = 1 - \frac{\sqrt{15}}{3}$$
$$= \frac{3 + \sqrt{15}}{3} \qquad = \frac{3 - \sqrt{15}}{3}$$

The solutions are $\dfrac{3 + \sqrt{15}}{3}$ and $\dfrac{3 - \sqrt{15}}{3}$.

You Try It 2

$$x^2 + 6x + 12 = 0$$
$$x^2 + 6x = -12$$
$$x^2 + 6x + 9 = -12 + 9$$
$$(x + 3)^2 = -3$$
$$\sqrt{(x + 3)^2} = \sqrt{-3}$$

$\sqrt{-3}$ is not a real number.

The quadratic equation has no real number solution.

You Try It 3 $\dfrac{x^2}{8} + x + 1 = 0$

$$8\left(\dfrac{x^2}{8} + x + 1\right) = 8(0)$$
$$x^2 + 8x + 8 = 0$$
$$x^2 + 8x = -8$$
$$x^2 + 8x + 16 = -8 + 16$$
$$(x + 4)^2 = 8$$
$$\sqrt{(x + 4)^2} = \sqrt{8}$$
$$x + 4 = \pm\sqrt{8} = \pm2\sqrt{2}$$

$$x + 4 = 2\sqrt{2} \qquad\qquad x + 4 = -2\sqrt{2}$$
$$x = -4 + 2\sqrt{2} \qquad\quad x = -4 - 2\sqrt{2}$$
$$\approx -4 + 2(1.414) \qquad \approx -4 - 2(1.414)$$
$$\approx -4 + 2.828 \qquad\quad \approx -4 - 2.828$$
$$\approx -1.172 \qquad\qquad\quad \approx -6.828$$

The solutions are approximately -1.172
and -6.828.

SECTION 12.3   *pages 463–464*

You Try It 1 $3x^2 + 4x - 4 = 0$
$a = 3, b = 4, c = -4$

$$x = \dfrac{-(4) \pm \sqrt{(4)^2 - 4(3)(-4)}}{2 \cdot 3}$$
$$= \dfrac{-4 \pm \sqrt{16 + 48}}{6}$$
$$= \dfrac{-4 \pm \sqrt{64}}{6} = \dfrac{-4 \pm 8}{6}$$

$$x = \dfrac{-4 + 8}{6} \qquad x = \dfrac{-4 - 8}{6}$$
$$= \dfrac{4}{6} = \dfrac{2}{3} \qquad\quad = \dfrac{-12}{6} = -2$$

The solutions are $\dfrac{2}{3}$ and -2.

You Try It 2 $\dfrac{x^2}{4} + \dfrac{x}{2} = \dfrac{1}{4}$

$$4\left(\dfrac{x^2}{4} + \dfrac{x}{2}\right) = 4\left(\dfrac{1}{4}\right)$$
$$x^2 + 2x = 1$$
$$x^2 + 2x - 1 = 0$$
$$a = 1, b = 2, c = -1$$
$$x = \dfrac{-(2) \pm \sqrt{(2)^2 - 4(1)(-1)}}{2 \cdot 1}$$
$$= \dfrac{-2 \pm \sqrt{4 + 4}}{2} = \dfrac{-2 \pm \sqrt{8}}{2}$$
$$= \dfrac{-2 \pm 2\sqrt{2}}{2} = -1 \pm \sqrt{2}$$

The solutions are $-1 + \sqrt{2}$ and
$-1 - \sqrt{2}$.

SECTION 12.4   *pages 467–468*

You Try It 1
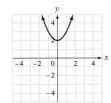

You Try It 2
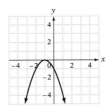

Content and Format © 1995 HMCo.

You Try It 1

Strategy
- This is a geometry problem.
- Width of the rectangle: w
  Length of the rectangle: $w + 2$
- Use the equation $A = l \cdot w$.

Solution

$A = l \cdot w$

$15 = (w + 2)w$

$15 = w^2 + 2w$

$0 = w^2 + 2w - 15$

$0 = (w + 5)(w - 3)$

$w + 5 = 0 \qquad\qquad w - 3 = 0$

$\qquad w = -5 \qquad\qquad\quad w = 3$

The solution -5 is not possible.
The width is 3 m.

ANSWERS to Chapter 1 Odd-Numbered Exercises

SECTION 1.1   *pages 5–6*

1. $8 > -6$ **3.** $-12 < 1$ **5.** $42 > 19$ **7.** $0 > -31$ **9.** $53 > -46$ **11.** false
13. true   **15.** false   **17.** true   **19.** false   **21.** 5   **23.** $-18, -23$ **25.** 21, 37
27. 46, 52   **29.** -23 **31.** -4 **33.** 9   **35.** 28   **37.** 14   **39.** -77 **41.** 0
43. 74   **45.** -82 **47.** -81 **49.** $>$ **51.** $<$ **53.** $>$ **55.** $<$ **57.** $-p \in \{19, 0, -28\}$
59. $-|x| \in \{-45, 0, -17\}$ **61.** $\{-15, -10\}$ **63.** positive   **65.** sometimes true
67. Golf, rummy, hearts.   **69.** Answers will vary.

SECTION 1.2   *pages 11–14*

1. -11 **3.** -5 **5.** -83 **7.** -46 **9.** 0   **11.** -5 **13.** 9   **15.** 1   **17.** -10
19. -18 **21.** -41 **23.** -12 **25.** 0   **27.** -34 **29.** 0   **31.** -61 **33.** 27   **35.** 8
37. -7 **39.** -9 **41.** 9   **43.** -3 **45.** 18   **47.** -9 **49.** 11   **51.** -18 **53.** 0
55. 2   **57.** -138 **59.** -8 **61.** -12 **63.** -20
65. The difference in elevation is 20,602 ft.   **67.** The difference in elevation is 30,314 ft.
69. The difference in temperature is 35°F.   **71.** No, the largest decrease in temperature is 19°F.
73. The temperature rose 49°F.   **75.** The total loss was $7973 million.
77. The difference in profit was $-\$3006$ million.   **79.** -3 **81.** sometimes true
83. No $10 - (-8) = 18$

SECTION 1.3   *pages 19–20*

1. 42   **3.** -28 **5.** 60   **7.** -253 **9.** -238 **11.** -114 **13.** -105 **15.** 252
17. -240 **19.** 96   **21.** -216 **23.** -315 **25.** 420   **27.** 2880   **29.** -2772 **31.** 0
33. -2 **35.** 8   **37.** -7 **39.** -12 **41.** -6 **43.** -7 **45.** 11   **47.** -14 **49.** 15
51. -16 **53.** 0   **55.** -29 **57.** undefined   **59.** -11 **61.** undefined
63. The average daily high temperature was -26°F.
65. The five day moving average was $\{-35, -30, -30, -25, -10, -5\}$.
67. The score for the exam was 74.   **69.** -5 **71.** Answers will vary.   **73.** negative

SECTION 1.4   *pages 29–34*

1. $\frac{1}{3}$ **3.** $\frac{4}{11}$ **5.** $\frac{2}{3}$ **7.** $\frac{3}{2}$ **9.** 0   **11.** $\frac{3}{5}$ **13.** $\frac{11}{15}$ **15.** $\frac{1}{2}$ **17.** $\frac{3}{16}$ **19.** $\frac{3}{5}$
21. 0.8   **23.** $0.8\overline{3}$ **25.** 0.875   **27.** $0.\overline{8}$ **29.** $0.58\overline{3}$ **31.** 0.5625   **33.** $0.3\overline{8}$
35. 0.05   **37.** 0.56   **39.** 0.225   **41.** $\frac{3}{4}$, 0.75   **43.** $\frac{16}{25}$, 0.64   **45.** $\frac{5}{4}$, 1.25   **47.** $\frac{19}{100}$, 0.19
49. $\frac{1}{20}$, 0.05   **51.** $\frac{19}{5}$, 3.8   **53.** $\frac{1}{50}$, 0.02   **55.** $\frac{41}{50}$, 0.82   **57.** $\frac{3}{70}$ **59.** $\frac{3}{8}$ **61.** $\frac{1}{400}$
63. $\frac{1}{16}$ **65.** $\frac{23}{400}$ **67.** 0.091   **69.** 0.167   **71.** 0.009   **73.** 0.0915   **75.** 0.1823
77. 0.0015   **79.** 0.0002   **81.** 15%   **83.** 5%   **85.** 17.5%   **87.** 115%   **89.** 0.8%
91. 54%   **93.** $33\frac{1}{3}$%   **95.** $45\frac{5}{11}$%   **97.** 87.5%   **99.** $166\frac{2}{3}$%   **101.** $\frac{13}{12}$ **103.** $-\frac{5}{24}$
105. $-\frac{19}{24}$ **107.** $\frac{5}{26}$ **109.** $\frac{11}{8}$ **111.** $\frac{1}{12}$ **113.** $\frac{7}{24}$ **115.** 0   **117.** $-\frac{7}{16}$ **119.** $\frac{11}{24}$
121. 1   **123.** $\frac{11}{8}$ **125.** 8.022   **127.** -38.8 **129.** -6.192 **131.** 13.355   **133.** 4.676
135. -10.03 **137.** -60.03 **139.** 11.56   **141.** -0.88 **143.** -4.73 **145.** $-\frac{3}{8}$
147. $\frac{1}{10}$ **149.** $-\frac{4}{9}$ **151.** $-\frac{7}{30}$ **153.** $\frac{15}{64}$ **155.** $-\frac{10}{9}$ **157.** $-\frac{147}{32}$ **159.** $\frac{25}{8}$ **161.** $\frac{2}{3}$
163. 4.164   **165.** 4.347   **167.** -4.028 **169.** -2.22 **171.** -1.104 **173.** 0.506
175. -0.2376 **177.** -274.44 **179.** -2.59 **181.** -5.11 **183.** -2060.55

185. a. San Antonio came closest to winning $\frac{2}{3}$ of its games.

    b. Minnesota and Dallas lost more than $\frac{3}{5}$ of their games.

    c. Dallas lost 84.1% of its games.

187. a. The difference in profit was $-\$30.8$ million.
    b. The difference in profit was $\$30.8$ million.

189. The total tax for gasoline was $335,506,050.   **191.** greater   **193.** yes, $c = \dfrac{a+b}{2}$

195. 48.8 months to 97.5 months

SECTION 1.5   *pages 39–40*

1. 36   **3.** -49 **5.** 9   **7.** 81   **9.** $\frac{1}{4}$ **11.** 0.09   **13.** 12   **15.** 0.216   **17.** -12
19. 16   **21.** -864 **23.** -1008 **25.** 3   **27.** $-77{,}760$ **29.** 9   **31.** 12   **33.** 1
35. 8   **37.** -16 **39.** 12   **41.** 13   **43.** -36 **45.** 13   **47.** 4   **49.** 15   **51.** -1
53. 4   **55.** 0.51   **57.** 1.7   **59.** $-\frac{1}{16}$ **61.** $\frac{17}{24}, \frac{33}{48}$

63. The step is incorrect. The order of operations agreement indicates that exponential expressions are to be evaluated before multiplication.

CHAPTER REVIEW   *pages 43–44*

1. $-4 < 2$ **2.** -4 **3.** 4   **4.** -5 **5.** $<$ **6.** $-|y| \in \{-7, 0, -9\}$ **7.** -6 **8.** -13
9. -4 **10.** 1   **11.** The score on the test was 98.   **12.** -42 **13.** 81   **14.** 8   **15.** -20
16. The difference in temperature is 396°C.   **17.** 0.28   **18.** $0.1\overline{3}$ **19.** $\frac{7}{12}$ **20.** -1.068

21. $-\frac{8}{15}$ **22.** -4.6224 **23.** $\frac{159}{200}$ **24.** 0.062   **25.** 62.5%   **26.** $54\frac{2}{7}\%$
27. a. The increase in earnings was 21.1.   **b.** The decrease in earnings was 28.6%.   **28.** -25
29. $\frac{16}{81}$ **30.** 10   **31.** 60   **32.** 1   **33.** 31

CHAPTER TEST   *pages 45–46*

1. $-2 > -40$ [1.1A]   **2.** 4 [1.1A]   **3.** 4 [1.1B]   **4.** -4 [1.1B]   **5.** $\frac{9}{20}$; 0.45 [1.4B]
6. -16 [1.2A]   **7.** -14 [1.2B]   **8.** 4 [1.2B]   **9.** -48 [1.3A]   **10.** 90 [1.3A]
11. $\frac{3}{8}$ [1.4B]   **12.** 17 [1.3B]   **13.** 17 [1.5B]
14. The average low temperature was -2°F. [1.3C]   **15.** 102.5% [1.4B]   **16.** $0.\overline{7}$ [1.4A]
17. $\frac{1}{15}$ [1.4C]   **18.** $83\frac{1}{3}\%$ [1.4B]   **19.** $-\frac{1}{2}$ [1.4D]   **20.** -5.3578 [1.4D]
21. -108 [1.5A]   **22.** 12 [1.5A]   **23.** 9 [1.5B]   **24.** 8 [1.5B]
25. The value in U.S. Dollars is $7958.81. [1.4E]

ANSWERS to Chapter 2 Odd-Numbered Exercises

SECTION 2.1   *pages 51–52*

1. $2x^2$, $5x$, -8 **3.** $-a^4$, 6 **5.** $7x^2y$ and $6xy^2$ **7.** 1, -9 **9.** 1, -4, -1 **11.** 10
13. 32   **15.** 21   **17.** 16   **19.** -9 **21.** 41   **23.** -7 **25.** 13   **27.** -15 **29.** 41
31. 1   **33.** 5   **35.** 1   **37.** 57   **39.** 5   **41.** 8   **43.** -3 **45.** -2 **47.** -4
49. 10   **51.** -25 **53.** Answers will vary.   **55.** -1 **57.** 4   **59.** 81
61. $n^x \geq x^4$ if $x \geq n + 1$

SECTION 2.2   *pages 59–62*

1. $14x$ **3.** $5a$ **5.** $-6y$ **7.** $-3b - 7$ **9.** $5a$ **11.** $-2ab$ **13.** $5xy$ **15.** 0

17. $-\frac{5}{6}x$ **19.** $-\frac{1}{24}x^2$ **21.** $11x$ **23.** $7a$ **25.** $-14x^2$ **27.** $-x + 3y$ **29.** $17x - 3y$

31. $-2a - 6b$ **33.** $-3x - 8y$ **35.** $-4x^2 - 2x$ **37.** $12x$ **39.** $-21a$ **41.** $6y$ **43.** $8x$

45. $-6a$ **47.** $12b$ **49.** $-15x^2$ **51.** x^2 **53.** a **55.** x **57.** n **59.** x **61.** y

63. $3x$ **65.** $-2x$ **67.** $-8a^2$ **69.** $8y$ **71.** $4y$ **73.** $-2x$ **75.** $6a$ **77.** $-x - 2$

79. $8x - 6$ **81.** $-2a - 14$ **83.** $-6y + 24$ **85.** $35 - 21b$ **87.** $2 - 5y$ **89.** $15x^2 + 6x$

91. $2y - 18$ **93.** $-15x - 30$ **95.** $-6x^2 - 28$ **97.** $-6y^2 + 21$ **99.** $3x^2 - 3y^2$

101. $-4x + 12y$ **103.** $-6a^2 + 7b^2$ **105.** $4x^2 - 12x + 20$ **107.** $x - 3y + 4$

109. $-12a^2 - 20a + 28$ **111.** $12x^2 - 9x + 12$ **113.** $10x^2 - 20xy - 5y^2$

115. $-8b^2 + 6b - 9$ **117.** $a - 7$ **119.** $-11x + 13$ **121.** $-4y - 4$ **123.** $-2x - 16$

125. $14y - 45$ **127.** $a + 7b$ **129.** $6x + 28$ **131.** $5x - 75$ **133.** $4x - 4$ **135.** $2x - 9$

137. **a.** false $8 \div 4 \neq 4 \div 8$ **b.** false $(8 \div 4) \div 2 \neq 8 \div (4 \div 2)$
   c. false $(7 - 5) - 1 \neq 7 - (5 - 1)$ **d.** false $6 - 3 \neq 3 - 6$ **e.** true

139. No, 0   **141.** Answers will vary.   **143.** Answers will vary.   **145.** **a.** yes   **b.** no

SECTION 2.3   *pages 67–70*

1. $8 + y$ **3.** $t + 10$ **5.** $z + 14$ **7.** $x^2 - 20$ **9.** $\frac{3}{4}n + 12$ **11.** $8 + \frac{n}{4}$ **13.** $3(y + 7)$

15. $t(t + 16)$ **17.** $\frac{1}{2}x^2 + 15$ **19.** $5n^3 + n^2$ **21.** $r - \frac{r}{3}$ **23.** $x^2 - (x + 17)$ **25.** $9(z + 4)$

27. $12 - x$ **29.** $\frac{2}{3}x$ **31.** $\frac{2x}{9}$ **33.** $11x - 8$ **35.** $(x + 2) - 9; x - 7$ **37.** $\frac{7}{5 + x}$

39. $5 + \frac{1}{2}(x + 3); \frac{1}{2}x + \frac{13}{2}$ **41.** $x + (2x - 4); 3x - 4$ **43.** $(x - 5)7; 7x - 35$ **45.** $\frac{2x + 5}{x}$

47. $x - (3x - 8); -2x + 8$ **49.** $x + 3x; 4x$ **51.** $(x + 6) + 5; x + 11$ **53.** $x - (x + 10); -10$

55. $\frac{1}{6}x + \frac{4}{9}x; \frac{11}{18}x$ **57.** $\frac{x}{3} + x; \frac{4}{3}x$ **59.** rate of jet plane: j rate of propeller plane: $\frac{1}{2}j$

61. diameter of basketball: x diameter of baseball: $\frac{1}{4}x$

63. lowest tax bracket: t highest tax bracket: $2t + 3$

65. larger container: L smaller container: $20 - L$ **67.** older model: x new model: $\frac{1}{2}x + 7$

69. length of wire: x length of side of square $\frac{1}{4}x$ **71.** $\frac{3}{5}x$ **73.** Answers will vary.

CHAPTER REVIEW   *pages 73–74*

1. -7 **2.** 29 **3.** -4 **4.** 79 **5.** $11x$ **6.** $-5y$ **7.** $8a - 4b$ **8.** $-4x^2 + 6x$

9. $9c - 5d$ **10.** $-9r + 8s$ **11.** $20x$ **12.** $36y$ **13.** $-6a$ **14.** $-42x^2$ **15.** $-5n$

16. $10x - 35$ **17.** $12y^2 + 8y - 10$ **18.** $-63 - 36x$ **19.** $3x^2 - 24x - 21$

20. $28a^2 - 8a + 12$ **21.** $-4x + 20$ **22.** $7x + 46$ **23.** $24y + 30$ **24.** $-90x + 25$

25. $4x$ **26.** $x - 6$ **27.** $\frac{2}{3}(x + 10)$ **28.** $x + 2x; 3x$ **29.** $3x + 5(x - 1); 8x - 5$

30. $2x - \frac{1}{2}x; \frac{3}{2}x$ **31.** number of calories in an apple: a number of calories in a candy

bar: $2a + 8$ **32.** number of ten-dollar bills: T number of five-dollar bills: $35 - T$

33. diameter of #30 wire: d diameter of #8 wire: $13d - 2$

34. number of American league cards: A number of National league cards: $5A$

CHAPTER TEST *pages 75–76*

1. 22 [2.1A]   **2.** 3 [2.1A]   **3.** $5x$ [2.2A]   **4.** y^2 [2.2A]   **5.** $-9x - 7y$ [2.2A]

6. $-2x - 5y$ [2.2C]   **7.** $2x$ [2.2B]   **8.** $3x$ [2.2B]   **9.** $36y$ [2.2B]   **10.** $-10a$ [2.2B]

11. $15 - 35b$ [2.2C]   **12.** $-4x + 8$ [2.2C]   **13.** $-6x^2 + 21y^2$ [2.2C]

Content and Format © 1995 HMCo.

14. $-10x^2 + 15x - 30$ [2.2C]   **15.** $-x + 6$ [2.2D]   **16.** $7x + 38$ [2.2D]

17. $-7x + 33$ [2.2D]   **18.** $2x + y$ [2.2D]   **19.** $a^2 - b^2$ [2.3A]   **20.** $x + 2x^2$ [2.3A]

21. $\frac{6}{x} - 3$ [2.3A]   **22.** $6 - 7b$; $-6b$ [2.3B]   **23.** $10(x - 3)$; $10x - 30$ [2.3B]

24. speed of return: s speed of fastball: $2s$ [2.3C]

25. length of shorter piece: x length of longer piece: $4x - 3$ [2.3C]

CUMULATIVE REVIEW *pages 77–78*

1. -7 [1.2A]   **2.** 5 [1.2B]   **3.** 24 [1.3A]   **4.** -5 [1.3B]   **5.** 1.25 [1.4A]

6. $\frac{11}{48}$ [1.4C]   **7.** $\frac{1}{6}$ [1.4D]   **8.** $\frac{1}{4}$ [1.4D]   **9.** 75% [1.4B]   **10.** -5 [1.5B]

11. $\frac{53}{48}$ [1.5B]   **12.** 16 [2.1A]   **13.** $5x^2$ [2.2A]   **14.** $-7a - 10b$ [2.2A]   **15.** $6a$ [2.2B]

16. $30b$ [2.2B]   **17.** $24 - 6x$ [2.2C]   **18.** $6y - 18$ [2.2C]   **19.** $\frac{3}{8}$ [1.4B]

20. 0.0105 [1.4B]   **21.** $-8x^2 + 12y^2$ [2.2C]   **22.** $-9y^2 + 9y + 21$ [2.2C]

23. $-7x + 14$ [2.2D]   **24.** $5x - 43$ [2.2D]   **25.** $17x - 24$ [2.2D]   **26.** $-3x + 21y$ [2.2D]

27. $\frac{1}{2}b + b$ [2.3A]   **28.** $\frac{10}{y - 2}$ [2.3A]   **29.** $8 - \frac{x}{12}$ [2.3A]   **30.** $x + (x + 2)$; $2x + 2$ [2.3B]

31. $(3 + x)5 + 12$; $27 + 5x$ [2.3B]   **32.** speed of normal drive: s speed of triple-speed Rom: $3s$

ANSWERS to Chapter 3 Odd-Numbered Exercises

SECTION 3.1   *pages 87–92*

1. yes   **3.** no   **5.** no   **7.** yes   **9.** yes   **11.** yes   **13.** no   **15.** yes   **17.** yes

19. yes   **21.** no   **23.** $y = 6$ **25.** $z = 16$ **27.** $x = 7$ **29.** $t = -2$ **31.** $x = 1$

33. $y = 0$ **35.** $x = 3$ **37.** $n = -10$ **39.** $x = -3$ **41.** $y = -14$ **43.** $t = 2$

45. $a = 11$ **47.** $n = -9$ **49.** $y = -1$ **51.** $x = -14$ **53.** $x = -5$ **55.** $m = -1$

57. $x = 1$ **59.** $y = -\frac{1}{2}$ **61.** $m = -\frac{3}{4}$ **63.** $x = \frac{1}{12}$ **65.** $x = -\frac{7}{12}$ **67.** $d = 0.6529$

69. $x = -0.283$ **71.** $z = 9.257$ **73.** $x = -3$ **75.** $b = 0$ **77.** $x = -2$ **79.** $x = 9$

81. $c = 80$ **83.** $w = -4$ **85.** $x = 0$ **87.** $y = 8$ **89.** $t = -7$ **91.** $x = 12$

93. $b = -18$ **95.** $x = 15$ **97.** $m = -20$ **99.** $x = 0$ **101.** $x = 15$ **103.** $x = 75$

105. $x = \frac{8}{3}$ **107.** $y = \frac{1}{3}$ **109.** $x = -\frac{1}{2}$ **111.** $n = -\frac{3}{2}$ **113.** $m = \frac{15}{7}$ **115.** $n = 4$

117. $y = 3$ **119.** $x = 4.745$ **121.** $a = 2.06$ **123.** $x = -2.13$ **125.** 28 **127.** 0.72

129. 64 **131.** 24% **133.** 7.2 **135.** 400 **137.** 9 **139.** 25% **141.** 5 **143.** 200%

145. 400 **147.** 7.7 **149.** 200 **151.** 400 **153.** 20 **155.** 80.34%

157. 19% of the students are in the fine arts college.

159. The number of playoff games was increased 40%.   **161.** The discount is $127.50.

163. At least 67 votes are required for an override.   **165.** The airline would sell 177 tickets.

167. $x = \frac{b}{a}$; no, $a \neq 0$ **169.** Answers will vary.   **171.** **a.** $\frac{6}{5}$ **b.** 2 **175.** 0

SECTION 3.2   *pages 95–98*

1. $x = 3$ **3.** $a = 6$ **5.** $x = -1$ **7.** $x = -3$ **9.** $w = 2$ **11.** $t = 2$ **13.** $a = 5$

15. $b = -3$ **17.** $x = 6$ **19.** $x = 3$ **21.** $a = 1$ **23.** $b = 6$ **25.** $m = -7$ **27.** $y = 0$

29. $x = \frac{3}{4}$ **31.** $x = \frac{4}{9}$ **33.** $x = \frac{1}{3}$ **35.** $w = -\frac{1}{2}$ **37.** $b = -\frac{3}{4}$ **39.** $a = \frac{1}{3}$

41. $x = -\frac{1}{6}$ **43.** $a = 1$ **45.** $x = 1$ **47.** $x = 0$ **49.** $x = \frac{13}{10}$ **51.** $a = \frac{2}{5}$

53. $x = -\dfrac{4}{3}$ **55.** $x = -\dfrac{3}{2}$ **57.** $m = 18$ **59.** $n = 8$ **61.** $b = -16$ **63.** $y = 25$

65. $c = 21$ **67.** $w = 15$ **69.** $x = -16$ **71.** $x = -21$ **73.** $x = \dfrac{15}{2}$ **75.** $y = -\dfrac{18}{5}$

77. $y = 2$ **79.** $z = 3$ **81.** $b = 1$ **83.** $m = -2$ **85.** 19 **87.** -1 **89.** -11
91. The initial velocity is 8 ft/s. **93.** The depreciated value will be \$38,000 after 2 years.
95. The approximate length is 31.8 in. to the nearest tenth.
97. The approximate population is 51,000 to the nearest thousand.
99. The value of y is 66°. **101.** $x = 385$
103. It is not an equation and cannot be solved. **107.** Distance and speed

SECTION 3.3 *pages 103–106*

1. $x = 2$ **3.** $x = 3$ **5.** $x = -1$ **7.** $x = 2$ **9.** $x = -2$ **11.** $b = -3$ **13.** $y = 0$

15. $x = -1$ **17.** $x = -3$ **19.** $x = -1$ **21.** $m = 4$ **23.** $b = \dfrac{2}{3}$ **25.** $x = \dfrac{5}{6}$

27. $m = \dfrac{3}{4}$ **29.** -17 **31.** 41 **33.** 8 **35.** $y = 1$ **37.** $x = 4$ **39.** $m = -1$

41. $x = -1$ **43.** $b = -\dfrac{2}{3}$ **45.** $a = \dfrac{4}{3}$ **47.** $x = \dfrac{1}{2}$ **49.** $y = -\dfrac{1}{3}$ **51.** $a = \dfrac{10}{3}$

53. $b = -\dfrac{1}{4}$ **55.** 0 **57.** -1 **59.** The force applied must be 25 lb.

61. The child must sit 10 ft from the fulcrum. **63.** The see-saw is not balanced.
65. There must be 350 televisions sold. **67.** There must be 1200 compact discs sold.
69. The measure of the complementary angles are 28° and 62°.
71. The supplementary angles are 48° and 132°. **73.** Answers will vary. **75.** yes
77. No solution **79.** -21

SECTION 3.4 *pages 111–114*

1. $x - 15 = 7; x = 22$ **3.** $7x = -21; x = -3$ **5.** $3x - 4 = 5; x = 3$
7. $4(2x + 3) = 12; x = 0$ **9.** $12 = 6(x - 3); x = 5$ **11.** $4x = 3(35 - x); x = 15$
13. $3x - 1 = 15 - x;$ 4 and 11 **15.** $2x - (14 - x) = 1;$ 5 and 9
17. $8 - 2x = 4(2 - x) - 2;$ -1 and 3 **19.** The consecutive odd integers are 15, 17, 19.
21. The consecutive odd integers are 11, 13, 15. **23.** The consecutive even integers are 4 and 6.
25. The consecutive even integers are 2, 4, 6. **27.** The consecutive odd integers are 3 and 5.
29. The consecutive even integers are 20, 22, 24. **31.** The consecutive integers are $-8, -7, -6$.
33. The consecutive integers are 15, 16, 17. **35.** The rating of the second computer is 15 mips.
37. The length of the patio is 15 ft. **39.** The basketball team scored 15 three-point baskets.
41. The measures of the angles in the triangle are 30°, 60°, and 90°.
43. The family used 515 kWh of electricity. **45.** $a = 120°$ $b = 120°$ $c = 60°$
47. $a = 140°$ $b = 40°$ $c = 40°$ **49.** Answers will vary. **51.** Identity **53.** Identity
55. Contradiction

CHAPTER REVIEW *pages 117–118*

1. no **2.** no **3.** $x = 21$ **4.** $a = \dfrac{5}{6}$ **5.** $x = 2.5$ **6.** $a = 20$ **7.** 250% **8.** 1600

9. $x = 7$ **10.** $x = -3$ **11.** $x = 6$ **12.** $x = -\dfrac{6}{7}$

13. In 15 years, the interest will be \$1800. **14.** \$28 **15.** The temperature is 37°C.

16. \$79.25 **17.** $x = 4$ **18.** $y = \dfrac{1}{3}$ **19.** $x = 4$ **20.** $x = -5$ **21.** $x = -2$ **22.** $x = 3$

23. $x = 10$ **24.** $x = -1$ **25.** A 24 lb force must be applied to the other end of the lever.
26. The fulcrum is 3 ft from the 25 lb force. **27.** $5n - 4 = 16; n = 4$
28. $4(n + 1) = n + (n + 2); n = -1$ **29.** $3x = 2(21 - x) - 2;$ 8, 13
30. The two lengths of wire are 14 in. and 21 in.
31. The consulting fee consisted of 8 h of consulting. **32.** The Eiffel Tower is 993 ft. tall.

CHAPTER TEST   *pages 119–120*

1. no [3.1A]   **2.** 0.04 [3.1D]   **3.** $x = -5$ [3.1B]   **4.** -12 [3.1C]   **5.** $x = -3$ [3.2A]

6. $x = 5$ [3.2A]   **7.** 200 calculators [3.2B]   **8.** $x = -5$ [3.3A]   **9.** $x = -\frac{1}{2}$ [3.3A]

10. $x = 2$ [3.3A]   **11.** $x = 2$ [3.3B]   **12.** $x = -\frac{1}{3}$ [3.3B]   **13.** $x = \frac{12}{11}$ [3.3B]

14. The final temperature is 60°C. [3.3C]   **15.** $3x - 15 = 27$; $x = 14$ [3.4A]
16. The three consecutive even integers are 10, 12, 14. [3.4A]
17. The two numbers are 8 and 10. [3.4A]   **18.** The time is 7 h. [3.4B]
19. The pieces measured 6 ft and 12 ft. [3.4B]
20. The new disk controller requires 15 milliseconds to access the data. [3.4B]

CUMULATIVE REVIEW   *pages 121–122*

1. 6 [1.2B]   **2.** -48 [1.3A]   **3.** $-\frac{19}{48}$ [1.4C]   **4.** -2 [1.4D]   **5.** 54 [1.5A]

6. 24 [1.5B]   **7.** 6 [2.1A]   **8.** $-17x$ [2.2A]   **9.** $-5a - 2b$ [2.2A]   **10.** $2x$ [2.2B]
11. $36y$ [2.2B]   **12.** $2x^2 + 6x - 4$ [2.2C]   **13.** $-4x + 14$ [2.2D]   **14.** $6x - 34$ [2.2D]
15. yes [3.1A]   **16.** no [3.1A]   **17.** 19.2 [3.1D]   **18.** $x = -25$ [3.1C]
19. $x = -3$ [3.2A]   **20.** $x = 3$ [3.2A]   **21.** $x = 13$ [3.3B]   **22.** $x = 2$ [3.3B]

23. $x = -3$ [3.3A]   **24.** $x = \frac{1}{2}$ [3.3A]   **25.** 250 cameras were produced. [3.2B]

26. The final temperature is 60°C. [3.3C]   **27.** $12 - 5x = -18$; $x = 6$ [3.4A]
28. $6x + 13 = 3x - 5$; $x = -6$ [3.4A]   **29.** The area of the garage is 600 ft². [3.4B]
30. The pieces measure 5 ft and 11 ft. [3.4B]

ANSWERS to Chapter 4 Odd-Numbered Exercises

SECTION 4.1   *pages 127–128*

1. The selling price is $35.   **3.** The markup rate is 87.5%.   **5.** The selling price is $196.
7. The markup rate is 40%.   **9.** The markup rate is 23.3%.   **11.** The sale price is $41.25.
13. The discount rate is 25%.   **15.** The discount rate is 40%.
17. The sale price per shirt is $14.45.   **19.** The discount rate to the nearest percent is 26%.
21. The markup is $18.   **23.** The regular price is $317.65.   **25.** The largest possible price is $96.

SECTION 4.2   *pages 131–132*

1. There must be an additional $5000 added.
3. There was $9000 invested at 7% and $6000 at 6.5%.
5. There was $2500 deposited in the mutual fund.
7. The university deposited $200,000 at 10% and $100,000 at 8.5%.
9. The mechanic invests $3000 in additional bonds.   **11.** The total amount invested was $32,000.
13. The total amount invested was $75,000.
15. $12,000 was invested in bonds, $21,000 was invested in a simple interest account, and $27,000 was invested in corporate bonds.
17. The value of the investment in two years will be $2916.

SECTION 4.3   *pages 137–140*

1. The mixture contains 2 lb of diet supplement and 3 lb of vitamin supplement.
3. The selling price is $6.98 per pound.
5. The combination contained 56 oz at $4.30 and 144 oz at $1.80.
7. The selling price is $2.90 per lb.   **9.** There must be 10 kg of hard candy.
11. The mixture contains 30 lb at $2.20 and 20 lb at $4.20.
13. There must be 8 kg of soil supplement.
15. The mixture contains 63 lb of walnuts and 37 lb of almonds.

17. The cereal costs $0.70 per pound.   **19.** There must be 9.6 lb of lima beans.
21. The solution must contain 20 ml of 13% acid and 30 ml of 18% acid.
23. There is a 50% concentration of silver.   **25.** There was 30 lb of 60% mixture used.
27. There is 0.74% of hydrocortizone.   **29.** The hair dye contains 100 ml of 7% and 200 ml of 4%.
31. There must be 25 oz of pure water added.   **33.** The percent concentration is 27%.
35. There must be 10 oz of pure bran flakes added.   **37.** The cost is $3.65 per ounce.
39. 10 oz of water must be evaporated to produce a 15% salt solution.
41. 75 g of pure water must be added to produce a 40% acid solution.

SECTION 4.4   *pages 143–144*

1. The plane flew 2 h at 105 mph and 3 h at 115 mph.   **3.** The sailboat traveled 36 mi.
5. The rate of the passenger train is 50 mph, and the rate of the freight train is 30 mph.
7. The rate of the cyclist is 16 mph
9. The rate of the first plane is 95 mph, and the rate of the second plane is 120 mph.
11. They will meet after 1 hour.
13. The second runner will overtake the first runner after 3 hours.
15. It took 20 min downstream.   **17.** The campers turned around at 10:15 A.M.
19. The van overtook the truck at 2:15 P.M.   **21.** It is impossible to average 60 mph.

SECTION 4.5   *pages 147–148*

1. The lengths are 50 ft, 50 ft, and 25 ft.   **3.** The lengths are 9 m, 9 m, and 3 m.
5. The length is 40 ft, and the width is 20 ft.   **7.** The lengths are 11 ft, 10 ft, and 12 ft.
9. The length is 130 ft, and the width is 39 ft.   **11.** The angles are 37°, 37°, and 106°.
13. The angles are 40°, 20°, and 120°.   **15.** The angles are 63°, 21°, and 96°.
17. The angles are 31°, 59°, and 90°.   **19.** The angles are 60°, 55°, and 65°.
21. The length is 9 cm and the width is 3 cm.   **23.** The length is 9 cm and the width is 4 cm.
25. The length of the rectangle is 16*x*.

CHAPTER REVIEW   *pages 151–152*

1. The cost of the curio cabinet is $671.25.   **2.** The sale price of the shoes is $41.25.
3. The regular price of the carpet sweeper is $32.   **4.** The discount rate is $33\frac{1}{3}$%.
5. The amount invested at 4% is $9600. The amount invested at 9% is $14,400.
6. The total amount invested is $650,000.
7. The amount invested at 6.75% is $1400. The amount invested at 9.45% is $1000.
8. The total investment was $23,333.33.
9. The mixture contains 7 qt of cranberry juice and 3 qt of apple juice.
10. The selling price of the mixture is $1.84 per pound.   **11.** The mixture is 14% butterfat.
12. One liter of pure water should be added.
13. The jet overtakes the propeller-driven plane 600 mi from the starting point.
14. The average speed on the winding road was 32 mph.
15. The sides measure 8 in., 12 in., and 15 in.   **16.** The length is 80 ft. The width is 20 ft.
17. The measures of the angles are 16°, 82°, and 82°.
18. The measures of the angles are 75°, 60°, and 45°.

CHAPTER TEST   *pages 153–154*

1. The cost was $350. [4.1A]   **2.** The discount rate was 20%. [4.1B]
3. The markup rate was $33\frac{1}{3}$% [4.1A]   **4.** The sale price is $250.25. [4.1B]
5. They should invest $18,000 at 6% and $36,000 at 9%. [4.2A]
6. The amount invested at 6% is $3500. The amount invested at 9% is $1500. [4.2A]
7. The amount invested at 7% is $4500. The amount invested at 9% is $3500. [4.2A]
8. The baker should use 10 lb and $0.70 and 5 lb at $0.40. [4.3A]

9. The selling price is $3.00 per pound. [4.3A]
10. 1.25 gal of water should be mixed with the salt solution. [4.3B]
11. The mixture is 32.5% silver. [4.3B] **12.** The mixture is 52% wild rice. [4.3B]
13. The rate of the snowmobile is 6 mph. [4.4A]
14. The distance between the airports is 360 mi. [4.4A]
15. The length is 6 ft, and the width is 4 ft. [4.5A]
16. The measures of the sides of the triangle are 5 ft, 8 ft, and 10 ft. [4.5A]
17. The length is 40 m, and the width is 16 m. [4.5A]
18. The measure of one of the equal angles is 70°. [4.5B]
19. The measures of the angles of the triangle are 36°, 36°, and 108°. [4.5B]
20. The measures of the angles of the triangle are 36°, 67°, and 77°. [4.5B]

CUMULATIVE REVIEW   *pages 155–156*

1. -36 [1.5B] **2.** $-\dfrac{1}{16}$ [1.5A] **3.** $\dfrac{49}{40}$ [1.5B] **4.** 9 [2.1A] **5.** $8a - 9b$ [2.2A]

6. $-9x^2 - 12x + 21$ [2.2C] **7.** $10x - 4$ [2.2D] **8.** Yes [3.1A] **9.** $x = -2$ [3.1B]

10. $x = -12$ [3.1C] **11.** $x = 2$ [3.2A] **12.** $x = 2$ [3.3B] **13.** $\dfrac{11}{20}$ [1.4B]

14. $\dfrac{2}{3}$ [1.4B] **15.** 4 and 6 [3.4A] **16.** 1.5 h [3.4B] **17.** 103% [1.4B]

18. 45% [1.4B] **19.** 15 [3.1D] **20.** 120 [3.1D]
21. The value of the investment was $4000. [3.1D]
22. An additional $4800 must be deposited. [4.2A] **23.** The markup rate is 40%. [4.1A]
24. The sale price is $37.20. [4.1B] **25.** 20 lb of oat flour must be used. [4.3A]
26. 25 g of pure gold must be used. [4.3B] **27.** The length is 12 ft, and the width is 10 ft. [4.5A]
28. The measure of one angle is 60°. [4.5B] **29.** The integers are $-1, 0, 1$. [3.4A]
30. The length of the track is 120 m. [4.4A]

ANSWERS to Chapter 5 Odd-Numbered Exercises

SECTION 5.1   *pages 161–162*

1. yes **3.** no **5.** yes **7.** yes **9.** Binomial **11.** Trinomial **13.** none
15. Binomial **17.** $-2x^2 + 3x$ **19.** $y^2 - 8$ **21.** $5x^2 + 7x + 20$ **23.** $x^3 + 2x^2 - 6x - 6$
25. $2a^3 - 3a^2 - 11a + 2$ **27.** $5x^2 + 8x$ **29.** $7x^2 + xy - 4y^2$ **31.** $3a^2 - 3a + 17$
33. $5x^3 + 10x^2 - x - 4$ **35.** $3r^3 + 2r^2 - 11r + 7$ **37.** $-2x^3 + 3x^2 + 10x + 11$ **39.** $4x$
41. $3y^2 - 4y - 2$ **43.** $-7x - 7$ **45.** $4x^3 + 3x^2 + 3x + 1$ **47.** $y^3 + 5y^2 - 2y - 4$
49. $-y^2 - 13xy$ **51.** $2x^2 - 3x - 1$ **53.** $-2x^3 + x^2 + 2$ **55.** $3a^3 - 2$
57. $4y^3 + 2y^2 + 2y - 4$ **59.** $x^2 + 9x - 11$ **61.** $-3x^2 - 4x - 3$ **63.** Answers will vary.
65. yes; $(2x^3 + 3x - 4) + (-2x^3 + 5x^2 - 6) = 5x^2 + 3x - 10$

SECTION 5.2   *pages 165–166*

1. $30x^3$ **3.** $-42c^6$ **5.** $9a^7$ **7.** x^3y^4 **9.** $-10x^9y$ **11.** $12x^7y^8$ **13.** $-6x^3y^5$
15. x^4y^5z **17.** $a^3b^5c^4$ **19.** $-30a^5b^8$ **21.** $6a^5b$ **23.** $40y^{10}z^6$ **25.** $x^3y^3z^2$
27. $-24a^3b^3c^3$ **29.** $8x^7yz^6$ **31.** $30x^6y^8$ **33.** $-36a^3b^2c^3$ **35.** x^{15} **37.** x^{14} **39.** x^8
41. y^{12} **43.** $-8x^6$ **45.** x^4y^6 **47.** $9x^4y^2$ **49.** $-243x^{15}y^{10}$ **51.** $-8x^7$ **53.** $24x^8y^7$
55. a^4b^6 **57.** $64x^{12}y^3$ **59.** $-18x^3y^4$ **61.** $-8a^7b^5$ **63.** $-54a^9b^3$ **65.** $12x^2$
67. $2x^6y^2 + 9x^4y^2$ **69.** 0 **71.** $17x^4y^8$ **73.** true **75.** false $(x^2)^5 = x^{2 \cdot 5} = x^{10}$
77. not equal $2^{(3^2)}$ is larger **79.** sometimes

SECTION 5.3   *pages 171–174*

1. $x^2 - 2x$ **3.** $-x^2 - 7x$ **5.** $3a^3 - 6a^2$ **7.** $-5x^4 + 5x^3$ **9.** $-3x^5 + 7x^3$
11. $12x^3 - 6x^2$ **13.** $6x^2 - 12x$ **15.** $3x^2 + 4x$ **17.** $-x^3y + xy^3$ **19.** $2x^4 - 3x^2 + 2x$
21. $2a^3 + 3a^2 + 2a$ **23.** $3x^6 - 3x^4 - 2x^2$ **25.** $-6y^4 - 12y^3 + 14y^2$ **27.** $-2a^3 - 6a^2 + 8a$
29. $6y^4 - 3y^3 + 6y^2$ **31.** $x^3y - 3x^2y^2 + xy^3$ **33.** $x^3 + 4x^2 + 5x + 2$
35. $a^3 - 6a^2 + 13a - 12$ **37.** $-2b^3 + 7b^2 + 19b - 20$ **39.** $-6x^3 + 31x^2 - 41x + 10$
41. $x^3 - 3x^2 + 5x - 15$ **43.** $x^4 - 4x^3 - 3x^2 + 14x - 8$ **45.** $15y^3 - 16y^2 - 70y + 16$
47. $5a^4 - 20a^3 - 5a^2 + 22a - 8$ **49.** $y^4 + 4y^3 + y^2 - 5y + 2$ **51.** $x^2 + 4x + 3$
53. $a^2 + a - 12$ **55.** $y^2 - 5y - 24$ **57.** $y^2 - 10y + 21$ **59.** $2x^2 + 15x + 7$
61. $3x^2 + 11x - 4$ **63.** $4x^2 - 31x + 21$ **65.** $3y^2 - 2y - 16$ **67.** $9x^2 + 54x + 77$
69. $21a^2 - 83a + 80$ **71.** $6a^2 - 25ab + 14b^2$ **73.** $2a^2 - 11ab - 63b^2$
75. $100a^2 - 100ab + 21b^2$ **77.** $15x^2 + 56xy + 48y^2$ **79.** $14x^2 - 97xy - 60y^2$
81. $56x^2 - 61xy + 15y^2$ **83.** $y^2 - 25$ **85.** $4x^2 - 9$ **87.** $x^2 + 2x + 1$ **89.** $9a^2 - 30a + 25$
91. $9x^2 - 49$ **93.** $4a^2 + 4ab + b^2$ **95.** $x^2 - 4xy + 4y^2$ **97.** $16 - 9y^2$
99. $25x^2 + 20xy + 4y^2$ **101.** $(2L^2 - 2L)$ ft^2 **103.** $(90x + 2025)$ ft^2
105. $(3.14x^2 + 18.84x + 28.26)$ cm^2 **107.** $4ab$ **109.** $x^4 + 2x^3 - 5x^2 - 6x + 9$
111. $a^3 + 9a^2 + 27a + 27$ **113.** $12x^2 - x - 20$ **115.** $7x^2 - 11x - 8$ **117.** 5

SECTION 5.4   *pages 181–184*

1. $\dfrac{1}{25}$ **3.** 64 **5.** $\dfrac{1}{27}$ **7.** 2 **9.** $\dfrac{1}{x^2}$ **11.** a^6 **13.** $\dfrac{4}{x^7}$ **15.** $\dfrac{2}{3z^2}$ **17.** $5b^8$ **19.** $\dfrac{x^2}{3}$

21. 1 **23.** -1 **25.** y^4 **27.** a^3 **29.** p^4 **31.** $2x^3$ **33.** $2k$ **35.** m^5n^2 **37.** $\dfrac{3r^2}{2}$

39. $-\dfrac{2a}{3}$ **41.** $\dfrac{1}{y^5}$ **43.** $\dfrac{1}{a^6}$ **45.** $\dfrac{1}{3x^3}$ **47.** $\dfrac{2}{3x^5}$ **49.** $\dfrac{y^4}{x^2}$ **51.** $\dfrac{2}{5m^3n^8}$ **53.** $\dfrac{1}{p^3q}$ **55.** $\dfrac{1}{2y^3}$

57. $\dfrac{7xz}{8y^3}$ **59.** $\dfrac{p^2}{2m^3}$ **61.** $-\dfrac{8x^3}{y^6}$ **63.** $\dfrac{9}{x^2y^4}$ **65.** $\dfrac{2}{x^4}$ **67.** $-\dfrac{5}{a^8}$ **69.** $-\dfrac{a^5}{8b^4}$ **71.** $\dfrac{10y^3}{x^4}$

73. $\dfrac{1}{2x^3}$ **75.** $\dfrac{3}{x^3}$ **77.** $\dfrac{1}{2x^2y^6}$ **79.** $\dfrac{1}{x^6y}$ **81.** $\dfrac{a^4}{y^{10}}$ **83.** $-\dfrac{1}{6x^3}$ **85.** $-\dfrac{a^2b}{6c^2}$ **87.** $-\dfrac{7b^6}{a^2}$

89. $\dfrac{t^4s^8}{4r^{12}}$ **91.** $\dfrac{125p^3}{27m^{15}n^6}$ **93.** 3.24×10^{-9} **95.** 3×10^{-18} **97.** 3.2×10^{16} **99.** 1.22×10^{-19}

101. 5.47×10^8 **103.** 0.000167 **105.** $68,000,000$ **107.** 0.0000305 **109.** 0.00000000102
111. $\$4.3 \times 10^{12}$ **113.** 5.98×10^{24} kg   **115.** 1.6×10^{-19} coulomb

117. $510,000,000,000,000$ m^2 **119.** $\dfrac{1}{4}, \dfrac{1}{2}, 1, 2, 4$ **121.** $4, 2, 1, \dfrac{1}{2}, \dfrac{1}{4}$ **123.** $\dfrac{1}{16}, \dfrac{1}{2}, 1, \dfrac{1}{2}, \dfrac{1}{16}$

125. false $\dfrac{1}{8a^3}$ **127.** true   **129.** false $\dfrac{1}{5}$ **131.** $\dfrac{x^6z^6}{4y^5}$ **133.** always positive

135. If $x = \dfrac{1}{3}$ then $3x - 1 = 0$ and 0^0 is undefined.

SECTION 5.5   *pages 187–188*

1. $2a - 5$ **3.** $6y + 4$ **5.** $x - 2$ **7.** $-x + 2$ **9.** $x^2 + 3x - 5$ **11.** $x^4 - 3x^2 - 1$

13. $xy + 2$ **15.** $-3y^3 + 5$ **17.** $3x - 2 + \dfrac{1}{x}$ **19.** $-3x + 7 - \dfrac{6}{x}$ **21.** $4a - 5 + 6b$

23. $9x + 6 - 3y$ **25.** $b - 7$ **27.** $y - 5$ **29.** $2y - 7$ **31.** $2y + 6 + \dfrac{25}{y - 3}$

33. $x - 2 + \dfrac{8}{x + 2}$ **35.** $3y - 5 + \dfrac{20}{y + 4}$ **37.** $6x - 12 + \dfrac{19}{x + 2}$ **39.** $b - 5 - \dfrac{24}{b - 3}$

41. $3x + 17 + \dfrac{64}{x - 4}$ **43.** $5y + 3 + \dfrac{1}{2y + 3}$ **45.** $4a + 1$ **47.** $2a + 9 + \dfrac{33}{3a - 1}$
49. $x^2 - 5x + 2$ **51.** $x^2 + 5$ **53.** $3ab$

CHAPTER REVIEW   *pages 191–192*

1. $21y^2 + 4y - 1$ **2.** $2x^3 + 9x^2 - 3x - 12$ **3.** $10a^2 + 12a - 22$ **4.** $2x^2 + 3x - 8$
5. $13y^3 - 12y^2 - 5y - 1$ **6.** $-2y^2 + y - 5$ **7.** $-20x^3y^5$ **8.** $x^4y^8z^4$ **9.** $-54a^{13}b^5c^7$
10. 64 **11.** $9x^4y^6$ **12.** $100a^{15}b^{13}$ **13.** $-8x^3 - 14x^2 + 18x$ **14.** $8a^3b^3 - 4a^2b^4 + 6ab^5$
15. $6y^3 + 17y^2 - 2y - 21$ **16.** $12b^5 - 4b^4 - 6b^3 - 8b^2 + 5$ **17.** $10a^2 + 31a - 63$
18. $8b^2 - 2b - 15$ **19.** $a^2 - 49$ **20.** $25y^2 - 70y + 49$ **21.** $(2w^2 - w)\ \text{ft}^2$

22. $(9x^2 - 12x + 4)\ \text{in.}^2$ **23.** $\dfrac{2x^3}{3}$ **24.** $-\dfrac{2a^3}{3b^3}$ **25.** $\dfrac{x^4y^6}{9}$ **26.** b^2 **27.** 1.27×10^{-7}

28. 0.0000000000032 **29.** $-4y + 8$ **30.** $4b^4 + 12b^2 - 1$ **31.** $2y - 9$

32. $b^2 + 5b + 2 + \dfrac{7}{b - 7}$

CHAPTER TEST   *pages 193–194*

1. $3x^3 + 6x^2 - 8x + 3$ [5.1A]   **2.** $-5a^3 + 3a^2 - 4a + 3$ [5.1B]   **3.** a^4b^7 [5.2A]
4. $-6x^3y^6$ [5.2A]   **5.** 3.02×10^{-9} [5.4B]   **6.** $-8a^6b^3$ [5.2B]   **7.** $4x^3 - 6x^2$ [5.3A]
8. $6y^4 - 9y^3 + 18y^2$ [5.3A]   **9.** $x^3 - 7x^2 + 17x - 15$ [5.3B]
10. $-4x^4 + 8x^3 - 3x^2 - 14x + 21$ [5.3B]   **11.** $a^2 + 3ab - 10b^2$ [5.3C]
12. $10x^2 - 43xy + 28y^2$ [5.3C]   **13.** $16y^2 - 9$ [5.3D]   **14.** $4x^2 - 20x + 25$ [5.3D]

15. $(3.14x^2 - 31.4x + 78.5)\ \text{m}^2$ [5.3E]   **16.** $-\dfrac{4}{x^6}$ [5.4A]   **17.** $-2x^3$ [5.4A]   **18.** $8ab^4$ [5.4A]

19. $\dfrac{9y^{10}}{x^{10}}$ [5.4A]   **20.** $4a - 7$ [5.5A]   **21.** $4x^4 - 2x^2 + 5$ [5.5A]   **22.** $4x - 1 + \dfrac{3}{x^2}$ [5.5A]

23. $x + 7$ [5.5B]   **24.** $x - 1 + \dfrac{2}{x + 1}$ [5.5B]   **25.** $2x + 3 + \dfrac{2}{2x - 3}$ [5.5B]

CUMULATIVE REVIEW   *pages 195–196*

1. $\dfrac{5}{144}$ [1.4C]   **2.** $\dfrac{5}{3}$ [1.5A]   **3.** $\dfrac{25}{11}$ [1.5B]   **4.** $-\dfrac{22}{9}$ [2.1A]   **5.** $5x - 3xy$ [2.2A]
6. $-9x$ [2.2B]   **7.** $-18x + 12$ [2.2D]   **8.** $x = -16$ [3.1C]   **9.** $x = -16$ [3.3A]
10. $x = 15$ [3.3B]   **11.** 22% [3.1D]   **12.** $4b^3 - 4b^2 - 8b - 4$ [5.1A]
13. $3y^3 + 2y^2 - 10y$ [5.1B]   **14.** a^9b^{15} [5.2B]   **15.** $-8x^3y^6$ [5.2A]
16. $6y^4 + 8y^3 - 16y^2$ [5.3A]   **17.** $10a^3 - 39a^2 + 20a - 21$ [5.3B]
18. $15b^2 - 31b + 14$ [5.3C]   **19.** $9b^2 + 12b + 4$ [5.3D]   **20.** $\dfrac{1}{2b^2}$ [5.4A]

21. 0.0000609 [5.4B]   **22.** $6a - 4 + \dfrac{2}{a^2}$ [5.5A]   **23.** $a - 7$ [5.5B

24. $8x - 2x = 18;\ x = 3$ [3.4A]   **25.** The selling price is \$43.20 [4.1A]
26. The resulting mixture is 28% orange juice [4.3B]
27. The car overtakes the cyclist 25 mi. from the starting point [4.4A]
28. The length is 15 m; the width is 6 m. [4.5A]
29. The consecutive odd integers are 21 and 23. [3.4A]   **30.** $[4x^2 + 12x + 9]$ square units [5.3E]

ANSWERS to Chapter 6 Odd-Numbered Exercises

SECTION 6.1   *pages 203–204*

1. $5(a + 1)$ **3.** $8(2 - a^2)$ **5.** $4(2x + 3)$ **7.** $6(5a - 1)$ **9.** $x(7x - 3)$ **11.** $a^2(3 + 5a^3)$
13. $y(14y + 11)$ **15.** $2x(x^3 - 2)$ **17.** $2x^2(5x^2 - 6)$ **19.** $4a^5(2a^3 - 1)$ **21.** $xy(xy - 1)$
23. $3xy(xy^3 - 2)$ **25.** $xy(x - y^2)$ **27.** $5y(y^2 - 4y + 2)$ **29.** $3y^2(y^2 - 3y - 2)$
31. $3y(y^2 - 3y + 8)$ **33.** $a^2(6a^3 - 3a - 2)$ **35.** $ab(2a - 5ab + 7b)$ **37.** $2b(2b^4 + 3b^2 - 6)$
39. $x^2(8y^2 - 4y + 1)$ **41.** $(y + 7)(a + z)$ **43.** $(3r + s)(a - b)$ **45.** $(t - 7)(m - 7)$
47. $(2y + 1)(4a - b)$ **49.** $(x + 2)(x + 2y)$ **51.** $(p - 2)(p - 3r)$ **53.** $(b - 4)(a + 6)$
55. $(z + y)(2z - 1)$ **57.** $(4v + 7)(2v - 3y)$ **59.** $(2x - 5)(x - 3y)$ **61.** $(3y - a)(y - 2)$
63. $(y + 1)(3x - y)$ **65.** $(t - 2)(t + 3s)$ **67. a.** $(2x + 5)(x + 3)$ **b.** $(x + 3)(2x + 5)$

69. 28 is the one perfect number between 20 and 30.
71. In the equation for the perimeter of a rectangle, when $l + w$ doubles, the perimeter doubles.

SECTION 6.2 *pages 209–212*

1. $(x + 1)(x + 2)$ **3.** $(x - 2)(x + 1)$ **5.** $(a + 4)(a - 3)$ **7.** $(a - 1)(a - 2)$
9. $(a + 2)(a - 1)$ **11.** $(b - 3)^2$ **13.** $(b + 8)(b - 1)$ **15.** $(y + 11)(y - 5)$
17. $(y - 3)(y - 2)$ **19.** $(z - 5)(z - 9)$ **21.** $(z - 20)(z + 8)$ **23.** $(p + 3)(p + 9)$
25. $(x + 10)^2$ **27.** $(b + 4)(b + 5)$ **29.** $(x + 3)(x - 14)$ **31.** $(b - 5)(b + 4)$
33. $(y - 17)(y + 3)$ **35.** $(p - 7)(p + 3)$ **37.** Nonfactorable over the integers
39. $(x - 5)(x - 15)$ **41.** $(x - 7)(x - 8)$ **43.** $(x + 8)(x - 7)$ **45.** $(a - 24)(a + 3)$
47. $(a - 3)(a - 12)$ **49.** $(z - 17)(z + 8)$ **51.** $(c - 10)(c + 9)$ **53.** $(z + 11)(z + 4)$
55. $(c + 2)(c + 17)$ **57.** $(x - 12)(x + 8)$ **59.** $(x - 8)(x - 14)$ **61.** $(b + 15)(b - 7)$
63. $(a - 12)(a + 3)$ **65.** $(b - 6)(b - 17)$ **67.** $(a + 3)(a + 24)$ **69.** $(x + 12)(x + 13)$
71. $(x - 16)(x + 6)$ **73.** $2(x + 1)(x + 2)$ **75.** $-(x + 2)(x - 9)$ **77.** $a(b + 5)(b - 3)$
79. $x(y - 2)(y - 3)$ **81.** $z(z - 4)(z - 3)$ **83.** $-3y(y - 2)(y - 3)$ **85.** $3(x + 4)(x - 3)$
87. $5(z - 7)(z + 4)$ **89.** $2a(a + 8)(a - 4)$ **91.** $(x - 3y)(x - 2y)$ **93.** $(a - 4b)(a - 5b)$
95. $(x - 7y)(x + 4y)$ **97.** Nonfactorable over the integers   **99.** $z^2(z - 5)(z - 7)$
101. $b^2(b - 10)(b - 12)$ **103.** $2y^2(y - 16)(y + 3)$ **105.** $-x^2(x + 8)(x - 1)$
107. $4y(x + 7)(x - 2)$ **109.** $c(c + 20)(c - 2)$ **111.** $-4x(x + 3)(x - 2)$
113. $(y - 8x)(y + x)$ **115.** $(y + 7z)(y - 3z)$ **117.** $(y - z)(y - 15z)$ **119.** $4y(x - 18)(x + 1)$
121. $4x(x + 8)(x - 5)$ **123.** $5z(z - 12)(z + 2)$ **125.** $5x(x + 2)(x + 4)$
127. $4(p - 15)(p + 8)$ **129.** $p^2(p + 8)(p - 7)$ **131.** $(a - 5b)^2$ **133.** $(x + 10y)(x - 6y)$
135. $6x(x - 5)(x + 4)$ **137.** $y(x - 9)(x + 6)$ **139.** $k = -36, -12, 12, 36$
141. $-22, -10, 10, 22$ **143.** $k = 6, 10, 12$ **145.** $k = 6, 10, 12$ **147.** $k = 4, 6$

SECTION 6.3 *pages 217–220*

1. $(2x + 1)(x + 1)$ **3.** $(2y + 1)(y + 3)$ **5.** $(2a - 1)(a - 1)$ **7.** $(2b - 1)(b - 5)$
9. $(2x - 1)(x + 1)$ **11.** $(2x + 1)(x - 3)$ **13.** $(2t - 5)(t + 2)$ **15.** $(3p - 1)(p - 5)$
17. $(4y - 1)(3y - 1)$ **19.** Nonfactorable over the integers   **21.** $(2t - 1)(3t - 4)$
23. $(8x + 1)(x + 4)$ **25.** Nonfactorable over the integers   **27.** $(4y + 5)(3y + 1)$
29. $(7a - 2)(a + 7)$ **31.** $(3b - 4)(b - 4)$ **33.** $(2z + 1)(z - 14)$ **35.** $(3p - 2)(p + 8)$
37. $2(2x + 1)(x + 1)$ **39.** $5(3y - 7)(y - 1)$ **41.** $x(2x - 1)(x - 5)$ **43.** $b(3a - 4)(a - 4)$
45. Nonfactorable over the integers   **47.** $-3x(x + 4)(x - 3)$ **49.** $4(4y - 1)(5y - 1)$
51. $z(4z + 1)(2z + 3)$ **53.** $y(3x + 2)(2x - 5)$ **55.** $5(2t - 5)(t + 2)$ **57.** $p(3p - 1)(p - 5)$
59. $2(13z - 3)(z + 4)$ **61.** $2y(5y - 2)(y - 4)$ **63.** $yz(4z - 3)(z + 2)$
65. $3a(2a + 3)(7a - 3)$ **67.** $y(3x - 5y)^2$ **69.** $xy(3x - 4y)^2$ **71.** $(3x - 4)(2x - 3)$
73. $(5b - 2)(b + 7)$ **75.** $(3a + 8)(2a - 3)$ **77.** $(4z + 3)(z + 2)$ **79.** $(2p + 5)(11p - 2)$
81. $(8y + 9)(y + 1)$ **83.** $(3t + 1)(6t - 5)$ **85.** $(6b - 1)(b + 12)$ **87.** $(3x + 2)^2$
89. $(3b - 2)(2b - 3)$ **91.** $(11b - 7)(3b + 5)$ **93.** $(3y - 4)(6y - 5)$ **95.** $(3a + 7)(5a - 3)$
97. $(2y - 5)(4y - 3)$ **99.** $(4z - 5)(2z + 3)$ **101.** Nonfactorable over the integers
103. $(2z - 5)(5z - 2)$ **105.** $(6z + 7)(6z + 5)$ **107.** $(3x - 2y)(x + y)$ **109.** $(3a - b)(a + 2$
111. $(4y - 3z)(y - 2z)$ **113.** $-(z - 7)(z + 4)$ **115.** $-(x - 1)(x + 8)$ **117.** $3(3x - 4)(x$
119. $4(2x - 3)(3x - 2)$ **121.** $a^2(7a - 1)(5a + 2)$ **123.** $5(3b - 2)(b - 7)$
125. $(3x - 5y)(x - 7y)$ **127.** $3(8y - 1)(9y + 1)$ **129.** $(x - 1)(x + 21)$
131. $(3a - 2b)(5a + 7b)$ **133.** $-z(z + 11)(z - 3)$ **135.** $(x + 3)(x - 2)$ **137.**
139. $(3a + 2)(a + 3)$ **141.** $k = -7, -5, 5, 7$ **143.** $k = -7, -5, 5, 7$ **145.** k

SECTION 6.4 *pages 225–228*

1. $(x - 2)(x + 2)$ **3.** $(a - 9)(a + 9)$ **5.** $(y + 1)^2$ **7.** $(a - 1)^2$
11. $(x^3 - 3)(x^3 + 3)$ **13.** Nonfactorable over the integers   **15.** $($
19. $(3x - 1)(3x + 1)$ **21.** $(1 + 8x)(1 - 8x)$ **23.** Nonfactorable ove
25. $(3a + 1)^2$ **27.** $(b^2 - 4a)(b^2 + 4a)$ **29.** $(2a - 5)^2$ **31.** $(3a - $
35. $(ab - 5)(ab + 5)$ **37.** $(5x - 1)(5x + 1)$ **39.** $(2a - 3b)^2$ **41.** $(2$

43. $\left(\dfrac{1}{x} - 2\right)\left(\dfrac{1}{x} + 2\right)$ **45.** $(3ab - 1)^2$ **47.** $2(2y - 1)(2y + 1)$ **49.** $3a(a + 1)^2$
51. $(m^2 + 16)(m + 4)(m - 4)$ **53.** $(9x + 4)(x + 1)$ **55.** $4y^2(2y + 3)^2$
57. $(y^4 + 9)(y^2 - 3)(y^2 + 3)$ **59.** $(5 - 2p)^2$ **61.** $(4x - 3 - y)(4x - 3 + y)$
63. $(x - 2 - y)(x - 2 + y)$ **65.** $5(x + 1)(x - 1)$ **67.** $x(x + 2)^2$ **69.** $x^2(x + 7)(x - 5)$
71. $5(b + 3)(b + 12)$ **73.** Nonfactorable over the integers   **75.** $2y(x + 11)(x - 3)$
77. $x(x^2 - 6x - 5)$ **79.** $3(y^2 - 12)$ **81.** $(2a + 1)(10a + 1)$ **83.** $y^2(x - 8)(x + 1)$
85. $5(2a - 3b)(a + b)$ **87.** $-2(x - 5)(x + 5)$ **89.** $b^2(a - 5)^2$ **91.** $ab(4a + b)(3a - b)$
93. $3a(2a - 1)^2$ **95.** $3(81 + a^2)$ **97.** $2a(2a - 5)(3a - 4)$ **99.** $a(2a + 5)^2$
101. $3b(3a - 1)^2$ **103.** $-6(x + 4)(x - 2)$ **105.** $x^2(x + y)(x - y)$ **107.** $2a(3a + 2)^2$
109. $-b(3a - 2)(2a + 1)$ **111.** $2x^2(2x - 3)(x - 8)$ **113.** $x^2(x + 5)(x - 5)$
115. $(a^2 + 4)(a + 2)(a - 2)$ **117.** $-3y^2(5 + 2y)(4y - 3)$ **119.** $2(x - 3)(2a - b)$
121. $(y + 1)(y - 1)(a - b)$ **123.** $(x - y)(a + b)(a - b)$ **125.** $-12, 12$ **127.** $-16, 16$
129. $-10, 10$ **131.** $(x + 2)(x^2 - 2x + 4)$ **133.** $(x + 4)(x^2 - 4x + 16)$
135. $(3y - 1)(9y^2 + 3y + 1)$ **137.** Since n or $n + 1$ is an even number, $4n(n + 1)$ is divisible by 8.

SECTION 6.5   *pages 233–236*

1. The solutions are -3 and -2.   **3.** The solutions are 7 and 3.   **5.** The solutions are 0 and 5.

7. The solutions are 0 and 9.   **9.** The solutions are 0 and $-\dfrac{3}{2}$.   **11.** The solutions are 0 and $\dfrac{2}{3}$.

13. The solutions are -2 and 5.   **15.** The solutions are 9 and -9.

17. The solutions are $\dfrac{7}{2}$ and $-\dfrac{7}{2}$.   **19.** The solutions are $\dfrac{1}{3}$ and $-\dfrac{1}{3}$.

21. The solutions are -2 and -4.   **23.** The solutions are -7 and 2.

25. The solutions are 2 and 3.   **27.** The solutions are -7 and 3.

29. The solutions are $-\dfrac{1}{2}$ and 5.   **31.** The solutions are $-\dfrac{1}{3}$ and $-\dfrac{1}{2}$.

33. The solutions are 0 and 3.   **35.** The solutions are 0 and 7.
37. The solutions are -1 and -4.   **39.** The solutions are 2 and 3.

41. The solutions are $\dfrac{1}{2}$ and -4.   **43.** The solutions are $\dfrac{1}{3}$ and 4.

45. The solutions are 3 and 9.   **47.** The solutions are 9 and -2.
49. The solutions are -1 and -2.   **51.** The solutions are -9 and 5.
53. The solutions are -7 and 4.   **55.** The solutions are -2 and -3.
57. The solutions are -8 and 9.   **59.** The solutions are 1 and 4.
61. The solutions are -5 and 2.   **63.** The solutions are 3 and 4.

65. The solutions are -4 and $-\dfrac{1}{3}$.   **67.** The number is -2.   **69.** The numbers are 3 and 5.

71. The numbers are 6 and 8.   **73.** The numbers are 8 and 15.
75. There will be 15 consecutive numbers.   **77.** There are 10 teams.
79. The object will hit 4 seconds later.   **81.** It will be 6 seconds later.
83. The length is 16 ft and the width is 11 ft.
85. The length of the side of the original square is 3 in.   **87.** The width of the border is 2 ft.
89. The increase in area is 138.16 in^2.   **91.** The solutions are -3 and 4.   **93.** $3n^2 = 3$ or 48

95. The solutions are $-\dfrac{3}{2}$ and -5.   **97.** The solutions are 0 and 9.

CHAPTER REVIEW   *pages 239–240*

1. $5x(x^2 + 2x + 7)$ **2.** $3ab(4a + b)$ **3.** $7y^3(2y^6 - 7y^3 + 1)$ **4.** $(x - 3)(4x + 5)$
5. $(2x + 5)(5x + 2y)$ **6.** $(3a - 5b)(7x + 2y)$ **7.** $(b - 3)(b - 10)$ **8.** $(c + 6)(c + 2)$
9. $(x - 4)(y + 9)$ **10.** $3(a + 2)(a - 7)$ **11.** $4x(x - 6)(x + 1)$ **12.** $n^2(n + 1)(n - 3)$
13. $(x - 7)(3x - 4)$ **14.** $(6y - 1)(2y + 3)$ **15.** Nonfactorable over the integers.
16. $(x - 2)(x - 5)$ **17.** $(2a + 5)(a - 12)$ **18.** $(6a - 5)(3a + 2)$ **19.** $(a^3 + 10)(a^3 - 10)$
20. $(3y^2 + 5z)(3y^2 - 5z)$ **21.** $(ab + 1)(ab - 1)$ **22.** $5(x + 2)(x - 3)$ **23.** $3(x + 6)^2$

24. $2b(3b - 4)(2b - 7)$ **25.** The solutions are $\frac{1}{4}$ and -7.   **26.** The solutions are -3 and 7.

27. The length is 100 yd. The width is 60 yd.

28. The distance between the screen and the projector is 20 ft.

29. The width of the frame is $\frac{3}{2}$ in. or 1.5 in. or $1\frac{1}{2}$ in.   **30.** The length of a side is 20 ft.

CHAPTER TEST   *pages 241–242*

1. $2x(3x^2 - 4x + 5)$ [6.1A]   **2.** $(b + 6)(a - 3)$ [6.1B]   **3.** $(p + 2)(p + 3)$ [6.2A]

4. $(a - 3)(a - 16)$ [6.2A]   **5.** $(x + 5)(x - 3)$ [6.2A]   **6.** $(x + 3)(x - 12)$ [6.2A]

7. $5(x^2 - 9x - 3)$ [6.1A]   **8.** $2y^2(y + 1)(y - 8)$ [6.2B]

9. Nonfactorable over the integers. [6.3A]   **10.** $(2x + 1)(3x + 8)$ [6.3A]

11. $4(x + 4)(2x - 3)$ [6.3B]   **12.** $3y^2(2x + 1)(x + 1)$ [6.3B]   **13.** $(x - 2)(a + b)$ [6.3B]

14. $(p + 1)(x - 1)$ [6.3B]   **15.** $(b + 4)(b - 4)$ [6.4A]   **16.** $(2x + 7y)(2x - 7y)$ [6.4A]

17. $(p + 6)^2$ [6.4A]   **18.** $(2a - 3b)^2$ [6.4A]   **19.** $3(a + 5)(a - 5)$ [6.4B]

20. $3(x + 2y)^2$ [6.4B]   **21.** The solutions are $\frac{3}{2}$ and -7. [6.5A]

22. The solutions are $\frac{1}{2}$ and $-\frac{1}{2}$ [6.5A]   **23.** The solutions are 3 and 5. [6.5A]

24. The two numbers are 3 and 7. [6.5B]   **25.** The length is 15 cm. The width is 6 cm. [6.5B]

CUMULATIVE REVIEW   *pages 243–244*

1. 7 [1.2B]   **2.** 4 [1.5B]   **3.** -7 [2.1A]   **4.** $15x^2$ [2.2B]   **5.** 12 [2.2D]

6. $x = \frac{2}{3}$ [3.1C]   **7.** $x = \frac{7}{4}$ [3.3A]   **8.** $x = 3$ [3.3B]   **9.** 45 [3.1D]   **10.** $9a^6b^4$ [5.2B]

11. $x^3 - 3x^2 - 6x + 8$ [5.3B]   **12.** $4x + 8 + \frac{21}{2x - 3}$ [5.5B]   **13.** $\frac{y^6}{x^8}$ [5.4A]

14. $(a - b)(3 - x)$ [6.1B]   **15.** $5xy^2(3 - 4y^2)$ [6.1A]   **16.** $(x - 7y)(x + 2y)$ [6.2A]

17. $(p - 10((p + 1)$ [6.2A]   **18.** $3a(2a + 5)(3a + 2)$ [6.3B]   **19.** $(6a - 7b)(6a + 7b)$ [6.4A]

20. $(2x + 7y)^2$ [6.4A]   **21.** $(3x - 2)(3x + 7)$ [6.3A]   **22.** $2(3x - 4y)^2$ [6.4B]

23. $(x - 3)(3y - 2)$ [6.1B]   **24.** The solutions are $\frac{2}{3}$ and -7. [6.5A]

25. The pieces measure 4 ft and 6 ft. [3.4B]   **26.** The discount rate is 40%. [4.1B]

27. $6500 more must be invested. [4.2A]   **28.** The distance to the resort was 168 mi. [4.4A]

29. The integers are 10, 12, and 14. [3.4A]   **30.** The length of the base is 12 in. [6.5B]

ANSWERS to Chapter 7 Odd-Numbered Exercises

SECTION 7.1   *pages 251–254*

1. $\frac{3}{4x}$ **3.** $\frac{1}{x + 3}$ **5.** -1 **7.** $\frac{2}{3y}$ **9.** $-\frac{3}{4x}$ **11.** $\frac{a}{b}$ **13.** $-\frac{2}{x}$ **15.** $\frac{y - 2}{y - 3}$ **17.** $\frac{x}{}$

19. $\frac{x + 4}{x - 3}$ **21.** $-\frac{x + 2}{x + 5}$ **23.** $\frac{2(x + 2)}{x + 3}$ **25.** $\frac{2x - 1}{2x + 3}$ **27.** $-\frac{x + 7}{x + 6}$ **29.** $\frac{35ab^2}{24x^2y}$

33. $\frac{3}{4}$ **35.** ab^2 **37.** $\frac{x^2(x - 1)}{y(x + 3)}$ **39.** $\frac{y(x - 1)}{x^2(x + 10)}$ **41.** $-ab^2$ **43.** $\frac{x + 5}{x + 4}$

47. $-\frac{n - 10}{n + 7}$ **49.** $\frac{x(x + 2)}{2(x - 1)}$ **51.** $-\frac{x + 2}{x - 6}$ **53.** $\frac{x + 5}{x - 12}$ **55.** $\frac{3y + 2}{3y + 1}$

61. $\frac{3a}{2}$ **63.** $\frac{x^2(x + 4)}{y^2(x + 2)}$ **65.** $\frac{x(x - 2)}{y(x - 6)}$ **67.** $-\frac{3by}{ax}$ **69.** $\frac{(x + 6)(x - 3)}{(x + 7)(x - 6)}$

75. $\frac{2n + 1}{2n - 3}$ **77.** $-\frac{3x + 1}{2x - 3}$ **79.** yes; $3 + 10^{-8}$ **81.** $5, -5$ **83.** $\frac{4x}{3y}$

SECTION 7.2 *pages 257–258*

1. $24x^3y^2$ **3.** $30x^4y^2$ **5.** $8x^2(x+2)$ **7.** $6x^2y(x+4)$ **9.** $36x(x+2)^2$ **11.** $6(x+1)^2$

13. $(x-1)(x+2)(x+3)$ **15.** $(2x+3)^2(x-5)$ **17.** $(x-1)(x-2)$

19. $(x-3)(x+2)(x+4)$ **21.** $(x+4)(x+1)(x-7)$ **23.** $(x-6)(x+6)(x+4)$

25. $(x-10)(x-8)(x+3)$ **27.** $(3x-2)(x-3)(x+2)$ **29.** $(x+2)(x-3)$

31. $(x-5)(x+1)$ **33.** $(x-3)(x-2)(x-1)(x-6)$ **35.** $\dfrac{5}{ab^2}$ $\dfrac{6b}{ab^2}$ **37.** $\dfrac{15y^2}{18x^2y}$ $\dfrac{14x}{18x^2y}$

39. $\dfrac{a(y+5)}{y^2(y+5)}$ $\dfrac{6y}{y^2(y+5)}$ **41.** $\dfrac{a^2(y+7)}{y(y+7)^2}$ $\dfrac{ay}{y(y+7)^2}$ **43.** $\dfrac{b}{y(y-4)}$ $\dfrac{-b^2y}{y(y-4)}$ **45.** $-\dfrac{3(y-7)}{(y-7)^2}$ $\dfrac{2}{(y-7)^2}$

47. $\dfrac{2y^2}{y^2(y-3)}$ $\dfrac{3}{y^2(y-3)}$ **49.** $\dfrac{x^2(x+4)}{(2x-1)(x+4)}$ $\dfrac{(x+1)(2x-1)}{(2x-1)(x+4)}$ **51.** $\dfrac{3x(x+5)}{(x-5)(x+5)}$ $\dfrac{4}{(x-5)(x+5)}$

53. $\dfrac{x-3}{(3x-2)(x+2)}$ $\dfrac{2(3x-2)}{(3x-2)(x+2)}$ **55.** $\dfrac{(x-1)(x+1)}{(x+5)(x-3)(x+1)}$ $\dfrac{x(x-3)}{(x+5)(x-3)(x+1)}$ **57.** $\dfrac{800}{10^5}$ $\dfrac{9}{10^5}$

59. $\dfrac{x(x^2-1)}{x^2-1}$ $\dfrac{x}{x^2-1}$ **61.** $\dfrac{3c(c-d)}{3(6c+d)(c+d)(c-d)}$ $\dfrac{d(6c+d)}{3(6c+d)(c+d)(c-d)}$

SECTION 7.3 *pages 263–266*

1. $\dfrac{11}{y^2}$ **3.** $\dfrac{-7}{x+4}$ **5.** $\dfrac{8x}{2x+3}$ **7.** $\dfrac{5x+7}{x-3}$ **9.** $\dfrac{2x-5}{x+9}$ **11.** $\dfrac{3x+4}{2x+7}$ **13.** $\dfrac{1}{x+5}$ **15.** $\dfrac{1}{x-6}$

17. $\dfrac{3}{2y-1}$ **19.** $\dfrac{1}{x-5}$ **21.** $\dfrac{4y+5x}{xy}$ **23.** $\dfrac{19}{2x}$ **25.** $\dfrac{5}{12x}$ **27.** $\dfrac{19x-12}{6x^2}$ **29.** $\dfrac{52y-35x}{20xy}$

31. $\dfrac{13x+2}{15x}$ **33.** $\dfrac{7}{24}$ **35.** $\dfrac{x+90}{45x}$ **37.** $\dfrac{x^2+2x+2}{2x^2}$ **39.** $\dfrac{2x^2+3x-10}{4x^2}$ **41.** $\dfrac{-x^2-4x+4}{x+4}$

43. $\dfrac{4x+7}{x-1}$ **45.** $\dfrac{4x^2+9x+9}{24x^2}$ **47.** $\dfrac{3x-1-2xy-3y}{xy^2}$ **49.** $\dfrac{20x^2+28x-12xy+9y}{24x^2y^2}$

51. $\dfrac{9x^2-3x-2xy-10y}{18xy^2}$ **53.** $\dfrac{7x-23}{(x-3)(x-4)}$ **55.** $\dfrac{-y-33}{(y+6)(y-3)}$ **57.** $\dfrac{3x^2+20x-8}{(x-4)(x+6)}$

59. $\dfrac{3(4x^2+5x-5)}{(x+5)(2x+3)}$ **61.** $\dfrac{-4x+5}{x-6}$ **63.** $\dfrac{2(y+2)}{(y-4)(y+4)}$ **65.** $\dfrac{-4x}{(x+1)^2}$ **67.** $\dfrac{2x-1}{1-x^2}$ **69.** $\dfrac{14}{(x-5)^2}$

71. $-\dfrac{2(x+7)}{(x-7)(x+6)}$ **73.** $\dfrac{x-4}{x-6}$ **75.** $\dfrac{2x+1}{x-1}$ **77.** $-\dfrac{3(x^2+8x+25)}{(x-3)(x+7)}$ **79.** $\dfrac{2}{3}; \dfrac{3}{4}; \dfrac{4}{5}; \dfrac{50}{51}; \dfrac{100}{101}; \dfrac{1000}{1001}$

81. 1 **83.** $\dfrac{x^2-9x+30}{(x+5)(x+1)}$

SECTION 7.4 *pages 269–270*

1. $\dfrac{x}{x-3}$ **3.** $\dfrac{2}{3}$ **5.** $\dfrac{y+3}{y-4}$ **7.** $\dfrac{4x+26}{5x+36}$ **9.** $\dfrac{x+2}{x+3}$ **11.** $\dfrac{x-6}{x+5}$ **13.** $-\dfrac{x-2}{x+1}$ **15.** $x-1$

17. $\dfrac{1}{2x-1}$ **19.** $\dfrac{x-3}{x+5}$ **21.** $\dfrac{x-7}{x-8}$ **23.** $\dfrac{2y-1}{2y+1}$ **25.** $\dfrac{x-2}{2x-5}$ **27.** $-\dfrac{x+1}{4x-3}$ **29.** $\dfrac{x+1}{2(5x-2)}$

31. $\dfrac{5}{3}$ **33.** $\dfrac{-1}{x-1}$ **35.** $\dfrac{y+4}{2(y-2)}$ **37.** $\dfrac{x+1}{x-1}$ **39.** $\dfrac{y^2+x^2}{xy}$

SECTION 7.5 *pages 273–274*

1. The solution is 3. **3.** The solution is 1. **5.** The solution is 9. **7.** The solution is 1.

The solution is $\dfrac{1}{4}$. **11.** The solution is 1. **13.** The solution is -3. **15.** The solution is $\dfrac{1}{2}$.

The solution is 8. **19.** The solution is 5. **21.** The solution is -1. **23.** The solution is 5.

The equation has no solution. **27.** The solutions are 2 and 4.

he solutions are 4 and $-\dfrac{3}{2}$. **31.** The solution is 3. **33.** The solution is 4.

solution is 0. **37.** The solution is $-\dfrac{2}{5}$. **39.** The solutions are 0 and $-\dfrac{2}{3}$.

Content and Format © 1995 HMCo.

SECTION 7.6   *pages 277–278*

1. The solution is 9.   **3.** The solution is 12.   **5.** The solution is 7.   **7.** The solution is 6.
9. The solution is 1.   **11.** The solution is -6.   **13.** The solution is 4.

15. The solution is $-\dfrac{2}{3}$.   **17.** There will be 20,000 voters voting in favor.

19. The building will need 140 air vents.   **21.** The shipment will be accepted.
23. It will take 18 min to print the document.   **25.** There are an estimated 75 elk in the preserve.
27. The first person's share of the winnings is $1.25 million.
29. The basketball player made 210 shots.

SECTION 7.7   *pages 281–282*

1. $y = -3x + 10$ **3.** $y = 4x - 3$ **5.** $y = -\dfrac{3}{2}x + 3$ **7.** $y = \dfrac{2}{5}x - 2$ **9.** $y = -\dfrac{2}{7}x + 2$

11. $y = -\dfrac{1}{3}x + 2$ **13.** $y = \dfrac{2}{9}x - 2$ **15.** $y = 2x + 5$ **17.** $x = -6y + 10$ **19.** $x = \dfrac{1}{2}y + 3$

21. $x = -\dfrac{3}{4}y + 3$ **23.** $x = 4y + 3$ **25.** $t = \dfrac{d}{r}$ **27.** $T = \dfrac{PV}{nR}$ **29.** $l = \dfrac{P - 2w}{2}$

31. $b_1 = \dfrac{2A - hb_2}{h}$ **33.** $h = \dfrac{3V}{A}$ **35.** $S = C - Rt$ **37.** $P = \dfrac{A}{1 + rt}$ **39.** $w = \dfrac{A}{S + 1}$

41. **a.** $S = \dfrac{F + BV}{B}$ **b.** The selling price per unit to break even is $180.

c. The selling price per unit to break even is $75.

SECTION 7.8   *pages 287–290*

1. It would take the experienced painter 6 h working alone.
3. It would take both skiploaders 3 h working together.
5. It would take 30 h with both computers working.
7. It would take 6 min with both air conditioners working.
9. It would take the second welder 15 h working alone.
11. It would take the second harvester 3 h working alone.
13. It would take the second mason 6 h.   **15.** It would take the second technician 3 h.
17. It would take one welder 40 h working alone.
19. It would take one machine 28 h working alone.   **21.** The camper hiked at a rate of 5 mph.
23. The rate of the jogger is 8 mph, and the rate of the cyclist is 20 mph.
25. The rate of the jet is 360 mph.   **27.** The rate of the second plane is 150 mph.
29. The rate of the car is 48 mph.   **31.** The rate of the first 9 mi was 3 mph.
33. The rate of the current is 2 mph.   **35.** The rate of the gulf current is 6 mph.
37. The rate of the trucker for the first 330 mi is 55 mph.
39. The less experienced helper can complete the job in 30 h.
41. The bus usually travels 60 mph.

CHAPTER REVIEW   *pages 293–294*

1. $\dfrac{2x^4}{3y^7}$ **2.** $-\dfrac{x + 6}{x + 3}$ **3.** $\dfrac{by^3}{6ax^2}$ **4.** $\dfrac{1}{x}$ **5.** $\dfrac{8x + 5}{3x - 4}$ **6.** $\dfrac{b^3y}{10ax}$ **7.** $\dfrac{1}{x^2}$ **8.** $\dfrac{(3y - 2)^2}{(y - 1)(y - 2)}$

9. $(2x - 1)(5x - 3)(4x - 1)$ **10.** $\dfrac{3x^2 - x}{(2x + 3)(6x - 1)(3x - 1)}, \dfrac{24x^3 - 4x^2}{(2x + 3)(6x - 1)(3x - 1)}$ **11.** $\dfrac{1}{x}$

12. $\dfrac{7x + 22}{60x}$ **13.** $\dfrac{2y - 3}{5y - 7}$ **14.** $\dfrac{3x - 1}{x - 5}$ **15.** $\dfrac{x}{x - 7}$ **16.** $x - 2$ **17.** $\dfrac{x - 2}{3x - 1}$

18. The solution is 2.   **19.** The equation has no solution.   **20.** The solut
21. The solution is 12.   **22.** The solution is 62.   **23.** The solution is

24. The pitcher's ERA is 1.35.   **25.** $y = -\dfrac{4}{9}x + 2$ **26.** $c = \dfrac{100m}{i}$

28. It would take 6 h to fill the pool using both hoses.   **29.** The rate of t
30. The rate of the wind is 20 mph.

CHAPTER TEST   *pages 295–296*

1. $\dfrac{2x^3}{3y^3}$ [7.1A]   **2.** $-\dfrac{x+5}{x+1}$ [7.1A]   **3.** $\dfrac{x+1}{x^3(x-2)}$ [7.1B]   **4.** $\dfrac{(x-5)(2x-1)}{(x+3)(2x+5)}$ [7.1B]

5. $\dfrac{x+5}{x+4}$ [7.1C]   **6.** $3(2x-1)(x+1)$ [7.2A]   **7.** $\dfrac{3(x+2)}{x(x+2)(x-2)}$ $\dfrac{x^2}{x(x+2)(x-2)}$ [7.2B]

8. $\dfrac{2}{x+5}$ [7.3A]   **9.** $\dfrac{5}{(2x-1)(3x+1)}$ [7.3B]   **10.** $\dfrac{x^2-4x+5}{(x-2)(x+3)}$ [7.3B]   **11.** $\dfrac{x-3}{x-2}$ [7.4A]

12. The solution is 2. [7.5A]   **13.** The equation has no solution [7.5A]

14. The solution is -1. [7.6A]   **15.** Two additional pounds of salt are required. [7.6B]

16. 54 sprinklers are needed [7.6B]   **17.** $y = \dfrac{3}{8}x - 2$ [7.7A]   **18.** $t = \dfrac{d-s}{t}$ [7.7A]

19. It would take 4 h to fill the pool with both pipes turned on. [7.8A]

20. The rate of the wind is 20 mph [7.8B]

CUMULATIVE REVIEW   *pages 297–298*

1. $\dfrac{31}{30}$ [1.5B]   **2.** 21 [2.1A]   **3.** $5x - 2y$ [2.2A]   **4.** $-8x + 26$ [2.2D]   **5.** $-\dfrac{9}{2}$ [3.2A]

6. $x = -12$ [3.3B]   **7.** 10 [1.4E]   **8.** a^3b^7 [5.2A]   **9.** $a^2 + ab - 12b^2$ [5.3C]

10. $3b^3 - b + 2$ [5.5A]   **11.** $x^2 + 2x + 4$ [5.5B]   **12.** $(4x + 1)(3x - 1)$ [6.3A]

13. $(y - 6)(y - 1)$ [6.2A]   **14.** $a(2a - 3)(a + 5)$ [6.3B]   **15.** $4(b - 5)(b + 5)$ [6.4B]

16. The solutions are -3 and $\dfrac{5}{2}$ [6.5A]   **17.** $\dfrac{2x^3}{3y^5}$ [7.1A]   **18.** $-\dfrac{x-2}{x+5}$ [7.1A]   **19.** 1 [7.1C]

20. $\dfrac{3}{(2x-1)(x+1)}$ [7.3B]   **21.** $\dfrac{x+3}{x+5}$ [7.4A]   **22.** The solution is 4. [7.5A]

23. The solution is 3. [7.6A]   **24.** $t = \dfrac{f-v}{a}$ [7.7A]   **25.** $5x - 13 = -8; x = 1$ [3.4A]

26. The 120 gram alloy is 70% silver. [4.5B]   **27.** The height is 6 in and the base is 10 in. [6.5B]

28. A policy of $5000 would cost $80. [7.6B]

29. Working together, it would take the pipes 6 min to fill the tank. [7.8A]

30. The rate of the current is 2 mph. [7.6B]

ANSWERS to Chapter 8 Odd-Numbered Exercises

SECTION 8.1   *pages 309–312*

1.   **3.** 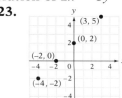   **5.**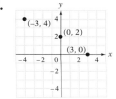

7. A is $(2, 3)$, B is $(4, 0)$, C is $(-4, -1)$ and D is $(-2, -2)$.

9. A is $(-2, 5)$, B is $(3, 4)$, C is $(0, 0)$ and D is $(-3, -2)$.

11.   **a.** The abscissa of point A is 2. The abscissa of point C is -4.

     b. The ordinate of point B is 1. The ordinate of point D is -3.

13. Yes, $(3, 4)$ is a solution of $y = -x + 7$.   **15.** No, $(-1, 2)$ is not a solution of $y = \dfrac{1}{2}x - 1$.

17. No, $(4, 1)$ is not a solution of $2x - 5y = 4$.   **19.** No, $(0, 4)$ is not a solution of $3x - 4y = -4$.

21. 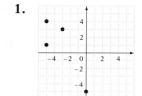   **23.** 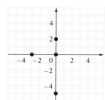   **25.** 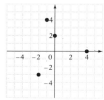   **27.**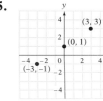

29. 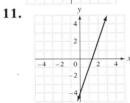 **31.**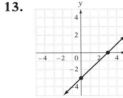

33. {(4, L) (6, W) (4, W) (2, W) (1, L) (6, W)} No, the relation is not a function.

35. {(11, 200) (9, 150) (7, 200) (12, 175) (12, 250)} No, the relation is not a function.   **37.** yes

39. no   **41.** yes   **43.** 8   **45.** 9   **47.** 2   **49.** −1   **51.** 22   **53.** $-\dfrac{3}{2}$ **55.** −7

57. The ordered pairs are being graphed in reverse order.

59. No. There are ordered pairs with the same first coordinates and different second coordinates.

SECTION 8.2   *pages 319–322*

1. 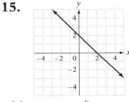 **3.** 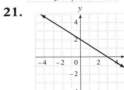 **5.** 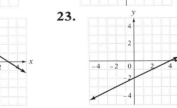 **7.**

9. 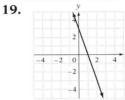 **11.** 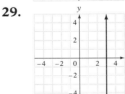 **13.** 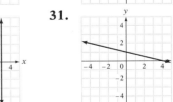 **15.**

17. 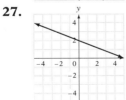 **19.** **21.** **23.**

25. 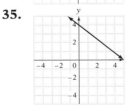 **27.** **29.** **31.**

33. 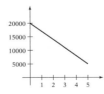 **35.** **37.** The value of the truck after 3 years is $10,400

39. The cost of designing a 3500-ft² home is $9775.   **41.** increase, 3 units, 3 unit

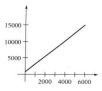

43. They are parallel lin

SECTION 8.3   *pages 329–332*

1. The x-intercept is $(3, 0)$ and the y-intercept is $(0, -3)$.
3. The x-intercept is $(2, 0)$, and the y-intercept is $(0, -6)$.
5. The x-intercept is $(10, 0)$, and the y-intercept is $(0, -2)$.
7. The x-intercept is $(-4, 0)$, and the y-intercept is $(0, 12)$.
9. The x-intercept is $(0, 0)$, and the y-intercept is $(0, 0)$.
11. The x-intercept is $(6, 0)$, and the y-intercept is $(0, 3)$.

13. 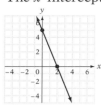 **15.** 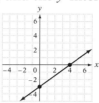 **17.** 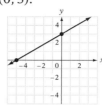 **19.** The slope is -2.

21. The slope is $\dfrac{1}{3}$.   **23.** The slope is $-\dfrac{5}{2}$.   **25.** The slope is $-\dfrac{1}{2}$.   **27.** The slope is -1.

29. The slope is undefined.   **31.** The line has zero slope.   **33.** The slope is $-\dfrac{1}{3}$.

35. The line has zero slope.

37.   **39.** 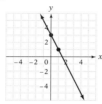 **41.** 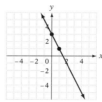 **43.**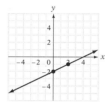

45. 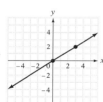 **47.** 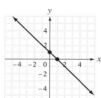 **49.** 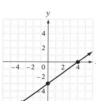 **51.**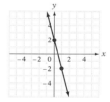

53. $m = 0.5$ For each additional foot a diver descends below the surface of the water, the pressure on the diver increases by 0.5 pound per square inch.

55. $m = -5000$ The median price of a home was decreasing \$5000 per month.   **57.** yes

59. $6\% = 0.06 = \dfrac{6}{100}$. A 6% grade means that the average slope of the road is $\dfrac{6}{100}$.

SECTION 8.4   *pages 337–340*

1. The equation of the line is $y = 2x + 2$.   **3.** The equation of the line is $y = -3x - 1$.

5. The equation of the line is $y = \dfrac{1}{3}x$.   **7.** The equation of the line is $y = \dfrac{3}{4}x - 5$.

9. The equation of the line is $y = -\dfrac{3}{5}x$.   **11.** The equation of the line is $y = \dfrac{1}{4}x + \dfrac{5}{2}$.

13. The equation of the line is $y = 2x - 3$.   **15.** The equation of the line is $y = -2x - 3$.

17. The equation of the line is $y = \dfrac{2}{3}x$.   **19.** The equation of the line is $y = \dfrac{1}{2}x + 2$.

21. The equation of the line is $y = -\dfrac{3}{4}x - 2$.   **23.** The equation of the line is $y = \dfrac{3}{4}x + \dfrac{5}{2}$.

25. The tread of the tire decreases 0.2 mm for each 1000 mi driven.

27. The revenue is increasing \$1.6 billion per year.   **29.** No, the third point is not on the line.

31. Yes, the third point is on the line.   **33.** The slope of the line is $-\dfrac{3}{2}$.   **35.** $n = -5$

37. The equation of the line is $y = -\dfrac{2}{3}x + \dfrac{5}{3}$.   **39.** To prevent division by zero.

CHAPTER REVIEW   *pages 343–344*

1. 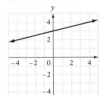   **b.** The abscissa of point A is -2.   **2.**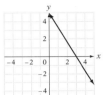
                          c. The ordinate of point B is -2.

3. Yes   **4.** $f(-1) = -1$

5. $\{(55, 95)\ (57, 101)\ (53, 94)\ (57, 98)\ (60, 100)\ (61, 105)\ (58, 97)\ (54, 95)\}$ No, the relation is not a function.

6. 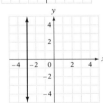   **7.**    **8.**   **9.**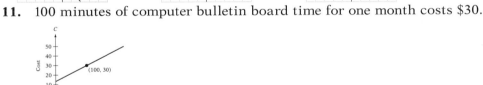

10.   **11.** 100 minutes of computer bulletin board time for one month costs $30.

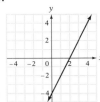

12. The x-intercept is $(8, 0)$, and the y-intercept is $(0, -12)$.   **13.** The slope is $\dfrac{7}{11}$.

14. The line has zero slope.

15. 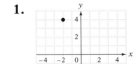   **16.** 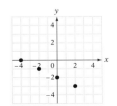   **17.**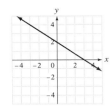

18. The equation of the line is $y = \dfrac{2}{3}x + 6$.   **19.** The equation of the line is $y = -\dfrac{5}{2}x + 16$.

20. The equation of the line is $y = -\dfrac{2}{3}x + \dfrac{11}{3}$.   **21.** The equation of the line is $y = -\dfrac{8}{3}x + $

22. As the age of the machine increases by one year, the down time per month increases 10.4 hours.

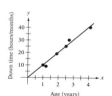

CHAPTER TEST   *pages 345–346*

1. The ordered-pair solution is $(3, -3)$.   [8.1B]

2. Yes, $y = \frac{1}{2}x - 3$ define y as a function of x.   [8.1C]   **3.**   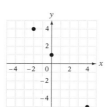   [8.1B]

4. $t(2) = 6$ [8.1D]   **5.** $f(-1) = 3$ [8.1D]

6. $[(3.5, 25), (4.0, 30), (5.2, 45), (5.0, 38), (4.0, 42), (6.3, 12), (5.4, 34)]$ No, the relation is not a function.   [8.1C]

7. 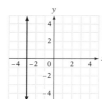   [8.2A]   **8.**   [8.2A]   **9.** 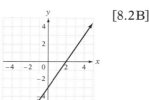   [8.2B]

10.   [8.2B]   **11.** 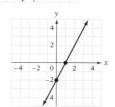   [8.3C]   **12.**   [8.3C]

13. After 1 s, the speed of the ball is 96 ft/s.   [8.2C]   **14.** The tuition is increasing \$1500 per year [8.3B].

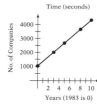

15. The number of mutual fund companies is increasing by 330 each year.   [8.4C]

16. The x-intercept is $(2, 0)$, and the y-intercept is $(0, -3)$.   [8.3A]

17. The x-intercept is $(-2, 0)$, and the y-intercept is $(0, 1)$.   [8.3A]   **18.** The slope is 2.   [8.3B]

19. The line has zero slope.   [8.3B]   **20.** The slope is undefined.   [8.3B]

21. The slope is $-\frac{2}{3}$.   [8.3C]   **22.** The equation of the line is $y = 3x - 1$.   [8.4A]

23. The equation of the line is $y = \frac{2}{3}x + 3$.   [8.4A]

24. The equation of the line is $y = -\frac{5}{8}x - \frac{7}{8}$.   [8.4B]

5. The equation of the line is $y = -\frac{2}{7}x - \frac{4}{7}$.   [8.4B]

ᴍULATIVE REVIEW   *pages 347–348*

ᴾ [1.5B]   **2.** $-\frac{5}{8}$ [2.1A]   **3.** $f(-2) = \frac{2}{3}$ [8.1D]   **4.** $x = \frac{3}{2}$ [3.2A]   **5.** $x = \frac{19}{18}$ [3.3B]

ᴬ]   **7.** $-32x^8y^7$ [5.2A]   **8.** $-3x^2$ [5.4A]   **9.** $x + 3$ [5.5B]

10. $5(x + 2)(x + 1)$ [6.2B]   **11.** $(a + 2)(x + y)$ [6.1A]

12. The solutions are 4 and -2.   [6.5A]   **13.** $\dfrac{x^3(x + 3)}{y(x + 2)}$ [7.1B]   **14.** $\dfrac{3}{x + 8}$ [7.3A]

15. $x = 2$ [7.5A]   **16.** $y = \dfrac{4}{5}x - 3$ [7.7A]   **17.** The ordered-pair solution is $(-2, -5)$ [8.1B]

18. The line has zero slope.   [8.3B]   **19.** The equation of the line is $y = \dfrac{1}{2}x - 2$.   [8.4A]

20. The equation of the line is $y = -3x + 2$.   [8.4A]
21. The equation of the line is $y = 2x + 2$.   [8.4A]

22. The equation of the line is $y = \dfrac{2}{3}x - 3$.   [8.4A]   **23.** The sale price is $62.30.   [4.1B]

24. The measure of the first angle is $46°$, the measure of the second angle is $43°$, and the measure of the third angle is $91°$.   [4.5B]   **25.** The value of the home is $110,000.   [7.6B]

26. It would take $3\dfrac{3}{4}$ h with both the electrician and the apprentice working.   [7.8A]

27. 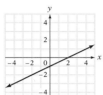   [8.2A]   **28.** 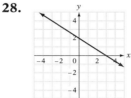   [8.3C]

ANSWERS to Chapter 9 Odd-Numbered Exercises

SECTION 9.1   *pages 355–358*

1. Yes, $(2, 3)$ is a solution of the system of equations.
3. Yes, $(1, -2)$ is a solution of the system of equations.
5. No, $(4, 3)$ is not a solution of the system of equations.
7. No, $(-1, 3)$ is not a solution of the system of equations.
9. No, $(0, 0)$ is not a solution of the system of equations.
11. Yes, $(2, -3)$ is a solution of the system of equations.
13. Yes, $(5, 2)$ is a solution of the system of equations.
15. Yes, $(-2, -3)$ is a solution of the system of equations.
17. No, $(0, -3)$ is not a solution of the system of equations.
19. 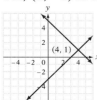   **21.** 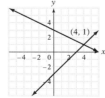   **23.**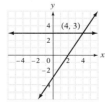

25. 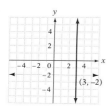   **27.** 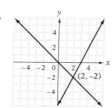   **29.**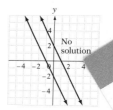

31.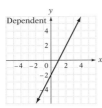

33.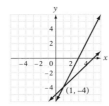

35.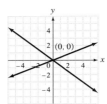

37.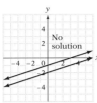

39. a. sometimes true   **b.** true   **c.** true   **d.** true
41. The graphs are parallel.   **43.** Answers will vary.
45. The solution of this system of equations is the date at which the Lipper index equals the S&P index and the return performance they both recorded.

SECTION 9.2   *pages 363–364*

1. The solution is $(2, 1)$.   **3.** The solution is $(4, 1)$.   **5.** The solution is $(-1, 1)$.
7. The system of equations is inconsistent and has no solution.
9. The system of equations is inconsistent and has no solution.   **11.** The solution is $\left(-\dfrac{3}{4}, -\dfrac{3}{4}\right)$.
13. The solution is $(1, 1)$.   **15.** The solution is $(2, 0)$.   **17.** The solution is $(1, -2)$.
19. The solution is $(0, 0)$.
21. The system of equations is dependent. The solutions are the ordered pairs that satisfy the equation $2x - y = 2$.   **23.** The solution is $(-4, -2)$.
25. The solution is $(10, 31)$.   **27.** The solution is $(3, -10)$.   **29.** The solution is $(-22, -5)$.
31. $k = 2$ **33.** $k = 2$ **35.** Answers will vary.
37. The assertion is not correct. The solution is $(0, 2)$.

SECTION 9.3   *pages 369–370*

1. The solution is $(5, -1)$.   **3.** The solution is $(1, 3)$.   **5.** The solution is $(1, 1)$.
7. The solution is $(3, -2)$.
9. The system is dependent. The solutions are the ordered pairs that satisfy the equation $2x - y = 1$.   **11.** The solution is $(3, 1)$.
13. The system is dependent. The solutions are the ordered pairs that satisfy the equation $2x - 3y = 1$.
15. The solution is $\left(-\dfrac{13}{17}, -\dfrac{24}{17}\right)$.   **17.** The solution is $(2, 0)$.   **19.** The solution is $(0, 0)$.
21. The solution is $(5, -2)$.   **23.** The solution is $\left(\dfrac{32}{19}, -\dfrac{9}{19}\right)$.   **25.** The solution is $(3, 4)$.
27. The solution is $(1, -1)$.
 9. The system is dependent. The solutions are the ordered pairs that satisfy the equation $5x + 15y = 20$.
The solution is $(3, 1)$.   **33.** The solution is $(-1, 2)$.   **35.** The solution is $(1, 1)$.
Answers will vary.   **39.** $A = 3, B = -1$
All real numbers, $k \neq 1$,   **b.** All real numbers, $k \neq \dfrac{3}{2}$,   **c.** All real numbers, $k \neq 4$.

SECTION 9.4   *pages 375–376*

1. The rate of the plane is 400 mph. The rate of the wind is 50 mph.
3. The rate of the boat is 7 mph. The rate of the current is 3 mph.
5. The rate of the plane is 125 mph. The rate of the wind is 25 mph.
7. The rate of the plane is 105 mph. The rate of the wind is 15 mph.
9. The word processing program costs $245. The spreadsheet program costs $325.
11. The dividend per share of the oil company stock is $0.25. The dividend per share of the movie company stock is $0.45.
13. There would have been 3 touchdowns and 4 field goals.
15. The measures of the two angles are 55° and 125°.   17. It is impossible to earn $600 in interest.
19. There are seven different combinations of nickels and dimes that could be in the bank.

CHAPTER REVIEW   *pages 379–380*

1. Yes, $(-1, -3)$ is a solution of the system.   2. No, $(-2, 0)$ is not a solution of the system.
3. 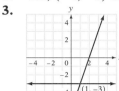   4. 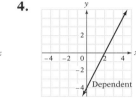   5.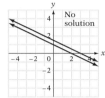

6. The solution is $(-1, 1)$.   7. The solution is $(1, 6)$.   8. The solution is $(1, 6)$.
9. The solution is $(-1, -3)$.   10. The system of equations is inconsistent and has no solution.
11. The system of equations is dependent. The solutions are the ordered pairs that satisfy the equation $4x + 3y = 12$.   12. The solution is $(-3, 1)$.   13. The solution is $(1, -5)$.
14. The solution is $\left(-\frac{5}{6}, \frac{1}{2}\right)$.   15. The system of equations is inconsistent and has no solution.
16. The system of equations is dependent. The solutions are the ordered pairs that satisfy the equation $3x + y = -2$.   17. The solution is $(-1, -2)$.   18. The solution is $\left(\frac{2}{3}, -\frac{1}{6}\right)$.
19. The rate of the canoeist in still water is 8 mph. The rate of the current is 2 mph.
20. The rate of the flight crew in calm air is 125 km/h. The rate of the wind is 15 km/h.
21. The rate of the plane in calm air is 105 mph. The rate of the wind is 15 mph.
22. The rate of the sculling team in calm water is 9 mph. The rate of the current is 3 mph.
23. There are 130 advertisements requiring 25¢ postage and 60 advertisements requiring 45¢ postage.
24. There are 350 bushels of lentils and 200 bushels of corn in the mixture.
25. There were 1300 shares at $6 per share and 200 shares at $25 per share purchased.

CHAPTER TEST   *pages 381–382*

1. Yes, $(-2, 3)$ is a solution of the system.   [9.1A]   2. Yes, $(1, -3)$ is a solution of the system.   [9.1A]
3. 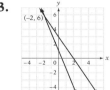   [9.1A]   4. The solution is $(3, 1)$.   [9.2A]   5. The solution is $(1, -1)$.   [9.2A]

6. The solution is $(2, -1)$.   [9.2A]   7. The solution is $\left(\frac{22}{7}, -\frac{5}{7}\right)$.   [9.2A]
8. The system is inconsistent and has no solution.   [9.2A]   9. The solution is $(2, 1)$.
10. The solution is $\left(\frac{1}{2}, -1\right)$.   [9.3A]
11. The system of equations is dependent. The solutions are the ordered pairs equation $x + 2y = 8$.   [9.3A]   12. The solution is $(2, -1)$.   [9.3A]
13. The solution is $(1, -2)$.   [9.3A]
14. The rate of the plane in calm air is 100 mph. The rate of the wind is 20 mph.
15. The price of a reserved-seat ticket is $10. The price of a general-admission tick

CUMULATIVE REVIEW   *pages 383–384*

1. $\frac{3}{2}$ [2.1A]   **2.** $x = -\frac{3}{2}$ [3.1C]   **3.** $f(2) = 7$ [3.3B]   **4.** $-6a^3 + 13a^2 - 9a + 2$ [5.3B]

5. $-2x^5y^2$ [5.4A]   **6.** $2b - 1 + \frac{1}{2b - 3}$ [5.5B]   **7.** $-\frac{4y}{x^3}$ [5.4A]

8. $4y^2(xy - 4)(xy + 4)$ [6.4B]   **9.** The solutions are 4 and -1.   [6.5A]   **10.** $x - 2$ [7.1C]

11. $\frac{x^2 + 2}{(x - 1)(x + 2)}$ [7.3B]   **12.** $\frac{x - 3}{x + 1}$ [7.4A]   **13.** $x = -\frac{1}{5}$ [7.5A]   **14.** $r = \frac{A - P}{Pt}$ [7.7A]

15. The x-intercept is $(6, 0)$, and the y-intercept is $(0, -4)$.   [8.3A]   **16.** The slope is $-\frac{7}{5}$.   [8.3B]

17. The equation of the line is $y = -\frac{3}{2}x$.   [8.4A]

18. Yes, $(2, 0)$ is a solution of the system of equations.   [9.1A]   **19.** The solution is $(-6, 1)$.   [9.2A]
20. The solution is $(4, -3)$.   [9.3A]
21. The amount invested at 9.6% is $3750. The amount invested at 7.2% is $5000.   [4.2A]
22. The rate of the freight train is 48 mph. The rate of the passenger train is 56 mph. [4.4A]
23. A side of the original square measures 8 in.   [6.5B]
24. The rate of the wind is 30 mph.   [7.8B]
25.   [8.2B]   **26.**   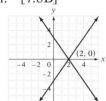   [9.1A]

27. The rate of the boat in calm water is 14 mph.   [9.4A]
28. 35.3% of the mixture is sugar.   [4.3B]

ANSWERS to Chapter 10 Odd-Numbered Exercises

SECTION 10.1   *pages 391–392*

1. $\{16, 17, 18, 19, 20, 21\}$ **3.** $\{9, 11, 13, 15, 17\}$ **5.** $\{b, c\}$ **7.** $A \cup B = \{3, 4, 5, 6\}$
9. $A \cup B = \{-10, -9, -8, 8, 9, 10\}$ **11.** $A \cup B = \{a, b, c, d, e, f\}$ **13.** $A \cup B = \{1, 3, 7, 9, 11, 13\}$
15. $A \cap B = \{4, 5\}$ **17.** $A \cap B = \varnothing$ **19.** $A \cap B = \{c, d, e\}$
21. $\{x \mid x > -5, x \text{ is a negative integer}\}$ **23.** $\{x \mid x > 30, x \in \text{integers}\}$
25. $\{x \mid x > 5, x \in \text{even integers}\}$ **27.** $\{x \mid x > 8, x \in \text{real numbers}\}$
29.   **31.**

33.   **35.**

37.   **39.** Answers will vary.

41. a. never true   **b.** always true   **c.** always true   **43. a.** yes   **b.** yes

SECTION 10.2   *pages 397–400*

1. $x < 2$ **3.** $x > 3$

5. $n \geq 3$ **7.** $x \leq 4$

9. $x \geq -1$ **11.** $y \geq -9$ **13.** $x < 12$ **15.** $x \geq 5$ **17.** $x < -11$

19. $x \leq 10$ **21.** $x \geq -6$ **23.** $x > 2$ **25.** $d < -\frac{1}{6}$ **27.** $x \geq -\frac{31}{24}$ **29.** $x < \frac{5}{8}$

31. $x < \dfrac{5}{4}$ **33.** $x > \dfrac{5}{24}$ **35.** $x < -3.8$ **37.** $x \leq -1.2$ **39.** $x < 5.6$ **41.** $x > -1.48$

43. $x < 4$ **45.** $y \geq 3$ **47.** $x \leq 1$

49. $x < -1$ **51.** $b < -4$ **53.** $y \leq 0$

55. $x > \dfrac{2}{7}$ **57.** $x \leq -\dfrac{5}{2}$ **59.** $x < 16$ **61.** $x \geq 16$ **63.** $x \geq -14$ **65.** $x \leq 21$

67. $x > 0$ **69.** $x \leq -\dfrac{12}{7}$ **71.** $x > \dfrac{2}{3}$ **73.** $x \leq \dfrac{2}{3}$ **75.** $x \leq 2.3$ **77.** $x < -3.2$

79. $x \leq 5$ **81.** $x < -5.4$

83. The possible diameters of a major league baseball will be in the range 2.87 in. $\leq d \leq$ 2.95 in.

85. A person with an annual income tax liability of \$3500 must pay $\geq$ \$3150.

87. A patient with a cholesterol level of 275 units should reduce his or her cholesterol level by $\geq$ 55 units.

89. The student must receive $\geq$ 78 on the fifth test to earn a B grade.

91. A dollar amount $>$ \$3125 will make the commission offer more attractive

93. $\{c \,|\, c > 0, c \in \text{real numbers}\}$ **95.** $\{c \,|\, c \in \text{real numbers}\}$ **97.** $\{c \,|\, c > 0, c \in \text{real numbers}\}$

SECTION 10.3 *pages 403–404*

1. $x < 4$ **3.** $x < -4$ **5.** $x \geq 1$ **7.** $x < 5$ **9.** $x < 0$ **11.** $x < 20$ **13.** $y \leq \dfrac{5}{2}$

15. $x < \dfrac{25}{11}$ **17.** $x > 11$ **19.** $n \leq \dfrac{11}{18}$ **21.** $x \geq 6$

23. The agent expects to sell at most \$20,000.

25. To exceed the \$10 fee, a person must use the service more than 60 min.

27. The maximum amount that can be added is 8 oz.

29. The distance to the ski area is greater than 76 mi.

31. a. never true **b.** sometimes true **c.** always true **d.** never true **33.** $d = \{1, 2\}$

35. $\{-1, 0, 1\}$ **37.** no solution

SECTION 10.4 *pages 407–408*

1. 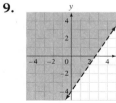 **3.** 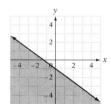 **5.** 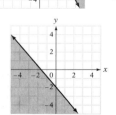 **7.**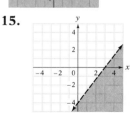

9. 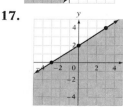 **11.** 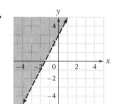 **13.** 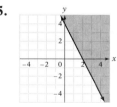 **15.**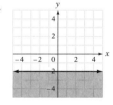

17. 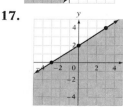 **19.** 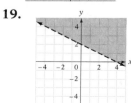 **21.** $y \geq 2x + 2$ **23.** $y \geq 2$

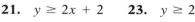

CHAPTER REVIEW   *pages 411–412*

1. $A = \{1, 3, 5, 7\}$ **2.** $A \cup B = \{2, 4, 6, 8, 10\}$ **3.** $A \cap B = \varnothing$ **4.** $A \cap B = \{1, 5, 9\}$
5. $\{x \mid x > -8, x \text{ is an odd integer}\}$ **6.** $\{x \mid x > 3, x \in \text{real numbers}\}$
7.   **8.** 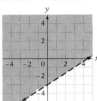   **9.**
10. $x > 2$ **11.** $x > 18$ **12.** $x \geq -3$ **13.** $x < -\dfrac{8}{9}$
14. $x > 31$; the smallest integer that will satisfy the inequality is 32.
15. The lowest score the student can receive is 72.   **16.** $x \geq -4$ **17.** $x \geq 4$ **18.** $x \geq 4$
19. $x > -18$ **20.** $x < \dfrac{1}{2}$ **21.** The minimum length is 24 ft.
22. Five or more residents make Florist B the more economical florist.
23.   **24.**   **25.**

CHAPTER TEST   *pages 413–414*

1. $A = \{4, 6, 8\}$ [10.1A]   **2.** $A \cap B = \{12\}$ [10.1A]
3. $\{x \mid x < 50, x \text{ is a positive integer}\}$ [10.1B]   **4.** $\{x \mid x > -23, x \in \text{real numbers}\}$ [10.1B]
5. [10.1C]   **6.** [10.1C]
7. $x < -3$ [10.2A]   **8.** $x > \dfrac{1}{8}$ [10.2A]
9. $x \geq 3$ [10.2B]   **10.** $x \geq -\dfrac{40}{3}$ [10.2B]
11. The child must grow at least 5 in. [10.2C]
12. The allowable diameters are between 0.0389 in. and 0.0395 in. [10.2C]   **13.** $x < -1$ [10.3A]
14. $x \geq -4$ [10.3A]   **15.** $x \leq -3$ [10.3A]   **16.** $x < -\dfrac{22}{7}$ [10.3A]
17. The maximum width is 11 ft. [10.3B]   **18.** The broker processed less than $75,000. [10.3B]
19. [10.4A]   **20.** [10.4A]

CUMULATIVE REVIEW   *pages 415–416*

. $40a - 28$ [2.2D]   **2.** $x = \dfrac{1}{8}$ [3.2A]   **3.** $x = 4$ [3.3B]   **4.** $-12a^7b^4$ [5.2B]

$-\dfrac{1}{b^4}$ [5.4A]   **6.** $4x - 2 - \dfrac{4}{4x - 1}$ [5.5B]   **7.** $f(-1) = 0$ [8.1D]

8. $3a^2(3x-1)(3x+1)$ [6.4B]   **9.** $\dfrac{1}{x+2}$ [7.1C]   **10.** $\dfrac{18a}{(2a-3)(a+3)}$ [7.3B]

11. $y=-\dfrac{5}{9}$ [7.5A]   **12.** $C=S+Rt$ [7.7A]   **13.** The slope is $-\dfrac{7}{3}$. [8.3B]

14. The equation of the line is $y=-\dfrac{3}{2}x-\dfrac{3}{2}$. [8.4A]   **15.** The solution is $(4,1)$. [9.2A]

16. The solution is $(1,-4)$. [9.3A]   **17.** $A\cup B=\{-10,-2,0,1,2\}$ [10.1A]

18. $\{x\mid x<48,\,x\in \text{real numbers}\}$ [10.1B]   **19.** 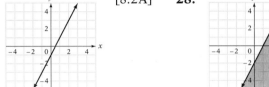 [10.1C]

20. $x>-2$ [10.2B]   **21.** $x<-15$ [10.2B]   **22.** $x>2$ [10.3A]

23. All integers less than or equal to -26 will satisfy the inequality. [10.2C]
24. The maximum number of miles is 359. [10.3B]   **25.** There are 5000 fish in the lake. [7.6B]
26. The measure of the first angle is 65°, the measure of the second angle is 35°, and the measure of the third angle is 80°. [3.5C]   **27.** [8.2A]   **28.** [10.4A]

ANSWERS to Chapter 11 Odd-Numbered Exercises

SECTION 11.1   *pages 423–424*

1. 4   **3.** 7   **5.** $4\sqrt{2}$ **7.** $2\sqrt{2}$ **9.** $18\sqrt{2}$ **11.** $10\sqrt{10}$ **13.** $\sqrt{15}$ **15.** $\sqrt{29}$
17. $-54\sqrt{2}$ **19.** $3\sqrt{5}$ **21.** 0   **23.** $48\sqrt{2}$ **25.** 15.492   **27.** 16.971   **29.** 16
31. 16.583   **33.** 15.652   **35.** 18.762   **37.** x^3 **39.** $y^7\sqrt{y}$ **41.** a^{10} **43.** x^2y^2 **45.** $2x^2$
47. $2x\sqrt{6}$ **49.** $2x^2\sqrt{15x}$ **51.** $7a^2b^4$ **53.** $3x^2y^3\sqrt{2xy}$ **55.** $2x^5y^3\sqrt{10xy}$ **57.** $4a^4b^5\sqrt{5a}$
59. $8ab\sqrt{b}$ **61.** x^3y **63.** $8a^2b^3\sqrt{5b}$ **65.** $6x^2y^3\sqrt{3y}$ **67.** $4x^3y\sqrt{2y}$
69. $5(a+4)$ or $5a+20$ **71.** $2(x+2)^2$ or $2x^2+8x+8$ **73.** $x+2$ **75.** $y+1$
77. no, $\sqrt{4+9}\neq\sqrt{4}+\sqrt{9}$ **79.** Answers will vary.   **81.** $\{3,4,5,6,7,8,9\}$
83. No. The $\sqrt{18}$ still contains a perfect-square factor. $6\sqrt{2}$ **85. a.** 2.9623   **b.** 2.0305

SECTION 11.2   *pages 427–428*

1. $3\sqrt{2}$ **3.** $-\sqrt{7}$ **5.** $-11\sqrt{11}$ **7.** $10\sqrt{x}$ **9.** $-2\sqrt{y}$ **11.** $-11\sqrt{3b}$ **13.** $2x\sqrt{2}$
15. $-3a\sqrt{3a}$ **17.** $-5\sqrt{xy}$ **19.** $8\sqrt{5}$ **21.** $8\sqrt{2}$ **23.** $15\sqrt{2}-10\sqrt{3}$ **25.** $\sqrt{x}$
27. $-12x\sqrt{3}$ **29.** $2xy\sqrt{x}-3xy\sqrt{y}$ **31.** $-9x\sqrt{3x}$ **33.** $-13y^2\sqrt{2y}$ **35.** $4a^2b^2\sqrt{ab}$
37. $7\sqrt{2}$ **39.** $6\sqrt{x}$ **41.** $-3\sqrt{y}$ **43.** $-45\sqrt{2}$ **45.** $13\sqrt{3}-12\sqrt{5}$ **47.** $32\sqrt{3}-3\sqrt{11}$
49. $6\sqrt{x}$ **51.** $-34\sqrt{3x}$ **53.** $10a\sqrt{3b}+10a\sqrt{5b}$ **55.** $-2xy\sqrt{3}$ **57.** $2ab\sqrt{ab}$ **59.** $5\sqrt{2}$
61. Answers will vary.   **63.** $-11\sqrt{2x+y}$ **65.** $13a\sqrt{a+1}$

SECTION 11.3   *pages 433–434*

1. 5   **3.** 6   **5.** x **7.** x^3y^2 **9.** $3ab^6\sqrt{2a}$ **11.** $12a^4b\sqrt{b}$ **13.** $2-\sqrt{6}$ **15.** $x-\sqrt{xy}$
17. $5\sqrt{2}-\sqrt{5x}$ **19.** $4-2\sqrt{10}$ **21.** $x-6\sqrt{x}+9$ **23.** $3a-3\sqrt{ab}$ **25.** $10abc$
27. $-2+2\sqrt{5}$ **29.** $16+10\sqrt{2}$ **31.** $6x+10\sqrt{x}-4$ **33.** $15x-22y\sqrt{x}+8y^2$
37. 4   **39.** 7   **41.** 3   **43.** $x\sqrt{5}$ **45.** $\dfrac{a^2}{7}$ **47.** $\dfrac{y\sqrt{3}}{3}$ **49.** $\sqrt{2y}$ **51.** $-$
53. $\dfrac{15-3\sqrt{5}}{20}$ **55.** $-\dfrac{7\sqrt{2}+49}{47}$ **57.** $-\dfrac{42-26\sqrt{3}}{11}$ **59.** $-\dfrac{20+7\sqrt{3}}{23}$ **61.** $\dfrac{6+5\sqrt{x}+}{4-x}$

63. $\dfrac{a - 5\sqrt{a} + 4}{2a - 2}$ **65.** $\dfrac{x + \sqrt{xy}}{x - y}$ **67.** Answers will vary.
69. a. true   **b.** true   **c.** false $x + 2\sqrt{x} + 1$ **d.** true

SECTION 11.4   *pages 439–440*

1. The solution is 25.   **3.** The solution is 144.   **5.** The solution is 5.
7. The equation has no solution.   **9.** The solution is -1.   **11.** The solution is 1.
13. The solution is 15.   **15.** The solution is $\dfrac{7}{3}$.   **17.** The solution is 45.

19. The equation has no solution.   **21.** The solution is 6.   **23.** The solution is $\dfrac{7}{5}$.

25. The pitcher's mound is more than halfway between home plate and second base.
27. The periscope would have to be 24 ft above the surface of the water.
29. The height of the TV screen is 19.8 in.   **31.** The length of the pendulum is 1.83 ft.
33. 3   **35.** 8   **37. a.** The perimeter of the triangle is 12 cm^2.
    b. The area of the triangle is 24 cm^2.
39. No, the Pythagorean Theorem is true only for right triangles.
41. The irrigation system can irrigate 3283.81 ft^2.

CHAPTER REVIEW   *pages 443–444*

1. 12   **2.** $20\sqrt{3}$ **3.** $-6\sqrt{30}$ **4.** $20\sqrt{10}$ **5.** $9a^2\sqrt{2ab}$ **6.** $2y^4\sqrt{6}$ **7.** $36x^8y^5\sqrt{3xy}$
8. $26\sqrt{3x}$ **9.** $7x^2y\sqrt{15xy}$ **10.** $18a\sqrt{5b} + 5a\sqrt{b}$ **11.** $-6x^3y^2\sqrt{2y}$ **12.** 3
13. $3a\sqrt{2} + 2a\sqrt{3}$ **14.** $8y + 10\sqrt{5y} - 15$ **15.** $7x^2y^4$ **16.** $\dfrac{16\sqrt{a}}{a}$ **17.** $\dfrac{8\sqrt{x} + 24}{x - 9}$
18. $-x\sqrt{3} - x\sqrt{5}$ **19.** The solution is 20.   **20.** The equation has no solution.

21. The equation has no solution.   **22.** The solution is $\dfrac{3}{5}$.

23. The explorer would weigh 144 lb on the surface of the earth.
24. The depth of the water is 100 ft.   **25.** The radius of the sharpest corner is 25 ft.

CHAPTER TEST   *pages 445–446*

1. $3\sqrt{5}$ [11.1A]   **2.** $5\sqrt{3}$ [11.1A]   **3.** $11x^4y$ [11.1B]   **4.** $6x^3y\sqrt{2x}$ [11.1B]
5. $4a^2b^5\sqrt{2ab}$ [11.1B]   **6.** $-5\sqrt{2}$ [11.2A]   **7.** $21\sqrt{2y} - 12\sqrt{2x}$ [11.2A]
8. $-2xy\sqrt{3xy} - 3xy\sqrt{xy}$ [11.2A]   **9.** $4x^2y^2\sqrt{5y}$ [11.3A]   **10.** $6x^2y\sqrt{y}$ [11.3A]
11. $a - \sqrt{ab}$ [11.3A]   **12.** $y + 2\sqrt{y} - 15$ [11.3A]   **13.** 9 [11.3B]   **14.** $7ab\sqrt{a}$ [11.3B]
15. $\sqrt{3} + 1$ [11.3B]   **16.** $\dfrac{17 - 8\sqrt{5}}{31}$ [11.3B]   **17.** The solution is 25. [11.4A]

18. The equation has no solution. [11.4A]   **19.** The larger integer is 51. [11.4B]
20. The length of the pendulum is 7.30 ft. [11.4B]

CUMULATIVE REVIEW   *pages 447–448*

1. $-\dfrac{1}{12}$ [1.5B]   **2.** $2x + 18$ [2.2D]   **3.** $x = \dfrac{1}{13}$ [3.3B]   **4.** $6x^5y^5$ [5.2A]

5. $-2b^2 + 1 - \dfrac{1}{3b^2}$ [5.5A]   **6.** $f(-3) = 1$ [8.1D]   **7.** $2a(a - 5)(a - 3)$ [6.3B]

8. $\dfrac{1}{4(x + 1)}$ [7.1B]   **9.** $\dfrac{x + 3}{x - 3}$ [7.3B]   **10.** $x = \dfrac{5}{3}$ [7.5A]

11. The equation of the line is $y = \dfrac{1}{2}x - 2$. [8.4A]   **12.** The solution is $(1, 1)$. [9.2A]

13. The solution is $(3, -2)$. [9.3A]   **14.** $x \le -\dfrac{9}{2}$ [10.3A]   **15.** $6\sqrt{3}$ [11.1A]

16. $-4\sqrt{2}$ [11.2A]   **17.** $4ab\sqrt{2ab} - 5ab\sqrt{ab}$ [11.2A]   **18.** $14a^5b^2\sqrt{2a}$ [11.3A]
͘. $3\sqrt{2} - x\sqrt{3}$ [11.3A]   **20.** 8 [11.3B]   **21.** $-6 - 3\sqrt{5}$ [11.3B]
    The solution is 6. [11.4A]   **23.** The book costs $24.50. [4.1A]

24. 56 oz of pure water must be added.   [4.3B]   **25.** The numbers are 13 and 8.   [6.5B]
26. Working alone, it would take the small pipe 48 h to fill the tank.   [7.8A]
27. The solution is $(2, -1)$.   [9.1B]   **28.**   [10.4A]

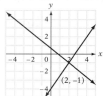   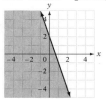

29. The smaller integer is 40.   [11.4B]   **30.** The building is 400 ft high.   [11.4B]

ANSWERS to Chapter 12 Odd-Numbered Exercises

SECTION 12.1   *pages 455–456*

1. The solutions are -5 and 3.   **3.** The solutions are 1 and 3.

5. The solutions are -1 and -2.   **7.** The solution is 3.   **9.** The solutions are 0 and $-\dfrac{2}{3}$.

11. The solutions are -2 and 5.   **13.** The solutions are $\dfrac{2}{3}$ and 1.

15. The solutions are -3 and $\dfrac{1}{3}$.   **17.** The solution is $\dfrac{2}{3}$.   **19.** The solutions are $-\dfrac{1}{2}$ and $\dfrac{2}{3}$.

21. The solution is $\dfrac{1}{2}$.   **23.** The solutions are -3 and 3.   **25.** The solutions are $-\dfrac{1}{2}$ and $\dfrac{1}{2}$.

27. The solutions are -3 and 5.   **29.** The solutions are 1 and 5.
31. The solutions are -6 and 6.   **33.** The solutions are -1 and 1.

35. The solutions are $-\dfrac{7}{2}$ and $\dfrac{7}{2}$.   **37.** The solutions are $-\dfrac{2}{3}$ and $\dfrac{2}{3}$.

39. The solutions are $-\dfrac{3}{4}$ and $\dfrac{3}{4}$.   **41.** The equation has no real number solution.

43. The solutions are $-2\sqrt{6}$ and $2\sqrt{6}$.   **45.** The solutions are -5 and 7.

47. The solutions are -7 and -3.   **49.** The solutions are $-\dfrac{1}{3}$ and $\dfrac{7}{3}$.

51. The solutions are $-\dfrac{2}{7}$ and $-\dfrac{12}{7}$.   **53.** The solutions are $4 + 2\sqrt{5}$ and $4 - 2\sqrt{5}$.

55. The equation has no real number solution.   **57.** The solutions are $-\dfrac{3}{4} + 2\sqrt{3}$ and $-\dfrac{3}{4} - 2\sqrt{3}$.

59. The solutions are $\sqrt{2}$ and $-\sqrt{2}$.   **61.** The solutions are 0 and $\dfrac{b}{a}$.

63. The solutions are 0 and 1.   **65.** The annual percentage rate is 8%.   **67.** Yes

SECTION 12.2   *pages 461–462*

1. The solutions are 1 and -3.   **3.** The solutions are 8 and -2.   **5.** The solution is 2.
7. The quadratic equation has no real number solution.   **9.** The solutions are -1 and -4.
11. The solutions are -8 and 1.   **13.** The solutions are $-2 + \sqrt{3}$ and $-2 - \sqrt{3}$.
15. The solutions are $-3 + \sqrt{14}$ and $-3 - \sqrt{14}$.   **17.** The solutions are $1 + \sqrt{2}$ and $1 - \sqrt{2}$.
19. The solutions are $\dfrac{-3 + \sqrt{13}}{2}$ and $\dfrac{-3 - \sqrt{13}}{2}$.   **21.** The solutions are 2 and 1.

23. The solutions are $\dfrac{-1 + \sqrt{13}}{2}$ and $\dfrac{-1 - \sqrt{13}}{2}$.   **25.** The solutions are $-5 + 4\sqrt{2}$ and $-5 - 4\sqrt{2}$.

27. The solutions are $\dfrac{3 + \sqrt{29}}{2}$ and $\dfrac{3 - \sqrt{29}}{2}$.   **29.** The solutions are $\dfrac{1 + \sqrt{17}}{2}$ and $\dfrac{1 - \sqrt{17}}{2}$.

31. The quadratic equation has no real number solution.   **33.** The solutions are 1 and $\dfrac{?}{?}$

35. The solutions are -3 and $\frac{1}{2}$.   **37.** The solutions are 2 and $\frac{3}{2}$.

39. The solutions are 1 and $-\frac{1}{2}$.   **41.** The solutions are -2 and $\frac{1}{3}$.

43. The solutions are -2 and $-\frac{2}{3}$.   **45.** The solutions are $\frac{1}{2}$ and $-\frac{3}{2}$.

47. The solutions are $\frac{1}{3}$ and $-\frac{3}{2}$.   **49.** The solutions are $-\frac{1}{2}$ and $\frac{4}{3}$.

51. The solutions are $\frac{1+\sqrt{2}}{2}$ and $\frac{1-\sqrt{2}}{2}$.   **53.** The solutions are approximately -4.193 and 1.193.

55. The solutions are approximately 2.766 and -1.266.
57. The solutions are approximately -1.652 and 0.152.   **59.** Answers will vary.
61. The solution is 7.   **63.** The solutions are $4 + \sqrt{7}$ and $4 - \sqrt{7}$.
65. The solutions are $6 + \sqrt{58}$ and $6 - \sqrt{58}$.   **67.** The solutions are 1 and 3.
69. The ball will hit the basket 2.13 s after it is released.

SECTION 12.3   *pages 465–466*

1. The solutions are -1 and 5.   **3.** The solutions are -3 and 5.   **5.** The solutions are -1 and 3.

7. The solutions are -5 and 1.   **9.** The solutions are $-\frac{1}{2}$ and 1.

11. The quadratic equation has no real number solution.   **13.** The solutions are 0 and 1.

15. The solutions are $-\frac{3}{2}$ and $\frac{3}{2}$.   **17.** The solutions are $-\frac{5}{2}$ and $\frac{3}{2}$.

19. The quadratic equation has no real number solution.
21. The solutions are $1 + \sqrt{6}$ and $1 - \sqrt{6}$.   **23.** The solutions are $-3 + \sqrt{10}$ and $-3 - \sqrt{10}$.

25. The solutions are $2 + \sqrt{13}$ and $2 - \sqrt{13}$.   **27.** The solutions are $\frac{-1+\sqrt{2}}{3}$ and $\frac{-1-\sqrt{2}}{3}$.

29. The solution is $-\frac{1}{2}$.   **31.** The quadratic equation has no real number solution.

33. The solutions are $\frac{-4+\sqrt{5}}{2}$ and $\frac{-4-\sqrt{5}}{2}$.   **35.** The solutions are $\frac{1+2\sqrt{3}}{2}$ and $\frac{1-2\sqrt{3}}{2}$.

37. The solutions are $\frac{-5+\sqrt{2}}{3}$ and $\frac{-5-\sqrt{2}}{3}$.   **39.** The solutions are approximately 5.690 and -3.690.

41. The solutions are approximately 7.690 and -1.690.
43. The solutions are approximately 4.590 and -1.090.
45. The solutions are approximately -2.118 and 0.118.
47. The solutions are approximately 1.105 and -0.905.
49. $a = 0$ causes the denominator of the quadratic formula to be equal to 0, and division by 0 is
    undefined.   **51. a.** false   **b.** false   **c.** false   **d.** true
53. The solutions are -3 and -4.   **55.** No real number solution.
57. The solutions are -1 and 11.   **59.** After 2 h the distance between the planes is 1000 mi.

SECTION 12.4   *pages 469–470*

1. 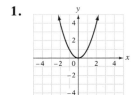   **3.** 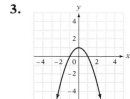   **5.** 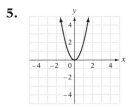   **7.**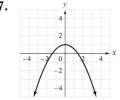

9. 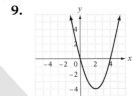   **11.** 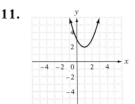   **13.** 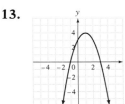   **15.**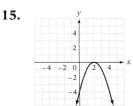

17. **19.** down   **21.** down   **23.** $y = 3x^2 + 6x + 5$

25. The x-intercepts are $(-3, 0)$ and $(3, 0)$.   **27.** The x-intercepts are $\left(-\dfrac{1}{2}, 0\right)$ and $(1, 0)$.

29. $g(2) = -3$ **31.** $P(-3) = 38$

SECTION 12.5   *pages 473–474*

1. The length is 6 ft, and the width is 4 ft.
3. The length of the pool is 100 ft, and the width is 50 ft.   **5.** The integers are 7 and 9.
7. The two integers are 5 and 7.   **9.** The integer is either 0 or 2.
11. The three odd integers are 9, 11, 13 or $-13, -11, -9$.
13. The height of the triangle is 10 ft. The length of the base is 4 ft.
15. The first computer alone would take 35 min. The second computer working alone would take 14 min.
17. The first engine alone would take 12 h. The second engine working alone would take 6 h.
19. The rate of the plane is 100 mph.
21. The four consecutive integers are 3, 4, 5, and 6 or $-6, -5, -4$, and -3.
23. The radius of the large pizza is 7 in.

CHAPTER REVIEW   *pages 477–478*

1. The solutions are $\dfrac{4}{3}$ and $-\dfrac{7}{2}$.   **2.** The solutions are 2 and $\dfrac{5}{12}$.

3. The solutions are -7 and -10.   **4.** The solutions are $-\dfrac{1}{3}$ and $-\dfrac{1}{2}$.

5. The solutions are $-\dfrac{5}{7}$ and $\dfrac{5}{7}$.   **6.** The equation has no real number solution.

7. The solutions are -9 and 1.   **8.** The solutions are $-2 - 2\sqrt{6}$ and $-2 + 2\sqrt{6}$.
9. The solutions are -1 and 2.   **10.** The solutions are -6 and 4.

11. The solutions are -4 and $\dfrac{3}{2}$.   **12.** The solutions are $2 - \sqrt{3}$ and $2 + \sqrt{3}$.

13. The equation has no real number solution.   **14.** The solutions are $\dfrac{-4 - \sqrt{23}}{2}$ and $\dfrac{-4 + \sqrt{23}}{2}$.

15. The solutions are -6 and 1.   **16.** The solutions are 1 and $\dfrac{3}{2}$.

17. The equation has no real number solution.   **18.** The solutions are $\dfrac{3 - \sqrt{29}}{2}$ and $\dfrac{3 + \sqrt{29}}{2}$.

19. The solutions are -2 and $-\dfrac{1}{2}$.   **20.**   **21.**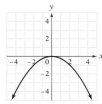

22. 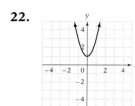   **23.** 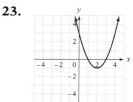   **24.**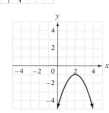

25. The rate of the hawk in calm air is 75 mph.

CHAPTER TEST   *pages 479–480*

1. The solutions are 6 and −1. [12.1A]   2. The solutions are −4 and $\frac{5}{3}$. [12.1A]

3. The solutions are 0 and 10. [12.1B]   4. The solutions are $-4 + 2\sqrt{5}$ and $-4 - 2\sqrt{5}$. [12.1B]

5. The solutions are $-2 + 2\sqrt{5}$ and $-2 - 2\sqrt{5}$. [12.2A]

6. The solutions are $\dfrac{-3 + \sqrt{41}}{2}$ and $\dfrac{-3 - \sqrt{41}}{2}$. [12.2A]

7. The solutions are $\dfrac{3 + \sqrt{7}}{2}$ and $\dfrac{3 - \sqrt{7}}{2}$. [12.2A]

8. The solutions are $\dfrac{-4 + \sqrt{22}}{2}$ and $\dfrac{-4 - \sqrt{22}}{2}$. [12.2A]

9. The solutions are $-2 + \sqrt{2}$ and $-2 - \sqrt{2}$. [12.3A]

10. The solutions are $\dfrac{3 + \sqrt{33}}{2}$ and $\dfrac{3 - \sqrt{33}}{2}$. [12.3A]   11. The solutions are $-\dfrac{1}{2}$ and 3. [12.3A]

12. The solutions are $\dfrac{1 + \sqrt{13}}{6}$ and $\dfrac{1 - \sqrt{13}}{6}$. [12.3A]   13. 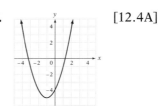   [12.4A]

14. The length is 8 ft. The width is 5 ft. [12.5A]
15. The rate of the boat in calm water is 11 mph. [12.5A]

CUMULATIVE REVIEW   *pages 481–482*

1. $-28x + 27$ [2.2D]   2. $x = \dfrac{3}{2}$ [3.1C]   3. $x = 3$ [3.3B]   4. $-12a^8b^4$ [5.2B]

5. $x + 2 - \dfrac{4}{x - 2}$ [5.5B]   6. $x(3x - 4)(x + 2)$ [6.3B]   7. $\dfrac{9x^2(x - 2)^2}{(2x - 3)^2}$ [7.1C]

8. $\dfrac{x + 2}{2(x + 1)}$ [7.3B]   9. $\dfrac{x - 4}{2x + 5}$ [7.4A]

10. The x-intercept is $(3, 0)$, and the y-intercept is $(0, -4)$. [8.3A]

11. The equation of the line is $y = -\dfrac{4}{3}x - 2$. [8.4A]   12. The solution is $(2, 1)$. [9.2A]

13. The solution is $(2, -2)$. [9.3A]   14. $x > \dfrac{1}{9}$ [10.3A]   15. $a - 2$ [11.3A]

16. $6ab\sqrt{a}$ [11.3B]   17. $\dfrac{-6 + 5\sqrt{3}}{13}$ [11.3B]   18. The solution is 5. [11.4A]

19. The solutions are $\dfrac{5}{2}$ and $\dfrac{1}{3}$. [12.1A]   20. The solutions are $5 + 3\sqrt{2}$ and $5 - 3\sqrt{2}$. [12.1B]

21. The solutions are $\dfrac{-7 + \sqrt{13}}{6}$ and $\dfrac{-7 - \sqrt{13}}{6}$. [12.2A]   22. The solutions are 2 and $-\dfrac{1}{2}$. [12.3A]

23. The selling price of the mixture is $2.25/lb. [4.3A]
24. 250 additional shares are required. [7.6B]
25. The rate of the plane in still air is 200 mph. The rate of the wind is 40 mph. [9.4A]
26. The student must receive a score of 77 or above. [10.2C]   27. The integer is −5 or 5. [12.5A]
28. The rate for the last 8 miles was 4 mph. [12.5A]
29. 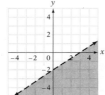   [10.4A]   30. 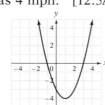   [12.4A]

FINAL EXAM   *pages 483–486*

1. -3 [1.1B]   **2.** -6 [1.2B]   **3.** -256 [1.5A]   **4.** -11 [1.5B]   **5.** $-\frac{15}{2}$ [2.1A]

6. $9x + 6y$ [2.2A]   **7.** $6z$ [2.2B]   **8.** $16x - 52$ [2.2D]   **9.** $x = -50$ [3.1C]

10. $x = -3$ [3.3B]   **11.** 12.5% [1.4B]   **12.** 15.2 [3.1D]   **13.** $-3x^2 - 3x + 8$ [5.1B]

14. $81x^4y^{12}$ [5.2B]   **15.** $6x^3 + 7x^2 - 7x - 6$ [5.3B]   **16.** $-\frac{x^4y}{2}$ [5.4A]

17. $\frac{3x}{y} - 4x^2 - \frac{5}{x}$ [5.4B]   **18.** $5x - 12 + \frac{23}{x+2}$ [5.5B]   **19.** $\frac{4y^6}{x^6}$ [5.4A]   **20.** $\frac{3}{4}$ [8.1D]

21. $(x - 6)(x + 1)$ [6.2A]   **22.** $(3x + 2)(2x - 3)$ [6.3A]   **23.** $4x(2x - 1)(x - 3)$ [6.3B]

24. $(5x - 4)(5x + 4)$ [6.4A]   **25.** $2(a + 3)(4 - x)$ [6.1B]   **26.** $3y(5 - 2x)(5 + 2x)$ [6.4B]

27. The solutions are $\frac{1}{2}$ and 3. [6.5A]   **28.** $\frac{2(x + 1)}{x - 1}$ [7.1B]   **29.** $\frac{-3x^2 + x - 25}{(2x - 5)(x + 3)}$ [7.3B]

30. $\frac{x^2 - 2x}{x - 1}$ [7.4A]   **31.** The solution is 2. [7.5A]   **32.** $a = b$ [7.7A]

33. The slope is $\frac{2}{3}$. [8.3B]   **34.** The equation of the line is $y = -\frac{2}{3}x - 2$. [8.4A]

35. The solution is $(6, 17)$. [9.2A]   **36.** The solution is $(2, -1)$. [9.3A]   **37.** $x \le -3$ [10.2A]

38. $y \ge \frac{5}{2}$ [10.3A]   **39.** $7x^3$ [11.1B]   **40.** $38\sqrt{3a}$ [11.2A]   **41.** $\sqrt{15} + 2\sqrt{3}$ [11.3B]

42. The solution is 2. [11.4A]   **43.** The solutions are -1 and $\frac{4}{3}$. [12.1A]

44. The solutions are $\frac{1 + \sqrt{5}}{4}$ and $\frac{1 - \sqrt{5}}{4}$. [12.3A]   **45.** $2x + 3(x - 2)$; $5x - 6$ [2.3C]

46. The original value is $3000. [4.1B]   **47.** The markup rate is 65% [4.1A]

48. An additional $6000 must be invested. [4.2A]   **49.** The mixture costs $3 per pound. [4.3A]

50. The percent concentration of the acid in the mixture is 36%. [4.3B]

51. In the first hour, the plane flew 215 km. [4.4A]

52. The measures of the angles are 50°, 60°, and 70°. [4.5B]

53. 4 or -4 [12.5A]   **54.** The length is 10 m. The width is 5 m. [6.5B]

55. Sixteen oz of dye are required for 120 oz of base paint. [7.6B]

56. Working together, it would take 0.6 h to prepare the dinner. [7.8A]

57. The rate of the boat in calm water is 15 mph. The rate of the current is 5 mph. [9.4A]

58. The rate of the wind is 25 mph. [12.5A]

59. 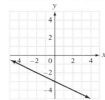   [8.3C]   **60.**   [12.4A]

Appendix: Guidelines for Using Graphing Calculators

TEXAS INSTRUMENTS TI-82

To evaluate an expression:

a. Press the $\boxed{Y=}$ key. A menu showing Y_1 through Y_2 will be displayed vertically with the cursor on Y_1. Press $\boxed{\text{CLEAR}}$, if necessary, to delete an unwanted expression.

b. Input the expression to be evaluated. For example, to input the expression $-3a^2b - 4c$, use the following keystrokes:

$\boxed{Y=}$ $\boxed{\text{CLEAR}}$ $\boxed{(-)}$ $\boxed{3}$ $\boxed{\text{ALPHA}}$ A $\boxed{\wedge}$ $\boxed{2}$ $\boxed{\text{ALPHA}}$ B $\boxed{-}$ $\boxed{4}$ $\boxed{\text{ALPHA}}$ C $\boxed{\text{2nd}}$ $\boxed{\text{QUIT}}$ (Note the difference between the keys for a *negative* sign $\boxed{(-)}$ and a *minus* sign $\boxed{-}$.)

c. Store the value of each variable that will be used in the expression. For example, to evaluate the expression above when $a = 3$, $b = -2$, and $c = -4$, use the following keystrokes:

$\boxed{3}$ $\boxed{\text{STO}\triangleright}$ $\boxed{\text{ALPHA}}$ A $\boxed{\text{ENTER}}$; $\boxed{(-)}$ $\boxed{2}$ $\boxed{\text{STO}\triangleright}$ $\boxed{\text{ALPHA}}$ B $\boxed{\text{ENTER}}$; $\boxed{(-)}$ $\boxed{4}$ $\boxed{\text{STO}\triangleright}$ $\boxed{\text{ALPHA}}$ C $\boxed{\text{ENTER}}$. These steps store the values of each variable.

d. Press $\boxed{\text{2nd}}$ $Y\text{-}VARS$ $\boxed{1}$ $\boxed{1}$ $\boxed{\text{ENTER}}$. The value of the expression, Y_1, for the given values is displayed, in the case, 70.

To graph a function:

a. Press the $\boxed{Y=}$ key. A menu showing Y_1 through Y_8 will be displayed vertically with the cursor on Y_1. Press $\boxed{\text{CLEAR}}$, if necessary, to delete an unwanted expression.

b. Input the expression for each function that is to be graphed. Press $\boxed{\text{X,T,}\theta}$ to input x. For example, to input $f(x) = x^3 + 2x^2 - 5x - 6$, use the following keystrokes:

$\boxed{Y=}$ $\boxed{\text{X,T,}\theta}$ $\boxed{\wedge}$ $\boxed{3}$ $\boxed{+}$ $\boxed{\text{X,T,}\theta}$ $\boxed{\wedge}$ $\boxed{2}$ $\boxed{-}$ $\boxed{5}$ $\boxed{\text{X,T,}\theta}$ $\boxed{-}$ $\boxed{6}$

c. Set the domain and range by pressing $\boxed{\text{WINDOW}}$. Enter the values for the minimum x-value (Xmin), maximum x-value (Xmax), distance between tic marks on the x-axis (Xscl), minimum y-value (Ymin), and maximum y-value (Ymax), and the distance between tic marks on the y-axis (Yscl). Press $\boxed{\text{2nd}}$ $\boxed{\text{QUIT}}$. Now press $\boxed{\text{GRAPH}}$. For the graph shown at the left, Xmin = -10, Xmax = 10, Xscl = 1, Ymin = -10, Ymax = 10, and Yscl = 1. This is called the standard viewing rectangle. Pressing $\boxed{\text{ZOOM}}$ $\boxed{6}$ is a quick way to set the calculator to the standard viewing rectangle. Note: This will also immediately graph the function in that window.

d. The equal sign has a black rectangle around it. This indicates that the function is *active* and will be graphed when the $\boxed{\text{GRAPH}}$ key is pressed. A function is deactivated by using the arrow keys. Move the cursor over the equal sign and press $\boxed{\text{ENTER}}$. When the cursor is moved to the right, the black rectangle will not be present and that equation will not be graphed.

e. Graphing some radical equations requires special care. To graph $f(x) = x^{2/3} - 1$, enter the following keystrokes:
$\boxed{Y=}$ $\boxed{\text{X,T,}\theta}$ $\boxed{\wedge}$ $\boxed{(}$ $\boxed{1}$ $\boxed{\div}$ $\boxed{3}$ $\boxed{)}$ $\boxed{)}$ $\boxed{\wedge}$ $\boxed{2}$ $\boxed{-}$ $\boxed{1}$. You are entering $x^{2/3}$ as $(x^{1/3})^2$. The graph is shown at the left.

To display the x-coordinate of rectangular coordinates as integers:

a. Set the range variables as follows: Xmin: -47, Xmax: 47, Xscl: 10, Ymin: -32, Ymax: 32, Yscl: 10.

b. Graph the function and use the trace key. Press $\boxed{\text{TRACE}}$ and then move the cursor with the $\boxed{\triangleleft}$ and $\boxed{\triangleright}$ keys. The values of x and $y = f(x)$ displayed on the bottom of the screen are the coordinates of a point on the graph.

To display the x-coordinate of rectangular coordinates in tenths:

a. Set the range variables as follows: $\boxed{\text{ZOOM}}$ $\boxed{4}$

b. Graph the function and use the trace key. Press $\boxed{\text{TRACE}}$ and then move the cursor with the $\boxed{\triangleleft}$ and $\boxed{\triangleright}$ keys. The values of x and $y = f(x)$ displayed on the bottom of the screen are the coordinates of a point on the graph.

To evaluate a function for a given value of x or to produce a pair of rectangular coordinates:

a. Input the equation; for example, input $Y_1 = 2X^3 - 3X + 2$

b. Press $\boxed{\text{2ND}}$ $\boxed{\text{QUIT}}$.

c. Input a value for x; for example, input 3 using the following keystrokes: 3 $\boxed{\text{STO}\triangleright}$ $\boxed{\text{X,T,}\theta}$ $\boxed{\text{ENTER}}$.

d. Press $\boxed{\text{2ND}}$ Y-$VARS$ $\boxed{1}$ $\boxed{1}$ $\boxed{\text{ENTER}}$. The value of the function, Y_1, for the given x-value is shown, in this case, 47. An ordered pair of the function is $(3, 47)$.

e. Repeat steps (c)–(d) to produce as many pairs as desired.

f. Repeat steps (c)–(d) to produce as many pairs as desired. The **TABLE** feature of **TI**-82 can also be used to determine ordered pairs.

To graph the sum (or difference) of two functions:

a. Input the first function as Y_1 and the second function as Y_2.

b. After the second function is entered, press $\boxed{\text{ENTER}}$ to move the cursor to Y_3.

c. Press $\boxed{\text{2ND}}$ Y-$VARS$ $\boxed{1}$ $\boxed{1}$ $\boxed{+}$ $\boxed{\text{2ND}}$ Y-$VARS$ $\boxed{2}$. Enter $\boxed{-}$ instead of $\boxed{+}$ to graph $Y_1 - Y_2$.

d. Graph Y_3, If you want to show only the graph of the sum (or difference), be sure that the black rectangle around the equal sign has been removed from Y_1 and Y_2.

ZOOM FEATURES OF THE TI-82

To zoom in or out on a graph:

a. There are two methods of using zoom. The first method uses the built-in features of the calculator. Move the cursor to a point on the graph that is of interest. Press $\boxed{\text{ZOOM}}$. The zoom menu will appear. Press $\boxed{2}$ $\boxed{\text{ENTER}}$ to zoom in on the graph by the amount shown under the Set Factors menu. The center of the new graph is the location at which you placed the cursor. Press $\boxed{3}$ $\boxed{\text{ENTER}}$ to zoom out on the graph by the amount shown under the Set Factors menu. (The Set Factors menu is accessed by pressing $\boxed{\text{ZOOM}}$ $\boxed{\triangleright}$ $\boxed{4}$).

b. The second method uses the ZBOX option under the ZOOM menu. To use this method, press $\boxed{\text{ZOOM}}$ $\boxed{1}$. A cursor will appear on the graph. Use the arrow keys to move the cursor to a portion of the graph that is of interest. Press $\boxed{\text{ENTER}}$. Now use the arrow keys to draw a box around the portion of the graph you wish to see. Press $\boxed{\text{ENTER}}$. The portion of the graph defined by the box will be drawn.

c. Pressing $\boxed{\text{ZOOM}}$ $\boxed{6}$ resets the RANGE to the standard 10×10 viewing window.

SOLVING EQUATIONS WITH THE TI-82

This discussion is based on the fact that the solution of an equation can be related to the x-intercepts of a graph. For instance, the solutions of the equation $x^3 = x + 1$ are the x-intercepts of the graph of $f(x) = x^3 - x - 1$, which are the zeros of f.

To solve $x^3 = x + 1$, rewrite the equation with all terms on one side. The equation is now $x^3 - x - 1 = 0$. Think of this equation as $Y_1 = x^3 - x - 1$. The x-intercepts of the graph of Y_1 are the solutions of the equation $x^3 = x + 1$.

a. Enter the function into Y_1.

b. Graph the equation. You may need to adjust the viewing window so that the x-intercepts are visible.

c. Press $\boxed{\text{2nd}}$ $CALC$ $\boxed{2}$.

d. Move the cursor to a point on the curve that is to the left of the x-intercept. Press $\boxed{\text{ENTER}}$.

e. Move the cursor to a point on the curve that is to the right of the x-intercept. Press $\boxed{\text{ENTER}}$.

f. Press $\boxed{\text{ENTER}}$.

g. The root is shown as the x-coordinate on the bottom of the screen, in this case, 1.324718. The **solve** feature under the MATH menu can also be used to find solutions of equations.

SOLVING SYSTEMS OF EQUATIONS IN TWO VARIABLES WITH THE TI-82

To solve a system of equations:

a. Solve each equation for y.

b. Enter the first equation as Y_1. For instance, let $Y_1 = x^2 - 1$.

c. Enter the second equation as Y_2. For instance, let $Y_2 = 1 - x$.

d. Graph both equations. (Note: The points of intersection must appear on the screen. It may be necessary to adjust the viewing window so that the point (or points) of intersection are displayed.)

e. Press $\boxed{\text{2nd}}$ $CALC$ $\boxed{5}$.

f. Move the cursor to the left of the first point of intersection. Press $\boxed{\text{ENTER}}$.

g. Move the cursor to the right of the first point of intersection. Press $\boxed{\text{ENTER}}$.

h. Press $\boxed{\text{ENTER}}$.

i. The first point of intersection is $(-2, 3)$.

j. Repeat this procedure for each point of intersection.

FINDING MINIMUM OR MAXIMUM VALUES OF A FUNCTION WITH THE TI-82

a. Enter the function into Y_1. The equation $Y_1 = x^3 - x - 1$ is used here.

b. Graph the equation. You may need to adjust the viewing window so that the x-intercepts are visible.

c. Press 2nd *CALC* 3 to determine a minimum value or press 2nd *CALC* 4 to determine a maximum value.

d. Move the cursor to a point on the curve that is to the left of the minimum (maximum). Press ENTER.

e. Move the cursor to a point on the curve that is to the right of the minimum (maximum). Press ENTER.

f. Press ENTER.

g. The minimum (maximum) is shown as the y-coordinate on the bottom of the screen, in this case the minimum $= -1.3849$ and the maximum $= -0.6150998$.

CASIO *fx-7700G*

To evaluate an expression:

a. For example, to input the expression $-3a^2b - 4c$, use the following keystrokes: (Note the difference between the keys for a *negative* sign $(-)$ and a *minus* sign $\boxed{-}$. To enter $(-)$, press SHIFT $(-)$).

$(-)$ 3 ALPHA *A* x^y 2 ALPHA *B* $\boxed{-}$ 4 ALPHA *C* SHIFT *F MEM* F1 1 EXE (The number, 1, entered here can be any number from 1 to 6.)

b. Store the value of each variable that will be used in the expression. For example, to evaluate the expression above when $a = 3$, $b = -2$, and $c = -4$, use the following keystrokes:

3 $\rightarrow$ ALPHA *A* EXE; SHIFT $(-)$ 2 $\rightarrow$ ALPHA *B* EXE; SHIFT $(-)$ 4 $\rightarrow$ ALPHA *C* EXE. These steps store the values of each variable.

c. To evaluate the expression, recall the expression from the function menu. Use the following keystrokes:

SHIFT *F MEM* F2 1 EXE. The value of the expression is displayed as 70.

To graph a function:

a. Ensure the calculator is in graphics mode.

Press MODE 1 MODE $+$ MODE SHIFT $+$

b. To graph $f(x) = x^3 + 2x^2 - 5x - 6$ use the following keystrokes:

GRAPH X,θ,T x^y 3 $+$ 2 X,θ,T SHIFT x^2 $-$ 5 X,θ,T $-$ 6 EXE

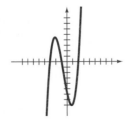

c. Set the domain and range by pressing RANGE. Enter the values for the minimum x-value (Xmin), maximum x-value (Xmax), X-scale (Xscl), minimum y-value (Ymin), maximum y-value (Ymax), and Y-scale (Yscl). Now press EXE. For the graph shown at the left, Xmin $= -10$, Xmax $= 10$, Xscl $= 1$, Ymin $= -10$, Ymax $= 10$, and Yscl $= 1$. Press the RANGE key until you return to the display of the graph. Press EXE.

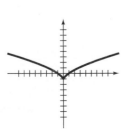

d. Graphing some radical equations requires special care. To graph $f(x) = x^{2/3} - 1$, enter the following keystrokes:

Graph (X,θ,T x^y (1 $\div$ 3)) x^y 2 $-$ 1. You are entering $x^{2/3}$ as $(x^{1/3})^2$. The graph is shown at the left.

e. If you graph a function and then return to the regular screen (for example, by pressing AC), you can view the graph again by pressing G ↔ T.

To display the x-coordinate of rectangular coordinates as integers:

a. Set the range variables as follows: Xmin: -47, Xmax: 47, Xscl: 10, Ymin: -31, Ymax: 31, Yscl: 10.

b. Graph the function and use the trace key. Press $\boxed{\text{F1}}$ and then use the cursor keys to move along the graph. The values of x and $y = f(x)$ displayed on the bottom of the screen are the coordinates of a point on the graph.

To display the x-coordinate of rectangular coordinates in tenths:

a. Set the range variables as follows: $\boxed{\text{RANGE}}$ $\boxed{\text{F1}}$ $\boxed{\text{RANGE}}$ $\boxed{\text{RANGE}}$

b. Graph the function and use the trace key. Press $\boxed{\text{F1}}$ and then use the cursor keys to move along the graph.

To produce a pair of rectangular coordinates:

a. Input the function; for example, input $3X - 4$.

b. Press $\boxed{\text{SHIFT}}$ $\boxed{\text{F}}$ *MEM* $\boxed{\text{F1}}$ $\boxed{\text{1}}$ $\boxed{\text{EXE}}$ $\boxed{\text{AC}}$ to store the function in f_1.

c. Input any value for x; for example, input 3 using the keystroke 3 $\boxed{\rightarrow}$ $\boxed{\text{X,}\theta\text{,T}}$ $\boxed{\text{EXE}}$ to store the value in x.

d. Press $\boxed{\text{SHIFT}}$ $\boxed{\text{F}}$ *MEM* $\boxed{\text{F2}}$ $\boxed{\text{1}}$ $\boxed{\text{EXE}}$ to find the corresponding value for the stored x-value; in this example you should get 5. The point is $(3, 5)$.

e. Repeat steps (c)–(d) to produce as many pairs as desired.

To graph the sum (or difference) of two functions:

a. Input the first function.

b. Press $\boxed{\text{SHIFT}}$ $\boxed{\text{F}}$ *MEM* $\boxed{\text{F1}}$ $\boxed{\text{1}}$ $\boxed{\text{EXE}}$ $\boxed{\text{AC}}$

c. Input the second function.

d. $\boxed{\text{SHIFT}}$ $\boxed{\text{F}}$ *MEM* $\boxed{\text{F1}}$ $\boxed{\text{2}}$ $\boxed{\text{EXE}}$ $\boxed{\text{AC}}$

e. $\boxed{\text{SHIFT}}$ $\boxed{\text{F3}}$ $\boxed{\text{1}}$ $\boxed{+}$ $\boxed{\text{F3}}$ $\boxed{\text{2}}$ $\boxed{\text{EXE}}$. Enter $\boxed{-}$ instead of $\boxed{+}$ to graph $f_1 - f_2$.

ZOOM FEATURES OF THE CASIO fx-7700G

To zoom in or out on a graph:

a. There are two methods of using zoom. The first method uses the built-in features of the calculator. Press $\boxed{\text{SHIFT}}$ $\boxed{\text{F1}}$. This activates the cursor. Use the arrow keys to move the cursor to a portion of the graph that is of interest. Press $\boxed{\text{F2}}$. The zoom menu will appear. Press $\boxed{\text{F3}}$ to zoom in on the graph by the amount shown by FCT (factor). The center of the new graph is the location at which you placed the cursor. Press $\boxed{\text{F4}}$ to zoom out on the graph by the amount shown in FCT.

b. The second method uses the BOX option under the ZOOM menu. If necessary, press $\boxed{\text{G} \leftrightarrow \text{T}}$ to view the graph. Press $\boxed{\text{F2}}$. Press $\boxed{\text{F1}}$ to select BOX. Use the arrow keys to move the cursor to a portion of the graph that is of interest. Press $\boxed{\text{EXE}}$. Now use the arrow keys to draw a box around the portion of the graph you wish to see. Press $\boxed{\text{EXE}}$. The portion of the graph defined by the box will be drawn.

SOLVING EQUATIONS WITH THE CASIO fx-7700G

This discussion is based on the fact that the solution of an equation can be related to the x-intercepts of a graph. For instance, the solutions of the equation $x^2 - x = 6$ are the x-intercepts of the graph of $f(x) = x^2 - x - 6$, which are the zeros of f.

To solve $x^3 = x + 1$, rewrite the equation with all terms on one side. The equation is now $x^3 - x - 1 = 0$. Think of this equation as $Y_1 = x^3 - x - 1$. The x-intercepts of the graph of Y are the solutions of the equation $x^3 = x + 1$.

a. Enter the function.

b. Graph the equation. You may need to adjust the viewing window so that the x-intercepts are visible.

c. Press $\boxed{F1}$ and move the cursor to the approximate intersection.

d. Press $\boxed{F2}$ $\boxed{F3}$.

e. Repeat steps (c) and (d) until you can approximate the x-intercept to the desired degree of accuracy. The root is 1.324718.

SOLVING SYSTEMS OF EQUATIONS IN TWO VARIABLES WITH THE CASIO fx-7700G

To solve a system of equations:

a. Solve each equation for y.

b. Enter the first equation as f_1. For instance, let $f_1 = x^2 - 1$.

c. Enter the second equation as f_2. For instance, let $f_2 = 1 - x$.

d. Graph both equations. (Note: The points of intersection must appear on the screen. It may be necessary to adjust the viewing window so that the point (or points) of intersection are displayed.)

e. Press $\boxed{F1}$ and the move the cursor to the approximate x-intercept.

f. Press $\boxed{F2}$ $\boxed{F3}$.

g. Repeat steps (e) and (f) until you can approximate the point of intersection to the desired degree of accuracy. The first point of intersection is $(-2, 3)$.

FINDING MINIMUM OR MAXIMUM VALUES OF A FUNCTION WITH THE CASIO fx-7700G

To find the minimum (maximum) value of a function:

a. Graph the function of the equation $Y = x^3 - x - 1$.

b. Press $\boxed{F1}$ and then move the cursor to the approximate minimum (maximum).

c. Press $\boxed{F2}$ $\boxed{F3}$.

d. Repeat steps (b) and (c) until you can approximate the minimum (maximum) value to the desired degree of accuracy. The minimum (maximum) is shown as the y-coordinate on the bottom of the screen, in this case the minimum = -1.3849 and the maximum = -0.6150998.

SHARP EL-9300

To evaluate an expression:

a. The *SOLVER* mode of the calculator is used to evaluate expressions. To enter solver mode, press $\boxed{2ndF}$ *SOLVER* $\boxed{CL}$. The expression $-3a^2b - 4c$ must be entered as the equation $-3a^2b - 4c = t$. The letter t can be any letter other than one used in the expression. When entering an expression in

SOLVER mode, the variables appear on the screen in lower case. Use the following keystrokes to input $-3a^2b - 4c = t$:

(−) 3 ALPHA *A* a^b 2 ▷ ALPHA *B* − 4 ALPHA *C* ALPHA = ALPHA *T* ENTER.

b. After you press ENTER, the variables used in the equation will be displayed on the screen. To evaluate the expression for $a = 3$, $b = -2$, and $c = -4$, input each value, pressing ENTER after each number. When the cursor moves to t, press ENTER. A small window will appear. Press ENTER.

c. Pressing ENTER again will allow you to evaluate the expression for new values of a, b, and c. Press ⊞ to return to normal operation.

To graph a function:

a. Press the 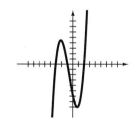 key. The screen will show Y1 = .

b. Input the expression for a function that is to be graphed. Press X/θ/T to enter x. For example, to input $f(x) = x^3 + 2x^2 - 5x - 6$, use the following keystrokes:

⌁ CL X/θ/T a^b 3 ▷ + 2 X/θ/T a^b 2 ▷ − 5 X/θ/T − 6 ⌁

c. Set the domain and range by pressing RANGE . Enter the values for the minimum x-value (Xmin), maximum x-value (Xmax), minimum y-value (Ymin), and maximum y-value (Ymax). Press ⌁ . For the graph shown at the left, enter Xmin = −10, Xmax = 10, Ymin = −10, and Ymax = 10. Press ⌁ .

d. The equal sign has a black rectangle around it. This indicates that the function is *active* and will be graphed when the ⌁ key is pressed. A function is deactivated by using the arrow keys. Move the cursor over the equal sign and press ENTER . When the cursor is moved to the right, the black rectangle will not be present and that equation will not be graphed. The EQTN key is used to display the equations that are currently stored.

e. Graphing some radical equations requires special care. To graph $f(x) = x^{2/3} - 1$, enter the following keystrokes:

⌁ (X/θ/T a^b (1 ÷ 3) ▷) a^b 2 ▷ − 1 ⌁ . You are entering the exponent in the form $(x^{1/3})^2$. The graph is shown at the left.

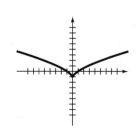

To display the x-coordinate of rectangular coordinates as integers:

a. Set the range variables as follows: Xmin: −47, Xmax: 47, Xscl: 10, Ymin: −31, Ymax: 31, Yscl: 10.

b. Graph the function and use the right and left arrow keys to trace the function. The values of x and $y = f(x)$ displayed on the bottom of the screen are the coordinates of a point on the graph.

To display the x-coordinate of rectangular coordinates in tenths:

a. Set the range variables as follows: Xmin: −4.7, Xmax: 4.7, Xscl: 1, Ymin: −3.1, Ymax: 3.1, Yscl: 1. This is accomplished by the following keystrokes:
RANGE MENU *A* ENTER

b. Graph the function and use the arrow keys to move along the graph of the function. The coordinates are displayed at the bottom of the screen.

To evaluate a function for a given value of x or to produce a pair of rectangular coordinates:

a. Enter *SOLVER* mode; press $\boxed{2ndF}$ *SOLVER* $\boxed{CL}$.

b. Input the expression; for instance, input $x^3 - 4x^2 + 1 = y$. Then press $\boxed{\text{ENTER}}$.

c. Move the cursor to the x variable (if it is not already there). Input any value for x; for example, input 3 and then press $\boxed{\text{ENTER}}$. The cursor will now be over y. Press $\boxed{\text{ENTER}}$ twice to evaluate the function. For this case, the value is -8. An ordered pair of the function is $(3, -8)$.

d. Repeat step (c) to produce as many pairs as desired.

To graph the sum (or difference) of two functions:

a. Input the first function as Y_1 and the second function as Y_2. Press $\boxed{\text{ENTER}}$ to move to Y_3.

b. Press $\boxed{\text{MATH}}$ E $\boxed{1}$ $\boxed{+}$ $\boxed{\text{MATH}}$ E $\boxed{2}$ $\boxed{\sim}$. To graph the difference $Y_1 - Y_2$, replace $\boxed{+}$ with $\boxed{-}$.

c. Graph Y_3. If you want to show only the graph of the sum (or difference), be sure that the black rectangle around the equal sign has been removed from Y_1 and Y_2.

ZOOM FEATURES OF THE SHARP EL-9300

To zoom in or out on a graph:

a. There are two methods of using zoom. The first method uses the built-in features of the calculator. Move the cursor to a point on the graph that is of interest. Press $\boxed{\text{ZOOM}}$. The zoom menu will appear. Press 2 to zoom in on the graph by the amount shown by FACTOR. The center of the new graph is the location at which you placed the cursor. Press 3 to zoom out on the graph by the amount shown in FACTOR.

b. The second method uses the BOX option under the ZOOM menu. To use this method, press $\boxed{\text{ZOOM}}$ 1. A cursor will appear on the graph. Use the arrow keys to move the cursor to a portion of the graph that is of interest. Press $\boxed{\text{ENTER}}$. Use the arrow keys to draw a box around the portion of the graph you wish to see. Press $\boxed{\text{ENTER}}$.

SOLVING EQUATIONS OR SYSTEMS OF EQUATIONS IN TWO VARIABLES WITH THE SHARP EL-9300

a. The x-intercept, y-intercept, and the point of intersection of the two graphs can be determined by using the JUMP command. Graph the functions of interest. Using the arrow keys, place the cursor on the graph of one of the functions. Press $\boxed{2nd}$ *JUMP*; the JUMP menu will appear. Press 1 to jump to the intersection of two graphs, press 4 to jump to the x-intercept, or press 5 to jump to the y-intercept. If there is more than one intercept or intersection, pressing $\boxed{2nd}$ *JUMP* again will allow you to find the remaining points. IMPORTANT: The intersection must be a point in the viewing window.

FINDING MAXIMUM AND MINIMUM VALUES OF A FUNCTION WITH THE SHARP EL-9300

a. The maximum and minimum values of a function can be determined by using the JUMP command. Graph the function of interest. Press $\boxed{\text{2nd}}$ *JUMP*; the JUMP menu will appear. Press 2 to jump to the minimum value or press 3 to jump to the maximum value of the function. If there is more than minimum or maximum, pressing $\boxed{\text{2nd}}$ *JUMP* again will allow you to find the remaining points. IMPORTANT: The minimum or maximum must be a point in the viewing window.

b. For the equation $Y_1 = x^3 - x - 1$, the minimum (maximum) is shown as the *y*-coordinate on the bottom of the screen, in this case the minimum $= -1.3849$ and the maximum $= -0.615099$.

Table of Square Roots

Decimal approximations have been rounded to the nearest thousandth.

Number	Square Root	Number	Square Root	Number	Square Root	Number	Square Root
1	1	51	7.141	101	10.050	151	12.288
2	1.414	52	7.211	102	10.100	152	12.329
3	1.732	53	7.280	103	10.149	153	12.369
4	2	54	7.348	104	10.198	154	12.410
5	2.236	55	7.416	105	10.247	155	12.450
6	2.449	56	7.483	106	10.296	156	12.490
7	2.646	57	7.550	107	10.344	157	12.530
8	2.828	58	7.616	108	10.392	158	12.570
9	3	59	7.681	109	10.440	159	12.610
10	3.162	60	7.746	110	10.488	160	12.649
11	3.317	61	7.810	111	10.536	161	12.689
12	3.464	62	7.874	112	10.583	162	12.728
13	3.606	63	7.937	113	10.630	163	12.767
14	3.742	64	8	114	10.677	164	12.806
15	3.873	65	8.062	115	10.724	165	12.845
16	4	66	8.124	116	10.770	166	12.884
17	4.123	67	8.185	117	10.817	167	12.923
18	4.243	68	8.246	118	10.863	168	12.961
19	4.359	69	8.307	119	10.909	169	13
20	4.472	70	8.367	120	10.954	170	13.038
21	4.583	71	8.426	121	11	171	13.077
22	4.690	72	8.485	122	11.045	172	13.115
23	4.796	73	8.544	123	11.091	173	13.153
24	4.899	74	8.602	124	11.136	174	13.191
25	5	75	8.660	125	11.180	175	13.229
26	5.099	76	8.718	126	11.225	176	13.267
27	5.196	77	8.775	127	11.269	177	13.304
28	5.292	78	8.832	128	11.314	178	13.342
29	5.385	79	8.888	129	11.358	179	13.379
30	5.477	80	8.944	130	11.402	180	13.416
31	5.568	81	9	131	11.446	181	13.454
32	5.657	82	9.055	132	11.489	182	13.491
33	5.745	83	9.110	133	11.533	183	13.528
34	5.831	84	9.165	134	11.576	184	13.565
35	5.916	85	9.220	135	11.619	185	13.601
36	6	86	9.274	136	11.662	186	13.638
37	6.083	87	9.327	137	11.705	187	13.675
38	6.164	88	9.381	138	11.747	188	13.711
39	6.245	89	9.434	139	11.790	189	13.748
40	6.325	90	9.487	140	11.832	190	13.784
41	6.403	91	9.539	141	11.874	191	13.820
42	6.481	92	9.592	142	11.916	192	13.856
43	6.557	93	9.644	143	11.958	193	13.892
44	6.633	94	9.695	144	12	194	13.928
45	6.708	95	9.747	145	12.042	195	13.964
46	6.782	96	9.798	146	12.083	196	14
47	6.856	97	9.849	147	12.124	197	14.036
48	6.928	98	9.899	148	12.166	198	14.071
49	7	99	9.950	149	12.207	199	14.107
50	7.071	100	10	150	12.247	200	14.142

Table of Symbols

$+$	add		$<$	is less than
$-$	subtract		$\leq$	is less than or equal to
$\cdot, \times, (a)(b)$	multiply		$>$	is greater than
$\frac{a}{b}, \div\ a\overline{)b}$	divide		$\geq$	is greater than or equal to
$(\ )$	parentheses, a grouping symbol		(a, b)	an ordered pair whose first component is a and whose second component is b
$[\ ]$	brackets, a grouping symbol			
π	pi, a number approximately equal to $\frac{22}{7}$ or 3.14		$^\circ$	degree (for angles)
			$\sqrt{a}$	the principal square root of a
$-a$	the opposite, or additive inverse, of a		$\varnothing, \{\ \}$	the empty set
$\frac{1}{a}$	the reciprocal, or multiplicative inverse, of a		$\lvert a \rvert$	the absolute value of a
			$\cup$	union of two sets
$=$	is equal to		$\cap$	intersection of two sets
$\approx$	is approximately equal to		$\in$	is an element of (for sets)
$\neq$	is not equal to			

Table of Measurement Abbreviations

U.S. Customary System

Length		Capacity		Weight		Area	
in.	inches	oz	ounces	oz	ounces	in^2	square inches
ft	feet	c	cups	lb	pounds	ft^2	square feet
yd	yards	qt	quarts				
mi	miles	gal	gallons				

Metric System

Length		Capacity		Weight/Mass		Area	
mm	millimeter (0.001 m)	ml	milliliter (0.001 L)	mg	milligram (0.001 g)	cm^2	square centimeters
cm	centimeter (0.01 m)	cl	centiliter (0.01 L)	cg	centigram (0.01 g)	m^2	square meters
dm	decimeter (0.1 m)	dl	deciliter (0.1 L)	dg	decigram (0.1 g)		
m	meter	L	liter	g	gram		
dam	decameter (10 m)	dal	decaliter (10 L)	dag	decagram (10 g)		
hm	hectometer (100 m)	hl	hectoliter (100 L)	hg	hectogram (100 g)		
km	kilometer (1000 m)	kl	kiloliter (1000 L)	kg	kilogram (1000 g)		

Time

h	hours	min	minutes	s	seconds

Glossary

abscissa The first number in an ordered pair. It measures a horizontal distance and is also called the first coordinate. (Sec. 8.1)

absolute value of a number Distance of a number from zero on the number line. (Sec. 1.1)

addition method Method of finding an exact solution of a system of linear equations wherein we use the Addition Property of Equations. (Sec. 9.3)

additive inverses Numbers that are the same distance from zero on the number line, but on opposite sides; also called opposites. (Sec. 1.1)

analytic geometry Geometry in which a coordinate system is used to study the relationships between variables. (Sec. 8.1)

arithmetic mean of values Average determined by calculating the sum of the values and then dividing that result by the number of values. (Sec. 1.3)

axes The two number lines that form a rectangular coordinate system; also called coordinate axes. (Sec. 8.1)

base In exponential notation, the factor that is taken the number of times shown by the exponent. (Sec. 1.5)

binomial A polynomial of two terms. (Sec. 5.1)

clearing denominators Removing denominators from an equation that contains fractions by multiplying each side of the equation by the LCM of the denominators. (Sec. 7.5)

combining like terms Using the Distributive Property to add the coefficients of like variable terms. (Sec. 2.2)

completing the square Adding to a binomial the constant term that makes it a perfect-square trinomial. (Sec. 12.2)

complex fraction A fraction whose numerator or denominator contains one or more fractions. (Sec. 7.4)

conjugates Binomial expressions that differ only in the sign of a term. The expressions $a + b$ and $a - b$ are conjugates. (Sec. 11.3)

consecutive even integers Even integers that follow one another in order. (Sec. 3.4)

consecutive integers Integers that follow one another in order. (Sec. 3.4)

consecutive odd integers Odd integers that follow one another in order. (Sec. 3.4)

constant term A term that includes no variable part; also called a constant. (Sec. 2.1)

coordinate axes The two number lines that form a rectangular coordinate system; also simply called axes. (Sec. 8.1)

coordinates of a point The numbers in the ordered pair that is associated with the point. (Sec. 8.1)

cost The price that a business pays for a product. (Sec. 4.1)

degree of a polynomial in one variable The largest exponent that appears on the variable. (Sec. 5.1)

dependent system of equations A system of equations that has an infinite number of solutions. (Sec. 9.1)

dependent variable In a function, the variable whose value depends on the value of another variable known as the independent variable. (Sec. 8.1)

discount The amount by which a retailer reduces the regular price of a product for a promotional sale. (Sec. 4.1)

discount rate The percent of the regular price that the discount represents. (Sec. 4.1)

domain The set of first coordinates of the ordered pairs in a relation. (Sec. 8.1)

element of a set One of the objects in a set. (Sec. 10.1)

empty set The set that contains no elements; also called the null set. (Sec. 10.1)

equation A statement of the equality of two mathematical expressions. (Sec. 3.1)

equivalent equations Equations that have the same solution. (Sec. 3.1)

evaluating the function Replacing x in $f(x)$ with some value and then simplifying the numerical expression that results. (Sec. 8.1)

evaluating the variable expression   Replacing each variable by its value and then simplifying the resulting numerical expression. (Sec. 2.1)

even integer   An integer that is divisible by 2. (Sec. 3.4)

exponent   In exponential notation, the elevated number that indicates how many times the factor occurs in the multiplication. (Sec. 1.5)

exponential form   The expression 2^5 is in exponential form. Compare *factored form.* (Sec. 1.5)

factor a polynomial   To write the polynomial as a product of other polynomials. (Sec. 6.1)

factor a trinomial of the form $ax^2 + bx + c$ To express the trinomial as the product of two binomials. (Sec. 6.2)

factored form   The expression $2 \cdot 2 \cdot 2 \cdot 2 \cdot 2$ is in factored form. Compare *exponential form.* (Sec. 1.5)

first coordinate   The first number in an ordered pair. It measures a horizontal distance and is also called the abscissa. (Sec. 8.1)

FOIL   A method of finding the product of two binomials; the letters stand for First, Outer, Inner, and Last. (Sec. 5.3)

formula   An equation that states rules about measurements. (Sec. 7.7)

fraction in simplest form   A fraction in which there are no common factors in the numerator and the denominator. (Sec. 1.4)

function   A relation in which no two ordered pairs that have the same first coordinate have different second coordinates. (Sec. 8.1)

functional notation   A function designated by $f(x)$, which is the value of the function at x. (Sec. 8.1)

graph of a relation   The graph of the ordered pairs that belong to the relation. (Sec. 8.1)

graph of an equation in two variables   A graph of the ordered-pair solution of an equation. (Sec. 8.2)

graph of an integer   A heavy dot directly above that number on the number line. (Sec. 1.1)

graph of an ordered pair   The dot drawn at the coordinates of the point in the plane. (Sec. 8.1)

graphing a point in the plane   Placing a dot at the location given by the ordered pair; also called plotting a point in the plane. (Sec. 8.1)

greater than   A number a is greater than another number b, written $a > b$, if a is to the right of b on the number line. (Sec. 1.1)

greater than or equal to   The symbol $\geq$ means "is greater than or equal to." (Sec. 1.1)

greatest common factor   The greatest common factor (GCF) of two or more integers is the greatest integer that is a factor of all the integers. The greatest common factor of two or more monomials is the product of the GCF of the coefficients and the common variable factors. (Sec. 6.1)

half-plane   The solution set of an inequality in two variables. (Sec. 10.4)

hypotenuse   In a right triangle, the side opposite the 90° angle. (Sec. 11.4)

inconsistent system of equations   A system of equations that has no solution. (Sec. 9.1)

independent system of equations   A system of equations that has one solution. (Sec. 9.1)

independent variable   In a function, the variable that varies independently and whose value determines the value of the dependent variable. (Sec. 8.1)

inequality   An expression that contains the symbol $>$, $<$, $\geq$ (is greater than or equal to), or $\leq$ (is less than or equal to). (Sec. 10.1)

integers   The numbers $\ldots, -3, -2, -1, 0, 1, 2, 3, \ldots$. (Sec. 1.1)

intersection of sets A and B   The set that contains the elements that are common to both A and B. (Sec. 10.1)

irrational number   The square root of a number that is not a perfect square. The decimal representation of an irrational number never repeats or terminates and can only be approximated. (Sec. 11.1)

isosceles triangle   A triangle that has two equal angles and two equal sides. (Sec. 4.5)

least common denominator   The smallest number that is a multiple of each denominator in question. (Sec. 1.4)

least common multiple (LCM)   The LCM of two or more numbers is the smallest number that contains the prime factorization of each number. (Sec. 7.2)

less than A number a is less than another number b, written $a < b$, if a is to the left of b on the number line. (Sec. 1.1)

less than or equal to The symbol $\leq$ means "is less than or equal to." (Sec. 1.1)

like terms Terms of a variable expression that have the same variable part. (Sec. 2.2)

line of best fit A line drawn to approximate data that are graphed as points in a coordinate system. (Sec. 8.4)

linear equation in two variables An equation of the form $y = mx + b$, where m and b are constants; also called a linear function. (Sec. 8.2)

linear function An equation of the form $y = mx + b$, where m and b are constants; also called a linear equation in two variables. (Sec. 8.2)

linear model A first-degree equation that is used to describe a relationship between quantities. (Sec. 8.4)

literal equation An equation that contains more than one variable. (Sec. 7.7)

markup The difference between selling price and cost. (Sec. 4.1)

markup rate The percent of the retailer's cost that the markup represents. (Sec. 4.1)

monomial A number, a variable, or a product of numbers and variables; a polynomial of one term. (Sec. 5.1)

multiplicative inverse of a number The reciprocal of a number. (Sec. 2.2)

natural numbers The numbers $1, 2, 3, \ldots$. (Sec. 1.1)

negative integers The integers to the left of zero on the number line. (Sec. 1.1)

negative slope A property of a line that slants downward to the right. (Sec. 8.3)

nonfactorable over the integers A polynomial that does not factor using only integers. (Sec. 6.2)

null set The set that contains no elements; also called the empty set. (Sec. 10.1)

numerical coefficient The number part of a variable term. When the numerical coefficient is 1 or -1, the 1 is usually not written. (Sec. 2.1)

odd integer An integer that is not divisible by 2. (Sec. 3.4)

opposite of a polynomial The polynomial created when the sign of each term of the original polynomial is changed. (Sec. 5.1)

opposites Numbers that are the same distance from zero on the number line, but on opposite sides; also called additive inverses. (Sec. 1.1)

ordered pair Pair of numbers, such as (a, b) that can be used to identify a point in the plane determined by the axes of a rectangular coordinate system. (Sec. 8.1)

ordinate The second number in an ordered pair. It measures a vertical distance and is also called the second coordinate. (Sec. 8.1)

origin The point of intersection of the two coordinate axes that form a rectangular coordinate system. (Sec. 8.1)

percent Parts of 100. (Sec. 1.4)

perfect square The square of an integer. (Sec. 11.1)

perfect-square trinomial A trinomial that is a product of a binomial and itself. (Sec. 6.4)

perimeter The distance around a geometric figure. (Sec. 4.5)

plane Flat surface determined by the intersection of two lines. (Sec. 8.1)

plotting a point in the plane Placing a dot at the location given by the ordered pair; also called graphing a point in the plane. (Sec. 8.1)

point–slope formula If (x_1, y_1) is a point on a line with slope m, then $y - y_1 = m(x - x_1)$. (Sec. 8.4)

polynomial A variable expression in which the terms are monomials. (Sec. 5.1)

positive integers The integers to the right of zero on the number line. (Sec. 1.1)

positive slope A property of a line that slants upward to the right. (Sec. 8.3)

prime polynomial A polynomial that is nonfactorable over the integers. (Sec. 6.2)

principal square root The positive square root of a number. (Sec. 11.1)

proportion An equation that states the equality of two ratios or rates. (Sec. 7.6)

quadrant One of the four regions into which the two axes of a rectangular coordinate system divide the plane. (Sec. 8.1)

quadratic equation An equation of the form $ax^2 + bx + c = 0$, where a, b, and c are constants and a is not equal to zero; also called a second-degree equation. (Sec. 12.1)

quadratic equation in two variables An equation of the form $y = ax^2 + bx + c$, where a is not equal to zero. (Sec. 12.4)

quadratic function A quadratic function is given by $f(x) = ax^2 + bx + c$, where a is not equal to zero. The graph of a quadratic function is a parabola. (Sec. 12.4)

radical The symbol $\sqrt{}$, which is used to indicate the positive, or principal, square root of a number. (Sec. 11.1)

radical equation An equation that contains a variable expression in a radicand. (Sec. 11.4)

radicand In a radical expression, the expression under the radical sign. (Sec. 11.1)

range The set of second coordinates of the ordered pairs in a relation. (Sec. 8.1)

rate The quotient of two quantities that have different units. (Sec. 7.6)

rate of work That part of a task that is completed in one unit of time. (Sec. 7.8)

ratio The quotient of two quantities that have the same unit. (Sec. 7.6)

rational expression A fraction in which the numerator or denominator is a polynomial. (Sec. 7.1)

rational number A number that can be written in the form a/b, where a and b are integers and b is not equal to zero. (Sec. 1.4)

rationalizing the denominator The procedure used to remove a radical from the denominator of a fraction. (Sec. 11.3)

real numbers The rational numbers and the irrational numbers. (Sec. 1.4)

reciprocal Interchanging the numerator and denominator of a rational number yields that number's reciprocal. (Sec. 2.2)

reciprocal of a fraction A fraction with the numerator and denominator interchanged. (Sec. 7.1)

rectangular coordinate system System formed by two number lines, one horizontal and one vertical, that intersect at the zero point of each line. (Sec. 8.1)

relation Any set of ordered pairs. (Sec. 8.1)

repeating decimal Decimal that is formed when dividing the numerator of its fractional counterpart by the denominator results in a decimal part wherein a block of digits repeat infinitely. (Sec. 1.4)

right angle An angle whose measure is 90 degrees. (Sec. 4.5)

right triangle A triangle that includes an angle of 90 degrees. (Sec. 4.5)

roster method Method of writing a set by enclosing a list of the elements in braces. (Sec. 10.1)

scatter diagram A graph of collected data as points in a coordinate system. (Sec. 8.4)

scientific notation Notation in which each number is expressed as the product of two factors, one a number between 1 and 10, and the other a power of ten. (Sec. 5.4)

second coordinate The second number in a ordered pair. It measures a vertical distance and is also called the ordinate. (Sec. 8.1)

second-degree equation An equation of the form $ax^2 + bx + c = 0$, where a, b. and c are constants and a is not equal to zero; also called a quadratic equation. (Sec. 12.1)

selling price The price for which a business sells a product to a customer. (Sec. 4.1)

set A collection of objects. (Sec. 10.1)

set-builder notation A method of designating a set that makes use of a variable and a certain property that only elements of that set possess. (Sec. 10.1)

simplest form A rational expression is in simplest form when the numerator and denominator have no common factors. (Sec. 7.1)

slope of a line A measure of the slant of a line. The symbol for slope is m. The formula for the slope is

$$m = \frac{y_2 - y_1}{x_2 - x_1}$$

where (x_1, y_1) and (x_2, y_2) are the coordinates of two points on the line and $x_1 = x_2$. (Sec. 8.3)

slope–intercept form The slope–intercept form of an equation of a straight line is $y = mx + b$. (Sec. 8.3)

solution of a system of equations in two variables An ordered pair that is a solution of each equation of the system. (Sec. 9.1)

solution of an equation A number that, when substituted for the variable, results in a true equation. (Sec. 3.1)

solution of an equation in two variables An ordered pair whose coordinates make the equation a true statement. (Sec. 8.1)

solution set of an inequality A set of numbers each element of which, when substituted for the variable, results in a true inequality. (Sec. 10.2)

solving an equation Finding a solution of the equation. (Sec. 3.1)

square root A square root of a positive number x is a number a for which $a^2 = x$. (Sec. 11.1)

standard form A quadratic equation is in standard form when the polynomial is in descending order and equal to zero. $ax^2 + bx + c = 0$ is in standard form. (Sec. 6.5)

substitution method Method of finding an exact solution of a system of linear equations wherein we use the Substitution Property of Equality. (Sec. 9.2)

system of equations Equations that are considered together. (Sec. 9.1)

terminating decimal Decimal that is formed when dividing the numerator of its fractional counterpart by the denominator results in a remainder of zero. (Sec. 1.4)

terms of a variable expression The addends of the expression. (Sec. 2.1)

trinomial A polynomial of three terms. (Sec. 5.1)

undefined slope A property of a vertical line. (Sec. 8.3)

uniform motion The motion of a moving object whose speed and direction do not change. (Sec. 4.4)

union of sets A and B The set that contains all elements of A and all elements of B. (Sec. 10.1)

value of the variable The number assigned to the variable. (Sec. 2.1)

variable A letter of the alphabet used to stand for a quantity that is unknown or can change. (Sec. 2.1)

variable expression An expression that contains one or more variables. (Sec. 2.1)

variable term A term composed of a numerical coefficient and a variable part. (Sec. 2.1)

x-coordinate The abscissa in an xy-coordinate system. (Sec. 8.1)

x-intercept The point at which a graph crosses the x-axis. (Sec. 8.3)

xy-coordinate system A rectangular coordinate system in which the horizontal axis is labeled x and the vertical axis is labeled y. (Sec. 8.1)

y-coordinate The ordinate in an xy-coordinate system. (Sec. 8.1)

y-intercept The point at which a graph crosses the y-axis. (Sec. 8.3)

zero slope A property of a horizontal line. (Sec. 8.3)

Index

Abscissa, 301
Absolute value, 4
Absolute value function, 308
Addition
  associative property of, 53
  commutative property of, 54
  of decimal numbers, 25
  of integers, 7–8
  order of operations, 36
  of polynomials, 159
  of radical expressions, 425–426
  of rational expressions, 259–262
  of rational numbers, 24–25
  verbal phrases for, 63
Addition method for solving
    systems of equations, 365–368
Addition Property of Equations,
    82
Addition Property of Inequalities,
    393–394
Addition Property of Zero, 54
Additive inverse, 4, 54
Adjacent angles, 98
Algebraic symbolism, 450
Al-Khwarisimi, 475
Alternate exterior angles,
    114
Alternate interior angles,
    114
Analytic geometry, 301
Angles, 98, 106, 114
  triangles, 146
Application problems
  angles, 98, 106
  area, 170, 173–174, 402
  averages, 18, 20
  break-even point, 105–106
  cost of production, 94
  depreciation, 97, 322, 326
  elevation, 13
  factoring, 231–232, 234–236
  geometry, 145–148
  inequalities, 396, 399–400,
    402–404
  investment and interest,
    129–132, 318
  levers, 102, 105
  linear equations, 318, 322, 336,
    339
  line of best fit, 336
  markup and discount, 125–128
  mixtures, 133–140
  percent, 85–86
  positive and negative numbers,
    10
  pressure, 94

profit and loss, 28, 33
  quadratic equations, 471–474
  radical expressions, 437–440
  rate and proportion, 276–278
  rate-of-wind or -current,
    371–372, 375, 471, 474
  sums of squares, 231, 234
  systems of equations in two
    variables, 371–376
  temperature, 10, 13–14, 18, 20,
    110
  triangles, 437–440, 441
  uniform motion, 141–144, 285,
    286, 289–290
  variable expressions, 66
  work, 283–284, 287–288
Area problems, 170, 173–174, 402
Arithmetic mean, 18
Associative Property of Addition,
    53
Associative Property of
    Multiplication, 55
Average, 18
Axes, 301

Babylonian mathematics, 198, 418
Base, 35
Basic percent equation, 85
Best fit, 336
Binomial(s), 159
  multiplication of, 168–170
  square of, 169, 221, 457
Binomial factors, 201
Break-even analysis, 282
Break-even point, 105–106, 282

Calculators
  graphing linear equations with,
    341
  graphing quadratic functions
    with, 476
  plus/minus (+/−) key, 41
  polynomial evaluation, 237
  solving systems of equations
    with, 377–378
Celsius temperature, 110
Chapter review, 43–45, 73–74,
    117–118, 151–152, 191,
    239–240, 293–294, 343–345,
    379–380, 411–412, 443–444,
    477–478
Chapter summary, 41–42, 72, 116,
    150, 190, 238, 292, 342, 378,
    410, 442, 476
Chapter test, 45–46, 75–76,
    119–120, 153–154, 192,

241–242, 295–296, 345–346,
    381–382, 413–414, 445–446,
    479–480
Circle, area of, 170
Circumference of the earth, 246
Clearing denominators, 271
Coefficient, 49
Combining like terms, 53
Common denominator, 24–25
  least common multiple, 256, 260
Common factors, 21, 199
Commutative Property of
    Addition, 54
Commutative Property of
    Multiplication, 55
Complementary angles, 106
Completing the square, 457–459
  geometric method, 475
  quadratic formula derivation,
    463
Complex fractions, 267–268, 291
Composite number, 71
Conditional equation, 114
Conjugates, 429
Consecutive integers, 107, 231
Constant, 49
Constant term, 49
Continued fractions, 291
Contradiction, 114
Convergent, 291
Coordinate(s), 301
Coordinate axes, 301
Coordinate system, 301–303
Corresponding angles, 114
Cost, 125
  of production, 94
Counterexamples, 149
Cubed, 35
Cubed of, 63
Cumulative review, 77–78,
    121–122, 155–156, 193–194,
    243–244, 297–298, 347–348,
    383–384, 415–416, 447–448,
    481–482
Current problems, 371–372, 375,
    471, 474

Decimal notation, 22
Decimal numbers, 22
  adding or subtracting, 25
  multiplying and dividing, 26
  percent as, 23
  rounding, 409
Decimal point, in scientific
    notation, 180
Degree of a polynomial, 159

Denominator
   common, 24
   least common multiple, 256
   rationalizing, 431
   clearing, 271
Dependent system of equations, 352, 353
Dependent variable, 308
Depreciation problems, 97, 322, 326
Descartes, René, 48, 301
Descending order, 159
Difference of two squares, 221
Discount, 126, 127–128
Discount rate, 126
Distance-rate problems, 371–372, 375, 471, 474
Distributive property, 53
   FOIL method, 168
   removing parentheses with, 100
   simplifying variable expressions with, 56–58
Division
   of decimal numbers, 26
   of exponential expressions, 177
   of inequalities, 394–395
   of integers, 16–17
   of monomials, 175–179
   one and, 17
   order of operations, 36
   of polynomials, 185–186
   of radical expressions, 430–431
   of rational expressions, 250
   of rational numbers, 26
   verbal phrases, 63
   zero and, 17
Domain of a relation, 305
Dotted line, 405

Earth, circumference of, 246
Egyptian mathematics, 2, 158, 386
Element of a set, 3
Elevation problems, 13
Empty set ($\emptyset$), 387
Equal, equivalent phrases, 107
Equality, Square Root Property of, 453
Equations, 81
   Addition Property of, 82
   basic percent, 85
   conditional, 114
   equivalent, 82
   linear, see Linear equations
   literal, 279–280
   Multiplication Property of, 83–84
   quadratic, see Quadratic equations
   solution of, 81
   solving, 82. See also Solving equations

squaring both sides of, 435
straight lines, 333–335
translating sentences into, 107–110
in two variables, solution of, 303
Equivalent equations, 82
Eratosthenes, 71, 246
Euler, Leonard, 291
Evaluating the function, 308
Even integers, 107
Exchange rates, 33
Exponent, 35, 178
   negative, 176
   scientific notation and, 180
   zero as, 175
Exponential expressions
   dividing, 177
   evaluation of, 35–36
   multiplying, 163
   quotient of, 175
   simplest form of, 177
   simplifying power of, 164
Exponential form, 35

Factor, 15
Factored form, 35
Factoring, 199
   application problems, 231–232, 234–236
   $ax^2 + bx + c$, 213–216
   completely, 207, 223
   difference of two squares, 221
   by grouping, 301, 215–216
   monomial from polynomial, 199–200
   nonfactorable over the integers, 206, 216
   perfect square trinomial, 221–223
   solving equations by, 229–230, 451–452
   using trial factors, 213–214
   $x^2 + bx + c$, 205–208
Fahrenheit temperature, 110
Final exam, 483–486
First coordinate of ordered pair, 301
FOIL method, 168, 205
Formula, 279
Fourth power, 35
Fractions
   addition of, 24
   ancient Egyptian, 2
   complex, 267–268, 291
   continued, 291
   division of, 26
   least common multiple, 255–256
   negative exponents on, 176
   percent as, 23

product of, 26, 248
ratio and proportion, 275
rationalizing the denominator, 431
reciprocal of, 250
simplest form of, 21
solving equations containing, 271–272
See also Rational expressions; Rational numbers
Function, 305–307
   evaluation of, 308
   value of, 308
Functional notation, 308

GCF, see Greatest common factor
Geometric method for completing the square, 475
Geometry problems, 145–148. See also Angles
Golden rectangle, 291
Grading scale, 306
Graph(s) and graphing, 301
   application problems, 336, 339
   $Ax + By = C$, 315–317
   equation in two variables, 303, 313–317
   horizontal line, 316
   inequalities, 389, 405–406
   integer, 3
   ordered pair, 301
   parabola, 465–466, 476
   points in the rectangular coordinate system, 301
   quadratic equations, 467–468, 476
   relation, 305
   slope-intercept equation, 327
   systems of equations, 351–354
   vertical line, 316
   x- and y-intercepts, 323
   $y = mx + b$, 313–315
Graphing calculator, 341, 378, 476
Greater than ($>$), 3
Greater than or equal to ($\geq$), 3
Greatest common factor (GCF), 199
Grouping method of factoring, 201, 215–216
Grouping symbols, order of operations, 36

Half-plane, 405
Horizon, distance to, 441
Horizontal line
   graph of, 316
   slope of, 325
Hypotenuse, 437

Identity, 114
Illumination intensity, 189

Inconsistent system of equations, 352
Independent system of equations, 352
Independent variable, 308
Inequalities
    Addition Property of, 393–394
    application problems, 396, 399–400, 402–404
    graphing, 389, 405–406
    Multiplication Property of, 394–395
    parentheses in, 401
    solution set of, 389
    solving, *see* Solving inequalities
Inequality symbols, 3–4
Input-output analysis, 350
Integers, 3, 107
    addition of, 7–8
    division of, 16–17
    multiplication of, 15–16
    nonfactorable polynomials over, 206, 216
    subtraction of, 8–9
Intensity of illumination, 189
Intercepts, 323
    slope-intercept equation, 327
Interest rate, 129
Intersection, 387
Inverse
    additive, 4, 54
    multiplicative, 55
Inverse Property of Addition, 54
Inverse Property of Multiplication, 55
Investment problems, 129–132, 318
Irrational numbers, 22, 419
Isosceles triangle, 146

LCM, *see* Least common multiple
Least common denominator, 25
Least common multiple (LCM), 24, 255–256, 260
Less than (<), 3
Less than or equal to (≤), 3
Lever problems, 102, 105
Like terms, 53
Linear equations, 313
    application problems, 318, 322, 336
    graphing calculator application, 341
    graph of $Ax + By = C$, 315–317
    graph of $y = mx + b$, 313–315
    systems of, *see* Systems of linear equations
Linear equations in two variables, 313
Linear function, 313
Linear inequalities, graphing, 405–406

Linear model, 336
Line of best fit, 336
Literal equations, 279–280
Lower half-plane, 405
Lumens, 189

m, 324–326. *See also* Slope
Magic square, 300
Markup, 125, 127
Markup rate, 125
Mersenne primes, 80
Mixture problems, 133–140
    percent mixtures, 135–136
    value mixtures, 133–134
Monomials, 159
    division and, 175–179, 185–186
    multiplication of, 163, 167
    simplifying powers of, 164
Moving average, 18
Multiplication
    associative property of, 55
    of binomials, 168–170
    commutative property of, 55
    of decimal numbers, 26
    exponential expression evaluation, 35
    of integers, 15–16
    inverse property of, 55
    of monomials, 163
    order of operations, 36
    of polynomials, 167–170
    of radical expressions, 429
    of rational expressions, 248–249
    of rational numbers, 26
    verbal phrases, 63
Multiplication Property of Equations, 83–84
Multiplication Property of Inequalities, 394–395
Multiplication Property of One, 55
Multiplication Property of Zero, 229
Multiplicative inverse, 55

Natural numbers, 3
Negative exponent, 176
Negative integers, 3
Negative numbers, 3
    multiplying inequality by, 395
    product of, 15–16
Negative slope, 325
Nielsen ratings, 115
Nonfactorable over the integers, 206, 216
Null set, (∅), 387
Number line, 3
    inequality solution set on, 389
    integer addition on, 7
Numerical coefficient, 49

Odd integers, 107
One
    division properties, 17
    multiplication property, 55
Opposite, 4, 54
    of polynomial, 160
Order of Operations Agreement, 36
Ordered pair, 301
    functions and relations, 305–307
    graph of, 301
Ordinate, 301
Origin, 301

Parabola, 465–466, 476
Parentheses, 15
    in inequalities, 401
    order of operations, 36
    solving equations containing, 100–101
Per, 326
Percent (%), 23
Percent equation, 85
Percent mixture problems, 135–136
Perfect square, 419, 421
Perfect square trinomial, 221–223, 457
Perimeter, 145–146, 147
Pi (π), 22, 291, 386
Plane, 301
Plot, 301
Point-slope formula, 333–334
Polynomials, 159
    addition of, 159
    $ax^2 + bx + c$, 213–216
    calculator applications, 237
    degree of, 159
    division of, 185–186
    factoring, *see* Factoring
    least common multiple, 255–256
    multiplication of, 167–170
    nonfactorable over the integers, 206, 216
    prime, 206, 216
    subtraction of, 160
    $x^2 + bx + c$, 205–208
    See also Binomial(s); Monomials

Positive integers, 3
Positive slope, 326
Powers, 35
    simplifying, 164
    verbal phrases, 63
    See also Exponent
Pressure problems, 94
Prime numbers, 71, 80

Prime polynomials, 206, 216
Principal square root, 419
Principle of Zero Products, 229, 451
Problems, *see* Application problems
Product, 15–16
  of fractions, 248
  simplifying powers of, 164
  of sum and difference of two terms, 169
  See also Multiplication
Product Property of Square Roots, 419
Profit and loss problems, 28, 33
Properties of Addition, 53–54
Properties of Multiplication, 55–56
Property of Squaring Both Sides of an Equation, 435
Proportion, 275–278
Pythagoras, 437
Pythagorean Theorem, 437

Quadrants, 301
Quadratic equations, 229, 467–468
  application problems, 471–474
  graphing, 467–468, 476
  solving by completing the square, 457–459
  solving by factoring, 451–452
  solving by quadratic formula, 463–464
  solving by taking square roots, 453–454
  standard form of, 229, 451
Quadratic equations in two variables, 467–468
Quadratic formula, 463–464
Quadratic function, 467
Quotient, *see* Division
Quotient Property of Square Roots, 430

Radical, 419
Radical equation, 435
Radical expressions
  addition and subtraction of, 425–426
  application problems, 437–440
  division of, 430–431
  multiplication of, 429
  rationalizing denominator, 431
  simplification of, 419–421
Radicand, 419
Range of a relation, 305
Rate, 275, 276–278
Rate-of-wind or -current problems, 371–372, 375, 471, 474

Rate of work, 283
Rating point, 115
Ratio, 275
Rational expressions, 247
  addition and subtraction of, 259–262
  division of, 250
  multiplication of, 248–249
  simplifying, 247–248
  solving equations, 435–436
  See also Fractions
Rationalizing the denominator, 431
Rational numbers, 21
  addition of, 24
  decimal notation, 22
  division of, 26
  multiplication of, 26
  subtraction of, 25
  See also Fractions
Real numbers, 22
Reciprocal, 55, 250
Rectangle
  area of, 170
  golden, 291
Rectangular coordinate system, 301–303
Relations, 305–306
Repeating decimal, 22
Rhetoric, 450
Rhind Papyrus, 158, 386
Right angle, 146
Right triangle, 146, 437
Roster method of writing sets, 387–388
Rounding, 409
Rule for Dividing Exponential Expressions, 177
Rule for Multiplying Exponential Expressions, 163
Rule for Negative Exponents on Fractional Expressions, 176
Rule for Simplifying the Power of an Exponential Expression, 164
Rule of Exponents, 178

Scatter diagram, 336
Scientific notation, 180
Second coordinate of ordered pair, 301
Second-degree equation, 451
Selling price, 125
Sentences, translating into equations, 107–110
Set, 3, 387
  element of, 3
  empty, 387
  roster method of writing, 387–388

union and intersection, 387
Set-builder notation, 388
Sieve of Eratosthenes, 71
Simple interest, 129
Simplest form
  of exponential expression, 177
  of fraction, 21
  of radical expression, 419
  of rational expression, 247
Simplification
  of complex fractions, 267
  of powers of monomials, 164
  of radical expressions, 419–421
  of rational expressions, 247–248
  of variable expressions, 53–58, 421
Slope, 324–326
  formula, 326
Slope–intercept equation, 327
Solid line, 405
Solution of an equation, 81
Solution of an equation in two variables, 303
Solution of application problems, 10
Solution of system of equations in two variables, 351
Solution set of an inequality, 389
Solving equations, 82
  $ax = b$, 83–84
  $ax + b = c$, 83
  $ax + b = cx + d$, 99–100
  by factoring, 229–230
  fractions and, 271–272
  geometry, 145–146
  investment and interest, 129–130
  markup and discount, 125–128
  mixtures, 133–136
  ordered pair solution, 303
  parentheses and, 100–101
  quadratic equations, *see* Quadratic equations
  rational expressions and, 435–436
  systems, *see* Systems of linear equations
  uniform motion, 141–142
  $x + a = b$, 82–83
  See also Application problems
Solving inequalities
  using Addition Property of Inequalities, 393–394
  general inequalities, 401–402
  using Multiplication Property of Inequalities, 394–395
Square, 63
  of binomial, 169, 221, 457
  perfect, 419, 421
  sums of, 231, 234

Squared, 35
Square root, 419
  Babylonian table, 418
  method for solving quadratic
    equation, 453–454
  of perfect square, 419, 421
  Product Property of, 419
  Quotient Property of, 430
  See also Radical expressions
Square Root Property of an
  Equality, 453
Squaring both sides of an
  equation, 435
Standard form of a quadratic
  equation, 229, 451
Straight line
  finding equations of, 333–335
  intercepts, 323
  point-slope formula, 333–334
  slope-intercept equation of, 327
  slope of, 324–326
Straight-line depreciation, 97
Strategy, for application problems,
  10
Substitution method for solving
  systems of equations,
  359–362
Subtraction
  Addition Property of Equations
    and, 82
  of decimal numbers, 25
  of integers, 8–9
  order of operations, 36
  of polynomials, 160
  of radical expressions, 425–426
  of rational expressions, 259–262
  of rational numbers, 25
  verbal phrases, 63
Sum and difference of two terms,
  product of, 169
Supplementary angles, 106
Systems of equations, 351
Systems of linear equations
  application problems, 371–376

calculator applications, 377
dependent, independent, and
  inconsistent, 352
solving
  addition method, 365–368
  graphing, 351–354
  substitution method, 359–362

Temperature problems, 10, 13–14,
  18, 20, 110
Terminating decimal, 22
Terms of a variable, 49
TRACE, 341
Transversal, 114
Trial factors, 213–214
Triangle problems, 146–148,
  437–440, 441
Trinomials, 159
  factoring by grouping, 215–216
  factoring using trial factors,
    213–214
  factors of, 205
  perfect square, 221–223, 457

Uniform motion problems,
  141–144, 285–286, 289–290
Union, 387
Unit fractions, 2
Upper half-plane, 405

Value mixture problems, 133–134
Value of a function, 308
Value of a variable, 49
Variable, 3, 49
  dependent and independent, 308
  history of, 48
  value of, 49
Variable expressions, 49
  application problems, 66
  evaluating, 50
  simplifying
    using addition properties,
      53–54

    using distributive property,
      56–58
    using multiplication
      properties, 55–56
  translating verbal expressions
    into, 63–66
Variable part of variable term, 49
Variable term, 49
Verbal expressions, translating
  into variable expressions,
  63–66
Vertical angles, 98
Vertical line
  graph of, 316
  slope of, 325
Viewing window, 341

Wind-chill temperature, 13–14
Wind problems, 371–372, 375, 474
Word problems, 124
  translating sentences into
    equations, 107–110
  See also Application problems
Work problems, 283–284, 287–288

x, as variable, 48
x-coordinate, 303
x-intercept, 323
xy coordinate system, 303

y, as variable, 48
y-coordinate, 303
y-intercept, 323

z, as variable, 48
Zero, 3
  addition property of, 54
  division properties, 17
  as an exponent, 175
  Multiplication Property of, 229
Zero Products, Principle of, 229,
  451
Zero slope, 325